SOUTH WESTERN
Algebra 1
AN INTEGRATED APPROACH

SOLUTIONS MANUAL

GERVER, SGROI, CARTER, HANSEN
MOLINA & WESTEGAARD

JOIN US ON THE INTERNET
WWW: http://www.thomson.com
EMAIL: findit@kiosk.thomson.com   A service of I(T)P®

SOUTH
WESTERN
EDUCATIONAL
PUBLISHING

I(T)P® An International Thomson Publishing Company

Cincinnati • Albany • Bonn • Boston • Detroit • London • Madrid • Melbourne • Mexico City • New York
Philadelphia • Pacific Grove • Paris • San Francisco • Singapore • Tokyo • Toronto • Washington

ISBN: 0-538-66464-9
  2 3 4 5 6 7 8 MZ 03 02 01 00 99 98
Printed in the United States of America

I(T)P
International Thomson Publishing

TABLE OF CONTENTS

JOIN US ON THE INTERNET

WWW: http://www.thomson.com
E-MAIL: findit@kiosk.thomson.com

Access South-Western Educational Publishing's complete catalog online through thomson.com. Internet users can search catalogs, examine subject-specific resource centers, and subscribe to electronic discussions lists. In addition, you'll find new product information and information on upcoming events in Mathematics.

For information on our products and services, point your web browser to http://www.swpco.com/swpco.html

For technical support, you may e-mail hotline_education@kdc.com

South-Western Algebra 1: An Integrated Approach Internet Connection is a web site which accompanies the chapter projects. You can access the Internet Connection at www.swpco.com/swpco/algebra1.html

A service of I(T)P®

Chapter 1   Data and Graphs

Data Activity, page 3

1. $25.9 - 11.4 = 14.5$ Pts

2. $(25.9 + 14.9 + 14.2 + 13.9 + 13.4 + 13.4 + 12.3 + 12.2 + 11.5 + 11.4) \div 10 = \dfrac{143.1}{10} = 14.3$

3. $(954,000) \cdot (14.9) = 14,214,600$

4. $14,215,000;\ 14,210,000;\ 14,000,000$

5. $(954,000) \cdot (13.4) - (954,000) \cdot (11.5) = 1,812,600$

6. ABC — 6 shows out of $10 = \dfrac{6}{10} = 0.60 = 60\%$

   NBC — 4 shows out of $10 = \dfrac{4}{10} = 0.40 = 40\%$

   CBS — 0 shows out of $10 = \dfrac{0}{10} = 0.00 = 0\%$

Lesson 1.1, pages 5–7

THINK BACK

1. DATA—Information that has been gathered about a sample or population.
   STATISTIC—Collecting, analyzing, and interpreting data.

2. Many reasons could apply such as: location of stores in city, number of CDs and cassettes sold per store, possibly a big country concert in July in this city, etc.

EXPLORE

3.

Music Type	Number of CDs and Tapes
Classical	$542 + 254 + 957 + 328 + 1307$ $= \mathbf{3388}$
Country	$3251 + 1389 + 2586 + 1654$ $+ 256 = \mathbf{9136}$
Easy Listening	$325 + 173 + 427 + 53 + 184$ $= \mathbf{1162}$
Jazz	$683 + 298 + 854 + 127 + 945$ $= \mathbf{2907}$
Rock	$2963 + 2108 + 2385 + 1489$ $+ 2723 = \mathbf{11,668}$
Other	$157 + 88 + 239 + 22 + 176$ $= \mathbf{682}$

4. Rock music is more popular with 11,668 CDs and tapes sold.

5. Answers will vary. Example include:
   What percentage of sales are in Rock music?
   $11,668 \div (3388 + 9136 + 1162 + 2907 + 11,668 + 682)$ $\approx 0.4031 \approx 40\%$
   What is the average number of CDs and tapes sold per store?
   $\dfrac{11,668}{5} \approx 2,334$

MAKE CONNECTIONS

6. Answers will depend on data collected.

7. Answers will depend on data collected.

8. Answers will depend on data collected.

9. Conclusions could be the same or different from individuals or groups depending on the distribution of favorite TV shows.

10. Answers will vary, such as: What type of show is most popular? Do differences exist between male and female students?

SUMMARIZE

11. Numerous answers such as: What things are being compared? How should we categorize data in columns and rows? A table is helpful to summarize and organize data.

12. Data must be organized correctly to allow interpretations to be made about the question asked. For instance, the CD example in the text showed country music most popular until the actual numbers and not the types most frequently sold per store were counted.

13. Most popular music in a city should be derived from a random sample in the city. Was this the only music chain in the city or are other outlets available? Many times a particular chain specializes in a certain type of music.

14. Tabulate the data as follows:

Tickets Sold, Both Adults and Children

	Saturday	Sunday	Monday	Total
Penguin	3266	1262	734	**5262**
Mark	2077	1622	1333	**5032**
Graduation	443	1143	277	**1863**
Dreams	1214	856	2236	**4306**
Water	1633	1382	1797	**4812**
Total	**8633**	**6265**	**6377**	**21,275**

a. "The Little Penguin" had the best attendance with 5262 tickets sold.

b. $\dfrac{5,262}{21,275} \approx 0.2473 \approx 25\%$

c. The most tickets were sold on Saturday, with 8633.

Lesson 1.2, pages 8–12

EXPLORE

1. Count the activities. Art, 6; Basketball, 9; Chess, 3; Dance, 7; Drama, 5; Swimming, 5

2. Chess could be cut since it was listed only 3 times. Basketball, with 9 responses, was most popular and should not be cut. Swimming was not as popular as some others, but costs to maintain facilities would still occur, so swimming should not be cut.

1. **Number of Hours Health Club Members Exercise per Week**

Number of Hours	Tally	Frequency
0	I I	2
1	I I I	3
2	⊬⊤	5
3	I I I	3
4	I I	2
5	I I	2
6	I	1
7	I I	2

2. $\dfrac{\text{frequency of 5 or more}}{\text{sum of frequencies}}$

$= \dfrac{2+1+2}{2+3+5+3+2+2+1+2} = \dfrac{5}{20} = 25\%$

3. Most frequent is 2 hours with a frequency of 5.

4. No, data is not available on the number of hours per visit.

5. **Number of Books Read During Summer Reading Program**

Number of Books	Tally	Frequency
0–4	I I I I	4
5–9	⊬⊤ I I	7
10–14	⊬⊤ I I I	8
15–19	I I I	3
20–24	I I I	3

6. $8 + 3 + 3 = 14$

7. $\dfrac{\text{frequency of fewer than 20}}{\text{sum of frequencies}}$

$= \dfrac{4+7+8+3}{4+7+8+3+3} = \dfrac{22}{25} = 88\%$

8. There are many possible answers, such as: What is the average number of books read?

Number of Movie Screens per Theater

Number of Screens	Tally	Frequency
2	I I I I	4
3	I I	2
4	I I I I	4
5		0
6	I	1
7	I	1
8	I I I	3
9		0
10	I I	2
11		0
12	I I I	3

1. $4 + 2 + 4 = 10$

2. $\dfrac{3}{4+2+4+1+1+3+2+3} = \dfrac{3}{20}$

Exam Scores of Students in Ms. Rivera's Algebra Class

Scores	Tally	Frequency
40–49	I I	2
50–59	I	1
60–69	I I I	3
70–79	⊬⊤ I	6
80–89	⊬⊤ ⊬⊤ I I	12
90–99	I I I I	4

3. $\dfrac{2+1}{2+1+3+6+12+4} = \dfrac{3}{28} \approx 11\%$

4. $\dfrac{12}{2+1+3+6+12+4} = \dfrac{12}{28} = \dfrac{3}{7}$

Circulation of Top 14 U.S. Magazines, 1993

Circulation	Tally	Frequency
3,000,000–3,999,999	I I	2
4,000,000–4,999,999	I I I	3
5,000,000–5,999,999	I I I	3
7,000,000–7,999,999	I	1
9,000,000–9,999,999	I	1
14,000,000–14,999,999	I	1
16,000,000–16,999,999	I	1
22,000,000–22,999,999	I I	2

5. 10 under 10 million, 2 over 20 million;
   $10 - 2 = 8$ more magazines

6. $\dfrac{1+1+1}{2+3+3+1+1+1+1+2} = \dfrac{3}{14} \approx 21\%$

7. Answers will vary, but a good rule is to use between 5 and 12 intervals. These will depend on the values and distribution of your data. You should have enough intervals to indicate differences, but not so many that there is no summarization of the data.

EXTEND

8.

Measurement	Frequency
0.365	1
0.366	1
0.367	1
0.368	2
0.369	4
0.370	2
0.371	4
0.372	2
0.373	2
0.374	1

9. 0.370; There is a total of 20 points, so count to the tenth point from the top or bottom to find 0.370.

10. $\dfrac{4+2+4}{1+1+1+2+4+2+4+2+2+1} = \dfrac{10}{20} = 50\%$

11. Answers will vary: 0.369 and 0.371 are the most frequent (modes), 0.370 is in the middle (median), and 0.3699 is the average (mean).

12. $5 + 6 + 8 + 0 + 4 + 2 + 1 = 26$

13. Greatest frequency is $6.00–$6.99 with 8.
    11 charge less than $6.00.
    7 charge more than $6.99.

14. There are many possible answers. One might be to charge between $5.00 and $5.99 as that is below the most common rate and will attract customers.

THINK CRITICALLY

15. Yes. The exact number he and every other boy made would be included in the table.

16. No. With a grouped table he can find out only how many boys made approximately the same number of free throws as he made.

Lesson 1.3, pages 13–17

EXPLORE

1. passing average of a player with 15 completions in 25 attempts; $\dfrac{15}{25} = 0.60$

2. percent of games won if you have 15 wins and 25 losses; $\dfrac{15}{15+25} \cdot 100\% = \dfrac{15}{40} \cdot 100\% = 37.5\%$

3. passing average of a player with 21 completions in 84 attempts; $\dfrac{21}{84} = 0.25$

4. percent of games won if you have 21 wins and 84 losses; $\dfrac{21}{21+84} \cdot 100\% = \dfrac{21}{103} \cdot 100\% = 20\%$

5. $A \# B = \dfrac{B}{A+B} \cdot 100\%$

6. $6 \cdot A + 3 \cdot B$

TRY THESE

1. $N - M = 10 - 6 = 4$

2. $L * M = 2 \times 6 = 12$

3. $M / L = 6 \div 2 = 3$

4. $6 * M - N = 6 \times 6 - 10 = 36 - 10 = 26$

5. $5 * 2 / N = 5 \times 2 \div 10 = 10 \div 10 = 1$

6. $4 * M / (L + N) = 4 \times 6 \div (2 + 10)$
   $= 4 \times 6 \div 12 = 24 \div 12 = 2$

7. Answers will vary such as $A = bh$. Letters A, b, and h are variables and each side of the equation represents an algebraic expression.

8. The total number of television sets sold at each store.

9. $9 + 21 + 16 = 46$ and $15 + 30 + 34 = 79$

10.

	A	B	C	D	E
1		Premier	Mark IV	Entertainer	Net Receipts
2	Plaza	9	21	16	19200 ◁ =699*B2+425*C2+249*D2
3	Mall	15	30	34	31701 ◁ =699*B3+425*C3+249*D3

PRACTICE

1. $R + N = 9 + 16 = 25$

2. $N - M = 16 - 5 = 11$

3. $7 * M = 7 \times 5 = 35$

4. $R / 3 = 9 \div 3 = 3$

5. $N + M - R = 16 + 5 - 9 = 12$

6. $15 + N + R = 15 + 16 + 9 = 40$

7. $N / (M - 3) = 16 \div (5 - 3) = 16 \div 2 = 8$

8. $10 * R / M = 10 \times 9 \div 5 = 90 \div 5 = 18$

9. $(4 * N) / (2 * M) = (4 \times 16) \div (2 \times 5)$
   $= 64 \div 10 = 6.4$

10.
&
11.

	A	B	C	D	E	F
1		G	PG	PG13	R	Total
2	Store 1	73	55	38	71	237 ◁ =B2+C2+D2+E2
3	Store 2	28	30	41	51	150 ◁ =B3+C3+D3+E3
4	Store 3	107	88	95	120	410 ◁ =B4+C4+D4+E4

12. $0.5 * 24 * 9 = 12 * 9 = 108$

13. $0.5 * 41 * 35 = 20.5 * 35 = 717.5$

14. $0.5 * 0.62 * 0.8 = 0.31 * 0.8 = 0.248$

15. Change Line 20 to **PRINT B * H**.

16. $1.8 * 45 + 32 = 81 + 32 = 113$

17. $1.8 * 100 + 32 = 180 + 32 = 212$

18. $1.8 * 0 + 32 = 0 + 32 = 32$

19. Answers will vary, but one such answer is to allow formulas to be input which require values that change with the problem such as in 12 to 18 above. Variables allow the use of one formula applied to many situations.

20. Multiply depth in feet by 0.44 lbs/in^2 and add 14.7 lbs/in^2.

21.

	A Trench or Basin	B Depth, ft	C Water Pressure, lb/in.^{2}	D Total Pressure, lb/in.^{2}
1				
2	Puerto Rico	28232	12422.08	12436.78
3	Japan	27599	12143.56	12158.26
4	Brazil	20076	8833.44	8848.14
5	Peru-Chile	26457	11641.08	11655.78
6	Eurasia	17881	7867.64	7882.34
7	Mariana	35840	15769.60	15784.30

EXTEND

22. & 23.

	A	B	C	D	E
1		G	PG	PG13	R
2	Store 1	73	55	38	71
3	Store 2	28	30	41	51
4	Store 3	107	88	95	120
5	Total	208	173	174	242

=B2+B3+B4 =C2+C3+C4 =D2+D3+D4 =E2+E3+E4

24. 10 INPUT B, H
20 PRINT (2 * B) + (2 * H)

25. 10 INPUT H, M, N
20 PRINT 0.5 * H * (M + N)

26. 10 INPUT L, W, H
20 PRINT 2 * (L * W + L * H + W * H)

THINK CRITICALLY

27–31 Answers will vary. Possible answers are:

27. B – A / 2, 2 * A

28. (B – A) * A, B / C * 3 * A

29. B * C, A * C * 2.5

30. (A + C) * B, B * B – 10

31. A * B * C, 2 * B * B

32. Yes, order of operations has multiplication performed before addition. $x = 2$, $y = 3$, $z = 4$
$x + y * z = 2 + 3 * 4 = 2 + 12 = 14$ and
$y * z + x = 3 * 4 + 2 = 12 + 2 = 14$.

33. No, order of operations has operations within parentheses performed before multiplication.
$x = 2$, $y = 3$, and $z = 4$,
$x * y + z = 2 * 3 + 4 = 6 + 4 = 10$ and
$x * (y + z) = 2 * (3 + 4) = 2 * 7 = 14$.

MIXED REVIEW

34.

Temp (°F)	Tally	Frequency
14 – 16	III	3
17 – 19	I	1
20 – 22		0
23 – 25	IIII	4
26 – 28	卌	5
29 – 31	卌 II	7
32 – 34	IIII	4
35 – 37	II	2
38 – 40	I	1
41 – 43	I	1

$$\frac{4+2+1+1}{3+1+0+4+5+7+4+2+1+1} = \frac{8}{28} \approx 29\%$$

35. $7 * (K + M) / (4 * L) = 7 * (5 + 3) / (4 * 8)$
$= 7 * 8 / 32 = 56 / 32 = 1.75$; C

Lesson 1.4, pages 18–20

THINK BACK

1. Answers will vary depending on data.

EXPLORE

2. Each symbol represents 2 students.
There are 12 symbols and 4 half-symbols.
$12 * 2 + 4 * 1 = 28$ students

3. 3 symbols and 1 half-symbol $= 3 * 2 + 1 = 7$ students

4. Rock column currently has 6 students. If each symbol is 3 students, then $\frac{6}{3} = 2$ symbols. Country column currently has 8 students. If each symbol represents 3 students, then $\frac{8}{3} = 2\frac{2}{3}$ symbols.

5. No, it would be difficult to divide this symbol into thirds.

6. No, with 1000 people, 500 symbols would be needed and that would be tedious to produce and difficult to read. Better to use one symbol for 200 or 500 people. Then only 5 or 2 symbols, respectively, are needed.

7. Answers will vary. Both types of graphs are visual and both allow for quick comparisons; both use a standard space or symbol to represent a certain number of items. In a pictograph you refer to the value of each symbol.

In a bar graph you measure the height of a column by referring to the numbers on the vertical axis.

8. Answers will vary. The scale should be chosen so that both the tallest and shortest bars can be shown.The greatest number on the scale should be equal to or slightly greater than the value of the tallest bar. The intervals should be chosen so that the reader can judge the height of each bar easily.

9. Answers will vary. Draw rectangles to enclose each stack of X's, then delete the X's. Draw a vertical axis for measuring the heights of the bars; count the X's and use as numbers for a bar graph.

10. Answers will vary. In bar graphs and line plots, the heights of the bars and stacks of X's immediately indicate the relative sizes of the categories of data. If the symbols are chosen carefully, however a pictograph can be more enjoyable to look at than a bar graph or line plot, and can give a stronger sense of the relevance of the data. A line plot may be inappropriate for very large data sets, but a pictograph can use symbols representing millions.

11. Answers will vary, but should refer to data in which there is a change in time. Examples: stacks represent scores on successive math quizzes, where the ups and downs of the line would show a student"s progress quiz by quiz; changes in temperature.

12. Largest sector is 28%. Full circle is 360°. Largest sector is 28% of $360° = 0.28 \cdot 360° = 100.8°$

13. Answers will vary. Instead of showing the number of people choosing each type of music, the circle graph shows the percent of all people surveyed who chose each type as their favorite. A circle graph shows how each category relates to the whole.

MAKE CONNECTIONS

14. Answers will vary depending on student's data.

15. Answers will vary depending on student's data.

16. Answers will vary depending on student's data.

SUMMARIZE

17. Answers will vary depending on student's data.

18. Answers will vary depending on student's justification.

19. The range is used to determine the value of a symbol of a pictograph, and the vertical scale of a bar graph or a line plot.

20. Answers will vary depending on student's data.

21. Answers will vary. Example: Choose one show and ask how many students watch it regularly; ask again each month; a line graph of the data will show increases and decreases in the popularity of the show among the students in the class.

22. Answers will vary. Pictographs, bar graphs, and line graphs tell the values of each category and allow the viewer to compare actual values. Circle graphs tell the percentages of the whole represented by each category and allow the viewer to compare their relative values.

23. Answers will vary depending on student's data.

Lesson 1.5, pages 21–26

TRY THESE

1. Convenience: people stopping in at campaign head-quarters are likely to be strong supporters of the mayor.

2. Cluster: mainly people with an interest in the services provided within the building will be surveyed.

3. Systematic: only people with jobs and cars will be surveyed.

4. $\dfrac{1}{4} = \dfrac{n}{8}$

$1 \cdot 8 = 4 \cdot n$

$\dfrac{8}{4} = n$

$2 = n$

5. $\dfrac{2}{3} = \dfrac{n}{12}$

$2 \cdot 12 = 3 \cdot n$

$24 = 3n$

$\dfrac{24}{3} = n$

$8 = n$

6. $\dfrac{n}{20} = \dfrac{3}{4}$

$4 \cdot n = 20 \cdot 3$

$4n = 60$

$n = \dfrac{60}{4}$

$n = 15$

7. $\dfrac{2}{7} = \dfrac{16}{n}$

$2 \cdot n = 7 \cdot 16$

$2n = 112$

$n = \dfrac{112}{2}$

$n = 56$

8. Let n represent voters for the mayor.

$\dfrac{75}{160} = \dfrac{n}{8240}$

$75 \cdot 8240 = 160 \cdot n$

$618,000 = 160n$

$\dfrac{618,000}{160} = n$

$3862.5 = n$

$3900 \approx n$

9. Answers will vary. Students should recognize that some polls are purposely biased. This might be done, for example, to suggest that a particular product is more popular than it actually is.

PRACTICE

1. Systematic: the only people surveyed will be those who enjoy popcorn.

2. Cluster: those surveyed, even if they don't personally like popcorn, are likely to be biased in favor of it because of its importance to their state's economy.

3. Convenience: since people who don't like popcorn are not likely to visit a popcorn processing plant, most of the people surveyed are likely to enjoy popcorn. People taking a tour may feel obligated to answer more positively about popcorn on the survey.

4. $\frac{1}{2} = \frac{n}{10}$

$2 \cdot n = 1 \cdot 10$

$n = \frac{10}{2} = 5$

5. $\frac{1}{3} = \frac{n}{6}$

$3 \cdot n = 1 \cdot 6$

$n = \frac{6}{3} = 2$

6. $\frac{n}{10} = \frac{2}{5}$

$5 \cdot n = 2 \cdot 10$

$n = \frac{20}{5} = 4$

7. $\frac{n}{8} = \frac{12}{32}$

$32 \cdot n = 8 \cdot 12$

$n = \frac{96}{32} = 3$

8. $\frac{7}{10} = \frac{49}{n}$

$7 \cdot n = 10 \cdot 49$

$n = \frac{490}{7} = 70$

9. $\frac{18}{81} = \frac{n}{117}$

$81 \cdot n = 18 \cdot 117$

$n = \frac{2106}{81} = 26$

10. $\frac{200}{750} = \frac{96}{n}$

$200 \cdot n = 750 \cdot 96$

$n = \frac{72,000}{200} = 360$

11. $\frac{0.75}{n} = \frac{0.375}{0.5}$

$0.375 \cdot n = 0.75 \cdot 0.5$

$n = \frac{0.375}{0.375} = 1$

12. Let n represent the adults in favor of smoke-free public buildings.

$\frac{415}{640} = \frac{n}{32,700}$

$640 \cdot n = 415 \cdot 32,700$

$640n = 13,570,500$

$n = \frac{13,570,500}{640} \approx 21,200$

13. Answers will vary. Students might want to know the size of the sample to judge whether it was sufficiently large to give accurate results. They might want to know the sampling method that was used in order to judge whether the survey was biased.

EXTEND

14. 28% of 300 = $0.28 \cdot 300 = 84$

15. 5% of 250 million
= $0.05 \cdot 250$ million
= 12.5 million

16. $\frac{(922 + 845 + 1047 + 606)}{684,000} = \frac{3420}{684,000} = 0.005 = 0.5\%$

17. $\frac{606}{(922 + 845 + 1047 + 606)} = \frac{606}{3420} = \frac{n}{684,000}$

$3420 \cdot n = 606 \cdot 384,000$

$3420n = 414,504,000$

$n = \frac{414,504,000}{3420} = 121,200$

18. Let n represent voters for Rivera.

$\frac{922}{3420} = \frac{n}{684,000}$

$3420 \cdot n = 922 \cdot 684,000$

$3420n = 630,648,000$

$n = \frac{630,648,000}{3420} = 184,400$

50% of the undecided vote:
$0.50(121,200) = 60,600$
90% of the registered voters voted:
$0.90(184,400 + 60,600) = 220,500$

THINK CRITICALLY

19. $\frac{527}{850} = 0.62 = 62\%$ favor term limits from survey results
maximum who favor term limits: $62\% + 3\% = 65\%$
minimum who favor term limits: $62\% - 3\% = 59\%$

20. maximum: 65% of 2 million = $0.65 \cdot 2$ million = 1.3 million
minimum: 59% of 2 million = $0.59 \cdot 2$ million = 1.18 million

ALGEBRA WORKS

1. 98% of 95 million = $0.98 \cdot 95$ million
= 93.1 million

2. Let n represent number of people watching TV.
$0.77 \cdot n = 125$ million
$n = \frac{125 \text{ million}}{0.77} \approx 162$ million people watching TV

3. 77% of 10,000 viewers = $0.77 \cdot 10,000$ viewers
= 7700 viewers

4. 2% of 125 million viewers who watched M*A*S*H
= $0.02 \cdot 125$ million = 2.5 million ≈ 3 million;
greatest number who watched:
= 125 million + 3 million = 128 million
Least number who watched:
= 125 million − 3 million = 122 million

5. Let n represent the number of people watching "News at Five."
$\frac{86}{500} = \frac{n}{240,000}$

$500 \cdot n = 86 \cdot 240,000$

$500n = 20,640,000$

$n = \frac{20,640,000}{500} = 41,280 \approx 41,300$

Lesson 1.6, pages 27–31

EXPLORE

1. Answers will vary. One possibility follows:

	Gold	Silver	Bronze	Total
Unified Team	45	38	29	112
United States	37	34	37	108
Germany	33	21	28	82
China	16	22	16	54

2. Answers will vary. A table displays the number of medals won by each country in rows and the country totals for each type of medal in columns. A table is easy to read and allows comparisons.

3. Using a table, variables could be assigned as they are for a spreadsheet, with columns lettered and rows numbered. In the table from answer 1, the number 21 would be assigned the variable C4.

TRY THESE

1. 9, 15, 7, 2, 6; 1×5

2. $\begin{bmatrix} 3+\ 7 & 5+16 & 9+20 & 19+\ 5 \\ 13+31 & 6+15 & 20+19 & 2+40 \end{bmatrix}$

$= \begin{bmatrix} 10 & 21 & 29 & 24 \\ 44 & 21 & 39 & 42 \end{bmatrix}$

3. $\begin{bmatrix} 7-\ 5 & 16-\ 9 & 20-13 & 5-\ 5 \\ 31-55 & 15-11 & 19-\ 4 & 40-19 \end{bmatrix}$

$= \begin{bmatrix} 2 & 7 & 7 & 0 \\ -24 & 4 & 15 & 21 \end{bmatrix}$

4. $\begin{bmatrix} 3+\ 7+\ 5 & 5+16+\ 9 & 9+20+13 & 19+\ 5+\ 5 \\ 13+31+55 & 6+15+11 & 20+19+\ 4 & 2+40+19 \end{bmatrix}$

$= \begin{bmatrix} 15 & 30 & 42 & 29 \\ 99 & 32 & 43 & 61 \end{bmatrix}$

5. $\mathbf{A} = \begin{bmatrix} 117 & 88 \\ 91 & 95 \end{bmatrix}$

$\mathbf{B} = \begin{bmatrix} 38 & 35 \\ 25 & 40 \end{bmatrix}$

$\mathbf{C} = \begin{bmatrix} 13 & 9 \\ 6 & 10 \end{bmatrix}$

6. $\mathbf{A} + \mathbf{B} - \mathbf{C}$

$= \begin{bmatrix} 117+38-13 & 88+35-\ 9 \\ 91+25-\ 6 & 95+40-10 \end{bmatrix}$

$= \begin{bmatrix} 142 & 114 \\ 110 & 125 \end{bmatrix}$

7. Answers will vary. Data on inventory can be written in a matrix. A second matrix can represent additions to inventory; a third, deletions from inventory. Combining the three will give a count of inventory on hand.

PRACTICE

1. 4, 5, 8, 3

2. 3×2

3. $\begin{bmatrix} 4+(-1) & 5+\ 6 \\ 8+\ 2 & 3+(-7) \end{bmatrix} = \begin{bmatrix} 3 & 11 \\ 10 & -4 \end{bmatrix}$

4. $\begin{bmatrix} 12+26 & 9+14 \\ 8+\ 9 & 6+12 \\ 20+\ 5 & 13+25 \end{bmatrix} = \begin{bmatrix} 38 & 23 \\ 17 & 18 \\ 25 & 38 \end{bmatrix}$

5. $\begin{bmatrix} 12-\ 9 & 9-5 \\ 8-\ 8 & 6-0 \\ 20-16 & 13-7 \end{bmatrix} = \begin{bmatrix} 3 & 4 \\ 0 & 6 \\ 4 & 6 \end{bmatrix}$

6. $\begin{bmatrix} 9+26-12 & 5+14-\ 9 \\ 8+\ 9-\ 8 & 0+12-\ 6 \\ 16+\ 5-20 & 7+25-13 \end{bmatrix} = \begin{bmatrix} 23 & 10 \\ 9 & 6 \\ 1 & 19 \end{bmatrix}$

7. $\mathbf{I} = \begin{bmatrix} 317 & 490 & 166 \\ 555 & 207 & 181 \end{bmatrix}$

$\mathbf{N} = \begin{bmatrix} 52 & 70 & 48 \\ 88 & 86 & 66 \end{bmatrix}$

$\mathbf{S} = \begin{bmatrix} 61 & 90 & 77 \\ 114 & 98 & 50 \end{bmatrix}$

8. $\mathbf{I} + \mathbf{N} + \mathbf{S}$

$= \begin{bmatrix} 317+52-\ 61 & 490+70-90 & 166+48-77 \\ 555+88-114 & 207+86-98 & 181+66-50 \end{bmatrix}$

$= \begin{bmatrix} 308 & 470 & 137 \\ 529 & 195 & 197 \end{bmatrix}$

9. tax = 8% of sticker price = 0.08 • price

$= \begin{bmatrix} 0.08 \cdot 12{,}400 & 0.08 \cdot 12{,}600 & 0.08 \cdot 12{,}000 \\ 0.08 \cdot 11{,}100 & 0.08 \cdot 11{,}400 & 0.08 \cdot 11{,}500 \\ 0.08 \cdot 14{,}800 & 0.08 \cdot 14{,}100 & 0.08 \cdot 14{,}400 \\ 0.08 \cdot 10{,}200 & 0.08 \cdot 10{,}900 & 0.08 \cdot 10{,}300 \end{bmatrix}$

$= \begin{bmatrix} 992 & 1008 & 960 \\ 888 & 912 & 920 \\ 1184 & 1128 & 1152 \\ 816 & 872 & 824 \end{bmatrix}$

10. $\begin{bmatrix} 12{,}400+\ 992 & 12{,}600+1008 & 12{,}000+\ 960 \\ 11{,}100+\ 888 & 11{,}400+\ 912 & 11{,}500+\ 920 \\ 14{,}800+1184 & 14{,}100+1128 & 14{,}400+1152 \\ 10{,}200+\ 816 & 10{,}900+\ 872 & 10{,}300+\ 824 \end{bmatrix}$

$= \begin{bmatrix} 13{,}392 & 13{,}608 & 12{,}960 \\ 11{,}988 & 12{,}312 & 12{,}420 \\ 15{,}984 & 15{,}228 & 15{,}552 \\ 11{,}016 & 11{,}772 & 11{,}124 \end{bmatrix}$

11. Answers will vary. One such possibility is to calculate retail prices.

EXTEND

12. $\mathbf{M} = \begin{bmatrix} 81-29 & 83-63 \\ 115-77 & 62-49 \end{bmatrix} = \begin{bmatrix} 52 & 20 \\ 38 & 13 \end{bmatrix}$

13. $\mathbf{M} = \begin{bmatrix} 19-\ 8 & 44-29 & 30-\ 6 \\ 62-43 & 91-60 & 55-27 \end{bmatrix} = \begin{bmatrix} 11 & 15 & 24 \\ 19 & 31 & 28 \end{bmatrix}$

THINK CRITICALLY

14. $\begin{bmatrix} 6+6 & 3+3 \\ 5+5 & 9+9 \end{bmatrix} = \begin{bmatrix} 12 & 6 \\ 10 & 18 \end{bmatrix}$

15. To double a matrix, multiply each element by 2.

16. Multiply each element by N.

17. Let S represent the sticker price matrix.
Total price $= 1.08S$.

$$1.08 \begin{bmatrix} 12{,}400 & 12{,}600 & 12{,}000 \\ 11{,}100 & 11{,}400 & 11{,}500 \\ 14{,}800 & 14{,}100 & 14{,}400 \\ 10{,}200 & 10{,}900 & 10{,}300 \end{bmatrix}$$

18. $A + B \overset{?}{=} B + A$

$$\begin{bmatrix} a+e & b+f \\ c+g & d+h \end{bmatrix} \overset{?}{=} \begin{bmatrix} e+a & f+b \\ g+c & h+d \end{bmatrix}$$

From the commutative property, corresponding elements are equal.
Thus, $A + B = B + A$.

19. $A + (B+C) \overset{?}{=} (A+B) + C$

$$\begin{bmatrix} a & b \\ c & d \end{bmatrix} + \begin{bmatrix} e+i & f+j \\ g+k & h+l \end{bmatrix} \overset{?}{=} \begin{bmatrix} a+e & b+f \\ c+g & d+h \end{bmatrix} + \begin{bmatrix} i & j \\ k & l \end{bmatrix}$$

$$\begin{bmatrix} a+e+i & b+f+j \\ c+g+k & d+h+l \end{bmatrix} \overset{?}{=} \begin{bmatrix} a+e+i & b+f+j \\ c+g+k & d+h+l \end{bmatrix}$$

Corresponding elements are equal.
Thus, $A + (B+C) = (A+B) + C$.

20. $A + D = A$

$D = A - A$

$$D = \begin{bmatrix} a & b \\ c & d \end{bmatrix} - \begin{bmatrix} a & b \\ c & d \end{bmatrix} = \begin{bmatrix} 0 & 0 \\ 0 & 0 \end{bmatrix}$$

Similarly for equations $B + D = B$ and $C + D = C$.

MIXED REVIEW

21. Let n represent the number of votes for Green.

$$\frac{30}{315} = \frac{n}{16{,}400}$$

$$315 \cdot n = 30 \cdot 16{,}400$$

$$315n = 492{,}000$$

$$n = \frac{492{,}000}{315} \approx 1562$$

Best answer is "B," about 1,600.

22. 2×3

23.
$$\begin{bmatrix} 114+151 & 91+114 & 172+198 \\ 88+156 & 65+79 & 80+80 \end{bmatrix}$$

$$= \begin{bmatrix} 265 & 205 & 370 \\ 244 & 144 & 160 \end{bmatrix}$$

24.
$$\begin{bmatrix} 151-114 & 114-91 & 198-172 \\ 156-88 & 79-65 & 80-80 \end{bmatrix}$$

$$= \begin{bmatrix} 37 & 23 & 26 \\ 68 & 14 & 0 \end{bmatrix}$$

Lesson 1.7, pages 32–37

EXPLORE

1. $\dfrac{100 + 100 + 100 + 99{,}700}{4} = 25{,}000$

Yes, if you assume the "average" is the mean.

2. The high average was due entirely to the huge attendance at Heavenly. At none of the other concerts was attendance even remotely close to 25,000. At most concerts 100 people attended.

3. Answers will vary. Possible answer: "Live Math National Tour Huge Failure! Disregarding Heavenly Rock Festival, average concert attendance 100!"

4. A small number of extreme values can affect the average.

5. Answers will vary. The tour went from Maine to California, so it was "National," but between Maine and California, it included only two states. The tour was not a "huge success."

TRY THESE

1. Mean:
$$\frac{98+77+89+93+75+81+77+88+78}{9} = \frac{756}{9} = 84$$
Median: 75, 77, 77, 78, **81**, 88, 89, 93, 98; median = 81
Mode: 77 with a frequency of 2.
Range: $98 - 75 = 23$

2. Mean:
$$\frac{5+9+3+1+0+4+5+2+1+8+7+6}{12} = \frac{51}{12} = 4.25$$
Median: 0, 1, 1, 2, 3, **4**, **5**, 5, 6, 7, 8, 9;
$$\text{median} = \frac{4+5}{2} = \frac{9}{2} = 4.5$$
Mode: 1 and 5 both with a frequency of 2.
Range: $9 - 0 = 9$

3. Mean: $\dfrac{188+260+174+206}{4} = \dfrac{828}{4} = 207$
Median: 174, **188**, **206**, 260;
$$\text{median} = \frac{188+206}{2} = \frac{394}{2} = 197$$
Mode: none
Range: $260 - 174 = 86$

4. Mean:
$$\frac{3.8+2.6+4.1+4.8+5.9+2.7+6.9}{7} = \frac{30.8}{7} = 4.4$$
Median: 2.6, 2.7, 3.8, **4.1**, 4.8, 5.9, 6.9;
median = 4.1
Mode: none
Range: $6.9 - 2.6 = 4.3$

5. Mean: $\dfrac{180+186+185+180+210}{5}=\dfrac{941}{5}=188.2$

Median: 180, 180, **185**, 186, 210;

median = 185

Mode: 180 with a frequency of 2.

Range: $210 - 180 = 30$

6. Answers will vary, accept any answer students can justify.

7. Mean:

$\dfrac{68+2(73)+75+3(80)+(3)85+(2)91}{12}=\dfrac{966}{12}=80.5$

Median: 68, 73, 73, 75, 80, **80**, **80**, 85, 85, 85, 91, 91;

median = $\dfrac{80+80}{2}=\dfrac{160}{2}=80$

Mode: 80 and 85 both with a frequency of 3.

Range: $91 - 68 = 23$

8. Answers will vary, accept any answer students can justify.

9. Answers will vary. For example: Mean can be thought of as the balance point for the data set. It is the average. Median is the middle point or average of two middle points for a set of sorted data. Mode is the point that occurs most frequently. See problems 1–7 for examples.

10.

Music Type	Frequency
Blues	2
Classical	2
Country	2
Folk	1
Jazz	2
Rap	2
Rock	4
Show music	1

Mode: Rock with a frequency of 4.

PRACTICE

1. Mean:

$\dfrac{23+36+25+28+21+30+32+23}{8}=\dfrac{218}{8}=27.25$

Median: 21, 23, 23, **25**, **28**, 30, 32, 36;

median = $\dfrac{25+28}{2}=\dfrac{53}{2}=26.5$

Mode: 23 with a frequency of 2.

Range: $36 - 21 = 15$

2. Mean: $\dfrac{9.5+9.9+10.3+8.4+7.7+10.0}{6}=\dfrac{55.8}{6}=9.3$

Median: 7.7, 8.4, **9.5**, **9.9**, 10.0, 10.3;

median = $\dfrac{9.5+9.9}{2}=\dfrac{19.4}{2}=9.7$

Mode: none

Range: $10.3 - 7.7 = 2.6$

3. Mean:

$\dfrac{280+295+235+210+230+235+195}{7}=\dfrac{1680}{7}=240$

Median: 195, 210, 230, **235**, 235, 280, 295;

median = 235

Mode: 235 with a frequency of 2.

Range: $295 - 195 = 100$

4. Mean: $\dfrac{5+5+7+6+7+5+7+7+8+5}{10}=\dfrac{62}{10}=6.2$

Median: 5, 5, 5, 5, **6**, **7**, 7, 7, 7, 8;

median = $\dfrac{6+7}{2}=\dfrac{13}{2}=6.5$

Mode: 5 and 7 both with a frequency of 4.

Range: $8 - 5 = 3$

5. Mean: $\dfrac{4237+4516+4444+4379}{4}=\dfrac{17,576}{4}=4394$

Median: 4237, **4379**, **4444**, 4516;

median = $\dfrac{4379+4444}{2}=\dfrac{8823}{2}=4411.5$

Mode: none

Range: $4516 - 4237 = 279$

6. Mean: $\dfrac{3\frac{1}{2}+5\frac{3}{4}+4\frac{3}{8}+3\frac{7}{8}}{4}=$

$\dfrac{\frac{28}{8}+\frac{46}{8}+\frac{35}{8}+\frac{31}{8}}{4}=$

$\dfrac{\frac{140}{8}}{4}=\dfrac{140}{8}\cdot\dfrac{1}{4}=\dfrac{140}{32}=4\frac{3}{8}$

Median: $3\frac{1}{2}, 3\frac{7}{8}, 4\frac{3}{8}, 5\frac{3}{4}$;

median = $\dfrac{3\frac{7}{8}+4\frac{3}{8}}{2}=\dfrac{8\frac{2}{8}}{2}=\dfrac{\frac{66}{8}}{2}$

$=\dfrac{66}{8}\cdot\dfrac{1}{2}=\dfrac{66}{16}=4\frac{1}{8}$

Mode: none

Range: $5\frac{3}{4}-3\frac{1}{2}=\dfrac{23}{4}-\dfrac{14}{4}=\dfrac{9}{4}=2\frac{1}{4}$

7. Mean:

$\dfrac{12,560+14,300+13,750+12,400+13,680+15,420}{6}=$

$\dfrac{82,110}{6}=13,685$

Median: 12,400, 12,560, **13,680**, **13,750**, 14,300, 15,420;

median = $\dfrac{13,680+13,750}{2}=$

$\dfrac{27,430}{2}=13,715$

Mode: none

Range: $15,420 - 12,400 = 3020$

8. Mean:

$$\frac{18{,}000 + 18{,}000 + 22{,}000 + 45{,}000}{4} = \frac{103{,}000}{4} = 25{,}750$$

Median: 18,000, **18,000**, **22,000**, 45,000;

$$\text{median} = \frac{18{,}000 + 22{,}000}{2} =$$

$$\frac{40{,}000}{2} = 20{,}000$$

Mode: 18,000 with a frequency of 2.
Range: $45{,}000 - 18{,}000 = 27{,}000$

9. Answers will vary, accept any answer students can justify.

10. Since 17,000 is smaller than any point in the set, mean will decrease. Similarly, median will also decrease since 18,000 is now middle point.

11. Mean: $\dfrac{390 + 5(400) + 402 + 3(404) + 3(410) + 440}{14} =$

$$\frac{5674}{14} \approx 405.3$$

Median: 390, 400, 400, 400, 400, 400, **402**, **404**, 404, 404, 410, 410, 410, 440;

$$\text{median} = \frac{402 + 404}{2} = \frac{806}{2} = 403$$

Mode: 400 with a frequency of 5.
Range: $440 - 390 = 50$

12. Answers will vary depending on student's data. One possible answer could be the grades on a difficult test in a large class. Median would best represent the data as a difficult test would probably have scores skewed to the right.

13. Mean:

$$\frac{210.5 + 212.4 + 210.0 + 209.8 + 213.5 + 212.6 + 210.5 + 211.5 + 210.8}{9}$$

$$= \frac{1901.6}{9} \approx 211.3$$

Median: 209.8, 210.0, 210.5, 210.5, **210.8**, 211.5, 212.4, 212.6, 213.5;
median = 210.8

Mode: 210.5
Range: $213.5 - 209.8 = 3.7$

EXTEND

14. Let n represent Michelle's total test score for quizzes 1-8.

Then $\dfrac{n}{8} = 89$

$$n = 712$$

Let q represent Michelle's 9th test score.

Then $\dfrac{712 + q}{9} = 90$

$$712 + q = 810$$

$$q = 98$$

Michelle must score 98.

15. Answers will vary. One solution is 18, 20, 20, 20, 22.

16. Median, since 50% were priced at $150,000 or less.

17. Mode, since it was most frequent.

18. Mean

19. Range, since this is how the data varies, not a measure of center.

20. Mode, since they were most frequent.

21. Either $14 or $15 since each of these occur twice.

22. Since $14 and $15 are modes now, any value not $12, $14, or $15.

23. $12 because than all values occur twice.

24. $= \dfrac{90 + 95 + 75 + 80 + 85 + 65 + 70 + 80 + 60 + x}{10} = 78$

$$700 + x = 780$$

$$x = 80$$

THINK CRITICALLY

25. Age 15: $\quad \dfrac{9}{360} = \dfrac{n}{240}$

$$360 \cdot n = 9 \cdot 240$$

$$360n = 2160$$

$$n = 6$$

Age 16: $\quad \dfrac{144}{360} = \dfrac{n}{240}$

$$360 \cdot n = 144 \cdot 240$$

$$360n = 34{,}560$$

$$n = 96$$

Age 17: $\quad \dfrac{180}{360} = \dfrac{n}{240}$

$$360 \cdot n = 180 \cdot 240$$

$$360n = 43{,}200$$

$$n = 120$$

Age 18: $\quad \dfrac{27}{360} = \dfrac{n}{240}$

$$360 \cdot n = 27 \cdot 240$$

$$360n = 6480$$

$$n = 18$$

26. Mean:

$$\frac{6(15) + 96(16) + 120(17) + 18(18)}{240} = \frac{3990}{240} = 16.625$$

Median is the average of the 120th and 121st data points when they are sorted. The first 6 data points are 15 years, the next 96 are 16 years, and the next 120 are 17 years.

$$\frac{17 + 17}{2} = \frac{34}{2} = 17$$

Mode: 17 with a frequency of 120.

27. Data item must be equal to one item already in data set.

28. Data item must be smaller than the original mean of the set.

29. Answers will vary; possible answers include:
$A = \{10, 20, 30, 40, 50\}$ and $B = \{10, 25, 25, 40, 50\}$.
Sum of the data must be 150 for both A and B. Largest and smallest elements of A and B must be the same.

MIXED REVIEW

30.
$$\frac{13}{26} = \frac{1}{m}$$
$$13 \cdot m = 26 \cdot 1$$
$$m = 2$$

31.
$$\frac{x}{3} = \frac{36}{27}$$
$$27 \cdot x = 3 \cdot 36$$
$$x = 4$$

32.
$$\frac{10}{15} = \frac{p}{12}$$
$$15 \cdot p = 10 \cdot 12$$
$$p = 8$$

33.
$$\frac{35}{y} = \frac{25}{30}$$
$$25 \cdot y = 35 \cdot 30$$
$$y = 42$$

34. 44, 56, 56, **66**, **75**, 77, 80, 81;
median $= \dfrac{66 + 75}{2} = \dfrac{141}{2} = 70.5$; A

ALGEBRA WORKS

1. Sales increased from 0 CDs to 434 million CDs in 9 years or about 48.2 million CDs per year. From 1991 to 1998 is 8 years, so sales will increase by $8 \cdot 48.2$ million ≈ 386.
$434 + 386 = 820$ million.
Let **D** represent the inventory matrix after shipments.

2. Then $\mathbf{A + B - C = D}$;
$D_{11} = 88{,}000 + 46{,}000 - 35{,}000 = 99{,}000$;
$D_{12} = 195{,}000 + 123{,}000 - 98{,}000 = 220{,}000$;
$D_{13} = 52{,}000 + 18{,}000 - 21{,}000 = 49{,}000$;
$D_{14} = 436{,}000 + 257{,}000 - 209{,}000 = 484{,}000$;
$D = [99{,}000 \quad 220{,}000 \quad 49{,}000 \quad 484{,}000]$

3a. E2 = A2 + B2 + C2 + D2
F2 = E2 / 4
E3 = A3 + B3 + C3 + D3
F3 = E3 / 4

b. 288,000; 72,000; 223,000; 55,750

Lesson 1.8, pages 38–41

EXPLORE/WORKING TOGETHER

1. Answers will vary depending on student's data.

2. Ratios will vary. Extensive sampling reveals that about 23.5% of the letters on a page of normal english prose are E's and T's. Students should see that the groups results are likely to be different, depending on the lines that each group chose to sample.

3. Since there are $5 \cdot 150{,}000 = 750{,}000$ letters in this novel, students should multiply 750,000 by the percentage they found in problem **2** above.

TRY THESE

1. $\dfrac{80}{80} = 1$; about 1

2. $\dfrac{0}{80} = 0$; about 0

3. $\dfrac{30}{80} = 0.375$; about 0.375

4. $\dfrac{56}{175} = 0.32$ or 32%

5. 32% of 250,000 $= 0.32 \cdot 250{,}000 = 80{,}000$

6. $\dfrac{8}{50} = \dfrac{4}{25} = 0.16$ or 16%

7. $\dfrac{42}{50} = \dfrac{21}{25} = 0.84$ or 84%

8. $\dfrac{50}{50} = 1$ or 100%

9. $\dfrac{0}{50} = 0$ or 0%

10. $\dfrac{151 + 21 + 31 + 2}{151 + 130 + 21 + 31 + 19 + 2} = \dfrac{205}{354} = 0.579$ or 57.9%

11. Answers will vary, but students should describe an experiment where they toss a folded index card and record the results. The more times the experiment is performed, the more reliable the results.

PRACTICE

1. $\dfrac{251}{365} = 0.688$ or 68.8%

2. $\dfrac{12}{75} = \dfrac{4}{25} = 0.16$ or 16%

3. $\dfrac{18}{75} = \dfrac{6}{25} = 0.24$ or 24%

4. $\dfrac{69}{75} = \dfrac{23}{25} = 0.92$ or 92%

5. $\dfrac{0}{75} = 0$ or 0%

6. $\dfrac{63 + 11}{248 + 90} = \dfrac{74}{338} \approx 0.219$ or 21.9%

7. 21.9% of 800,000 = 0.219 • 800,000 = 175,200

8. Answers will vary. On 75% of the days in the past when weather conditions were the same as today's conditions, it has rained.

9. $\dfrac{145+114}{136+145+109+114} = \dfrac{259}{504} \approx 0.514$ or 51.4%

10. $\dfrac{142+136+109+122}{142+136+109+122+165+145+114+102} =$ $\dfrac{509}{1035} \approx 0.492$ or 49.2%

11. $\dfrac{122+102}{1035} = \dfrac{224}{1035} \approx 0.216$ or 21.6%

EXTEND

12. Let f = number of families surveyed.
$$0.28f = 105$$
$$f = \dfrac{105}{0.28} = 375 \text{ families}$$

13. Answers will vary. Possible answer: a coin that was tossed 200 times came up tails 90 times. What is the experimental probability of tossing a tail?

THINK CRITICALLY

14. Answers will vary. The likelihood of tossing a head is around 50% on every toss, regardless of what happened before. The coin does not "know" that it has come up heads 20 times in a row.

Lesson 1.9, pages 42–45

EXPLORE THE PROBLEM

1. The final scores for all Super Bowls are given. The standard used to decide if a game was exciting is given (winning margin of 7 or fewer points).

2. The winning margin is the difference between the winner's and loser's final scores.

3. Answers will vary. What percent of Super Bowl games ended with a winning margin of 7 or fewer points?

4. The percent of Super Bowl games through 1995 that ended with a winning margin of 7 points or less.

5. No, data are organized sufficiently.

6. Subtract loser's score from winner's score to obtain winning margin.

7. Count the number of winning margins of 7 or less and divide by 29 games.

8. Use a calculator for the arithmetic.

9. $\dfrac{7}{29} \approx 0.24$ or 24%

10. Answers will vary. $\dfrac{7}{29}$ is a little less than $\dfrac{1}{4}\left(\dfrac{7}{28}\right)$ and 24% is a little less than 25%.

INVESTIGATE FURTHER

11. Answers will vary. Find the average winning margin for the years 1967 to 1995 by adding all the winning margins and dividing by 29. Find the average winning margin for 1985 to 1995 by adding all the winning margins for those years and dividing by 11. Subtract the smaller from the greater average value.

12. winning margin from 1967 to 1995:
$$\dfrac{490}{20} \approx 16.9; \ 16.9 \text{ points}$$
winning margin from 1985 to 1995:
$$\dfrac{247}{11} \approx 22.5; \ 22.5 \text{ points}$$
$22.5 - 16.9 = 5.6$; The difference is 5.6 points.

13. Answers will vary. Students could find the total points for 1985–1998 first, record the total, and then continue adding to find the total for all games. This saves some work and reduces the chance of error.

14. Answers will vary, such as: Which team has made more trips to the Super Bowl than any other? What is the average number of points they scored per game? What percentage of Super Bowls did they win?

APPLY THE STRATEGY

15. 2 • 21 = 42; 42 weeks

16. $\dfrac{58+4+42+7+5+139+8+9+2+12+30+21}{12} =$ $\dfrac{337}{12} \approx 28; \ 28 \text{ weeks}$

17. $\dfrac{5}{12} = 0.416 \approx 42\%$

18. 30 • 150 • $16.94 = $76,230

19. $\dfrac{4,500,000}{139} \approx 32,374$ copies/week

20. 4,500,000 copies • $2\dfrac{3}{8}$ in./copy = 10,687,500 in.
10,687,500 in. ÷ 12 in./ft = 890,625 ft
890,625 ft ÷ 5280 ft/mi ≈ 168.7 mi

21. Each book has area = 7.5" • 10" = 75 in.^{2}
4,500,000 • 75 in.^{2} = 337,500,000 in.^{2}
337,500,000 in.^{2} ÷ 144 in.^{2}/ft^2
= 2,343,750 ft^2

REVIEW PROBLEM SOLVING STRATEGIES

1a. Job takes 3 days, so they get $\dfrac{1}{3}$ done per day.

b. There are 3 workers, so $\dfrac{1}{3}$ of $\dfrac{1}{3} = \dfrac{1}{9}$ is amount each worker completes per day.

c. 1 job $- \dfrac{1}{3}$ complete $= \dfrac{2}{3}$ of job remaining

d. $\dfrac{1}{9} + \dfrac{1}{9} = \dfrac{2}{9}$ done per day by Leon and Yvonne

e. $\dfrac{\frac{2}{3}\text{ remaining}}{\frac{2}{9}\text{ done per day}} = \dfrac{2}{3} \cdot \dfrac{9}{2} = 3$ more days to complete job

f. Length of road was not needed.

2a. No. Since 1 is the multiplicative identity, another number would have to be repeated.

b. 4, because 5 or more would have a 2 digit product when multiplied by a number greater than one.

c. 6 or 8; none of the other numbers is a product of distinct factors.

d.
$$9 - 5 = 4$$
$$\times$$
$$6 \div 3 = 2$$
$$=$$
$$7 + 1 = 8 \quad \text{(or } 1 + 7)$$

3. three steps;

step 1: ↑ ↑ ↑(↑ ↑)↓

step 2: ↑(↓ ↓)↑ ↑ ↓

step 3: ↑ ↓(↑ ↓)↑ ↓

other steps possible:

step 1: ↑ ↑(↓ ↑)↓ ↓

step 2: ↑(↓ ↑)↑ ↓ ↓

step 3: ↑ ↓ ↑(↓ ↑)↓

Chapter Review, pages 46–47

1. a   **2.** d   **3.** e   **4.** b   **5.** c

6. Answers will vary. Graphs will vary depending on type of graph drawn and whether items of data are shown individually or grouped by intervals.

Sample answers:

Price	Tally	Frequency
$15–19	III	3
$20–24	LHT	5
$25–29	IIII	4
$30–34	LHT	5
$35–39	III	3

Price	Tally	Frequency
$17	I	1
$18	II	2
$20	II	2
$24	III	3
$25	IIII	4
$32	LHT	5
$35	I	1
$38	II	2

7. $\dfrac{8}{20} = 0.4$ or 40%

8. Answers will vary.

	A	B	C	D
1		Regal	Ovenmaster	Total
2	Campus	38	52	90 ⟵ =B2+C2
3	Eastwood	49	70	119 ⟵ =B3+C3
4	Central	25	41	66 ⟵ =B4+C4

9. $179 \cdot 5 + 229 \cdot 9 = 895 + 2061 = 2956$

10. convenience sampling

11.
$$\dfrac{650}{3278} = \dfrac{n}{90,546}$$
$$3278 \cdot n = 650 \cdot 90,546$$
$$3278n = 58,854,900$$
$$n = \dfrac{58,854,900}{3278} \approx 18,000$$

12. $\begin{bmatrix} 7+3 & 4+3 & 9+8 & 6+4 \\ 2+2 & 9+0 & 1+0 & 5+5 \end{bmatrix}$

$= \begin{bmatrix} 10 & 7 & 17 & 10 \\ 4 & 9 & 1 & 10 \end{bmatrix}$

13. $\begin{bmatrix} 16-3 & 9-3 & 18-8 & 7-4 \\ 25-2 & 30-0 & 2-0 & 15-5 \end{bmatrix}$

$= \begin{bmatrix} 13 & 6 & 10 & 3 \\ 23 & 30 & 2 & 10 \end{bmatrix}$

14. $\begin{bmatrix} 7-3+16 & 4-3+9 & 9-8+18 & 6-4+7 \\ 2-2+25 & 9-0+30 & 1-0+2 & 5-5+15 \end{bmatrix}$

$= \begin{bmatrix} 20 & 10 & 19 & 9 \\ 25 & 39 & 3 & 15 \end{bmatrix}$

15. Mean:

$$\dfrac{84+80+78+80+79+81+84+86+82+83}{10}$$
$$= \dfrac{817}{10} = 81.7$$

Median: 78, 79, 80, 80, **81**, **82**, 83, 84, 84, 86;

$$\text{median} = \dfrac{81+82}{2} = \dfrac{163}{2} = 31.5$$

Mode: 80 and 84 both with a frequency of 2.

Range: $86 - 78 = 8$

16. Mean: $\dfrac{2+4(2)+12(3)+9(4)+2(5)+6}{2+4+12+9+2+1} = \dfrac{98}{30} \approx 3.3$

Median: Median is the average of the 15th and 16th points, which are both 3.

$$\text{median} = \dfrac{3+3}{2} = 3.$$

Mode: 3 with a frequency of 12.

Range: $6 - 1 = 5$

17. $\dfrac{152}{152+124+96+578}=\dfrac{152}{950}=0.16$ or 16%

18. $\dfrac{152+96}{950}=\dfrac{248}{950}\approx 0.261$ or 26.1%

Chapter Assessment, page 48

1. $10-5=5$

2. $\dfrac{12\cdot 5}{6}=\dfrac{60}{6}=10$

3. $3\cdot 10+5=30+5=35$

4. $\begin{bmatrix} 12-10 & 9-9 \\ 7-4 & 14-12 \\ 8-5 & 20-15 \end{bmatrix}=\begin{bmatrix} 2 & 0 \\ 3 & 2 \\ 3 & 5 \end{bmatrix}$

5. $\begin{bmatrix} 3+12 & 7+9 \\ 8+7 & 2+14 \\ 5+8 & 6+20 \end{bmatrix}=\begin{bmatrix} 15 & 16 \\ 15 & 16 \\ 13 & 26 \end{bmatrix}$

6. $\begin{bmatrix} 3+10-12 & 7+9-9 \\ 8+4-7 & 2+12-14 \\ 5+5-8 & 6+15-20 \end{bmatrix}=\begin{bmatrix} 1 & 7 \\ 5 & 0 \\ 2 & 1 \end{bmatrix}$

7. Mean:

$$\dfrac{16,999+14,490+17,249+24,309+18,160+17,399}{6}$$

$$=\dfrac{108,606}{6}=18,101$$

Median: 14,490, 16,999, **17,249**, **17,399**, 18,160, 24,309;

$$\text{median}=\dfrac{17,249+17,399}{2}=$$

$$\dfrac{34,648}{2}=17,324$$

Mode: none
Range: $24,309-14,490=9819$
So, C is true.

8. $\dfrac{12}{30}=\dfrac{2}{5}=0.4$ or 40%

9. $\dfrac{12}{30}=\dfrac{2}{5}=0.4$ or 40%

10. C

11. A

12. B

13. Answer should describe using the survey findings and the number of registered voters to write and solve a proportion.

14.

mi/gal	Tally	Frequency
24	II	2
25	IIII	4
27	IIII	4
30	IIII	5
32	III	3
38	I	1
39	I	1

mi/gal	Tally	Frequency
21–25	IIII I	6
26–30	IIII IIII	9
31–35	III	3
36–40	II	2

15. $\dfrac{5}{20}=\dfrac{1}{4}=0.25$ or 25%

16.

	A	B	C
1		Posters	T-shirts
2	Stand 1	235	258
3	Stand 2	456	379
4	Stand 3	329	402
5	Total	1020	1039

17. $I=\begin{bmatrix} 400 & 400 \\ 500 & 600 \\ 500 & 600 \end{bmatrix}$ $S=\begin{bmatrix} 235 & 258 \\ 456 & 379 \\ 329 & 402 \end{bmatrix}$ $D=\begin{bmatrix} 300 & 200 \\ 600 & 400 \\ 500 & 400 \end{bmatrix}$

$I-S+D=$

$$\begin{bmatrix} 400-235+300 & 400-258+200 \\ 500-456+600 & 600-379+400 \\ 500-329+500 & 600-402+400 \end{bmatrix}$$

$$=\begin{bmatrix} 465 & 342 \\ 644 & 621 \\ 671 & 598 \end{bmatrix}$$

18. $\dfrac{456}{456+379}=\dfrac{456}{835}\approx 0.546$ or 54.6%

Cumulative Review, page 50

1.

$$\dfrac{3}{5}=\dfrac{n}{35}$$

$$5\cdot n=3\cdot 35$$

$$5n=105$$

$$n=\dfrac{105}{5}=21$$

2. $\dfrac{4}{t} = \dfrac{10}{15}$

$10 \cdot t = 4 \cdot 15$

$10t = 60$

$t = \dfrac{60}{10} = 6$

3. $36 \cdot \dfrac{7}{9} \;\square\; \dfrac{3}{4} \cdot 36$

$28 \;\square\; 27$

$>$

4. $0.05 \cdot 80 \;\square\; 0.80 \cdot 5$

$4 \;\square\; 4$

$=$

5. $100 \cdot 4\dfrac{63}{100} \;\square\; 4\dfrac{7}{10} \cdot 100$

$463 \;\square\; 470$

$<$

6. Winter

7. 7

8. $\dfrac{9}{27} = \dfrac{1}{3} = 0.33$ or 33%

9. $8 \cdot 5 = 40$

10. $\dfrac{6}{8-5} = \dfrac{6}{3} = 2$

11. $4 \cdot 6 + 5 - 8 = 24 + 5 - 8 = 29 - 8 = 21$

12. $\begin{bmatrix} 3+6-5 & 7+3-8 \\ 4+1-2 & 2+9-3 \end{bmatrix} = \begin{bmatrix} 4 & 2 \\ 3 & 8 \end{bmatrix}$

13. No. To add matrices, their dimensions must be the same, and 2×3 is not the same as 3×2.

14. Tallest bar is at 5, so mode is 5

15. $\dfrac{3+1+1}{1+2+5+9+8+3+1+1} = \dfrac{5}{30} = \dfrac{1}{6}$

16. I True, since 17 of 30 tossed 5 or 6 heads.

II False, since it is possible to get 0 or 10 heads.

III True, since the coin is fair.
Correct answer is C.

17. No. 90% is his current experimental probability, but it is very unlikely that Joey would continue to toss 9 more heads in every 10 tosses.

18. $\dfrac{38}{50} = \dfrac{n}{475}$

$50 \cdot n = 38 \cdot 475$

$50n = 18{,}050$

$n = \dfrac{18{,}050}{50} = 361$

Standardized Test, page 51

1. C; $0.75 \cdot n = 36$

$n = \dfrac{36}{0.75} = 48$

2. E; First four prime numbers are 2, 3, 5, 7.
LCM $= 2 \cdot 3 \cdot 5 \cdot 7 = 210$

3. C; $\dfrac{3}{5} \cdot 60 = \dfrac{180}{5} = 36$ calls made;

$60 - 36 = 24$ calls left to make.

4. A; $15\% = \dfrac{15}{100} = \dfrac{A}{1+A+4+7+5}$

$100A = 15(A + 17)$

$100A = 15A + 255$

$85A = 255$

$A = 3$

5. C; Volume of 2 cm cube is $2 \cdot 2 \cdot 2 = 8$;
Volume of 8 cm cube is $8 \cdot 8 \cdot 8 = 512$;

$\dfrac{512}{8} = 64$ cubes needed that are 2 cm

6. B; $(0.32 + 2.4 + 6.85) - 3.5 = 9.57 - 3.5 = 6.07$

7. E; Area $= \dfrac{1}{2} \cdot b \cdot h$; $16 = \dfrac{1}{2} \cdot 2h \cdot h$

$16 = h^2$

$4 = h$

$b = 2h = 2 \cdot 4 = 8$

8. A; From the first row, second column: $x - 2 = 4$, so $x = 6$. From the second row, first column: $7 - x = y$, since $x = 6$, $7 - 6 = y$, $1 = y$.

9. E;

I True, freshman + sophomores + juniors + seniors $=$ $15\% + 5\% + 25\% = 70\%$.

II False, since largest sector is 30% and $0.30 \times 360° = 108°$.

III True, 30% of 800 is $0.30 \cdot 800 = 240$.

10. B; Let n represent the number of numbers in the set.

$16 = \dfrac{96}{n}$

$16n = 96$

$n = 6$

Chapter 2   Variables, Expressions, and Real Numbers

Data Activity, pages 52–53

Let x represent the number of tornadoes in 1990.

1. Let x represent the number of tornadoes in 1990.

$x = 2 \cdot 684 - 235$

$x = 1368 - 235$

$x = 1133$

2. $\dfrac{866 + 783 + 1046 + 931 + 907 + 684 + 764 + 656 + 702 + 856}{10}$

$= \dfrac{8195}{10} \approx 820$

Range: $1046 - 656 = 390$

3. 1989 to 1990, increase of 277

4a. Answers will vary. Number has increased to over 1100.

  b. Answers will vary. Looking at 1990 to 1993, prediction would be about 1100 or more.

5. Pictographs will vary. Students may round numbers to the nearest 25 or 50, and let one symbol represent 100 tornadoes.

Lesson 2.1, pages 55–58

THINK BACK

1. $= 26 * A2$

2. $= A2 / 15$

3. $26 \cdot 145 = \$3770$

4. $145 \div 15 = 9.\overline{6}$ or 9 teams. No, each team does not get exactly 15 members. 9 teams of 15 is 135 people, so 10 are left over. Give one additional person to each team and assign the last person randomly to any team. There are now 8 teams with 16 members and 1 with 17 members.

EXPLORE

5. -4 **6.** $-x$ **7.** $x + 2$

8. $y + 3$ **9.** $6 + x$ **10.** $3x + 1$

11. $2y + 3$ **12.** $-3x$ **13.** $4 - 2y$

14.    **15.**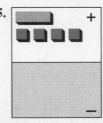

16. 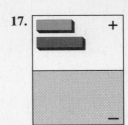   **17.**

18. 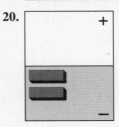   **19.**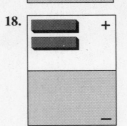

20.    **21.**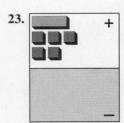

MAKE CONNECTIONS

22. 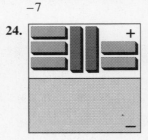   **23.**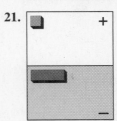

-7 $x + 5$

24. 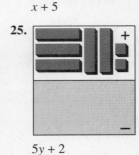   **25.**

$5x + 2y$ $5y + 2$

SUMMARIZE

26. Answers will vary but should be similar to text. One side of the Basic Mat represents positive values and the other side represents negative values. Different variables and numbers are represented by blocks of various colors and sizes. Expressions are formed using the various combinations of blocks and various placements on the Basic Mat.

27a.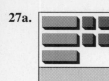

b.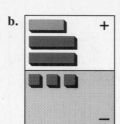

c.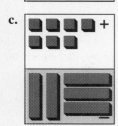

28. Remove the y block and replace it with 3 unit blocks. There are now 9 unit blocks on the positive side of the Basic Mat so the value of $y + 6 = 9$.

Lesson 2.2, pages 59–64

EXPLORE/WORKING TOGETHER

1. Joan added 40 and 88 first to get 128, then divided by 4 to get 32.

2. Leah divided 88 by 4 first to get 22, then add this to 40 to get 62.

3. They did the operations in a different order.

4. Answers will vary, but Leah used the correct order, which will be explained later in this section.

5. Answers will vary depending on calculator, but those with algebraic logic should give 62 as an answer.

TRY THESE

1. $9 \cdot 4 - 3 \cdot 8 = 36 - 24 = 12$

2. $8 + 12 \div 4 = 8 + 3 = 11$

3. $7 + 2^3 = 7 + 8 = 15$

4. $2 + 3 \cdot 5^2 - 6 \cdot 2 = 2 + 3 \cdot 25 - 6 \cdot 2 = 2 + 75 - 12 = 65$

5. $10 + 6 \div 4 \cdot 2 = 10 + 1.5 \cdot 2 = 10 + 3 = 13$

6. $(10 + 6) \div 4 \cdot 2 = 16 \div 4 \cdot 2 = 4 \cdot 2 = 8$

7. $(10 + 6) \div (4 \cdot 2) = 16 \div 8 = 2$

8. $10 + [6 \div (4 \cdot 2)] = 10 + [6 \div 8] = 10 + \frac{3}{4} = 10\frac{3}{4}$

9. $10 - 1 = 9$

10. $2 \cdot 10^2 = 2 \cdot 100 = 200$

11. $3 \cdot 10 + 10 = 30 + 10 = 40$

12. $\frac{10}{2} = 5$

13.

x	$5x + 2$
3	$5 \cdot 3 + 2 = 15 + 2 = 17$
5	$5 \cdot 5 + 2 = 25 + 2 = 27$
8	$5 \cdot 8 + 2 = 40 + 2 = 42$
10	$5 \cdot 10 + 2 = 50 + 2 = 52$

14. degree 6; 4 terms; constant is 7

15. degree 2; 3 terms; constant is -1

16. degree 1; 2 terms; constant is 24

17. 4   **18.** -7 **19.** 1   **20.** -1

21. $-\frac{3}{4}$

22. 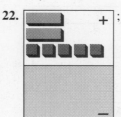 ; Let $x = 3$, then: 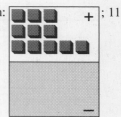 ; 11

23. octagon has 8 sides of length s, $8s - 4$

24. Answers will vary, but should include substituting a value for the variable in the expression and evaluating the expression using the order of operations.

PRACTICE

1. $8 + 3 \cdot 9 = 8 + 27 = 35$

2. $34 - 10 \div 2 = 34 - 5 = 29$

3. $9 \cdot 2 + 5 \cdot 3 = 18 + 15 = 33$

4. $9 \cdot (2 + 5) \cdot 3 = 9 \cdot 7 \cdot 3 = 189$

5. $42 - 5 \cdot 5 = 42 - 25 = 17$

6. $79 - 3 \cdot 8 = 79 - 24 = 55$

7. $7 + 81 \div 9 + 3 = 7 + 9 + 3 = 19$

8. $(12 + 15) \div 3 + 20 = 27 \div 3 + 20 = 9 + 20 = 29$

9. $200 \div 40 \cdot 8 = 5 \cdot 8 = 40$

10. $(25 - 5) \cdot 5 + 5 = 20 \cdot 5 + 5 = 100 + 5 = 105$

11. $36 \div 18 \div 2 = 2 \div 2 = 1$

12. $\frac{1}{4}(6 \cdot 8) + 2 \cdot 12 = \frac{1}{4}(48) + 2 \cdot 12 = 12 + 24 = 36$

13. $7^2 - 2 \cdot 8 = 49 - 2 \cdot 8 = 49 - 16 = 33$

14. $8^2 - 5 \cdot 4 = 64 - 5 \cdot 4 = 64 - 20 = 44$

15. $3 \cdot (4^2 + 2) = 3 \cdot (16 + 2) = 3 \cdot 18 = 54$

16. $2 \cdot 5^3 + 10 = 2 \cdot 125 + 10 = 250 + 10 = 260$

17. $3^4 - 2^3 = 81 - 8 = 73$

18. $4^3 - 4^2 = 64 - 16 = 48$

19. $10 \cdot 2^5 + 4 \cdot 2 + 36 \div 4 = 10 \cdot 32 + 4 \cdot 2 + 36 \div 4$
$= 320 + 8 + 9 = 337$

20. $2 \cdot 10^3 + 2 \cdot 30 + 200 \div 5 = 2 \cdot 1000 + 2 \cdot 30 + 200 \div 5$
$= 2000 + 60 + 40 = 2100$

21. $3.50 + x$

22. $3 \cdot 5 - 1 = 15 - 1 = 14$

23. $5^2 + 23 = 25 + 23 = 48$

24. $4 \cdot 5^3 = 4 \cdot 125 = 500$

25. $5 \cdot 5 - 2 = 25 - 2 = 23$

26. $\dfrac{3 \cdot 5}{10} = \dfrac{15}{10} = \dfrac{3}{2}$

27.

y	$3y - 2$
3	$3 \cdot 3 - 2 = 9 - 2 = 7$
4	$3 \cdot 4 - 2 = 12 - 2 = 10$
5	$3 \cdot 5 - 2 = 15 - 2 = 13$
6	$3 \cdot 6 - 2 = 18 - 2 = 16$

28.

a	$5a^2 + 6$
4	$5 \cdot 4^2 + 6 = 5 \cdot 16 + 6 = 80 + 6 = 86$
5	$5 \cdot 5^2 + 6 = 5 \cdot 25 + 6 = 125 + 6 = 131$
6	$5 \cdot 6^2 + 6 = 5 \cdot 36 + 6 = 180 + 6 = 186$
10	$5 \cdot 10^2 + 6 = 5 \cdot 100 + 6 = 500 + 6 = 506$

29.

g	$\frac{g}{4}$
4	$\frac{4}{4} = 1$
8	$\frac{8}{4} = 2$
12	$\frac{12}{4} = 3$
20	$\frac{20}{4} = 5$

30.

k	$\frac{k}{3}$
3	$\frac{3}{3} = 1$
6	$\frac{6}{3} = 2$
18	$\frac{18}{3} = 6$
24	$\frac{24}{3} = 8$

31. degree 4; 4 terms; constant is -3

32. degree 3; 4 terms; constant is 12

33. degree 6; 6 terms; constant is 10

34. degree 8; 2 terms; no constant term

35. 9   **36.** 1   **37.** -3 **38.** -1

39. $\dfrac{y}{8}$

40. Answer should include dealing with at least one unknown represented by a variable. For example, if x represents the number of friends amd a movie costs \$4, $4x$ represents the amount of money it will cost. If popcorn costs \$2, the total cost would be $4x + 2x$, or $6x$.

EXTEND

41. $2 \cdot 3 + 3 \cdot 4 = 6 + 12 = 18$

42. $3^2 + 4^2 = 9 + 13 = 25$

43. $3 \cdot 3^2 - 4 = 3 \cdot 9 - 4 = 27 - 4 = 23$

44. $5 \cdot 3 + 4^4 = 5 \cdot 3 + 256 = 15 + 256 = 271$

45. $12^4 + 7^5 = 20{,}736 + 16{,}807 = 37{,}543$

46. $125 + 28 \cdot 45 = 125 + 1260 = 1385$

47. $(2.5)^3 = 15.625; 15.6$

48. $(3.04)^4 = 85.40717056; 85.4$

49. $452 + 1360 \div 8 = 452 + 170 = 622$

50. $674 + 2943 \div 4 = 674 + 735.75 = 1409.75; 1409.8$

51. $450 + 0.02x$

52. $y + 0.04y$

53. $69.089 + 2.238 \cdot 42 = 69.089 + 93.996 = 163.085;$
163.1 cm or $\dfrac{163.1}{2.54} \times 64" = 5' \, 4"$.

THINK CRITICALLY

54. For $3x$ to be divisible by 6, $3x \div 6$ must have no remainder. Therefore x must be even.

55. $2^4 \stackrel{?}{=} 4^2$, $16 = 16$, true; no, $a^b \neq b^a$; example: $2^5 \stackrel{?}{=} 5^2$, $32 \neq 25$

56. 5 places; the number of places in a product equals the sum of the number of places in the factors; 2.48832

Lesson 2.3, pages 65–72

EXPLORE

1. $10°$ **2.** $2°$

3. shaded part moves up 5 short lines along scale

4. shaded part moves down 4 short lines along scale

5. by 6 short lines or $12°$

6. 6 short lines above $0°$ or $12°$

TRY THESE

1. rational; it is the ratio of integers

2. rational; it is a terminating decimal and may be written as $\dfrac{9}{10}$

3. irrational; 3 is not a perfect square

4. rational; it is a nonterminating, repeating decimal and may be written as $\dfrac{25}{99}$

5. irrational; 11 is not a perfect square

6. A   **7.** E   **8.** F

9. D   **10.** H   **11.** G

12. $-4 > -4.8$ **13.** $-2 < 0$ **14.** $-129 < -40$

15. $3 + 12 \div 3 \,\square\, 6$
$\quad 3 + 4 \,\square\, 6$
$\quad\quad 7 > 6$

16. $|-3| = 3$

17. $|9| = 9$

18. $\left| -\dfrac{1}{2} \right| = \dfrac{1}{2}$

19. $|4.9| = 4.9$

20. $-|-101| = -101$

21. Answers will vary. Possible answer: When modeling -4, you use 4 unit blocks. Their place on the mat determines the sign. The absolute value is the number of blocks.

22. Answers will vary, such as saving and borrowing money, bowling above and below average in a league, gaining and losing weight, etc.

23. Fairbanks; since $-4 > -20$

1. rational; may be written as $\frac{21}{4}$, a ratio of integers

2. rational; may be written as $\frac{28}{9}$, a ratio of integers

3. irrational; nonterminating and nonrepeating decimal

4. rational; may be written as $\frac{-21}{1}$, a ratio of integers

5. irrational; 5 is not a perfect square

6. irrational; 17 is not a perfect square

7. rational; may be written as $\frac{62}{10}$, a ratio of integers

8. rational; may be written as $-\frac{91}{10}$, a ratio of integers

9. rational; may be written as $\frac{7}{1}$, a ratio of integers

10. rational; may be written as $\frac{9}{1}$, a ratio of integers

11. rational; is a nonterminating, repeating decimal and may be written as $\frac{4}{9}$

12. rational; is a nonterminating, repeating decimal and may be written as $\frac{91}{99}$

13. C 14. F 15. H

16. J 17. E 18. B

19. A 20. D 21. G

22. I 23. $-12 < -4$ 24. $-20 > -21$

25. $-16 < 2$ 26. $-11 < 7$ 27. $-3 < -4$

28. $-1 < 0$ 29. $5.1 > -6.22$ 30. $10.2 > -4.3$

31. $|-2| = 2$ 32. $|-11| = 11$ 33. $\left|7\frac{2}{3}\right| = 7\frac{2}{3}$

34. $\left|-4\frac{3}{4}\right| = 4\frac{3}{4}$ 35. $-|-2.3| = -2.3$ 36. $-|4.7| = -4.7$

37. Answers will vary, but should include that every real number can be placed in only one location on the real number line. Also important is that the number line is the only way to show all real numbers.

38. Point B 39. Point E 40. Point D

41. 2.63 lb is the smallest weight still greater than or equal to $2\frac{1}{2}$ or 2.5 lb.

EXTEND

42. $-\sqrt{13}, -3.6, -3\frac{1}{3}, -3.31, \sqrt{9}, 3.2, \sqrt{11}, 3\frac{3}{8}$

43. $\sqrt{3}, 0.51, 0.5, \frac{9}{20}, -\frac{3}{10}, -0.98, -1.1, -\sqrt{4}$

44. $2 \cdot 5 + 36 \div 4 \ \square \ 4 \cdot 2 + 16 \div 8$

 $\qquad 10 + 9 \ \square \ 8 + 2$

 $\qquad\quad 19 \ > \ 10$

45. $-4|-3| \ \square \ -|-5.2|$

 $\quad -4 \cdot 3 \ \square \ -5.2$

 $\qquad -12 \ < \ -5.2$

46. $3 + 6 \cdot 7 \ \square \ 5^2 + 2 \cdot 10$

 $\quad 3 + 42 \ \square \ 25 + 20$

 $\qquad 45 \ = \ 45$

47. $6 + 3 \cdot 9 \ \square \ 4 + 8^2$

 $\quad 6 + 27 \ \square \ 4 + 64$

 $\qquad 33 \ < \ 68$

48. $-|-14| \ \square \ 2|-7|$

 $\quad -14 \ \square \ 2 \cdot 7$

 $\quad -14 \ < \ 14$

49. $|-4| \cdot |6| \ \square \ -|4| \cdot |6|$

 $\quad 4 \cdot 6 \ \square \ -4 \cdot 6$

 $\qquad 24 \ > \ -24$

50.

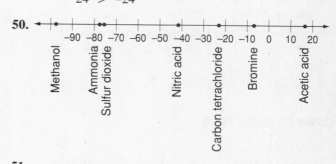

51.

THINK CRITICALLY

52. false; The absolute value of a number is never negative. If the number is positive, its absolute value is also positive.

53. true; The numbers on a number line decrease as you move to the left. However, the absolute value of numbers less than 0 increases as you move to the left.

MIXED REVIEW

54. $0.15x$ **55.** $\frac{x}{3}$

56. $\frac{[75-(9+13)]}{75} = 0.70\overline{6}$ or $70.\overline{6}\%$; B

57. $-33 < -12$ **58.** $-2 > -4$

59. $-6 > -10$ **60.** $-5 < 3$

ALGEBRA WORKS

1. 30 in.

2. Mercury will rise since the increased pressure will push more mercury into the tube.

3. Mercury will drop since the decreased pressure will allow more mercury to seep out of the tube into the container.

4. Laraville and Wiltshire

Lesson 2.4, pages 73–76

THINK BACK

1. Answers will vary.   **2.** Answers will vary.

3. $8 - 3 = 5$ points won

EXPLORE

4a.

     b. $9 - 4 = 5$ points won, overall score is not affected

     c. 5; Remove four blocks from both the positive and negative side.

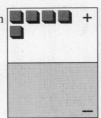

     5a. -6 **b.** 8   **c.** -4 **d.** -5

     6. In all four cases the addends were either both positive or both negative.

     7. -6 **8.** 3   **9.** 0   **10.** 0

     11. 0   **12.** 0   **13.** 0

     14. When you add the pair of opposites, the sum is zero.

     15. 0; 17 and -17 are opposites so the sum must be zero.

     16. yes; 5 and -5

     17. no; Total value is the same since zero was removed.

19. $-3 + 4 = 1$

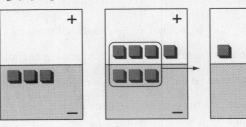

20. $3 + (-4) = -1$

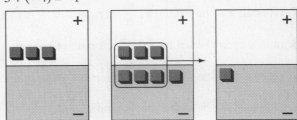

21. $5 + (-8) = -3$

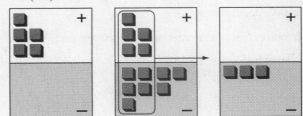

22. $-5 + 8 = 3$

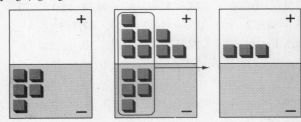

23. $-6 + 3 = -3$

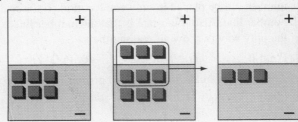

24. $6 + (-3) = 3$

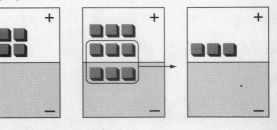

25. $-1 + 7 = 6$

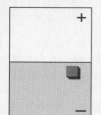 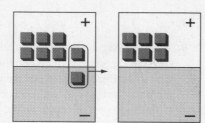 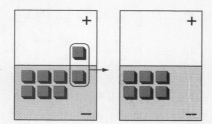

26. $1 + (-7) = -6$

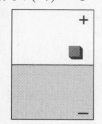 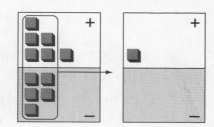

27. $6 + (-5) = 1$

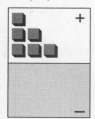

28. $3 + (-2) = 1$ **29.** $4 + (-1) = 3$

30. $1 + (-6) = -5$ **31.** $0 + (-5) = -5$

32. $6 - 4 = 2$ **33.** $-7 - (-3) = -4$

34. $9 - 8 = 1$ **35.** $4 - 3 = 1$

36. $-8 - (-5) = -3$ **37.** $-12 - (-8) = -4$

38a.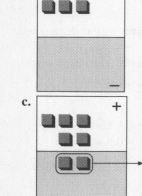

b. Add the zero pair +2 and −2 so there are 2 unit blocks on the negative side.

c.

d. 5

e. $3 - (-2) = 3 + 2 = 5$

39. $-4 - (-6) = 2$

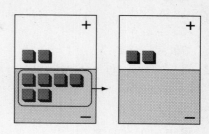 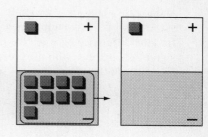 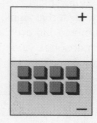

40. $-8 - (-9) = 1$

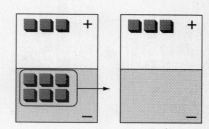

41. $-3 - (-6) = 3$

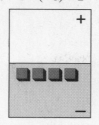 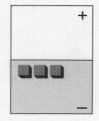 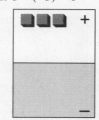

42. $3 - (-3) = 6$

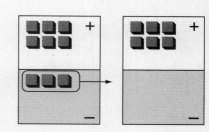

43. $3 - (-4) = 7$

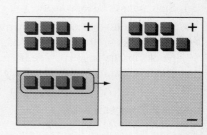

44. $6 - (-4) = 10$

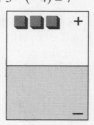 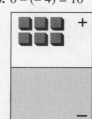 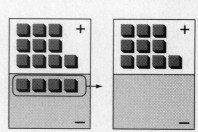

MAKE CONNECTIONS

45a. $5 - 3 = 2$ **b.** $5 + (-3) = 2$

46a. $-2 - 6 = -8$ **b.** $-2 + (-6) = -8$

47a. $-4 + 3 = -1$ **b.** $-4 - (-3) = -1$

48. The answers for a and b are the same

49. adding its opposite

SUMMARIZE

50. If two numbers have the same sign, add the numbers and use the same sign. Example: $2 + 3 = 5$ and $-2 + -3 = -5$. If two numbers have different signs, find the difference between the numbers and use the sign of the number with the greater absolute value. Example: $36 + (-3) = 33$ and $-36 + 3 = -33$.

51. $-11 + 9 = 2$

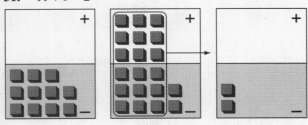

$-7 - (-9) = 2$

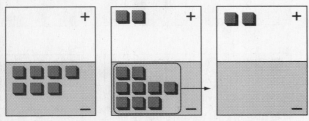

52. Answers should indicate that to subtract two integers, you add the opposite of the second integer and then use the rules for addition started in number 50 above.

53. Statement is false; consider $-1 + (-2) = -3$.

54. Place 3 unit blocks on the positive side of a mat, then 4 unit blocks on the negative side, then 2 more unit blocks on the negative side. Remove 3 unit pairs. The result is 3 unit blocks on the negative side, -3.

Lesson 2.5, pages 77–83

EXPLORE/WORKING TOGETHER

1. Answers will vary.   **2.** $9; 12 + (-3) = 9$

3. $-9; -4 + (-5) = -9$

TRY THESE

1. $6 + (-7) = -1$ **2.** $10 + (-3) = 7$

3. $-\frac{7}{11} + \left(-\frac{2}{11}\right) = -\frac{9}{11}$

4. $-\frac{4}{5} + \left(-\frac{9}{10}\right) = -\frac{8}{10} + \left(-\frac{9}{10}\right) = -\frac{17}{10}$

5. $-0.1 + 0.8 = 0.7$ **6.** $-1.5 + 0.7 = -0.8$

7. $-\frac{1}{4} + \frac{5}{6} = -\frac{3}{12} + \frac{10}{12} = \frac{7}{12}$ **8.** $-0.6 + 2 = 1.4$

9. When the signs are the same, add the number, and give the sum the same sign. When the signs are different, find the difference between the absolute values of the numbers and give the difference the sign of the number with the greater absolute value. Examples will vary.

10. $5 - 2 = 5 + (-2) = 3$ **11.** $11 - 4 = 11 + (-4) = 7$

12. $-\frac{8}{9} - \left(-\frac{1}{3}\right) = -\frac{8}{9} + \frac{1}{3} = -\frac{8}{9} + \frac{3}{9} = -\frac{5}{9}$

13. $-\frac{7}{8} - \left(-\frac{3}{4}\right) = -\frac{7}{8} + \frac{3}{4} = -\frac{7}{8} + \frac{6}{8} = -\frac{1}{8}$

14. $-6 - 2 = -6 + (-2) = -8$

15. $-8 - 1 = -8 + (-1) = -9$

16. $0.6 - (-1.1) = 0.6 + 1.1 = 1.7$

17. $4 - (-0.5) = 4 + 0.5 = 4.5$

18. $136° - (-126.9°) = 136° + 126.9° = 262.9°$

19. $45 + 75 - 80 - 32 - 16 =$
$45 + 75 + (-80) + (-32) + (-16) = -\$8$

20. $-4 + 2 - (-7) = -4 + 2 + 7 = 5$

21. $8 - 10 + (-8) = 8 + (-10) + (-8) = -10$

22. $1.4 - 2 + 0.7 = 1.4 + (-2) + 0.7 = 0.1$

23. $7 - 1.1 + (-0.3) = 7 + (-1.1) + (-0.3) = 5.6$

24. $-\frac{1}{2} + \left(-\frac{2}{3}\right) + \left(-\frac{3}{4}\right) = -\frac{6}{12} + \left(-\frac{8}{12}\right) + \left(-\frac{9}{12}\right) =$
$-\frac{23}{12} = -1\frac{11}{12}$

25. $-\frac{1}{4} + \frac{2}{5} - \left(-\frac{7}{10}\right) = -\frac{1}{4} + \frac{2}{5} + \frac{7}{10} = -\frac{5}{20} + \frac{8}{20} + \frac{14}{20} = \frac{17}{20}$

PRACTICE

1. Explanations will vary. Students may make analogies such as: Deducting \$50 from your account is the same as adding a negative charge or \$50 to your account.

2. $12 + (-6) = 6$ **3.** $14 + (-9) = 5$

4. $-6 + (-8) = -14$ **5.** $-9 + (-8) = -17$

6. $-1.5 + 0.2 = -1.3$ **7.** $-1 + 0.4 = -0.6$

8. $-1 + 3.2 = 2.2$ **9.** $-0.5 + 2 = 1.5$

10. $-2.1 + (-3.9) = -6.0$ **11.** $-4 + (-2.3) = -6.3$

12. $-3 + 1.7 = -1.3$ **13.** $-31 + 19 = -12$

14. $\frac{5}{6} + \left(-\frac{2}{3}\right) = \frac{5}{6} + \left(-\frac{4}{6}\right) = \frac{1}{6}$

15. $\frac{2}{5} + \left(-\frac{7}{10}\right) = \frac{4}{10} + \left(-\frac{7}{10}\right) = -\frac{3}{10}$

16. $3\frac{1}{6} + \left(-5\frac{1}{2}\right) = 3\frac{1}{6} + \left(-5\frac{3}{6}\right) = -2\frac{2}{6} = -2\frac{1}{3}$

17. $5 + \left(-2\frac{1}{3}\right) = 2\frac{2}{3}$

18. $-\frac{3}{7} + \left(-\frac{5}{7}\right) = -\frac{8}{7} = -1\frac{1}{7}$

19. $-\frac{2}{3} + \left(-\frac{1}{2}\right) = -\frac{4}{6} + \left(-\frac{3}{6}\right) = -\frac{7}{6} = -1\frac{1}{6}$

20. $-\frac{1}{4} + \frac{7}{10} = -\frac{5}{20} + \frac{14}{20} = \frac{9}{20}$

21. $-\frac{2}{9} + \frac{2}{3} = -\frac{2}{9} + \frac{6}{9} = \frac{4}{9}$

22. $29{,}028 - (-1299) = 30{,}327$ ft

23. $-5 - (-1) = -5 + 1 = -4$

24. $-9 - (-3) = -9 + 3 = -6$

25. $-3 - 8 = -3 + (-8) = -11$

26. $7 - (-3) = 7 + 3 = 10$

27. $12 - 18 = 12 + (-18) = -6$

28. $17 - (-9) = 17 + 9 = 26$

29. $-8 - 1.3 = -8 + (-1.3) = -9.3$

30. $-5.1 - (-9) = -5.1 + 9 = 3.9$

31. $-0.6 - (-0.7) = -0.6 + 0.7 = 0.1$

32. $6.2 - (-4.5) = 6.2 + 4.5 = 10.7$

33. $0.9 - (-0.03) = 0.9 + 0.03 = 0.93$

34. $-0.4 - 0.7 = -0.4 + (-0.7) = -1.1$

35. $\frac{5}{6} - \left(-\frac{3}{8}\right) = \frac{5}{6} + \frac{3}{8} = \frac{20}{24} + \frac{9}{24} = \frac{29}{24} = 1\frac{5}{24}$

36. $2 - \left(-1\frac{1}{2}\right) = 2 + 1\frac{1}{2} = 3\frac{1}{2}$

37. $-\frac{1}{2} - \frac{3}{4} = -\frac{1}{2} + \left(-\frac{3}{4}\right) = -\frac{2}{4} + \left(-\frac{3}{4}\right) = -\frac{5}{4} = -1\frac{1}{4}$

38. $\frac{1}{2} - \frac{7}{8} = \frac{1}{2} + \left(-\frac{7}{8}\right) = \frac{4}{8} + \left(-\frac{7}{8}\right) = -\frac{3}{8}$

39. $\frac{4}{5} - \frac{9}{10} = \frac{4}{5} + \left(-\frac{9}{10}\right) = \frac{8}{10} + \left(-\frac{9}{10}\right) = -\frac{1}{10}$

40. $-\frac{3}{5} - \left(-\frac{1}{5}\right) = -\frac{3}{5} + \frac{1}{5} = -\frac{2}{5}$

41. $1948 - (-800) = 1948 + 800 = 2748$;
2747 years since there is no year zero.

42. $-3 + 5 - 8 = -3 + 5 + (-8) = -6$

43. $-1 + 4 - 10 = -1 + 4 + (-10) = -7$

44. $36 - 40 - (-2) = 36 + (-40) + 2 = -2$

45. $51 - 12 - (-5) = 51 + (-12) + 5 = 44$

46. $-\frac{8}{9} + \frac{4}{9} - \frac{1}{3} - \left(-\frac{2}{3}\right) =$

$-\frac{8}{9} + \frac{4}{9} + \left(-\frac{1}{3}\right) + \frac{2}{3} =$

$-\frac{8}{9} + \frac{4}{9} + \left(-\frac{3}{9}\right) + \frac{6}{9} = -\frac{1}{9}$

47. $\frac{7}{10} + \left(-\frac{3}{5}\right) - \frac{1}{10} - \left(-\frac{2}{5}\right) =$

$\frac{7}{10} + \left(-\frac{3}{5}\right) + \left(-\frac{1}{10}\right) + \frac{2}{5} =$

$\frac{7}{10} + \left(-\frac{6}{10}\right) + \left(-\frac{1}{10}\right) + \frac{4}{10} = \frac{4}{10} = \frac{2}{5}$

48. $\frac{1}{4} + \frac{5}{6} - \frac{1}{3} - \frac{3}{4} =$

$\frac{1}{4} + \frac{5}{6} + \left(-\frac{1}{3}\right) + \left(-\frac{3}{4}\right) =$

$\frac{3}{12} + \frac{10}{12} + \left(-\frac{4}{12}\right) + \left(-\frac{9}{12}\right) = \frac{0}{12} = 0$

49. $-1.2 + 0.83 - 0.3 = -1.2 + 0.83 + (-0.3) = -0.67$

50. $-42.9 + 2.3 - 11.05 = -42.9 + 2.3 + (-11.05) = -51.65$

51. $20 + 36 - 21 + 3 + 5 - 15 =$
$20 + 36 + (-21) + 3 + 5 + (-15) = 28$ yard line

52. $16 \cdot 2 - 5^2 + (-13) = 16 \cdot 2 - 25 + (-13) =$
$32 - 25 + (-13) = 7 + (-13) = -6$

53. $12 \cdot 3 - 7^2 - (-3) = 12 \cdot 3 - 49 - (-3) =$
$36 - 49 - (-3) = 36 + (-49) + 3 = -10$

54. $-14 + 2^3 - 11 = -14 + 8 - 11 = -14 + 8 + (-11) = -17$

55. $-20 + 3^3 - 5 = -20 + 27 - 5 = -20 + 27 + (-5) = 2$

56. $120 \div 40 - 6^3 + 1 = 120 \div 40 - 216 + 1 =$
$3 - 216 + 1 = 3 + (-216) + 1 = -212$

57. $-12 + 36 \div 6 - 8 = -12 + 6 - 8 = -12 + 6 + (-8) = -14$

58. $|4 - 9| = |-5| = 5$

59. $|6| - |13| = 6 - 13 = 6 + (-13) = -7$

60. $|-(3 - 7)| = |-(-4)| = |4| = 4$

61. $-|-5 - 8| = -|-3| = -3$

62. $53\frac{1}{2} - 58\frac{1}{4} = 53\frac{1}{2} + \left(-58\frac{1}{4}\right) = 53\frac{2}{4} + \left(-58\frac{1}{4}\right) = -4\frac{3}{4}$

63. $12\frac{3}{8} - \left(-1\frac{1}{2}\right) = 12\frac{3}{8} + 1\frac{1}{2} = 12\frac{3}{8} + 1\frac{4}{8} = 13\frac{7}{8}$

64. Alaska with $100° - (-80°) = 100° + 80° = 180°$ difference

65. $-24° - (-40°) = -24° + 40° = 16°$ difference

66. The opposite is $-a - b$. Explanations may show worked-out examples that substitute numbers for a and b. Students may also show that $a + b - a - b = 0$ and, therefore, $-a - b$ is the opposite of $a + b$.

67. The opposite is $b - a$ or $-a + b$.. Explanations may show worked-out examples that substitute numbers for a and b. Students may also show that $a - b - a + b = 0$ and, therefore, $-a + b$ is the opposite of $a - b$.

68. always; $(a - b) - c = [a + (-b)] + (-c) =$
$a + (-b) + (-c) = a + (-c)(-b) = [a + (-c)] + (-b) =$
$(a - c) - b.$

69. Answers will vary. Possible answers: $4 + (-3) = 1$ and $2 + (-6) = -4$. If you want a positive sum, the number with the larger absolute value should have a positive sign; to get a negative sum, the number with the larger absolute value should have a negative sign.

Lesson 2.6, pages 84–87

THINK BACK

1. Pattern is add 5 to get 22, 27, 32.

2. Pattern is add 4 to get 20, 24, 28.

3. Pattern is subtract 2 to get $-3, -5, -7$.

EXPLORE

4.

									9	9	18	27	36	45	54	63	72	81
									8	8	16	24	32	40	48	56	64	72
									7	7	14	21	28	35	42	49	56	63
									6	6	12	18	24	30	36	42	48	54
									5	5	10	15	20	25	30	35	40	45
									4	4	8	12	16	20	24	28	32	36
									3	3	6	9	12	15	18	21	24	27
									2	2	4	6	8	10	12	14	16	18
									1	1	2	3	4	5	6	7	8	9
-9	-8	-7	-6	-5	-4	-3	-2	-1		1	2	3	4	5	6	7	8	9
									-1									
									-2									
									-3									
									-4									
									-5									
									-6									
									-7									
									-8									
									-9									

5. Each number is one less than the previous one.

6. zero

7.

-81	-72	-63	-54	-45	-36	-27	-18	-9	9	9	18	27	36	45	54	63	72	81
-72	-64	-56	-48	-40	-32	-24	-16	-8	8	8	16	24	32	40	48	56	64	72
-63	-56	-49	-42	-35	-28	-21	-14	-7	7	7	14	21	28	35	42	49	56	63
-54	-48	-42	-36	-30	-24	-18	-12	-6	6	6	12	18	24	30	36	42	48	54
-45	-40	-35	-30	-25	-20	-15	-10	-5	5	5	10	15	20	25	30	35	40	45
-36	-32	-28	-24	-20	-16	-12	-8	-4	4	4	8	12	16	20	24	28	32	36
-27	-24	-21	-18	-15	-12	-9	-6	-3	3	3	6	9	12	15	18	21	24	27
-18	-16	-14	-12	-10	-8	-6	-4	-2	2	2	4	6	8	10	12	14	16	18
-9	-8	-7	-6	-5	-4	-3	-2	-1	1	1	2	3	4	5	6	7	8	9
-9	-8	-7	-6	-5	-4	-3	-2	-1		1	2	3	4	5	6	7	8	9
									-1									
									-2									
									-3									
									-4									
									-5									
									-6									
									-7									
									-8									
									-9									

8. 15   **9.** -15 **10.** -48 **11.** -36

12. They are all positive.   **13.** They are all negative.

14. Each number is 3 less than the number above it.

15.

-81	-72	-63	-54	-45	-36	-27	-18	-9	9	9	18	27	36	45	54	63	72	81
-72	-64	-56	-48	-40	-32	-24	-16	-8	8	8	16	24	32	40	48	56	64	72
-63	-56	-49	-42	-35	-28	-21	-14	-7	7	7	14	21	28	35	42	49	56	63
-54	-48	-42	-36	-30	-24	-18	-12	-6	6	6	12	18	24	30	36	42	48	54
-45	-40	-35	-30	-25	-20	-15	-10	-5	5	5	10	15	20	25	30	35	40	45
-36	-32	-28	-24	-20	-16	-12	-8	-4	4	4	8	12	16	20	24	28	32	36
-27	-24	-21	-18	-15	-12	-9	-6	-3	3	3	6	9	12	15	18	21	24	27
-18	-16	-14	-12	-10	-8	-6	-4	-2	2	2	4	6	8	10	12	14	16	18
-9	-8	-7	-6	-5	-4	-3	-2	-1	1	1	2	3	4	5	6	7	8	9
-9	-8	-7	-6	-5	-4	-3	-2	-1		1	2	3	4	5	6	7	8	9
									-1	-1	-2	-3	-4	-5	-6	-7	-8	-9
									-2	-2	-4	-6	-8	-10	-12	-14	-16	-18
									-3	-3	-6	-9	-12	-15	-18	-21	-24	-27
									-4	-4	-8	-12	-16	-20	-24	-28	-32	-36
									-5	-5	-10	-15	-20	-25	-30	-35	-40	-45
									-6	-6	-12	-18	-24	-30	-36	-42	-48	-54
									-7	-7	-14	-21	-28	-35	-42	-49	-56	-63
									-8	-8	-16	-24	-32	-40	-48	-56	-64	-72
									-9	-9	-18	-27	-36	-45	-54	-63	-72	-81

16. -24 **17.** -10 **18.** -72 **19.** -21

20. They are all negative.

21. Each number is 5 more than the number above it.

22.

-81	-72	-63	-54	-45	-36	-27	-18	-9	9	9	18	27	36	45	54	63	72	81
-72	-64	-56	-48	-40	-32	-24	-16	-8	8	8	16	24	32	40	48	56	64	72
-63	-56	-49	-42	-35	-28	-21	-14	-7	7	7	14	21	28	35	42	49	56	63
-54	-48	-42	-36	-30	-24	-18	-12	-6	6	6	12	18	24	30	36	42	48	54
-45	-40	-35	-30	-25	-20	-15	-10	-5	5	5	10	15	20	25	30	35	40	45
-36	-32	-28	-24	-20	-16	-12	-8	-4	4	4	8	12	16	20	24	28	32	36
-27	-24	-21	-18	-15	-12	-9	-6	-3	3	3	6	9	12	15	18	21	24	27
-18	-16	-14	-12	-10	-8	-6	-4	-2	2	2	4	6	8	10	12	14	16	18
-9	-8	-7	-6	-5	-4	-3	-2	-1	1	1	2	3	4	5	6	7	8	9
-9	-8	-7	-6	-5	-4	-3	-2	-1		1	2	3	4	5	6	7	8	9
9	8	7	6	5	4	3	2	1	-1	-1	-2	-3	-4	-5	-6	-7	-8	-9
18	16	14	12	10	8	6	4	2	-2	-2	-4	-6	-8	-10	-12	-14	-16	-18
27	24	21	18	15	12	9	6	3	-3	-3	-6	-9	-12	-15	-18	-21	-24	-27
36	32	28	24	20	16	12	8	4	-4	-4	-8	-12	-16	-20	-24	-28	-32	-36
45	40	35	30	25	20	15	10	5	-5	-5	-10	-15	-20	-25	-30	-35	-40	-45
54	48	42	36	30	24	18	12	6	-6	-6	-12	-18	-24	-30	-36	-42	-48	-54
63	56	49	42	35	28	21	14	7	-7	-7	-14	-21	-28	-35	-42	-49	-56	-63
72	64	56	48	40	32	24	16	8	-8	-8	-16	-24	-32	-40	-48	-56	-64	-72
81	72	63	54	45	36	27	18	9	-9	-9	-18	-27	-36	-45	-54	-63	-72	-81

23. Each number is 2 more than the one to its right.

24. yes; Quadrant III is correctly filled in.

25. 20   **26.** 16   **27.** 56   **28.** 18

29. The signs are all positive.

30. Product of two positives or two negatives is positive. Product of a positive and a negative is negative.

31. yes; The rules about the signs of the products do not depend on the order of the factors.

32. yes; possible example:

$$2[(-3)(-5)]=[2(-3)](-5)$$
$$2[15]=[-6](-5)$$
$$30=30$$

33. Find 9 on the horizontal axis. Locate -36 in that column. Then locate the corresponding factor, -4.

34. $-12 \div 4 = -3$

35. $-45 \div 5 = -9$

36. $16 \div (-4) = -4$

37. $8 \div (-4) = -2$

38. When the signs are different, the quotient is negative. When the signs are the same, the quotient is positive.

MAKE CONNECTIONS

39. The sign of each quardrant is the sign of the product of the numbers that would appear in that quadrant.

40. $-2(4)$

41. $-2(4) = -8$

42.
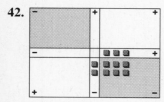

43. There are 6 blocks in Quadrant IV, so the result is -6

44. $4(-5) = -20$

45. $-3(-4) = 12$

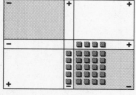
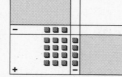

46. $(-5)(2) = -10$

47. $-2(-4) = 8$

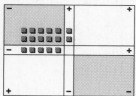
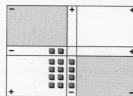

48. Take the number of cubes equal to the dividend. Form a rectangular arrangement using the divisor as one side. Place the cubes in a quadrant that has the same sign as the dividend. The other dimension of the rectangle tells you the quotient.

$-12 \div 3 = -4$

49. $-14 \div (-7) = 2$

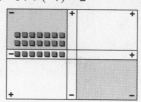

50. $-18 \div 9 = -2$

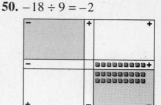

51. $-35 \div 7 = -5$

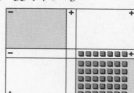

52. $32 \div (-8) = -4$

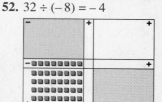

53. The product of a positive and a negative integer is negative.

54. $-4(-50) = 200$

55. The product of two negative integers is positive.

SUMARIZE

56. Answers will vary but should include a statement indicating that the prduct will be in the row containing one factor and the column containing the other factor.

57. Answers will vary but should include a statement indicating that one integer is used to place blocks on the horizontal axis, the other for the vertical axis. Form a rectangle of blocks using these as guidelines. Then count the blocks in the rectangle. The Quadrant indicates if the product is positive or negative.

58. $-24 \div 3 = -8$; Pictures will vary depending on which model student chose.

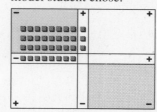

59. no; $(36 \div 6) \div 2 = 3 \neq 36 \div (6 \div 2) = 12$

60. $(-1)^2 = 1$; $(-1)^3 = -1$; $(-1)^4 = 1$; $(-1)^5 = -1$; $(-1)^{201} = -1$; Products of 2 negative numbers are positive and odd powers always have one negative left over so are negative.

Lesson 2.7, pages 88–94

EXPLORE

1a. $n = 6(-7)$

b. $n = -42$

2a. 8

b. $52 \div 6.5 = 8$

3a. 8

b. $-52 \div (-6.5) = 8$

1. Factors with like signs have a positive product, unlike signs have a negative product.

2. $-6(3) = -18$

3. $8(-0.4) = -3.2$

4. $\left(-\frac{1}{2}\right)\left(-\frac{5}{9}\right) = \frac{5}{18}$

5. $\left(-6\frac{1}{3}\right)(-4) = 24\frac{4}{3} = 25\frac{1}{3}$

6. $-\frac{1}{11}$

7. $-\frac{4}{3}$

8. $0.8 = \frac{8}{10}$, reciprocal is $\frac{10}{8} = \frac{5}{4} = 1\frac{1}{4}$

9. $-3\frac{2}{5} = -\frac{17}{5}$, reciprocal is $-\frac{5}{17}$

10. $72 \div (-8) = -9$

11. $-9.3 \div (-6) = 1.55$

12. $-\frac{2}{3} \div \frac{3}{4} = -\frac{2}{3} \cdot \frac{4}{3} = -\frac{8}{9}$

13. $-4 \div \left(-1\frac{1}{7}\right) = -4 \cdot \left(-\frac{7}{8}\right) = \frac{28}{8} = 3\frac{1}{2}$

14. $\frac{[2000 + 1500 + (-3600) + 2200]}{4} = \frac{2100}{4} = \$525$

15. $-3(2) = -6$

PRACTICE

1. $9(-11) = -99$

2. $-3(14) = -42$

3. $-5(-21) = 105$

4. $-25(-4) = 100$

5. $0.6(-3.4) = -2.04$

6. $1.5(-9) = -13.5$

7. $-2.7(-0.001) = 0.0027$

8. $-3.2(-5) = 16$

9. $-\frac{3}{4} \cdot \frac{7}{12} = -\frac{21}{48} = -\frac{7}{16}$

10. $-\frac{1}{9} \cdot \left(-\frac{3}{8}\right) = \frac{3}{72} = \frac{1}{24}$

11. $1\frac{1}{2} \cdot \left(-2\frac{3}{4}\right) = \frac{3}{2} \cdot \left(-\frac{11}{4}\right) = -\frac{33}{8} = -4\frac{1}{8}$

12. $-15 \cdot \left(-3\frac{1}{5}\right) = -15 \cdot \left(-\frac{16}{5}\right) = \frac{240}{5} = 48$

13. $-\frac{1}{21}$

14. $\frac{9}{8}$

15. $-3\frac{1}{6} = -\frac{19}{6}$, reciprocal is $-\frac{6}{19}$

16. $-0.02 = -\frac{2}{100}$, reciprocal is $-\frac{100}{2} = -50$

17. Answers will vary, such as:

   a. $2(-4) = -8$; $5(-3) = -15$; $-7(7) = -49$; $-8\left(\frac{1}{2}\right) = -4$

   b. $6(8) = 48$; $-3(-2) = 6$; $5(7) = 35$; $-\frac{2}{3}\left(-\frac{5}{7}\right) = \frac{10}{21}$

   c. $3\left(\frac{1}{3}\right) = 1$; $-4\left(-\frac{1}{4}\right) = 1$; $-8\left(-\frac{1}{8}\right) = 1$; $6\left(\frac{1}{6}\right) = 1$

18. $144 \div (-12) = -12$

19. $-150 \div 6 = -25$

20. $-201 \div (-3) = 67$

21. $-289 \div (-17) = 17$

22. $-2.4 \div 0.5 = -2.4 \div \frac{1}{2} = -2.4 \cdot 2 = -4.8$

23. $2.8 \div -70 = 2.8 \cdot \left(-\frac{1}{70}\right) = -\frac{2.8}{70} = -0.04$

24. $-\frac{8}{1.6} = -8 \div 1.6 = -8 \div \frac{16}{10} = -8 \cdot \frac{10}{16} = -\frac{80}{16} = -5$

25. $\frac{-200}{-2.5} = -200 \div (-2.5) = -200 \div \left(-\frac{5}{2}\right)$
   $= -200 \cdot \left(-\frac{2}{5}\right) = \frac{400}{5} = 80$

26. $\frac{5}{8} \div \left(-\frac{3}{4}\right) = \frac{5}{8} \cdot \left(-\frac{4}{3}\right) = -\frac{20}{24} = -\frac{5}{6}$

27. $-\frac{6}{7} \div \frac{2}{5} = \frac{6}{7} \cdot \frac{5}{2} = -\frac{30}{14} = -\frac{15}{7} = -2\frac{1}{7}$

28. $2\frac{1}{2} \div \left(-\frac{1}{10}\right) = \frac{5}{2} \cdot \left(-\frac{10}{1}\right) = -\frac{50}{2} = -25$

29. $-6\frac{1}{4} \div (-5) = -\frac{25}{4} \cdot \left(-\frac{1}{5}\right) = \frac{25}{20} = \frac{5}{4} = 1\frac{1}{4}$

30. $\frac{-5.7 + (-3.7) + (-8) + (-15.7)}{4} = \frac{-33.1}{4} = -8.275°$

31a. $1125m - 100m - 5m$

   b. $1125(4) - 100(4) - 5(4) = 4500 - 400 - 20 = \$4080$

32. Answers will vary.

EXTEND

33. $4 - 3(-5) = 4 - (-15) = 4 + 15 = 19$

34. $10 + (-7)(-3) = 10 + 21 = 31$

35. $-8 + (-2)(-9) = -8 + 18 = 10$

36. $(-7)^2 + (-9)(-2) = 49 + (-9)(-2) = 49 + 18 = 67$

37. $(-2)^3 - (3)(-7) = -8 - (-21) = -8 + 21 = 13$

38. $3(-4) + 5(-3) = -12 + (-15) = -27$

39. $-7(-4) + 9(-6) = 28 + (-54) = -26$

40. $-6(2^2) - 11(-3) = -6(4) - 11(-3) =$
   $-24 - (-33) = -24 + 33 = 9$

41. $-4(3^2) - 2(-8) = -4(9) - 2(-8) =$
   $-36 - (-16) = -36 + 16 = -20$

42. $2,735,000 + (-0.3\% \text{ of } 2,735,000) =$
   $2,735,000 + (-0.003)(2,735,000) =$
   $2,735,000 + (-8205) = 2,726,795$ people

43. 142; Start with 2, multiply by -3 and add 1.

THINK CRITICALLY

44a. $(b - c)$ is negative, $(a - c)$ is negative, product is positive

   b. a is negative, $(c - b)$ is positive, product is negative

45. no; $1 \div 2 = \frac{1}{2}$, not an integer

46. a and b are opposite signs, so ab is negative and equals $-|ab|$

47. a and b are both negative, so ab is positive and equals $|ab|$

48. a and b are opposite signs, so $a \div b$ is negative and equals $-\left|\dfrac{a}{b}\right|$

49. a and b are both negative, so $a \div b$ is positive and equals $\left|\dfrac{a}{b}\right|$

PROJECT CONNECTION

1.

	M	T	W	Th	F	S
High	69	71	71	72	71	70
Low	55	52	55	54	54	53
Average	62	61.5	63	63	62.5	61.5

2a. $63° - 2.5° = 63° + (-2.5°) = 60.5°$

b. Answers will vary. Both highs and lows seem to be decreasing. A possible solution is a high of 69°F and a low of 52°F for the desired 60.5°F average.

3. Answers will vary with student's data.

ALGEBRA WORKS

1. $-10°$F   **2.** about 50%   **3.** methanol

4. Answers will vary. Possible factors: price, availability, climate in your area

5. Find out the lowest possible temperature in your area. Use the graph to find the volume that will ensure protection to that point.

6. Antifreeze also raises the boiling point of water so that the radiator does not boil over when the air temperature is high.

Lesson 2.8, pages 95–101

EXPLORE

1.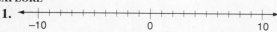

a. 3   **b.** 4.5   **c.** 1   **d.** 3.5

2. They are the same in every case.

3. $m = \dfrac{(a+b)}{2}$

TRY THESE

1. $5 + 4(9 - 11) = 5 + 4(-2) = 5 + (-8) = -3$

2. $12 \div (-4) + 2(3 + 4) = 12 \div (-4) + 2(7) = -3 + 14 = 11$

3. $5 \cdot 4^2 - (3 - 6)^2 = 5 \cdot 4^2 - (-3)^2 = 5 \cdot 16 - 9 = 80 - 9 = 71$

4. $2.3 + 5(1.4 + 8.3) = 2.3 + 5(9.7) = 2.3 + 48.5 = 50.8$

5. $0.2(1 - 0.7) + (-0.1)^3 = 0.2(0.3) + (-0.1)^3 = 0.2(0.3) + (-0.001) = 0.06 + (-0.001) = 0.059$

6. $\dfrac{2}{3}\left(\dfrac{3(4+8)}{5}\right) = \dfrac{2}{3}\left(\dfrac{3(12)}{5}\right) = \dfrac{2}{3}\left(\dfrac{36}{5}\right) = \dfrac{72}{15} = \dfrac{24}{5}$

7. $3 + [1 - 2(8 + 7)] = 3 + [1 - 2(15)] = 3 + [1 - 30] = 3 + (-29) = -26$

8. $4 - [(5 - 2) - 12] = 4 - [3 - 12] = 4 - (-9) = 4 + 9 = 13$

9. $100 \div [-4 + 3(2 - 9)] = 100 \div [-4 + 3(-7)] = 100 \div [-4 - 21] = 100 \div (-25) = -4$

10. $9(x + 2) = 9x + 9(2) = 9x + 18$

11. $(b - 3)4 = 4b - 3(4) = 4b - 12$

12. $-6(y + 2) = -6y - 6(2) = -6y - 12$

13. $(m - 8)(-2) = -2m - 8(-2) = -2m + 16$

14. yes   **15.** no   **16.** yes   **17.** no

18. $3x + 2(5x - 1) = 3x + 2(5x) + 2(-1) = 3x + 10x - 2 = 13x - 2$

19. $9 - 4(x - 8) = 9 - 4(x) - 4(-8) = 9 - 4x + 32 = 41 - 4x$

20. $7x - 2(8 + 3x) = 7x - 2(8) - 2(3x) = 7x - 16 - 6x = x - 16$

21. Figure 1: $3 \cdot 7$; figure 2: $3 \cdot 5 + 3 \cdot 2$; distributive property is shown

22. Combining like terms means adding coefficients of like terms. Like terms have the same variable base with the same exponent. Examples will vary. Possible examples: $4x - 10x = -6x$; $3x^2 + 2x^2 = 5x^2$

23. $C = \dfrac{5}{9}(23° - 32°) = \dfrac{5}{9}(-9°) = \dfrac{-45}{9} = -5°C$

PRACTICE

1. $3 - 6(7 - 11) = 3 - 6(-4) = 3 - (-24) = 3 + 24 = 27$

2. $5 + 2(9 - 16) = 5 + 2(-7) = 5 + (-14) = -9$

3. $(9 - 4)^2 - 7(8) = (5)^2 - 7(8) = 25 - 7(8) = 25 - 56 = -31$

4. $(2 + 1)^3 - 5(-4) = (3)^3 - 5(-4) = 27 - 5(-4) = 27 - (-20) = 27 + 20 = 47$

5. $4.3 + (9 - 3.02) = 4.3 + 5.98 = 10.28$

6. $7.7 - (5.12 + 2.4) = 7.7 - (7.52) = 0.18$

7. $(0.3 - 0.5)^2 - 2(0.9) = (-0.2)^2 - 2(0.9) = 0.04 - 2(0.9) = 0.04 - 1.8 = -1.76$

8. $(3.9 - 2.5)^2 + 5(0.01) = (1.4)^2 + 5(0.01) = 1.96 + 5(0.01) = 1.96 + 0.05 = 2.01$

9. $0.1(4.5 - 1.2) - 0.01(6 - 2.2) = 0.1(3.3) - 0.01(3.8) = 0.33 - 0.038 = 0.292$

10. Answers will vary. To determine whether or not a calculator follows the order of operations, enter an expression like $3 + 2 \times 5$. If the display reads 13, it does; if it reads 25, it does not follow the order of operations. To use the calculator to evaluate such an expression, enter the numbers and operations in the correct order. For example, enter $2 \times 5 =$, then $+ 3$.

(Solution continues on next page.)

The expression $9 - 3(5 + 2)$ involves parentheses. A calculator that follows the order of operations will not give the correct answer (-12) unless you enter the parentheses when you enter the expression. To evaluate this expression on a calculator that does not follow the order of operations, you must work backwards. Find $5 + 2 =$, then multiply that answer, 7, by -3. Then add the result, -21, and 9. You will get -12.

11. $\frac{3}{10}\left(\frac{1+5}{2 \cdot 5}\right) + \frac{1}{5} = \frac{3}{10}\left(\frac{6}{10}\right) + \frac{1}{5} =$

$\frac{18}{100} + \frac{1}{5} = \frac{18}{100} + \frac{20}{100} = \frac{38}{100} = \frac{19}{50}$

12. $\frac{4}{5}\left(\frac{3-5}{2 \cdot 4}\right) + 1\frac{1}{2} = \frac{4}{5}\left(\frac{-2}{8}\right) + 1\frac{1}{2} =$

$\frac{-8}{40} + 1\frac{1}{2} = \frac{-2}{10} + \frac{15}{10} = \frac{13}{10} = 1\frac{3}{10}$

13. $15 - [10 - (6 - 11)] = 15 - [10 - (-5)] =$
$15 - [10 + 5] = 15 - [15] = 0$

14. $-9 + [16 - (3 - 5)] = -9 + [16 - (-2)] =$
$-9 + [16 + 2] = -9 + [18] = 9$

15. $7 - [4(1 - 8) + 2] = 7 - [4(-7) + 2] =$
$7 - [-28 + 2] = 7 - [-26] = 7 + 26 = 33$

16. $-8 - [3(2 - 7) + 6] = -8 - [3(-5) + 6] =$
$-8 - [-15 + 6] = -8 - [-9] = -8 + 9 = 1$

17. $14 - [6 - (2 + 1)]^3 = 14 - [6 - 3]^3 =$
$14 - [3]^3 = 14 - 27 = -13$

18. $1 + [3 - 2(1 - 3)]^2 = 1 + [3 - 2(-2)]^2 =$
$1 + [3 - (-4)]^2 = 1 + [7]^2 = 1 + 49 = 50$

19. $5x + 4 - 3x - 11 = 5x - 3x + 4 - 11 =$
$(5 - 3)x + (4 - 11) = 2x - 7$

20. $-6a + 9b - 3a - 2b = -6a - 3a + 9b - 2b =$
$(-6 - 3)a + (9 - 2)b = -9a + 7b$

21. $15 - 3(8c + 2) = 15 - 3(8c) - 3(2) =$
$15 - 24c - 6 = 9 - 24c$

22. $-5(3x + 2) - 19 = -5(3x) - 5(2) - 19 =$
$-15x - 10 - 19 = -15x - 29$

23. $7a - 3b + 2c + 10a + b = 7a + 10a - 3b + b + 2c =$
$(7 + 10)a + (-3 + 1)b + 2c = 17a - 2b + 2c$

24. $x - y + 3z + 5y - 4x = x - 4x - y + 5y + 3z =$
$(1 - 4)x + (-1 + 5)y + 3z = -3x + 4y + 3z$

25. $3(r + 2) - (r + s) = 3(r) + 3(2) - r - s =$
$3r + 6 - r - s = 3r - r + 6 - s = (3 - 1)r + 6 - s =$
$2r + 6 - s$

26. $15(x + 3) = 15x + 15(3) = 15x + 45$ square units

27. $P = 2l + 2w = 2(l + w)$

28. $D = \frac{AB + C}{B} =$ total fixed and variable costs $\div$ number of mugs

EXTEND

29. $-5(-1)^2 = -5(1) = -5$

30. $(-5(-1))^2 = (5)^2 = 25$

31. $2(-1)^2 - 3(-1 + 2) = 2(-1)^2 - 3(-1) - 3(2) =$
$2(1) - 3(-1) - 3(2) = 2 + 3 - 6 = -1$

32. $4(-1)^3 - 5(2(-1) + 1) = 4(-1)^3 - 5(-2 + 1) =$
$4(-1)^3 - 5(-2) - 5(1) = -4 + 10 - 5 = 1$

33. $-6(3)(-4) = -18(-4) = 72$

34. $4(3)^2(-4) = 4(9)(-4) = 36(-4) = -144$

35. $5(3) + 2(-4) - 3(3) - 8(-4) = 15 + (-8) - 9 - (-32) =$
$15 + (-8) + (-9) + 32 = 30$

36. $9(3)^2 - 3(3)(-4) + 2(-4)^2 = 9(9) - 3(3)(-4) + 2(16) =$
$81 + 36 + 32 = 149$

37. If $x + y = x$, then $y = 0$ so $y(3x + 5) = 0(3x + 5) = 0$

38. If $xy = y$, then $x = 1$ so $x^{10} = 1$ (unless $y = 0$, then it could be any value)

39. $8 \cdot 3 + 8 \cdot 8 = 8 \cdot 3 + 8 \cdot n$
$n = 8$

40. $-6 \cdot 4 + 13 \cdot 4 = 6 \cdot n + 13 \cdot 4$
$n = 4$

41. $(7 + 1) \cdot 2 + 4 = 8 \cdot 2 + 4 = 16 + 4 = 20$

42. $15 - [20 \div (-3 - 1)] = 15 - [20 \div -4] =$
$15 - [-5] = 15 + [5] = 20$

43. $(10.45 + 10\sqrt{11} - 11)(33 - 10) =$
$(10\sqrt{11} - 0.55)(23) = 750.2$

44a. $Q = \frac{F}{S - v} = \frac{40,000}{(2.00 - 1.20)} = \frac{40,000}{0.80} = 50,000$ units

b. decrease in F or v, or an increase in s

THINK CRITICALLY

45. yes; Show that $\frac{(a + b)}{c} = \frac{a}{c} + \frac{b}{c}$ by using the definition of division $\frac{(a + b)}{c} = \frac{a}{c} + \frac{b}{c}$ and by using the distributive property for multiplication.

46. yes; $2n + 1$ is odd if n is an integer because $2n$ is even. The square of an odd number will always be an odd number.

Justification:

$(2n + 1)(2n + 1) = 2n(2n + 1) + 1(2n + 1)$
$= 4n^2 + 2n + 2n + 1$
$= 4n^2 + 4n + 1$
$= \text{even} + \text{even} + 1 = \text{odd}$

47. no; possible counterexample:

$5 + (3 \cdot 4) \neq (5 \cdot 3) + (5 \cdot 4)$

MIXED REVIEW

48. Both A and B are 3×3, so both are square matrices.

49. $\begin{bmatrix} 3+2 & 5+1 & -6+0 \\ 8+5 & -2+2 & 3+2 \\ -1+1 & 4+2 & 4+1 \end{bmatrix} = \begin{bmatrix} 5 & 6 & -6 \\ 13 & 0 & 5 \\ 0 & 6 & 5 \end{bmatrix}$

50. $\begin{bmatrix} 3-2 & 5-1 & -6-0 \\ 8-5 & -2-2 & 3-2 \\ -1-1 & 4-2 & 4-1 \end{bmatrix} = \begin{bmatrix} 1 & 4 & -6 \\ 3 & -4 & 1 \\ -2 & 2 & 3 \end{bmatrix}$

51. D   **52.** F   **53.** C   **54.** G

55. $\dfrac{-3+0+5+(-2)}{4} = \dfrac{0}{4} = 0$; B

AlgebraWorks

1. $-4°$ F

2. Between 5 and 10 mph since 10 mph is twice the speed at 5 mph.

3. $-13° - (-3°) = -10°$ F

4. Indicates relative temperature so you know how to dress for outdoor activity.

Lesson 2.9, pages 102–105

Problem

$59(101) = 5959$
$78(101) = 7878$
$94(101) = 9494$

Explore the Problem

1. The product of a two digit number and 101 is the two digit number written twice, side by side.

2. They will follow the pattern.

3.
```
    1 0 1
  ×   b a
  ─────────
    a 0 a
  b 0 b
  ─────────
  b a b a
```

4. $315(101) = 31{,}815$
$283(101) = 28{,}583$
$452(101) = 45{,}652$
$426(101) = 43{,}026$
$715(101) = 72{,}215$
$832(101) = 84{,}032$

5. For a three-digit number cba, the product is of the form $cb(a+c)ba$ if $a+c < 9$. If $a+c > 9$, then the thousands digit is $b+1$ and the hundreds digit is the ones digit of $a+c$.
Justification:
```
        c b a
      × 1 0 1
    ───────────
        c b a
    c b a 0 0
    ───────────
    c b(a+c)b a
```

6. Product is of the form $cbacba$.
Justification:
```
          c b a
      × 1 0 0 1
    ─────────────
          c b a
    c b a 0 0 0
    ─────────────
    c b a c b a
```

Investigate Further

7. $2m + 2n$ **8.** $2(m + n)$

9. It is the sum of two whole numbers and is divisible by 2.

10. $2m + 1$; $2n + 1$

11. $(2m+1)+(2n+1)$
$= 2m+1+2n+1$ associative property
$= 2m+2n+2$ commutative property
$= 2(m+n+1)$ distributive property
$2(m+n+1)$ represents the sum of any two odd numbers. It is divisible by 2 and so is even.

12. The sum of an odd number and an even number is odd.
$= 2m+1+2n$
$= 2m+2n+1$
$= 2(m+n)+1$
Let $m+n = A$; Then $2(m+n) + 1 = 2A + 1$, which is the form of an odd number.

13. The paragraph should include the fact that specific examples support but do not establish a general conclusion. General statements show that something is true for all cases. The variables used in algebra allow such statements to be made because a variable is used to represent any number.

Apply the Strategy

14. Addends have the same digits in reverse order. Sums are all divisible by 11.

15a. $10a + b$; $10b + a$

b. $(10a + b) + (10b + a)$

c. $(10a+b)+(10b+a)$
$= 10a+b+10b+a$ associative property
$= 10a+a+10b+b$ commutative property
$= 11a+11b$ Combine like terms.
$= 11(a+b)$ distributive property
$11(a+b)$ represents a whole number divisible by 11.

16a. No solution, this is an activity.

b. The nonnegative difference for any two-digit whole number and the two-digit whole number formed by reversing its digits is divisible by 9.

c. $(10b + a) - (10a + b)$

$= 10b + a - 10a - b$	definition of subtraction and distributive property
$= 10b - b + a - 10a$	commutative property
$= 9b - 9a$	Combine like terms.
$= 9(b - a)$	distributive property

17. Since there are three integers, either two must be even or two must be odd.

Case 1: Assume two of the integers are even.
Then, if $A = 2m$ and $B = 2n$, $A - B = 2m - 2n$.

Using distributive property, $A - B = 2(m - n)$, which is divisible by 2 and, therefore, is even. Then $(A - B)(B - C)(C - A)$ is even because the factor $A - B$ is even.

Case 2: Assume two of the integers are odd.
Then, $A = 2m + 1$ and $B = 2n + 1$, and
$$A - B = (2m + 1) - (2n + 1)$$
$$= 2m + 1 - 2n - 1$$
$$= 2m - 2n$$
$$= 2(m - n)$$

Thus, $A - B = 2(m - n)$, which is divisible by 2 and, therefore, is even. Then $(A - B)(B - C)(C - A)$ is even because the factor $A - B$ is even.

18. Answers will vary by student, one might be 1, 2, 6, 24, 120, 720, ... $= n!$

REVIEW PROBLEM SOLVING STRATEGIES

SUM + GAME = SUM GAME!

a. no

b. All the numbers are divisible by 3.; 100 is not divisible by 3, and so a sum of 100 cannot be produced by addends that are all 3's or multiples of 3.

c. Include a limited number of caps with numbers that are not multiples of 3.

EASY AS XYZ

a. z

b. z; Adding z to a number does not change the number.

c. y; $x + y = z$ and z is the identity element. The sum of a number and its opposite equals the identity element.

d. Work from inside out.
$z - (y - x)$
$z - x \qquad x + x = y \rightarrow x = y - x$
$y \qquad y + x = z \rightarrow y = z - x$

PATH PUZZLE

a. 3 paths

b. 8 paths;

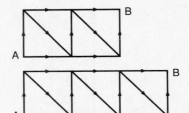

c. 21 paths;

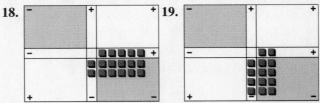

d. The increase in the number of paths from 1 square to 2 squares is 5. Add 5 to the number of paths in 2 squares to get the next increase, 13. The number of paths in 2 squares is 8, and $8 + 13 = 21$, the number of paths in 3 squares. Add 13 to the number of paths in 3 squares to get the next increase, 34. Add 34 to the 21 paths in 4 squares. Continue in this way to get total paths for 6 square, 377. The number of paths together with increases are the Fibonacci numbers.

square	1	2	3	4	5	6
	3	8	21	55	144	377
number	\	/ \	/ \	/ \	/ \	/
of paths	5	13	34	89	233	

Chapter Review, pages 106–107

1. f   **2.** b   **3.** d   **4.** c

5. e   **6.** a   **7.** -2 **8.** $4x$

9. $5 - 2y$ **10.** $2(6) - 5 = 12 - 5 = 7$

11. $3(6)^2 + 4 = 3(36) + 4 = 108 + 4 = 112$

12. $\dfrac{5(6)}{3} = \dfrac{30}{3} = 10$

13., 14., 15.

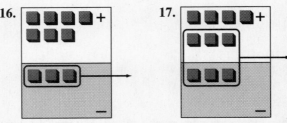

16.

17.

18.

19.

20. $4\frac{1}{2} + \left(-6\frac{1}{4}\right) = \frac{9}{2} + \left(-\frac{25}{4}\right) = \frac{18}{4} + \left(-\frac{25}{4}\right) = -\frac{7}{4} = -1\frac{3}{4}$

21. $7 - (-2.8) = 7 + 2.8 = 9.8$

22. $-8 - 5 + (-4) = -8 + (-5) + (-4) = -13 + (-4) = -17$

23. $3\frac{1}{3} - \left(-9\frac{1}{6}\right) = 3\frac{1}{3} + 9\frac{1}{6} =$

$\frac{10}{3} + \frac{55}{6} = \frac{20}{6} + \frac{55}{6} = \frac{75}{6} = 12\frac{3}{6} = 12\frac{1}{2}$

24. $8 + (-3.4) - 5 = 8 + (-3.4) + (-5) = 4.6 + (-5) = -0.4$

25. $-47 + 8.7 + (-5.7) = -38.3 + (-5.7) = -44$

26. $(-7)(20) = -140$

27. $\frac{2}{3} \div \left(-\frac{1}{4}\right) = \frac{2}{3} \cdot \left(-\frac{4}{1}\right) = -\frac{8}{3} = -2\frac{2}{3}$

28. $-400 \div (-25) = 16$

29. $(-5)(-8)(-4) = 40(-4) = -160$

30. $\left(-\frac{8}{9}\right) \div \left(5\frac{1}{3}\right) = \left(-\frac{8}{9}\right) \div \left(\frac{16}{3}\right) = \left(-\frac{8}{9}\right) \cdot \left(\frac{3}{16}\right) = -\frac{24}{144} = -\frac{1}{6}$

31. $-5780 \div 17 = -340$

32. $(3-5)^2 + 4(-2+1) = (-2)^2 + 4(-1) =$
$4 + 4(-1) = 4 + (-4) = 0$

33. $-2(4c - 7) - 10 = -2(4c) - 2(-7) - 10 =$
$-8c - (-14) - 10 = -8c + 4$

34. $4x + 2y - 5z - 3x + 2z = 4x - 3x + 2y - 5z + 2z =$
$(4 - 3)x + 2y + (-5 + 2)z = x + 2y - 3z$

35. Let n and $n + 1$ represent two consecutive whole numbers. Then $n + (n + 1) = (n + n) + 1 = 2n + 1$. $2n$ represents an even number, so $2n + 1$ represents an odd number.

Chapter Assessment, pages 108-109

1. $20 - 15 \div 5 = 20 - 3 = 17$

2. $(4^2 + 8) \div 4 \cdot 2 = (16 + 8) \div 4 \cdot 2 =$
$24 \div 4 \cdot 2 = 6 \cdot 2 = 12$

3. $-1 - (-5) = -1 + 5 = 4$

4. $(6)(-0.4) = -2.4$

5. $-\frac{1}{2} \div \left(-\frac{2}{3}\right) = -\frac{1}{2} \cdot \left(-\frac{3}{2}\right) = \frac{3}{4}$

6. $\frac{3}{4}\left(\frac{2(5-7)}{5}\right) = \frac{3}{4}\left(\frac{2(-2)}{5}\right) = \frac{3}{4}\left(\frac{-4}{5}\right) = \frac{-12}{20} = -\frac{3}{5}$

7. $6\left(\frac{1}{8} - \frac{1}{3}\right) = 6\left(\frac{3}{24} - \frac{8}{24}\right) = 6\left(-\frac{5}{24}\right) = -\frac{30}{24} = -1\frac{1}{4}$

8. $\frac{7}{8}\left(\frac{2(4-9)}{15}\right) = \frac{7}{8}\left(\frac{2(-5)}{15}\right) = \frac{7}{8}\left(\frac{-10}{15}\right)$
$= \frac{7}{8}\left(-\frac{2}{3}\right) = \frac{-14}{24} = -\frac{7}{12}$

9. $-7.3 + 2.5 = -4.8$

10. $-4(b + 2) = -4b - 8$

11. $-3 + (-2) - (-4) + 8 = -5 - (-4) + 8 =$
$-5 + 4 + 8 = -1 + 8 = 7$

12. $4.5 + (0.2)(0.5 - 5) = 4.5 + (0.2)(-4.5) =$
$4.5 + (-0.9) = 3.6$

13. $10 \div (-5) - (3 - 7)^2 = 10 \div (-5) - (-4)^2 =$
$10 \div (-5) - 16 = -2 - 16 = -18$

14. $-20 \div 5 \div 4 - 3^2 = -20 \div 5 \div 4 - 9 =$
$-4 \div 4 - 9 = -1 - 9 = -10$

15. $3x + 2(x - 7) = 3x + 2(x) + 2(-7) =$
$3x + 2x - 14 = 5x - 14$

16. $7s + 5t - 3u - s + 4u = 7s - s + 5t - 3u + 4u =$
$(7 - 1)s + 5t + (-3 + 4)u = 6s + 5t + u$

17. $\frac{2^2}{20} = \frac{4}{20} = \frac{1}{5}$

18. $4(-7) - 12 = -28 - 12 = -40$

19. $2(7 - (-3)) + 4 = 2(7 + 3) + 4 = 2(10) = 20 + 4 = 24$

20. $\frac{0.8^2}{4} - \frac{0.8}{5} = \frac{0.64}{4} - \frac{0.8}{5} = 0.16 - 0.16 = 0$

21. $\frac{28(-4)}{3(-4) - 2} = \frac{-112}{-12 - 2} = \frac{-112}{-14} = 8$

22. Use the order of operations to evaluate $-2 \cdot 3 + 5$. Compare this value to -3 and determine which is farther to the left or right on the number line. Since $-2 \cdot 3 + 5 = -1$ and -1 is to the right of -3, use $>$ to write $-2 \cdot 3 + 5 > -3$.

23. $-|2.5| = -2.5; |-2.5| = 2.5; -|-2.5| = -2.5;$ I and III are true.; B

24. A   **25.** G   **26.** E   **27.** C

28. D, since 27 is not a perfect square.

29. $t - 15$

30. $4(c + 4) = 4c + 4(4) = 4c + 16$

31. $50 + 18h$

32. Area = Length $\cdot$ Width = $l(l - 5) = l^2 - 5l$

33. $\left[-\frac{5}{8} + 1\frac{1}{2} + 1 + \left(-\frac{3}{4}\right) + (-2)\right] \div 5 =$
$\left[-\frac{5}{8} + \frac{12}{8} + \frac{8}{8} + \left(-\frac{6}{8}\right) + \left(-\frac{16}{8}\right)\right] \cdot \left(\frac{1}{5}\right) =$
$\left(-\frac{7}{8}\right) \cdot \left(\frac{1}{5}\right) = -\frac{7}{40}$

34. Let $2m$ and $2n$ represent the two even numbers.; $2m(2n) = 4mn$; 4 is divisible by 2, therefore, $4mn$ is also divisible by 2 and represents an even number.

Cumulative Review, page 110

1. $5 + (-8) = -3$

2. $-4 - 9 = -4 + (-9) = -13$

3. $-13 + 24 = 11$

4. $-8 - (-2) = -8 + 2 = -6$

5. $-7 \cdot 8 = -56$

6. $-32 \div (-4) = 8$

7. $20 \div (-10) = -2$

8. $(-3)(-6) = 18$

9.

Number	Tally	Frequency
1	II	2
2	IIII	5
3	III	3
4	II	2
5	II	2
6	I	1

10. 2 with a frequency of 5

11. 1, 1, 2, 2, 2, 2, 2, **3**, 3, 3, 4, 4, 5, 5, 6; 3

12. $\frac{5}{15} = \frac{1}{3}$

13. $3 + 4 \cdot 5 = 3 + 20 = 23$

14. $7 - 12 \div 4 + 2 = 7 - 3 + 2 = 4 + 2 = 6$

15. $4 + 2^3 \div 2 = 4 + 8 \div 2 = 4 + 4 = 8$

16. $12 \div (4 + 2) - 7 = 12 \div 6 - 7 = 2 - 7 = -5$

17. commutative property of addition

18. multiplicative inverse property

19. associative property of multiplication

20. additive identity property

21. distributive property of multiplication over addition

22. $-14 \ \square \ -5$
      $<$

23. $\sqrt{17} \ \square \ |-5|$
     $\sqrt{5} \ \square \ 5$
        $<$

24. $3^2 \ \square \ 2^3$
   $9 \ \square \ 8$
    $>$

25. $-6 + 8 \ \square \ 8 - 6$
     $2 \ \square \ 2$
      $=$

26. $-4.5 \ \square \ -4\frac{1}{3}$
   $-4.5 \ \square \ -4.3\overline{3}$
      $<$

27. $\frac{-12}{-2} \ \square \ \sqrt{25}$
    $6 \ \square \ 5$
     $>$

28. $-|4 \cdot (-3)| = -|-12| = -12; \ (-2)(-3)(-4) = 6(-4) =$
$-24; \ -(-1)^{17} = 1; \ \frac{-8 + (-4)}{-2} = \frac{-12}{-2} = 6;$
II and IV are positive.; D

29. $\begin{bmatrix} 6+10-5 & 12+4-13 \\ 9+6-7 & 4+7-6 \end{bmatrix} = \begin{bmatrix} 11 & 3 \\ 8 & 5 \end{bmatrix}$

30. $\begin{bmatrix} 3+(-2) & -1+(-9) \\ -2+6 & 8+(-8) \end{bmatrix} = \begin{bmatrix} 1 & -10 \\ 4 & 0 \end{bmatrix}$

31. $\begin{bmatrix} -3+(-4)-2 & 12+(-4)-8 \\ 2+3-(-2) & 0+(-2)-(-6) \end{bmatrix} = \begin{bmatrix} -9 & 0 \\ 7 & 4 \end{bmatrix}$

32. Coefficients are $3, -2, -1, 5$.

33. $-\frac{2}{5} + \frac{7}{10} = -\frac{4}{10} + \frac{7}{10} = \frac{3}{10}$

34. $12.6 - 24.25 = 12.6 + (-24.25) = -11.65$

35. $\frac{3}{7} \div \left(-\frac{9}{14}\right) = \frac{3}{7} \cdot \left(-\frac{14}{9}\right) = -\frac{42}{63} = -\frac{2}{3}$

36. $-9.2 \cdot (-8.43) = 77.556$

37. Since 2 is a factor of $2n$, $2n$ must be even. Adding 1 to an even number produces an odd number.

Standardized Test, page 111

1. $-|3.2| + 5.8 = -3.2 + 5.8 = 2.6$

2. $\frac{50 - 8}{50} = \frac{42}{50} = 0.84$

3. $3(-3)^2 - 2(-3) + 4 = 3(9) - 2(-3) + 4 =$
$27 - (-6) + 4 = 27 + 6 + 4 = 37$

4. $-3 - (-2.46) + (-0.7) = -3 + 2.46 + (-0.7) =$
$-0.54 + (-0.7) = -1.24$; opposite is 1.24

5. $\quad -3(h + t) = -3h - 21$
$\quad -3h - 3t = -3h - 21$
$\quad\quad\quad -3t = -21$
$\quad\quad\quad\quad\ t = 7$

6. $p = -0.2 = -\frac{2}{10} = -\frac{1}{5}; \ \left|\frac{1}{p}\right| = \left|\frac{1}{-\frac{1}{5}}\right| = \left|-\frac{5}{1}\right| = 5$

7. $64° - (-29°) = 64° + 29° = 93°$

8. $-\frac{-b}{6} = -\left[-\left(\frac{-3}{4}\right)\left(\frac{2}{5}\right)\right] \div 6 = -\left[-\left(\frac{-6}{20}\right)\right] \div 6 =$
$-\left(\frac{3}{10}\right) \div 6 = -\frac{3}{10} \cdot \frac{1}{6} = -\frac{3}{60} = -\frac{1}{20}$

9. $27 - 10 + 2 = 17 + 2 = 19$

10. Smallest is 12; range is 20 so largest is $12 + 20 = 36$; mode $= x =$ other two points.
Therefore mean $= \dfrac{12 + x + x + 32}{4} = 19$
$\quad\quad\quad\quad\quad\quad 44 + 2x = 76$
$\quad\quad\quad\quad\quad\quad\quad\ 2x = 32$
$\quad\quad\quad\quad\quad\quad\quad\quad x = 16$

11. $0.36(2 + 1 + c + 4 + 3 + 5 + 3 + 1) = 2 + 1 + c$
$\quad\quad\quad 0.36(19 + c) = 3 + c$
$\quad\quad\quad 6.84 + 0.36c = 3 + c$
$\quad\quad\quad\quad\quad\quad 3.84 = 0.64c$
$\quad\quad\quad\quad\quad\quad\quad\ 6 = c$

12. 100, between 92 and 93

13. $\frac{3}{7} = \frac{n}{12}$
$7 \cdot n = 3 \cdot 12$
$n = \frac{36}{7} \approx 5.14$

14. $238 + 576 - c = 198$
$\quad\quad\ 814 - c = 198$
$\quad\quad\quad\quad 616 = c$

Data Activity, page 113

1. $$\frac{422.6 + 454.8 + 494.1 + 546.0 + 604.3 + 675.0 + 751.8}{7}$$

   $= 564.1 =$ mean

   range $= 751.8 - 422.6 = 329.2$

2. 85 to 86: $454.8 - 422.6 = 32.2$
   86 to 87: $494.1 - 454.8 = 39.3$
   87 to 88: $546.0 - 494.1 = 51.9$
   88 to 89: $604.3 - 546.0 = 58.3$
   89 to 90: $675.0 - 604.3 = 70.7$
   90 to 91: $751.8 - 675.0 = 76.8$
   Greatest change occurs from 1990 to 1991.

3. 85 to 86: $\frac{32.2}{422.6} \approx 7.6\%$

   86 to 87: $\frac{39.3}{454.8} \approx 8.6\%$

   87 to 88: $\frac{51.9}{494.1} \approx 10.5\%$

   88 to 89: $\frac{58.3}{546.0} \approx 10.7\%$

   89 to 90: $\frac{70.7}{604.3} \approx 11.7\%$

   90 to 91: $\frac{76.8}{675} \approx 11.4\%$

4. Answers will vary. Based on dollar differences, it appears as if the rate of increase of expenditures is increasing. However, using the percent results, the rate of increase slowed during the 1987–1989 period and decreased for 1990–1991.

5. Answers will vary. Using an average increase of 10.1%, the expenditures would be $827.73, $911.33, and $1,003.37.

6. Answers will vary. Exact numbers are easy to determine from the table; both line and bar graphs show change visually. The slope of segments in a line graph gives information on rate of change as well as whether quantity is increasing or decreasing.

U.S. Expenditures for Health, 1985–1991

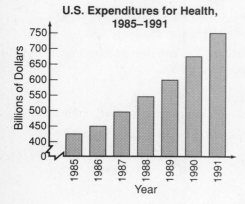

U.S. Expenditures for Health, 1985–1991

Lesson 3.1, pages 115–119

TRY THESE

1. 16; $16 + 5 \stackrel{?}{=} 21$, $21 = 21$ ✔

2. 27; $27 \div 3 \stackrel{?}{=} 9$, $9 = 9$ ✔

3. 5; $6 \cdot 5 \stackrel{?}{=} 30$, $30 = 30$ ✔

4. 16; $32 - 16 \stackrel{?}{=} 16$, $16 = 16$ ✔

5. 28; $28 - 11 \stackrel{?}{=} 17$, $17 = 17$ ✔

6. 48; $\frac{1}{4}(48) \stackrel{?}{=} 12$, $12 = 12$ ✔

7. 49; $\frac{49}{7} \stackrel{?}{=} 7$, $7 = 7$ ✔

8. 11; $8 + 11 \stackrel{?}{=} 19$, $19 = 19$ ✔

9. 4; $9 \stackrel{?}{=} 36 \div 4$, $9 = 9$ ✔

10. 14; $10 \stackrel{?}{=} 14 - 4$, $10 = 10$ ✔

11. -5; $-2 \stackrel{?}{=} -5 + 3$, $-2 = -2$ ✔

12. 3; $3 - 5 \stackrel{?}{=} -2$, $-2 = -2$ ✔

13. 8; $2 \cdot 8 + 6 \stackrel{?}{=} 22$, $22 = 22$ ✔

14. 12; $45 + 3 \cdot 12 \stackrel{?}{=} 81$, $81 = 81$ ✔

15. 3; $60 \div (4 \cdot 3) \stackrel{?}{=} 5$, $60 \div 12 \stackrel{?}{=} 5$, $5 = 5$ ✔

16. 6; $7(6 + 9) \stackrel{?}{=} 105$, $7(15) \stackrel{?}{=} 105$, $105 = 105$ ✔

17. 9; $81 \div (3 \cdot 9) \stackrel{?}{=} 3$, $81 \div 27 \stackrel{?}{=} 3$, $3 = 3$ ✔

18. 4; $64 \div (12 - 4) \stackrel{?}{=} 8$, $12 \div 8 \stackrel{?}{=} 8$, $8 = 8$ ✔

19. 6; $9 \cdot 6 + 42 \stackrel{?}{=} 96$, $54 + 42 \stackrel{?}{=} 96$, $96 = 96$ ✔

20. 16; $3 \cdot 16 \div 4 \stackrel{?}{=} 12$, $48 \div 4 \stackrel{?}{=} 12$, $12 = 12$ ✔

21. 2.1; $13.7 - 2 \cdot 2.1 \stackrel{?}{=} 9.5$, $13.7 - 4.2 \stackrel{?}{=} 9.5$, $9.5 = 9.5$ ✔

22. Answers will vary. If the equation involves basic number facts or other "friendly" numbers for which mental math can be used, the cover-up method is more efficient. Otherwise, guess-and-check can be used.

23. Move right 8 places, so $t = 8$.

24. Move left 6 places, so $b = -6$.

25. Move right 3 places, so $x = 3$.

26. Moved left in problem 24 and right in problem 23, so got negative and positive answers respectively.

27. $A = \dfrac{bh}{2}$; $56 = \dfrac{14 \cdot h}{2}$, $14h = 112$, $h = 8$; Methods will vary.

PRACTICE

We do not have an addition rule yet so students are using mental math or guess-and-check.

1. 40; $4 - 17 \overset{?}{=} 23$, $23 = 23$ ✔

2. 27; $38 + 27 \overset{?}{=} 65$, $65 = 65$ ✔

3. 480; $\dfrac{480}{4} \overset{?}{=} 120$, $120 = 120$ ✔

4. 47; $61 - 47 \overset{?}{=} 14$, $14 = 14$ ✔

5. 75; $\left(\dfrac{1}{3}\right) \cdot 75 \overset{?}{=} 25$, $25 = 25$

6. 3; $\dfrac{81}{3} \overset{?}{=} 27$, $27 = 27$ ✔

7. 14; $59 \overset{?}{=} 73 - 14$, $59 = 59$ ✔

8. 17; $51 \overset{?}{=} 3 \cdot 17$, $51 = 51$ ✔

9. 9.1; $23.9 - 9.1 \overset{?}{=} 14.8$, $14.8 = 14.8$ ✔

10. 94; $37 \overset{?}{=} 94 - 57$, $37 = 37$ ✔

11. 52; $\dfrac{156}{52} \overset{?}{=} 3$, $3 = 3$ ✔

12. 100; $\left(\dfrac{3}{4}\right) \cdot 100 \overset{?}{=} 75$, $75 = 75$ ✔

13. 10; $2 \cdot 10 - 1 \overset{?}{=} 19$, $19 = 19$ ✔

14. -15; $12 + (-15) \overset{?}{=} -3$, $-3 = -3$ ✔

15. 5; $-17 + 5 \overset{?}{=} -12$, $-12 = -12$ ✔

16. 4; $2.1 \cdot 4 \overset{?}{=} 8.4$, $8.4 = 8.4$ ✔

17. 8; $-14 + 8 \overset{?}{=} -6$, $-6 = -6$ ✔

18. 20; $2(20 - 9) \overset{?}{=} 22$, $2(11) \overset{?}{=} 22$, $22 = 22$ ✔

19. 38; $\dfrac{38+4}{6} \overset{?}{=} 7$, $\dfrac{42}{6} \overset{?}{=} 7$, $7 = 7$ ✔

20. 5; $\dfrac{45.5}{5} \overset{?}{=} 9.1$, $9.1 = 9.1$ ✔

21. Answers will vary but should be equivalent to: $13 - t = 12.5$; $t = 0.5$ g.

22. Answers will vary but should be equivalent to: $75.25 + m = 125.99$, $m = \$50.74$.

23. Answers will vary. Possible response: I can solve equations using mental math or guess-and-check. I can also cover up terms or use a number line.

In Problems 24–32, students are still using mental math or guess and check, but now a calculator may be available for use.

EXTEND

24. 7.7   25. 38   26. -16 27. 5

28. 5   29. 6   30. 12   31. -81

32. 34.6

33. Answers will vary. One such might be to guess a number such as 7 and check your answer. Since $23.2 - 7 = 16.2$ which is too large, then 7 was too small. Guess again using a larger value until 7.7 is found.

34. Greater, since $x - 9$ is a positive value.

35. Less, since $-16 + m$ must equal -28 which is smaller than -16.

36. Greater, since $36 - p$ is a negative value.

37. Less, since $7 + 2a$ must equal 1 which is smaller than 7.

38. Greater, since $18 - x$ is a negative value.

39. Less, since $\dfrac{r}{-3}$ must equal 12 which is positive.

40. Answers will vary, but may be similar to the above.

41. 29   42. 7

43. Let x represent the number of species whose population increased.
$$29 + 7 + x = 40$$
$$36 + x = 40$$
$$x = 4$$

44. $2.59 + 3t = 6.16$

45. No, since $3 \cdot 2.00 = 6.00$ which is almost the total cost of the bill.

46. try: $\$1.25$; $2.59 + 3(1.25) = 2.59 + 3.75 = 6.34$, too big;
try: $\$1.20$; $2.59 + 3(1.20) = 2.59 + 3.60 = 6.19$, still too big, but close;
try $\$1.19$; $2.59 + 3(1.19) = 2.59 + 3.57 = 6.16$, therefore $t = \$1.19$

47. possible answer: $\$5.75$ is approximately $\$6.00$; $15 \cdot 6 = 90$; $\$89.95$ is approximately $\$90$.

48. Answers will vary. You can round decimal numbers to whole numbers to estimate.

THINK CRITICALLY

49. The variable must represent a negative number.

50. It could be either, e.g. $x - 5 = -1$, $x = 4$ or $x - 5 = -10$, $x = -5$.

51. If $x + 7 = 29$, $x = 22$, so $x - 16 = 22 + 16 = 38$.

52. If $2y - 23 = 45$, $y = 34$, so $2y - 35 = 2(34) - 35 = 68 - 35 = 33$.

53. If $\frac{4w}{6} + 13 = 50$, $\frac{4w}{6} = 37$, $4w = 222$, $w = 55.5$;
$\frac{4w}{6} + 40 = \frac{4 \cdot 55.5}{6} + 40 = 77$.

MIXED REVIEW

54. Mean:
$$\frac{1 \cdot 23 + 1 \cdot 17 + 2 \cdot 9 + 2 \cdot 6 + 3 \cdot 5 + 4 \cdot 2}{1 + 1 + 2 + 2 + 3 + 4} = \frac{93}{13} \approx 7.2$$
Median: 23, 17, 9, 9, 6, 6, **5**, 5, 5, 2, 2, 2, 2; median = 5
Mode: 2 with a frequency of 4.
Range: $23 - 2 = 21$

55. $8^2 - 4 \cdot 6 = 64 - 4 \cdot 6 = 64 - 24 = 40$; C

Lesson 3.2, pages 120–123

THINK BACK

1. a, c; b may be changed to show zero by adding 3 positive unit blocks or by removing 3 negative unit blocks; d may be changed to show zero by adding 1 negative x-block or by removing the positive x-block; There are many other ways to make zero.

2. Remove pairs of blocks that make zero. The simplified mat will have one positive x-block and four positive unit blocks.

EXPLORE

3a. $x + 5 = 12$ **b.** $x - 1 = -3$ **c.** $2 - x = 7$

4. Answers will vary.

5a.

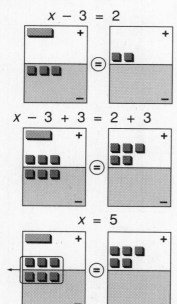

b.

$x + 6 = -10$

$x + 6 - 6 = -10 - 6$

$x = -16$

c.

$x - 4 = -10$

$x - 4 + 4 = -10 + 4$

$x = -6$

6. Answers will vary.

7. Adam's work is incorrect. Explanations will vary. Students might say that Adam added 7 negative unit blocks to the left side and 7 positive unit blocks to the right side. He then removed 7 zero pairs of unit blocks from the left side and 3 zero pairs of unit blocks from the right side. He should have added 7 negative unit blocks to both sides. Then, after removing 7 zero pairs of unit blocks from the left side, he would have x alone on the left and 10 negative unit blocks on the right.

(Solution continues on next page.)

$$x + 7 = -3$$

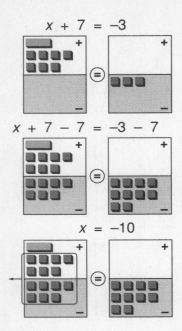

$$x + 7 - 7 = -3 - 7$$

$$x = -10$$

MAKE CONNECTIONS

8. $x - 2 = -7$

$x - 2 + 2 = -7 + 2$

$x = -5$

9. $x + 5 = -7$

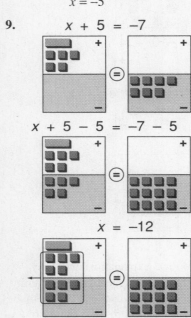

$$x + 5 - 5 = -7 - 5$$

$$x = -12$$

10. $x - 4 = -3$

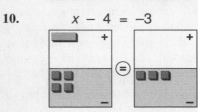

$$x - 4 + 4 = -3 + 4$$

$$x = 1$$

11. $x - (-3) = 8$

$$x - (-3) - 3 = 8 - 3$$

$$x = 5$$

12. $x + 2 = 6$

$$x + 2 - 2 = 6 - 2$$

$$x = 4$$

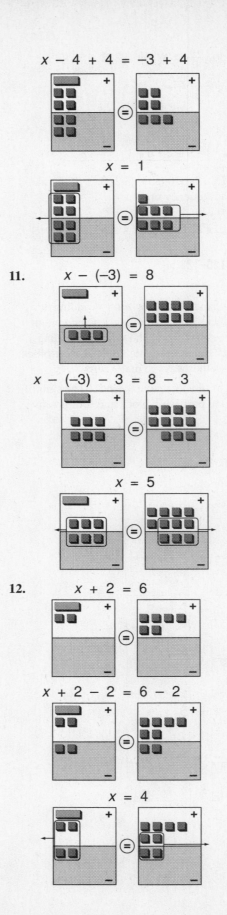

13.

$x - 3 = 2$

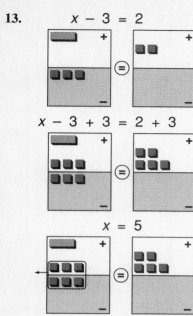

$x - 3 + 3 = 2 + 3$

$x = 5$

14.

$x - (-4) = 6$

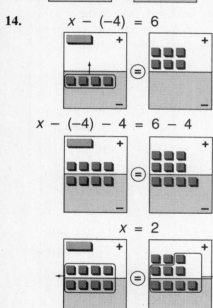

$x - (-4) - 4 = 6 - 4$

$x = 2$

SUMMARIZE

15. To get x alone on one side, count the unit blocks on the side that has x. Then add the opposite of that number of blocks to both sides. This allows you to remove all the unit blocks from the x side because you have formed pairs of blocks that make zero. Now x is alone on one side. The value of the other side is the solution.

16.

$x - 5 = -3$

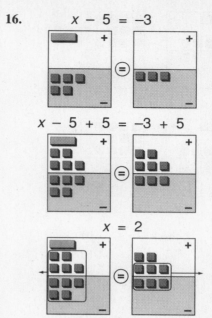

$x - 5 + 5 = -3 + 5$

$x = 2$

17. Examples will vary. In an equation such as $x - 7 = 0$, 7 positive blocks will be added to both sides but only the left side will need to be simplified.

18. Answers will vary. Add 5 unit blocks to both positive quadrants. Remove 5 zero pairs from the left side. So, $-x = 7$. Add an x-block to both sides. Remove zero pairs of x-blocks from the left side. This shows $0 = x + 7$. Add 7 unit blocks to both negative quadrants. Remove 7 zero pairs from the right side. So, $-7 = x$, or $x = -7$.

Lesson 3.3, pages 124–128

EXPLORE

1. Greater since in Figure A it tips the scale in that direction.

2. In Figure B, the scale is now in balance, it took 3 lbs less weight on the left to balance.

3. $x - 3 = 12$

4. Yes, since equal weight was added to both sides.

5. There would be x lbs on the left and 15 lbs on the right side.

TRY THESE

1. -3 **2.** 4

3. 17   **4.** 5

5. 6   **6.** 20

7. $c - (-4) = c + 4$; -4 **8.** $w - (-17) = w + 17$; -17

9. $x - 4 = 7;$

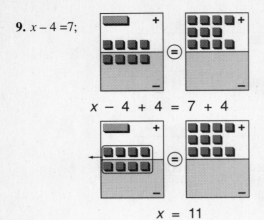

$$x - 4 + 4 = 7 + 4$$

$$x = 11$$

10.
$$r - 8 = 10 \qquad \text{check:} \quad r - 8 = 10$$
$$r - 8 + 8 = 10 + 8 \qquad 8 - 8 \stackrel{?}{=} 10$$
$$r = 18 \qquad\qquad 10 = 10 \checkmark$$

11.
$$a - 11 = -16 \qquad \text{check:} \quad a - 11 = -6$$
$$a - 11 + 11 = -11 + 11 \qquad -5 - 11 \stackrel{?}{=} -6$$
$$a = -5 \qquad\qquad -6 = -6 \checkmark$$

12.
$$z + 14 = -2 \qquad \text{check:} \quad z + 14 = -2$$
$$z + 14 - 14 = -2 - 14 \qquad -16 + 14 \stackrel{?}{=} -2$$
$$z = -16 \qquad\qquad -2 = -2 \checkmark$$

13.
$$g + 19 = 12 \qquad \text{check:} \quad g + 19 = 12$$
$$g + 19 - 19 = 12 - 19 \qquad -7 + 19 \stackrel{?}{=} 12$$
$$g = -7 \qquad\qquad 12 = 12 \checkmark$$

14.
$$x - (-16) = 8 \qquad \text{check:} \quad x - (-16) = 8$$
$$x + 16 = 8 \qquad -8 + 16 \stackrel{?}{=} 8$$
$$x + 16 - 16 = 8 - 16 \qquad 8 = 8 \checkmark$$
$$x = -8$$

15.
$$p - (-3) = 19 \qquad \text{check:} \quad p - (-3) = 19$$
$$p + 3 = 19 \qquad 16 + 3 \stackrel{?}{=} 19$$
$$p + 3 - 3 = 19 - 3 \qquad 19 = 19 \checkmark$$
$$p = 16$$

16.
$$y + 3 = 0 \qquad \text{check:} \quad y + 3 = 0$$
$$y + 3 - 3 = 0 - 3 \qquad -3 + 3 \stackrel{?}{=} 0$$
$$y = -3 \qquad\qquad 0 = 0 \checkmark$$

17.
$$k - 47 = 0 \qquad \text{check:} \quad k - 47 = 0$$
$$k - 47 + 47 = 0 + 47 \qquad 47 - 47 \stackrel{?}{=} 0$$
$$k = 47 \qquad\qquad 0 = 0 \checkmark$$

18.
$$n + \frac{1}{2} = 3 \qquad \text{check:} \quad n + \frac{1}{2} = 3$$
$$n + \frac{1}{2} - \frac{1}{2} = 3 - \frac{1}{2} \qquad 2\frac{1}{2} + \frac{1}{2} \stackrel{?}{=} 3$$
$$n = 2\frac{1}{2} \qquad\qquad 3 = 3 \checkmark$$

19.
$$f + \frac{3}{4} = 5 \qquad \text{check:} \quad f + \frac{3}{4} = 5$$
$$f + \frac{3}{4} - \frac{3}{4} = 5 - \frac{3}{4} \qquad 4\frac{1}{4} + \frac{3}{4} \stackrel{?}{=} 5$$
$$f = 4\frac{1}{4} \qquad\qquad 5 = 5 \checkmark$$

20.
$$(c - 7) + 15 = 26 \qquad \text{check:} \quad (c - 7) + 15 = 26$$
$$c - 7 + 15 = 26 \qquad (18 - 7) + 15 \stackrel{?}{=} 26$$
$$c + 8 = 26 \qquad 11 + 15 \stackrel{?}{=} 26$$
$$c + 8 - 8 = 26 - 8 \qquad 26 = 26 \checkmark$$
$$c = 18$$

21.
$$(b + 1) - 12 = 9 \qquad \text{check:} \quad (b + 1) - 12 = 9$$
$$b + 1 - 12 = 9 \qquad (20 + 1) - 12 \stackrel{?}{=} 9$$
$$b - 11 = 9 \qquad 21 - 12 \stackrel{?}{=} 9$$
$$b - 11 + 11 = 9 + 11 \qquad 9 = 9 \checkmark$$
$$b = 20$$

22. Answers will vary but might be to isolate the variable on one side of the equation by adding the opposite of the constant that is on the same side of the equation as the variable to both sides of the equation.

PRACTICE

1.
$$b + 15 = 40 \qquad \text{check:} \quad b + 15 = 40$$
$$b + 15 - 15 = 40 - 15 \qquad 25 + 15 \stackrel{?}{=} 40$$
$$b = 25 \qquad\qquad 40 = 40 \checkmark$$

2.
$$z + 28 = 79 \qquad \text{check:} \quad z + 28 = 79$$
$$z + 28 - 28 = 79 - 28 \qquad 51 + 28 \stackrel{?}{=} 79$$
$$z = 51 \qquad\qquad 79 = 79 \checkmark$$

3.
$$c + 10 = -10 \qquad \text{check:} \quad c + 10 = -10$$
$$c + 10 - 10 = -10 - 10 \qquad -20 + 10 \stackrel{?}{=} -10$$
$$c = -20 \qquad\qquad -10 = -10 \checkmark$$

4.
$$j + 8 = -5 \qquad \text{check:} \quad j + 8 = -5$$
$$j + 8 - 8 = -5 - 8 \qquad -13 + 8 \stackrel{?}{=} -5$$
$$j = -13 \qquad\qquad -5 = -5 \checkmark$$

5.
$$x - 9 = 9 \qquad \text{check:} \quad x - 9 = 9$$
$$x - 9 + 9 = 9 + 9 \qquad 18 - 9 \stackrel{?}{=} 9$$
$$x = 18 \qquad\qquad 9 = 9 \checkmark$$

6.
$$b - 41 = 42 \qquad \text{check:} \quad b - 41 = 42$$
$$b - 41 + 41 = 42 + 41 \qquad 83 - 41 \stackrel{?}{=} 42$$
$$b = 83 \qquad\qquad 42 = 42 \checkmark$$

7.
$$w - (-2) = -1 \qquad \text{check:} \quad w - (-2) = -1$$
$$w + 2 = -1 \qquad -3 - (-2) \stackrel{?}{=} -1$$
$$w + 2 - 2 = -1 - 2 \qquad -3 + 2 \stackrel{?}{=} -1$$
$$w = -3 \qquad\qquad -1 = -1 \checkmark$$

8. $k - (-18) = -3$ check: $k - (-18) = -3$
$\quad k + 18 = -3$ $\quad -21 - (-18) \overset{?}{=} -3$
$\quad k + 18 - 18 = -3 - 18$ $\quad -21 + 18 \overset{?}{=} -3$
$\quad\quad k = -21$ $\quad\quad -3 = -3$ ✔

9. $\quad 31 + x = 17$ check: $31 + x = 17$
$\quad 31 + x - 31 = 17 - 31$ $\quad 31 + (-14) \overset{?}{=} 17$
$\quad\quad x = -14$ $\quad 31 - 14 \overset{?}{=} 17$
$\quad\quad\quad 17 = 17$ ✔

10. $\quad 50 + p = 13$ check: $50 + p = 13$
$\quad 50 + p - 50 = 13 - 50$ $\quad 50 + (-37) \overset{?}{=} 13$
$\quad\quad p = -37$ $\quad 50 - 37 = 13$
$\quad\quad\quad 13 = 13$ ✔

11. $\quad 31 + x = -17$ check: $31 + (-48) = -17$
$\quad 31 + x - 31 = -17 - 31$ $\quad 31 - 48 \overset{?}{=} -17$
$\quad\quad x = -48$ $\quad\quad -17 = -17$ ✔

12. $\quad 37 + g = -25$ check: $37 + g = -25$
$\quad 37 + g - 37 = -25 - 37$ $\quad 37 + (-62) \overset{?}{=} -25$
$\quad\quad g = -62$ $\quad 37 - 62 \overset{?}{=} -25$
$\quad\quad\quad -25 = -25$ ✔

13. Equations will vary, possible response:
Let t represent temperature at 11 a.m.
$\quad t + 14° = 71°$ check: $t + 14° = 71°$
$\quad t + 14° - 14° = 71° - 14°$ $\quad 57° + 14° \overset{?}{=} 71°$
$\quad\quad t = 57°$ $\quad\quad 71° = 71°$ ✔

14. $\quad e + (-7) = 12$ check: $e + (-7) = 12$
$\quad e + (-7) + 7 = 12 + 7$ $\quad 19 + (-7) \overset{?}{=} 12$
$\quad\quad e = 19$ $\quad\quad 12 = 12$ ✔

15. $\quad f + (-15) = -8$ check: $f + (-15) = -8$
$\quad f + (-15) + 15 = -8 + 15$ $\quad 7 + (-15) \overset{?}{=} -8$
$\quad\quad f = 7$ $\quad 7 - 15 \overset{?}{=} -8$
$\quad\quad\quad -8 = -8$ ✔

16. $\quad -4 + n = -7$ check: $-4 + n = -7$
$\quad -4 + n + 4 = -7 + 4$ $\quad -4 + (-3) \overset{?}{=} -7$
$\quad\quad n = -3$ $\quad -4 - 3 \overset{?}{=} -7$
$\quad\quad\quad -7 = -7$ ✔

17. $\quad -9 + h = -9$ check: $-9 + h = -9$
$\quad -9 + h + 9 = -9 + 9$ $\quad -9 + 0 \overset{?}{=} -9$
$\quad\quad h = 0$ $\quad\quad -9 = -9$ ✔

18. $\quad r + 3.5 = -6.5$ check: $r + 3.5 = -6.5$
$\quad r + 3.5 - 3.5 = -6.5 - 3.5$ $\quad -10 + 3.5 \overset{?}{=} -6.5$
$\quad\quad r = -10$ $\quad\quad -6.5 = -6.5$ ✔

19. $\quad m + 2.6 = -1$ check: $m + 2.6 = -1$
$\quad m + 2.6 - 2.6 = -1 - 2.6$ $\quad -3.6 + 2.6 \overset{?}{=} -1$
$\quad\quad m = -3.6$ $\quad\quad -1 = -1$ ✔

20. $\quad x + \dfrac{5}{8} = \dfrac{7}{8}$ check: $x + \dfrac{5}{8} = \dfrac{7}{8}$
$\quad x + \dfrac{5}{8} - \dfrac{5}{8} = \dfrac{7}{8} - \dfrac{5}{8}$ $\quad \dfrac{2}{8} + \dfrac{5}{8} \overset{?}{=} \dfrac{7}{8}$
$\quad\quad x = \dfrac{2}{8} = \dfrac{1}{4}$ $\quad\quad \dfrac{7}{8} = \dfrac{7}{8}$ ✔

21. $\quad g + \dfrac{7}{12} = 2$ check: $g + \dfrac{7}{12} = 2$
$\quad g + \dfrac{7}{12} - \dfrac{7}{12} = 2 - \dfrac{7}{12}$ $\quad 1\dfrac{5}{12} + \dfrac{7}{12} \overset{?}{=} 2$
$\quad\quad g = 1\dfrac{5}{12}$ $\quad\quad 2 = 2$ ✔

22. $5 + (2 + y) = 8$ check: $5 + (2 + y) = 8$
$\quad 5 + 2 + y = 8$ $\quad 5 + (2 + 1) \overset{?}{=} 8$
$\quad\quad 7 + y = 8$ $\quad 5 + 3 \overset{?}{=} 8$
$\quad 7 + y - 7 = 8 - 7$ $\quad\quad 8 = 8$ ✔
$\quad\quad y = 1$

23. $7 + (4 + a) = 13$ check: $7 + (4 + a) = 13$
$\quad 7 + 4 + a = 13$ $\quad 7 + (4 + 2) \overset{?}{=} 13$
$\quad\quad 11 + a = 13$ $\quad 7 + 6 \overset{?}{=} 13$
$\quad 11 + a - 11 = 13 - 11$ $\quad\quad 13 = 13$ ✔
$\quad\quad a = 2$

24. $\quad 3 = -2 - (5 - c)$ check: $3 = -2 - (5 - c)$
$\quad 3 = -2 - 5 + c$ $\quad 3 \overset{?}{=} -2 - (5 - 10)$
$\quad 3 = -7 + c$ $\quad 3 \overset{?}{=} -2 - (-5)$
$\quad 3 + 7 = -7 + c + 7$ $\quad 3 \overset{?}{=} -2 + 5$ ✔
$\quad\quad 10 = c$

25. $\quad 16 = 9 - (1 - b)$ check: $16 = 9 - (1 - b)$
$\quad 16 = 9 - 1 + b$ $\quad 16 \overset{?}{=} 9 - (1 - 8)$
$\quad 16 = 8 + b$ $\quad 16 \overset{?}{=} 9 - (-7)$
$\quad 16 - 8 = 8 + b - 8$ $\quad 16 \overset{?}{=} 9 + 7$
$\quad\quad 8 = b$ $\quad 16 = 16$ ✔

26. Let c represent the change in temperature.
$\quad 168° - 105° = c$
$\quad\quad 63° = c$

27. Answers will vary, one answer might be: after spending \$32 at the mall, you still have \$48. How much money did you start with?

EXTEND

28. Add 2,394 to 339 to get 2,733. Check by subtracting 2,394 from 2,733 to see if you get 339.

29. estimate: -12;
$$t + 3.024 = -9.74$$
$$t + 3.024 - 3.024 = -9.74 - 3.024$$
$$t = -12.764$$
check: $t + 3.024 = -9.74$
$$-12.764 + 3.024 \stackrel{?}{=} -9.74$$
$$-9.74 = -9.74 \ \checkmark$$

30. estimate: -86;
$$-2.6015 = a + 83.3$$
$$-2.6015 - 83.3 = a + 83.3 - 83.3$$
$$-85.9015 = a$$
check: $-2.6015 = a + 83.3$
$$-2.6015 \stackrel{?}{=} -85.9015 + 83.3$$
$$-2.6015 = -2.6015 \ \checkmark$$

31. estimate: 5;
$$x - 0.1008 = 4.992$$
$$x - 0.1008 + 0.1008 = 4.992 + 0.1008$$
$$x = 5.0928$$
check: $x - 0.1008 = 4.992$
$$5.0928 - 0.1008 \stackrel{?}{=} 4.992$$
$$4.992 = 4.992 \ \checkmark$$

32. estimate: 24;
$$-18.9 + n = 5.01$$
$$-18.9 + n + 18.9 = 5.01 + 18.9$$
$$n = 23.91$$
check: $-18.9 + n = 5.01$
$$-18.9 + 23.91 \stackrel{?}{=} 5.01$$
$$5.01 = 5.01 \ \checkmark$$

33. Let C represent the RDA of calories for a 35-year-old male.
$$3000 = C + 300$$
$$3000 - 300 = C + 300 - 300$$
$$2700 = C$$

34. Let C represent the RDA of calories for a 55-year-old woman.
$$C = 2800 - 1000$$
$$C = 1800$$

35. Let r represent the number of calories an hour of roller-skating consumes.
$$220 = r - 130$$
$$220 + 130 = r - 130 + 130$$
$$350 = r$$

THINK CRITICALLY

36a. x must decrease if the sum is to still equal b.

b. b must decrease to keep the sum equal.

37a. x must decrease if the difference is to still equal b.

b. b must increase to keep the difference equal.

38.
$$z + 9 = -5 \qquad\qquad z - 3 = -14 - 3$$
$$+9 - 9 = -5 - 9 \qquad z - 3 = -17$$
$$z = -14$$

39.
$$c - 1\frac{1}{2} = 2\frac{1}{4} \qquad\qquad 2c = 2\left(3\frac{3}{4}\right)$$
$$c - 1\frac{1}{2} + 1\frac{1}{2} = 2\frac{1}{4} + 1\frac{1}{2} \qquad 2c = 2\left(\frac{15}{4}\right) = 7\frac{1}{2}$$
$$c = 3\frac{3}{4}$$

40. Answers will vary. Let p represent the regular price.
$p - 59 = 177$, $p = \$236$;
What is the regular price of the machine?

41. Answers will vary. Let w represent the width.
$w + 25.5 = 53$, $w = 27.5$;
What is the width of the pool?

PROJECT CONNECTION

1. $8 \cdot 9 = 72$ calories from fat
$$\frac{72}{140} \approx 51\% \text{ of caloric value from fat}$$

2. 30% of $370 = 0.30 \cdot 370 = 111$ calories;
111 calories $\div$ 9 grams/cal = 12 g

3. Answers will vary.

Lesson 3.4, pages 129–133

EXPLORE

1a. $4x$; 12; 12 people are to be divided into 4 teams, each having the same (unknown) number of people; the 12 unit cubes on the right represent the total number of people, and the 4 x-blocks on the left represent the four teams. x-blocks are used on the left because the number of people on each team is unknown

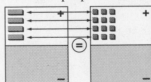

b. $4x = 12$
$$x = 3$$

2.

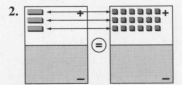

$$3x = 18$$
$$x = 6$$

1. Divide both sides by 7. 2. Divide both sides by −5.

3. Multiply both sides by 3. 4. Multiply both sides by −4.

5. Multiply both sides by $\frac{3}{2}$. 6. Multiply both sides by $\frac{4}{3}$.

7. Divide both sides by 9. 8. Divide both sides by −6.

9. $3x = -9$ 10. $2x = 6$

11. $\dfrac{a}{3} = 5$

$3 \cdot \dfrac{a}{3} = 5 \cdot 3$

$a = 15$

12. $12 = \dfrac{x}{5}$

$5 \cdot 12 = \dfrac{x}{5} \cdot 5$

$60 = x$

13. $5b = 65$

$\dfrac{5b}{5} = \dfrac{65}{5}$

$b = 13$

14. $36 = 9c$

$\dfrac{36}{9} = \dfrac{9c}{9}$

$4 = c$

15. $9.2b = 32.2$

$\dfrac{9.2b}{9.2} = \dfrac{32.2}{9.2}$

$b = \dfrac{32.2}{9.2}$

16. $22.1 = 3.4c$

$\dfrac{22.1}{3.4} = \dfrac{3.4c}{3.4}$

$\dfrac{22.1}{3.4} = c$

17. $\dfrac{d}{28.5} = 7.5$

$28.5\left(\dfrac{d}{28.5}\right) = 7.5(28.5)$

$d = 7.5(28.5)$

18. $\dfrac{f}{16} = 6.4$

$16\left(\dfrac{f}{16}\right) = 6.4(16)$

$f = 6.4(16)$

1. Divide both sides by 6. 2. Divide both sides by 12.

3. Divide both sides by −4. 4. Divide both sides by 8.

5. Divide both sides by −7. 6. Divide both sides −9.

7. Multiply both sides by 5. 8. Multiply both sides by $\frac{3}{2}$.

9. Multiply both sides by 12. 10. Multiply both sides by −10.

11. Multiply both sides by $\frac{4}{3}$. 12. Multiply both sides by $-\frac{5}{2}$.

13. $-4f = -480$ check: $-4f = -480$

$\dfrac{-4f}{-4} = \dfrac{-480}{-4}$ $-4(120) \overset{?}{=} -480$

$f = 120$ $-480 = -480$ ✔

14. $80c = 560$ check: $80c = 560$

$\dfrac{80c}{80} = \dfrac{560}{80}$ $80(7) \overset{?}{=} 560$

$c = 7$ $560 = 560$ ✔

15. $\dfrac{2}{3}x = 8$ check: $\dfrac{2}{3}x = 8$

$\dfrac{3}{2}\left(\dfrac{2}{3}x\right) = \dfrac{3}{2}(8)$ $\dfrac{2}{3}(12) \overset{?}{=} 8$

$x = \dfrac{24}{2} = 12$ $8 = 8$ ✔

16. $-\dfrac{1}{4}y = 7$ check: $-\dfrac{1}{4}y = 7$

$-4\left(-\dfrac{1}{4}x\right) = -4(7)$ $-\dfrac{1}{4}(28) \overset{?}{=} 7$

$y = -28$ $7 = 7$ ✔

17. $-5x = 90$ check: $-5x = 90$

$\dfrac{-5x}{-5} = \dfrac{90}{-5}$ $-5(-18) \overset{?}{=} 90$

$x = -18$ $90 = 90$ ✔

18. $\dfrac{3}{4}h = -6$ check: $\dfrac{3}{4}h = -6$

$\dfrac{4}{3}\left(\dfrac{3}{4}h\right) = \dfrac{4}{3}(-6)$ $\dfrac{3}{4}(-8) \overset{?}{=} -6$

$h = \dfrac{-24}{3} = -8$ $-6 = -6$ ✔

19. $6y = -102$ check: $6y = -102$

$\dfrac{6y}{6} = \dfrac{-102}{6}$ $6(-17) \overset{?}{=} -102$

$y = -17$ $-102 = -102$ ✔

20. $3.2p = 14.4$ check: $3.2p = 14.4$

$\dfrac{3.2p}{3.2} = \dfrac{14.4}{3.2}$ $3.2(4.5) \overset{?}{=} 14.4$

$p = 4.5$ $14.4 = 14.4$ ✔

21. $\dfrac{z}{6} = 13$ check: $\dfrac{z}{6} = 13$

$6\left(\dfrac{z}{6}\right) = 6(13)$ $\dfrac{78}{6} \overset{?}{=} 13$

$z = 78$ $13 = 13$ ✔

22. $2.6y = 24.7$ check: $2.6y = 24.7$

$\dfrac{2.6y}{2.6} = \dfrac{24.7}{2.6}$ $2.6(9.5) \overset{?}{=} 24.7$

$y = 9.5$ $24.7 = 24.7$ ✔

23. $\dfrac{2}{5}z = 1.2$ check: $\dfrac{2}{5}z = 1.2$

$\dfrac{5}{2}\left(\dfrac{2}{5}z\right) = \dfrac{5}{2}(1.2)$ $\dfrac{2}{5}(3) \overset{?}{=} 1.2$

$z = \dfrac{6}{2} = 3$ $1.2 = 1.2$ ✔

24. $\dfrac{2}{7}w = 3.4$ check: $\dfrac{2}{7}w = 3.4$

$\dfrac{7}{2}\left(\dfrac{2}{7}w\right) = \dfrac{7}{2}(3.4)$ $\dfrac{2}{7}(1.9) \overset{?}{=} 3.4$

$w = \dfrac{23.8}{2} = 11.9$ $3.4 = 3.4$ ✔

25. $\dfrac{-c}{13.6} = 3.5$ check: $\dfrac{-c}{13.6} = 3.5$

$-13.6\left(\dfrac{-c}{13.6}\right) = -13.6(3.5)$ $\dfrac{-(-47.6)}{13.6} \overset{?}{=} 3.5$

$c = -47.6$ $3.5 = 3.5$ ✔

26. $\dfrac{-j}{3.4} = 5.5$ check: $\dfrac{-j}{3.4} = 5.5$

$-3.4\left(\dfrac{-j}{3.4}\right) = -3.4(5.5)$ $\dfrac{-(-18.7)}{3.4} \stackrel{?}{=} 5.5$

$j = -18.7$ $5.5 = 5.5$ ✔

27. $5.4g = 13.5$ check: $5.4g = 13.5$

$\dfrac{5.4g}{5.4} = \dfrac{13.5}{5.4}$ $5.4(2.5) \stackrel{?}{=} 13.5$

$g = 2.5$ $13.5 = 13.5$ ✔

28. $3.8q = -9.5$ check: $3.8q = -9.5$

$\dfrac{3.8q}{3.8} = \dfrac{-9.5}{3.8}$ $3.8(-2.5) \stackrel{?}{=} 9.5$

$q = -2.5$ $9.5 = 9.5$ ✔

29. Let x represent number of teams that can be formed.

$6x = 1008$

$\dfrac{6x}{6} = \dfrac{1008}{6}$

$x = 168$ teams

30. Let x represent number of games sold.

$19.95x = 1017.45$; estimate: 50 games sold;

check: $19.95 \cdot 50 = \$997.50$;

$19.95x = 1017.45$

$\dfrac{19.95x}{19.95} = \dfrac{1017.45}{19.95}$

$x = 51$ games sold

31. Answers will vary but should include the fact that both sides of the equation are to be divided by -3.

32. Answers will vary, possible answers:

$\dfrac{x}{7} = 5$ and $3x = 96$

EXTEND

33. $-4(-5) = 8n$ check: $-4(-5) = 8n$

$20 = 8n$ $-4(-5) \stackrel{?}{=} 8(2.5)$

$\dfrac{20}{8} = \dfrac{8n}{8}$ $20 = 20$ ✔

$2.5 = n$

34. $\dfrac{k}{3} = 2(6 - 8 + 12)$ check: $\dfrac{k}{3} = 2(6 - 8 + 12)$

$\dfrac{k}{3} = 2(10)$ $\dfrac{60}{3} \stackrel{?}{=} 2(6 - 8 + 12)$

$\dfrac{k}{3} = 20$ $20 \stackrel{?}{=} 2(10)$

$3\left(\dfrac{k}{3}\right) = 3(20)$ $20 = 20$ ✔

$k = 60$

35. $\dfrac{p}{4 - 10} = \dfrac{4 + 5}{8 - 5}$ check: $\dfrac{p}{4 - 10} = \dfrac{4 + 5}{8 - 5}$

$\dfrac{p}{-6} = \dfrac{9}{3}$ $\dfrac{-18}{4 - 10} \stackrel{?}{=} \dfrac{4 + 5}{8 - 5}$

$\dfrac{p}{-6} = 3$ $\dfrac{-18}{-6} \stackrel{?}{=} \dfrac{9}{3}$

$-6\left(\dfrac{p}{-6}\right) = -6(3)$ $3 = 3$ ✔

$p = -18$

36. $\dfrac{64.8}{3} = -2.4x$ check: $\dfrac{64.8}{3} = -2.4x$

$21.6 = -2.4x$ $\dfrac{64.8}{3} \stackrel{?}{=} (-2.4)(-9)$

$\dfrac{21.6}{-2.4} = \dfrac{-2.4x}{-2.4}$ $21.6 = 21.6$ ✔

$-9 = x$

37. $6 + 18 \div 9 = (9 - 29)y$ check: $6 + 18 \div 9 = (9 - 29)y$

$6 + 18 \div 9 = -20y$ $6 + 18 \div 9 \stackrel{?}{=} (9 - 29)(-0.4)$

$6 + 2 = -20y$ $6 + 2 \stackrel{?}{=} -20(-0.4)$

$8 = -20y$ $8 = 8$ ✔

$\dfrac{8}{-20} = \dfrac{-20y}{-20}$

$-0.4 = y$

38. $4(-2.1) = \dfrac{n}{12(-3.5)}$ check: $4(-2.1) = \dfrac{n}{12(-3.5)}$

$-8.4 = \dfrac{n}{-42}$ $4(-2.1) \stackrel{?}{=} \dfrac{352.8}{12(-3.5)}$

$-42(-8.4) = -42\left(\dfrac{n}{-42}\right)$ $-8.4 = -8.4$ ✔

$352.8 = n$

39. $56.8g = 1209.84$ check: $56.8g = 1209.84$

$\dfrac{56.8g}{56.8} = \dfrac{1209.84}{56.8}$ $56.8(21.3) \stackrel{?}{=} 1209.84$

$g = 21.3$ $1209.84 = 1209.84$ ✔

40. $13.6 = \dfrac{e}{-9.7}$ check: $13.6 = \dfrac{e}{-9.7}$

$-9.7(13.6) = -9.7\left(\dfrac{e}{-9.7}\right)$ $13.6 \stackrel{?}{=} \dfrac{-131.92}{-9.7}$

$-131.92 = e$ $13.6 = 13.6$ ✔

41. $\dfrac{h}{123} = 0.88$ check: $\dfrac{h}{123} = 0.88$

$123\left(\dfrac{h}{123}\right) = 123(0.88)$ $\dfrac{108.24}{12.3} \stackrel{?}{=} 0.88$

$h = 108.24$ $108.24 = 108.24$ ✔

42.

$$8.036 = \frac{y}{-0.45}$$

check: $8.036 = \frac{-y}{-0.45}$

$$-0.45(8.036) = -0.45\left(\frac{y}{-0.45}\right)$$

$8.036 \stackrel{?}{=} \frac{-3.62}{-0.45}$

$$-3.62 = y$$

$-3.62 = -3.62$ ✔

43. $2142 = -2520p$ check: $2142 = -2520p$

$$\frac{2142}{-2520} = \frac{-2520p}{-2520}$$

$2142 \stackrel{?}{=} -2520(-0.85)$

$$-0.85 = p$$

$2142 = 2142$ ✔

44. $2.7x = 8.1(-9.3)$ check: $2.7x = 8.1(-9.3)$

$2.7x = -75.33$ $2.7(-27.9) \stackrel{?}{=} 8.1(-9.3)$

$$\frac{2.7x}{2.7} = \frac{-75.33}{2.7}$$

$-75.33 = -75.33$ ✔

$$x = -27.9$$

45.

$$-1 = \frac{3}{4}x$$

check: $-1 = \frac{3}{4}x$

$$\frac{4}{3}(-1) = \frac{4}{3}\left(\frac{3}{4}x\right)$$

$-1 \stackrel{?}{=} \frac{3}{4}\left(-\frac{4}{3}\right)$

$$-\frac{4}{3} = x$$

$-1 = -1$ ✔

$$-1.33 = x$$

46.

$$-\frac{x}{2} = 1\frac{1}{4}$$

check: $-\frac{x}{2} = 1\frac{1}{4}$

$$-2\left(-\frac{x}{2}\right) = -2\left(1\frac{1}{4}\right)$$

$-\frac{-2.5}{2} \stackrel{?}{=} 1\frac{1}{4}$

$$x = -2\frac{1}{2} = -2.5$$

$1.25 \stackrel{?}{=} 1.25$ ✔

47.

$$-\frac{3}{4}x = -\frac{2}{3}$$

check: $-\frac{3}{4}x = -\frac{2}{3}$

$$-\frac{4}{3}\left(-\frac{3}{4}x\right) = -\frac{4}{3}\left(-\frac{2}{3}\right)$$

$-\frac{3}{4}(0.89) \stackrel{?}{=} -\frac{2}{3}$

$$x = \frac{8}{9}$$

$-\frac{2.67}{4} \stackrel{?}{=} -\frac{2}{3}$

$$x = 0.89$$

$-0.67 = -0.67$ ✔

48. The correct equation is b.

$$\frac{x}{3} = 39.15$$

$$3\left(\frac{x}{3}\right) = 3(39.15)$$

$$x = 117.45$$

49. The correct answer is b.

$$s = \$4968.80 - \frac{3}{4}(\$4968.80)$$

$$s = \$4968.80 - (\$3726.60)$$

$$s = \$4968.80 - \$3726.60$$

$$s = \$1242.20$$

50. Area = Length • Width

$$7200 = 120 \cdot w$$

$$\frac{7200}{120} = \frac{120w}{120}$$ The width is 60 ft.

$$60 = w$$

51. perimeter of square = 4 • length of a side;

$$72 = 4x$$

$$\frac{72}{4} = \frac{4x}{4}$$

$18 = x$ The length of one side is 18 in.

52. $d = 14,494 - (-282)$

$d = 14,494 + 282$

$d = 14,776$ The difference in elevation is 14,776 ft.

53. $6.875g = 11$

$$\frac{6.875g}{6.875} = \frac{11}{6.875}$$

$g = 1.6$ The depth of the Grand Canyon is 1.6 km.

54. $0.45d = 12.6$

$$\frac{0.45d}{0.45} = \frac{12.6}{0.45}$$

$d = 28$ The experiment lasted 28 days.

55. $11,612m = 36,000,000$

$$\frac{11,612m}{11,612} = \frac{36,000,000}{11,612}$$

$m \approx 3100$ Mercury's diameter is 3100 mi.

THINK CRITICALLY

56a. one solution

b. Infinite number since any number times zero equals zero.

c. No, they have different solution sets.

d. No, since this results in an equation with an infinite number of solutions. It is not equivalent to the original equation.

MIXED REVIEW

57. $8 - 3r + 17 + 5r - 2$

$(-3r + 5r) + (8 + 17 - 2)$

$2r + 23$ Correct answer is D.

58. $0.4c = 2.6$ check: $0.4c = 2.6$

$$\frac{0.4c}{0.4} = \frac{2.6}{0.4}$$

$0.4(6.5) \stackrel{?}{=} 2.6$

$$c = 6.5$$

$2.6 = 2.6$ ✔

59.

$$\frac{d}{3} = 18$$

check: $\frac{d}{3} = 18$

$$3\left(\frac{d}{3}\right) = 3(18)$$

$\frac{54}{3} \stackrel{?}{=} 18$

$$d = 54$$

$18 = 18$ ✔

60.

$$f - 10 = 3.5 \qquad \text{check:} \quad f - 10 = 3.5$$
$$f - 10 + 10 = 3.5 + 10 \qquad 13.5 - 10 \overset{?}{=} 3.5$$
$$f = 13.5 \qquad 3.5 = 3.5 \; \text{✔}$$

61.

$$28.7 = 16.2 + c \qquad \text{check:} \quad 28.7 = 16.2 + c$$
$$28.7 - 16.2 = 16.2 + c - 16.2 \qquad 28.7 \overset{?}{=} 16.2 + 12.5$$
$$12.5 = c \qquad 28.7 = 28.7 \; \text{✔}$$

62.

$$(-4)^2 = 8 \cdot x$$
$$16 = 8x$$
$$\frac{16}{8} = \frac{8x}{8}$$
$$2 = x$$

63.

$$\frac{72}{2} = x + 15$$
$$72 = 2(x + 15)$$
$$72 = 2x + 30$$
$$72 - 30 = 2x + 30 - 30$$
$$42 = 2x$$
$$\frac{42}{2} = \frac{2x}{2}$$
$$21 = x$$

ALGEBRA WORKS

1. Not unless you know the male's height. For example, a male 6'1" tall could be medium-framed, but a 5'6" male with a 3" elbow measurement would be large-framed.

2. Answers will vary but might look like:

$$2\frac{3}{4} - 2\frac{1}{2} = d$$
$$\frac{1}{4} = d \qquad \text{My elbow is } \frac{1}{4}" \text{ larger.}$$

3. $x = 2\frac{5}{8} + \frac{1}{4}$

$$x = 2\frac{5}{8} + \frac{2}{8} = 2\frac{7}{8} \qquad \text{Theo's measurement is } 2\frac{7}{8}".$$

Lesson 3.5, pages 135–138

EXPLORE

1. Answers will vary. Students may estimate that $79.95 is $10 off of the original price, or about 11%; the other discounts are much greater.

2. Successive discounts of 10% and 25% are not equivalent to a 35% discount. A convenient price to try is $100: 10% off gives $90, then 25% off gives $67.50, but a 35% discount gives $65.00.

3.

$$\frac{1}{3} = \frac{p}{100}$$
$$3p = 100$$
$$p = 33\frac{1}{3}\%$$

4. Answers will vary. Students may conclude that sale price E is lowest.

TRY THESE

1. $0.40 \cdot 120 = 48$

2. $0.15 \cdot 260 = 39$

3. $0.09 \cdot 90 = 8.1$

4. $0.25 \cdot 56 = 14$

5. $0.84 \cdot 155 = 130.2$

6. $0.03 \cdot 30 = 0.9$

7. $x \cdot 70 = 20$
$$x = \frac{20}{70} \approx 29\%$$

8. $x \cdot 134 = 45$
$$x = \frac{45}{134} \approx 34\%$$

9. $x \cdot 18 = 30$
$$x = \frac{30}{18} \approx 167\%$$

10. $x \cdot 200 = 32$
$$x = \frac{32}{200} \approx 16\%$$

11.
$$64 = 0.80 \cdot x$$
$$\frac{64}{0.80} = x$$
$$80 = x$$

12.
$$78 = 0.16 \cdot x$$
$$\frac{78}{0.16} = x$$
$$487.5 = x$$

13.
$$130 = 1.25 \cdot x$$
$$\frac{130}{1.25} = x$$
$$104 = x$$

14.
$$90 = 2.00 \cdot x$$
$$\frac{90}{2.00} = x$$
$$45 = x$$

15. Let r represent the interest rate paid.
$$540 + 540 \cdot r = 564.30$$
$$540 + 540 \cdot r - 540 = 564.30 - 540$$
$$540 \cdot r = 24.30$$
$$r = \frac{24.30}{540} = 0.045 = 4.5\% = 4\frac{1}{2}\%$$

16. Let p represent the original price.
$$p - 0.25p = 11.96$$
$$0.75p = 11.96$$
$$p = \frac{11.96}{0.75} \approx \$15.95$$

17. Answers will vary. Possible answer: Divide 158,000,000 by 1000, find 3% of 158,000, and then multiply the result by 1000.

PRACTICE

1. $0.23 \cdot 45 \approx 10.4$

2. $0.80x = 35$
$$x = \frac{35}{0.80} \approx 43.8$$

3. $x \cdot 150 = 250$
$$x = \frac{250}{150} \approx 166.7\%$$

4. $1.25 \cdot 36 = 45$

5. $x \cdot 120 = 75$

$x = \dfrac{75}{120} = 62.5\%$

6. $x \cdot 7 = 28$

$x = \dfrac{28}{7} = 400\%$

7. $2.30 \cdot 230 = 529$

8. $x \cdot 50 = 14$

$x = \dfrac{14}{50} = 28\%$

9. $x \cdot 300 = 190$

$x = \dfrac{190}{300} \approx 63.3\%$

10. $0.56 \cdot 150 = 84$

11. $0.08 \cdot 32 \approx 2.6$

12. $x \cdot 12 = 40$

$x = \dfrac{40}{12} \approx 333.3\%$

13. $\dfrac{135 - 79}{79} = \dfrac{56}{79} \approx 0.71$ or 71%

14. $170 + 0.02 \cdot 170 = 170 + 3.4 = 173.4$ tons

15. Answers will vary.
Possible answer:
Let w represent the standard weight.

5000% of $w = 190$

$50 \cdot w = 190$

$w = \dfrac{190}{50} = 3.8$ g

EXTEND

16. $0.005 \cdot 120 = 0.6$

17. $x \cdot 12.6 = 3$

$x = \dfrac{3}{12.6} \approx 23.8\%$

18. $0.20x = 45.8$

$x = \dfrac{45.8}{0.20} = 229$

19. $0.139 \cdot 15 \approx 2.1$

20. $x \cdot 49 = 3.8$

$x = \dfrac{3.8}{49} \approx 7.8\%$

21. $1.505 \cdot 34 \approx 51.2$

22. $0.125 \cdot 246 \approx 30.8$

23. $x \cdot 0.155 = 11.6$

$x = \dfrac{11.6}{0.155} \approx 74.8$

24. Alabama, since $11,000 + 0.04(11,000) =$
$11,000 + 440 = 11,440.$

25. Solution 1:
Mississippi: $23,500 + 0.07(23,500) = 25,145;$
Colorado: $23,500 + 0.03(23,500) = 24,205;$
Price difference: $25,145 - 24,205 = \$940;$
Solution 2:
Similarly, difference in sales tax is 4%,
4% of $23,500 = 0.04(23,500) = \$940.$

26. Alabama: $17,990 + 0.04(17,990) = \$18,709.60;$
Colorado: $17,990 + 0.03(17,990) = \$18,529.70;$
Maine: $17,990 + 0.05(17,990) = \$18,889.50;$

THINK CRITICALLY

27. True, both equal 50.

28. True, both equal 4.5.

29. false; Example: start with $10 item.
first discount: $\$10 - 0.20 \cdot 10 = \$8;$
second discount: $\$8 - 0.20 \cdot 8 = \$6.40;$
A 40% discount would yield $\$10 - 0.40 \cdot 10 = \$6.00.$

30. True, both equal $18.55.

31. Information is inconsistent. 3 g is 4% of 75 g, but 7 g is 15% of 46.7 g therefore giving a different RDA for protein.

MIXED REVIEW

	1	2	3	4	5	6
1	2	3	4	5	6	7
2	3	4	5	6	7	8
3	4	5	6	7	8	9
4	5	6	7	8	9	10
5	6	7	8	9	10	11
6	7	8	9	10	11	12

32. $\dfrac{18}{36} = \dfrac{1}{2}$ **33.** $\dfrac{6}{36} = \dfrac{1}{6}$ **34.** $\dfrac{15}{36} = \dfrac{5}{12}$

35. $\dfrac{15}{36} = \dfrac{5}{12}$ **36.** $\dfrac{21}{36} = \dfrac{7}{12}$ **37.** $\dfrac{12}{36} = \dfrac{1}{3}$

38. $|\,23.5\,| + |\,-14\,| = 23.5 + 14 = 37.5$; B

ALGEBRA WORKS

1a. Walking 4 miles in 1 hour, 5 days a week or 20 miles burns 2000 calories. Let x represent calories burned walking 1 mile in 15 minutes.

$20x = 2000$

$x = 100$ calories

b. Swimming 30 minutes, 6 days a week or 180 minutes burns 2000 calories. Let x represent the calories burned swimming 30 minutes.

$6x = 2000$

$x \approx 333.33$ calories

c. Playing tennis 1 hour, 5 days a week or 5 hours burns 2000 calories. Let x represent calories burned playing tennis for 1 hour.

$5x = 2000$

$x = 400$ calories

d. Jogging 3 miles in 30 minutes, 6 days a week or 18 miles burns 2000 calories. Let x represent calories burned jogging 1 mile in 10 minutes.

$18x = 2000$

$x \approx 111.11$ calories

2. Answers will vary. Possible answer:
day 1: swim 30 min;
day 2: jog 3 mi;
day 3: play tennis for 1 h;
day 4: walk 3 mi in 45 min;
day 5: swim for 1 h

Lesson 3.6, pages 140–145

EXPLORE

1. $2x + 5 = 41$; guess: 18; check:
$2x + 5 = 41$, $2(18) + 5 \overset{?}{=} 41$, $41 = 41$ ✔

2. x represents the weight in ounces of each bag.

3. yes

4. $2x + 5 = 20 + 20 + 1$
$2x + 5 = 41$

5. $2x + 5 = 41$
$2(18) + 5 \overset{?}{=} 41$
$36 + 5 \overset{?}{=} 41$
$41 = 41$ ✔; Yes, $x = 18$ solves the equation.

TRY THESE

1. Add b to both sides.

2. Subtract 4 from both sides.

3. Add y to both sides.

4. Multiply both sides by 7.

5. $3m - 12 = 15$ check: $3m - 12 = 15$
$3m - 12 + 12 = 15 + 12$ $3(9) - 12 \overset{?}{=} 15$
$\dfrac{3m}{3} = \dfrac{27}{3}$ $27 - 12 \overset{?}{=} 15$
$m = 9$ $15 = 15$ ✔

6. $\dfrac{r}{8} - 7 = -1$ check: $\dfrac{r}{8} - 7 = -1$
$\dfrac{r}{8} - 7 + 7 = -1 + 7$ $\dfrac{48}{8} - 7 \overset{?}{=} -1$
$8\left(\dfrac{r}{8}\right) = 8(6)$ $6 - 7 \overset{?}{=} -1$
$r = 48$ $-1 = -1$ ✔

7. $\dfrac{z}{3} + 70 = 98$ check: $\dfrac{z}{3} + 70 = 98$
$\dfrac{z}{3} + 70 - 70 = 98 - 70$ $\dfrac{84}{3} + 70 \overset{?}{=} 98$
$3\left(\dfrac{z}{3}\right) = 3(28)$ $28 + 70 \overset{?}{=} 98$
$z = 84$ $98 = 98$ ✔

8. $9w + 6.25 = 123.25$
$9w + 6.25 - 6.25 = 123.25 - 6.25s$
$\dfrac{9w}{9} = \dfrac{117}{9}$
$w = 13$
check : $9w + 6.25 = 123.25$
$9(13) + 6.25 \overset{?}{=} 123.25$
$117 + 6.25 \overset{?}{=} 123.25$
$123.25 = 123.25$ ✔

9. $\dfrac{a}{2.5} - 3.8 = 8.2$ check: $\dfrac{a}{2.5} - 3.8 = 8.2$
$\dfrac{a}{2.5} - 3.8 + 3.8 = 8.2 + 3.8$ $\dfrac{30}{2.5} - 3.8 \overset{?}{=} 8.2$
$2.5\left(\dfrac{a}{2.5}\right) = 2.5(12)$ $12 - 3.8 \overset{?}{=} 8.2$
$a = 30$ $8.2 = 8.2$ ✔

10. $\dfrac{-6f}{2} + 1.9 = -8.6$ check: $\dfrac{-6f}{2} + 1.9 = -8.6$
$-3f + 1.9 - 1.9 = -8.6 - 1.9$ $\dfrac{-6(3.5)}{2} + 1.9 \overset{?}{=} -8.6$
$\dfrac{-3f}{-3} = \dfrac{-10.5}{-3}$ $\dfrac{-21}{2} + 1.9 \overset{?}{=} -8.6$
$f = 3.5$ $-8.6 = -8.6$ ✔

11. $-3x + 6 = 30$ check: $-3x + 6 = 30$
$-3x + 6 - 6 = 30 - 6$ $-3(-8) + 6 \overset{?}{=} 30$
$\dfrac{-3x}{-3} = \dfrac{24}{-3}$ $24 + 6 \overset{?}{=} 30$
$x = -8$ $30 = 30$ ✔

12. $8y - 15 = -87$ check: $-8y - 15 = -87$
$-8y - 15 + 15 = -87 + 15$ $-8(9) - 15 \overset{?}{=} -87$
$\dfrac{-8y}{-8} = \dfrac{-72}{-8}$ $-72 - 15 \overset{?}{=} -87$
$y = 9$ $-87 = -87$ ✔

13. $4(x - 8) = -4$ check: $4(x - 8) = -4$
$\dfrac{4(x - 8)}{4} = \dfrac{-4}{4}$ $4(7 - 8) \overset{?}{=} -4$
$x - 8 + 8 = -1 + 8$ $4(-1) \overset{?}{=} -4$
$x = 7$ $-4 = -4$ ✔

14.
$$5(15+h) = 65$$
$$\frac{5(15+h)}{5} = \frac{65}{5}$$
$$15 + h - 15 = 13 - 15$$
$$h = -2$$

check: $5(15+h) = 65$
$$5(15 + (-2)) \stackrel{?}{=} 65$$
$$5(13) \stackrel{?}{=} 65$$
$$65 = 65 \;\checkmark$$

15.
$$-2(3-x) = -14$$
$$\frac{-2(3-x)}{-2} = \frac{-14}{-2}$$
$$3 - x - 3 = 7 - 3$$
$$-1(-x) = (4)(-1)$$
$$x = -4$$

check: $-2(3-x) = -14$
$$-2(3 - (-4)) \stackrel{?}{=} -14$$
$$-2(3 + 4) \stackrel{?}{=} -14$$
$$-2(7) = -14$$
$$-14 = -14 \;\checkmark$$

16.
$$(c-3) - 2(8-c) = -37$$
$$c - 3 - 16 + 2c = -37$$
$$3c - 19 + 19 = -37 + 19$$
$$\frac{3c}{3} = \frac{-18}{3}$$
$$c = -6$$
check:
$$(c-3) - 2(8-c) = -37$$
$$(-6-3) - 2[8 - (-6)] \stackrel{?}{=} -37$$
$$-9 - 2(8+6) \stackrel{?}{=} -37$$
$$-9 - 2(14) \stackrel{?}{=} -37$$
$$-37 = -37 \;\checkmark$$

17.

$$3x + 4 = -2$$
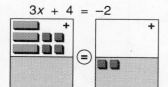

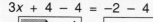

$$3x + 4 - 4 = -2 - 4$$
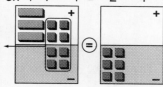

$$x = -2$$
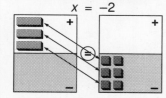

18. perimeter = 2 • length + 2 • width
Let w represent the width; $67 + w$ represent the length.
$$346 = 2(67 + w) + 2(w)$$
$$346 = 134 + 2w + 2w$$
$$346 - 134 = 134 + 4w - 134$$
$$\frac{212}{4} = \frac{4w}{4}$$
$$53 = w$$
length = $67 + w = 67 + 53 = 120$; 120 yd • 53 yd

19. Let x represent the number of years the car's value drops.
$$14,000 - 1800x = 6800$$
$$14,000 - 1800x - 14,000 = 6800 - 14,000$$
$$\frac{-1800x}{-1800} = \frac{-7200}{-1800}$$
$$x = 4$$
The car's value will drop to $6800 in 4 years.

20.
$$m + 2m = 3600$$
$$\frac{3m}{3} = \frac{3600}{3}$$
$$m = 1200 \text{ miles from Dallas to Cleveland;}$$
$$2m = 2 \cdot 1200;$$
$$2 \cdot 1200 = 2400 \text{ miles from Cleveland to Seattle}$$

21. Answers will vary but should include identifying a variable to represent the unknown quantity, set up an equation to model the quantitative description given in the problem, finally use the properties of equations to solve for the variable and answer the question.

PRACTICE

1. Solution is 3, since $3(3) - 9 = 9 - 9 = 0$.

2. Solution is -3, since $\frac{2(-3)}{2} = 5 = \frac{-6}{2} = 5 = -3 = 5 = 2$.

3. Solution is 3, since $-17 + 2(3) = -17 + 6 = -11$.

4. Solution is -3, since $2.5(-3) + 6 = -7.5 + 6 = -1.5$.

5. Solution is 3, since $-4(3) - 8 = -12 - 8 = -20$.

6. Solution is -3, since $25 - 8.5(-3) = 24 + 25.5 = 50 \cdot 5$

7.
$$14 = 8r - 58$$
$$14 + 58 = 8r - 58 + 58$$
$$\frac{72}{8} = \frac{8r}{8}$$
$$9 = r$$
check: $14 = 8r - 58$
$$14 \stackrel{?}{=} 8(9) - 58$$
$$14 \stackrel{?}{=} 72 - 58$$
$$14 = 14 \;\checkmark$$

8.
$$3.9w - 46.8 = 0$$
$$3.9w - 46.8 + 46.8 = 0 + 46.8$$
$$\frac{3.9w}{3.9} = \frac{46.8}{3.9}$$
$$w = 12$$
check: $3.9w - 46.8 = 0$
$$3.9(12) - 46.8 \stackrel{?}{=} 0$$
$$46.8 - 46.8 \stackrel{?}{=} 0$$
$$0 = 0 \;\checkmark$$

9.
$$\frac{1}{3}x - 15 = 10$$
$$\frac{1}{3}x - 15 + 15 = 10 + 15$$
$$3\left(\frac{1}{3}x\right) = 3(25)$$
$$x = 75$$
check: $\frac{1}{3}x - 15 = 10$
$$\frac{1}{3}(75) - 15 \stackrel{?}{=} 10$$
$$25 - 15 \stackrel{?}{=} 10$$
$$10 = 10 \;\checkmark$$

10. $\dfrac{a}{11}+16=25$ check: $\dfrac{a}{11}+16=25$

$\dfrac{a}{11}+16-16=25-16$ $\dfrac{99}{11}+16\overset{?}{=}25$

$11\left(\dfrac{a}{11}\right)=11(9)$ $9+16\overset{?}{=}25$

$a=99$ $25=25$ ✔

11. $-35-4y=5$ check: $-35-4y=5$

$-35-4y+35=5+35$ $-35-4(-10)\overset{?}{=}5$

$\dfrac{-4y}{-4}=\dfrac{40}{-4}$ $-35+40\overset{?}{=}5$

$y=-10$ $5=5$ ✔

12. $6(3-e)=42$ check: $6(3-e)=42$

$\dfrac{6(3-e)}{6}=\dfrac{42}{6}$ $6(3-(-4))\overset{?}{=}42$

$3-e-3=7-3$ $6(7)\overset{?}{=}42$

$-1(-e)=-1(4)$ $42=42$ ✔

$e=-4$

13. $13x-7x=24$ check: $13x-7x=24$

$6x=24$ $13(4)-7(4)\overset{?}{=}24$

$\dfrac{6x}{6}=\dfrac{24}{6}$ $52-28\overset{?}{=}24$

$x=4$ $24=24$ ✔

14. $-8e-3e=22$ check: $-8e-3e=22$

$-11e=22$ $-8(-2)-3(-2)\overset{?}{=}22$

$\dfrac{-11e}{-11}=\dfrac{22}{-11}$ $16+6=22$

$e=-2$ $22=22$ ✔

15. $23-\dfrac{1}{2}h=17.5$ check: $23-\dfrac{1}{2}h=17.5$

$23-\dfrac{1}{2}h-23=17.5-23$ $23-\dfrac{1}{2}(11)\overset{?}{=}17.5$

$-2\left(-\dfrac{1}{2}h\right)=-2(-5.5)$ $23-5.5\overset{?}{=}17.5$

$h=11$ $17.5=17.5$ ✔

16. $15(2+x)-3x=114$ check: $15(2+x)-3x=114$

$30+15x-3x=114y$ $15(2+7)-3(7)\overset{?}{=}114$

$30+12x-30=114-30$ $15(9)-3(7)\overset{?}{=}114$

$\dfrac{12x}{12}=\dfrac{84}{12}$ $135-21\overset{?}{=}114$

$x=7$ $114=114$ ✔

17. $-x+(9-2x)=39$ check: $-x+(9-2x)=39$

$-3x+9-9=39-9$ $-(-10)+[9-2(-10)]\overset{?}{=}39$

$\dfrac{-3x}{-3}=\dfrac{30}{-3}$ $10+29\overset{?}{=}39$

$x=-10$ $39=39$ ✔

18. $-6=-\dfrac{1}{4}(a+4)$ check: $-6=-\dfrac{1}{4}(a+4)$

$-4(-6)=-4\left(-\dfrac{1}{4}(a+4)\right)$ $-6\overset{?}{=}-\dfrac{1}{4}(20+4)$

$24-4=a+4-4$ $-6\overset{?}{=}-\dfrac{1}{4}(24)$

$24=a+4$ $-6=-6$ ✔

$20=a$

19. $4c-5(c-3)=12.5$ check: $4c-5(c-3)=12.5$

$4c-5c+15=12.5$ $4(2.5)-5(2.5-3)\overset{?}{=}12.5$

$-c+15-15=12.5-15$ $10-5(-0.5)\overset{?}{=}12.5$

$-1(-c)=-1(-2.5)$ $10+2.5\overset{?}{=}12.5$

$c=2.5$ $12.5=12.3$ ✔

20. $19-14x=6.4$ check: $19-14x=6.4$

$19-14x-19=6.4-19$ $19-14(0.9)\overset{?}{=}6.4$

$\dfrac{-14x}{-14}=\dfrac{-12.6}{-14}$ $19-12.6\overset{?}{=}6.4$

$x=0.9$ $6.4=6.4$ ✔

21. $2(x-8)-3(x+2)=-24$

$2x-16-3x-6=-24$

$-x-22+22=-24+22$

$-1(-x)=-1(-2)$

$x=2$

check: $2(x-8)-3(x+2)=-24$

$2(2-8)-3(2+2)\overset{?}{=}-24$

$2(-6)-3(4)\overset{?}{=}-24$

$-24=-24$ ✔

22. $\dfrac{5y}{4}-9=-49$ check: $\dfrac{5y}{4}-9=-49$

$\dfrac{5y}{4}-9+9=-49+9$ $\dfrac{5(-32)}{4}-9\overset{?}{=}-49$

$\dfrac{4}{5}\left(\dfrac{5y}{4}\right)=\dfrac{4}{5}(-40)$ $-40-9\overset{?}{=}-49$

$y=-32$ $-49=-49$ ✔

23. $\dfrac{x}{7}-3=5$ check: $\dfrac{x}{7}-3=5$

$\dfrac{x}{7}-3+3=5+3$ $\dfrac{56}{7}-3\overset{?}{=}5$

$7\left(\dfrac{x}{7}\right)=7(8)$ $8-3\overset{?}{=}5$

$x=56$ $5=5$ ✔

24.
$$\frac{2}{3}(z+1) = 4$$

$$\frac{3}{2}\left(\frac{2}{3}(z+1)\right) = \frac{3}{2}(4)$$

$$z + 1 - 1 = 6 - 1$$

$$z = 5$$

check: $\frac{2}{3}(z+1) = 4$

$$\frac{2}{3}(5+1) \stackrel{?}{=} 4$$

$$\frac{2}{3}(6) \stackrel{?}{=} 4$$

$$4 = 4 ✔$$

25. $6m - 3(m+4) = 6$

$$6m - 3m - 12 = 6$$

$$3m - 12 + 12 = 6 + 12$$

$$\frac{3m}{3} = \frac{18}{3}$$

$$m = 6$$

check: $6m - 3(m+4) = 6$

$$6(6) - 3(6+4) \stackrel{?}{=} 6$$

$$36 - 30 \stackrel{?}{=} 6$$

$$6 = 6 ✔$$

26.
$$8.9c + 2.3 = -16.39$$

$$8.9c + 2.3 - 2.3 = -16.39 - 2.3$$

$$\frac{8.9c}{8.9} = \frac{-18.69}{8.9}$$

$$c = -2.1$$

check: $8.9c + 2.3 = -16.39$

$$8.9(-2.1) + 2.3 \stackrel{?}{=} -16.39$$

$$-18.69 + 2.3 \stackrel{?}{=} -16.39$$

$$-16.39 = -16.39 ✔$$

27. $1.7a - (a+3) = 3.3$

$$1.7a - a - 3 = 3.3$$

$$0.7a - 3 + 3 = 3.3 + 3$$

$$\frac{0.7a}{0.7} = \frac{6.3}{0.7}$$

$$a = 9$$

check: $1.7a - (a+3) = 3.3$

$$1.7(9) - (9+3) \stackrel{?}{=} 3.3$$

$$15.3 - 12 \stackrel{?}{=} 3.3$$

$$3.3 = 3.3 ✔$$

28.
$$10.25 = 1.25 + 0.75(m-1)$$

$$10.25 = 1.25 + 0.75m - 0.75$$

$$10.25 - 0.50 = 0.50 + 0.75m - 0.50$$

$$\frac{9.75}{0.75} = \frac{0.75m}{0.75}$$

$$13 = m$$

check: $10.25 = 1.25 + 0.75(m-1)$

$$10.25 \stackrel{?}{=} 1.25 + 0.75(13-1)$$

$$10.25 \stackrel{?}{=} 1.25 + 0.75(12)$$

$$10.25 = 10.25 ✔$$

29. The passenger travelled 13 miles.

$$36 = \frac{400 - 5e}{10}$$

$$10(36) = 10\left(\frac{400 - 5e}{10}\right)$$

$$360 - 400 = 400 - 5e - 400$$

$$\frac{-40}{-5} = \frac{-5e}{-5}$$

$$8 = e$$

He made 8 errors

check: $36 = \frac{400 - 5c}{10}$

$$36 \stackrel{?}{=} \frac{400 - 5(8)}{10}$$

$$36 \stackrel{?}{=} \frac{400 - 40}{10}$$

$$36 = 36 ✔$$

30. Answers will vary by student and equation. The term(s) with the variable needs to be isolated on one side of the equation either by the addition and subtraction property or the multiplication and division property.

EXTEND

31.
$$(4.5 - w)3 + \frac{w}{2} = 21$$

$$13.5 - 3w + \frac{w}{2} = 21$$

$$13.5 - \frac{5w}{2} - 13.5 = 21 - 13.5$$

$$-\frac{2}{5}\left(-\frac{5w}{2}\right) = -\frac{2}{5}(7.5)$$

$$w = -3$$

check: $(4.5 - w)3 + \frac{w}{2} = 21$

$$(4.5 - (-3))3 + \frac{-3}{2} \stackrel{?}{=} 21$$

$$(4.5 + 3)3 - \frac{3}{2} \stackrel{?}{=} 21$$

$$22.5 - \frac{3}{2} \stackrel{?}{=} 21$$

$$\frac{45}{3} - \frac{3}{2} \stackrel{?}{=} 21$$

$$21 = 21 ✔$$

32.
$$-87 = \frac{7}{2}(x - 6) + 2x$$

$$-87 = \frac{7x}{2} - 21 + 2x$$

$$-87 + 21 = \frac{11x}{2} - 21 + 21$$

$$\frac{2}{11}(-66) = \frac{2}{11}\left(\frac{11x}{2}\right)$$

$$-12 = x$$

(Solution continues on next page.)

check: $-87 = \frac{7}{2}(x - 6) + 2x$

$-87 \overset{?}{=} \frac{7}{2}(-12 - 6) + 2(-12)$

$-87 \overset{?}{=} \frac{7}{2}(-18) - 24$

$-87 \overset{?}{=} -63 - 24$

$-87 = -87 \checkmark$

33. $-14\frac{2}{3} = \frac{4}{9}(-9y - 6)$ check: $-14\frac{2}{3} = \frac{4}{9}(-9y - 6)$

$-14\frac{2}{3} = -4y\frac{-24}{9}$ $\qquad -14\frac{2}{3} \overset{?}{=} \frac{4}{9}(-9(3) - 6)$

$-14\frac{2}{3} + \frac{8}{3} = -4y - \frac{8}{3} + \frac{8}{3}$ $\qquad -14\frac{2}{3} \overset{?}{=} -12 - \frac{24}{9}$

$\frac{-12}{-4} = \frac{-4y}{-4}$ $\qquad -14\frac{2}{3} = -14\frac{2}{3} \checkmark$

$3 = x$

34. average $= \frac{273}{3} = 91;$

$x + (x + 5) + (x - 2) = 273$

$3x + 3 - 3 = 273 - 3$

$\frac{3x}{3} = \frac{270}{3}$

$x = 90$

$x + 5 = 90 + 5 = 95;$

$x - 2 = 90 - 2 = 88;$

check: $x + (x + 5) + (x - 2) = 273$

$90 + 95 + 88 \overset{?}{=} 273$

$273 = 273 \checkmark$

The test scores were 90, 95, and 88.

35. Student's spinners will vary. Possible spinner:

Let x represent probability of an odd number.
Then $3x$ represent the probability of an even number.

$x + 3x = 1$ $\qquad$ check: $x + 3x = 1$

$4x = 1$ $\qquad \frac{1}{4} + 3\left(\frac{1}{4}\right) \overset{?}{=} 1$

$x = \frac{1}{4}$ $\qquad 1 = 1 \checkmark$

$p(\text{odd}) = x = \frac{1}{4} = 0.25;$

$p(\text{even}) = 3x = 3 \cdot \frac{1}{4} = \frac{3}{4} = 0.75$

36. cost $= 0.13(250) + 0.09(400 - 250)$

cost $= 0.13(250) + 0.09(150)$

cost $= 32.50 + 13.50$

cost $= \$46.00 > 40.00;$ Do not buy the refrigerator.

37. $(6x - 10) + (3x + 10) + (3x + 10) + (6x - 10) = 360°$

$6x + 3x + 3x + 6x - 10 + 10 + 10 - 10 = 360°$

$\frac{18x}{18} = \frac{360}{18}$

$x = 20°$

check:

$6x - 10 = 6(20) - 10 = 110°;\ 3x + 10 = 3(20) + 10 = 70°;$

$(6x - 10) + (3x + 10) + (3x + 10) + (6x - 10) = 360°$

$110° + 70° + 70° + 110° \overset{?}{=} 360$

$360 = 360 \checkmark$

Angles are 70°, 70°, 110°, 110°.

38. Answers will vary such as:

$6 + x \div 4 = 4;$ answer is -8

$18 - x \cdot 2 = 4;$ answer is 7

THINK CRITICALLY

39. P represents profit.; x represents number of sweaters sold per month.; $19x$ is the store's income; \$1300 is the store's expenses

$1778 = 19x - 1300$

$1778 + 1300 = 19x - 1300 + 1300$

$\frac{3078}{19} = \frac{19x}{19}$

$162 = x$

check: $n + (n + 1) + (n + 2) = 111$

$36 + 37 + 38 \overset{?}{=} 11$

$111 = 111 \checkmark$

The store sold 162 sweaters.

40. $n + (n + 1) + (r + 2) = 111$

$3n + 3 - 3 = 111 - 3$

$\frac{3n}{3} = \frac{108}{3}$

$n \overset{?}{=} 36;\ n + 1 = 37;\ n + 2 = 38$

check: $n + (n + 1) + (n + 2) = 111$

$36 + 37 + 38 \overset{?}{=} 111$

$111 = 111 \checkmark$

41. $n + (n + 2) + (n + 4) = 87$

$3n + 6 - 6 = 87 - 6$

$\frac{3n}{3} = \frac{81}{3}$

$n = 27;\ n + 2 = 29;\ n + 4 = 31$

check: $n + (n + 2) + (n + 4) \overset{?}{=} 87$

$27 + 29 + 31 \overset{?}{=} 87$

$87 = 87 \checkmark$

42. No; If you begin with an even integer n and count by two's then n, $n + 2$, $n + 4$ are consecutive even integers. $n + (n + 2) + (n + 4) = 100$, $3n + 6 = 100$, $3n = 94$ has no integer solutions since 94 is not divisible by 3.

43. $8m - 3m = 18$, $5m = 18$; $6m + 4m = 10m$; $10m = 2(5m)$ $= 2(18) = 36$

44. $\dfrac{9 - r}{2} = 14$, $9 - r = 28$; $2(9 - r) = 2(28) = 56$

45. $5(a + 3) - 2(a + 3) = 7$, $5a + 15 - 2a - 6 = 7$, $3a + 9 = 7$, $3(a + 3) = 7$; $12(a + 3) = 4 \cdot 3(a + 3) = 4(7) = 28$

46. Answers will vary, one such equation is
If $3(x - 4) - 5(x - 4) = 1$, find the value of $2(x - 4)$.
Solution: $3(x - 4) - 5(x - 4) = 1$, $3x - 12 - 5x + 20 = 1$, $-2x + 8 = 1$, $2(x - 4) = -1$; $2(x - 4) = -1$

Lesson 3.7, pages 146–150

EXPLORE

1. Answers will vary but should include that there is three times an unknown weight plus 6 pounds on one side that balances with 8 pounds less than 5 times that unknown weight.

2. $3x + 6 = 5x - 8$

3. Remove 3 of the 5 x's from the right side.

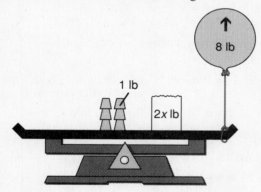

4. Adding 8 pounds of weight to the right side.

5. Add 8 pounds.

6. $14 = 2x$ **7.** $x = 7$

TRY THESE

1.
$$8a + 12 = 11a$$
$$8a + 12 - 8a = 11a - 8a$$
$$\frac{12}{3} = \frac{3a}{3}$$
$$4 = a$$

2.
$$-3c - 4 = -5c + 6$$
$$-3c - 4 + 5c = -5c + 6 + 5c$$
$$2c - 4 + 4 = 6 + 4$$
$$\frac{2c}{2} = \frac{10}{2}$$
$$c = 5$$

3.
$$6(r - 3) = 4r - 4$$
$$6r - 18 = 4r - 4$$
$$6r - 18 - 4r = 4r - 4 - 4r$$
$$2r - 18 + 18 = -4 + 18$$
$$\frac{2r}{2} = \frac{14}{2}$$
$$r = 7$$

4.
$$9b + 8 = 10b$$
$$9b + 8 - 9b = 10b - 9b$$
$$8 = b$$
check: $9b + 8 = 10b$
$9(8) + 8 \overset{?}{=} 10(8)$
$80 = 80$ ✔

5.
$$6f + 24 = -2f$$
$$6f + 24 - 6f = -2f - 6f$$
$$\frac{24}{-8} = \frac{-8f}{-8}$$
$$-3 = f$$
check: $5f + 24 = -2f$
$6(-3) + 24 \overset{?}{=} -2(-3)$
$6 = 6$ ✔

6.
$$-8 + 3h = 2(h - 5)$$
$$-8 + 3h - 2h = 2h - 10 - 2h$$
$$-8 + h + 8 = -10 + 8$$
$$h = -2$$
check: $-8 + 3h = 2(h - 5)$
$-8 + 3(-2) \overset{?}{=} 2((-2) - 5)$
$-8 - 6 \overset{?}{=} 2(-7)$
$-14 = -14$ ✔

7.
$$-3(z + 4) = -4z$$
$$-3z - 12 + 3z = -4z + 3z$$
$$-1(-12) = -1(-z)$$
$$12 = z$$
check: $-3(z + 4) = -4z$
$-3(12 + 4) \overset{?}{=} -4(12)$
$-36 - 12 \overset{?}{=} -4(12)$
$-48 = -48$ ✔

8.
$$24 - 9t = -13t + 8$$
$$24 - 9t + 13t = -13t + 8 + 13t$$
$$24 + 4t - 24 = 8 - 24$$
$$\frac{4t}{4} = \frac{-16}{4}$$
$$t = -4$$
check: $24 - 9t = -13t + 8$
$24 - 9(-4) \overset{?}{=} -13(-4) + 8$
$24 + 36 \overset{?}{=} 52 + 8$
$60 = 60$ ✔

9.
$$7x = 10(x - 1.5)$$
$$7x - 10x = 10x - 15 - 10x$$
$$\frac{-3x}{-3} = \frac{-15}{-3}$$
$$x = 5$$
check: $7x = 10(x - 1.5)$
$7(5) \overset{?}{=} 10(5 - 1.5)$
$35 \overset{?}{=} 10(3.5)$
$35 \overset{?}{=} 35$ ✔

10.

$$61 - 13r = -3r + 1$$
$$61 - 13r + 13r = -3r + 1 + 13r$$
$$61 - 1 = 10r + 1 - 1$$
$$\frac{60}{10} = \frac{10r}{10}$$
$$6 = r$$

check: $61 - 13r = 3r + 1$
$$61 - 13(6) \overset{?}{=} -3(6) + 1$$
$$61 - 78 \overset{?}{=} -18 + 1$$
$$-17 = -17 ✔$$

11.

$$-n = 5n - 72$$
$$-n - 5n = 5n - 72 - 5n$$
$$\frac{-6n}{-6} = \frac{-72}{-6}$$
$$n = 12$$

check: $-n = 5n - 72$
$$-12 \overset{?}{=} 5(12) - 72$$
$$-12 \overset{?}{=} 60 - 72$$
$$-12 = -12 ✔$$

12.

$$\frac{1}{2}(3c - 4) = c + 8$$
$$2\left(\frac{1}{2}(3c - 4)\right) = 2(c + 8)$$
$$3c - 4 - 2c = 2c + 16 - 2c$$
$$c - 4 + 4 = 16 + 4$$
$$c = 20$$

check: $\frac{1}{2}(3c - 4) = c + 8$
$$\frac{1}{2}(3(20) - 4) \overset{?}{=} 20 + 8$$
$$\frac{1}{2}(56) \overset{?}{=} 20 + 8$$
$$28 = 28 ✔$$

13.

$$3(5.5 - m) = 8m$$
$$16.5 - 3m + 3m = 8m + 3m$$
$$\frac{16.5}{11} = \frac{11m}{11}$$
$$1.5 = m$$

check: $3(5.5 - m) = 8m$
$$3(5.5 - 1.5) \overset{?}{=} 8(1.5)$$
$$3(4) \overset{?}{=} 12$$
$$12 = 12 ✔$$

14.

$$e - 17 = -4e + 24.25$$
$$e - 17 + 4e = -4e + 24.25 + 4e$$
$$5e - 17 + 17 = 24.25 + 17$$
$$\frac{5e}{5} = \frac{41.25}{5}$$
$$e = 8.25$$

check: $e - 17 = -4c + 24.25$
$$8.25 - 17 \overset{?}{=} -4(8.25) + 24.25$$
$$-8.75 \overset{?}{=} -33 + 24.25$$
$$-8.75 = -8.75 ✔$$

15.

$$\frac{1}{4}x = 2x + 17.5$$
$$\frac{1}{4}x - 2x = 2x + 17.5 - 2x$$
$$-\frac{4}{7}\left(-\frac{7}{4}x\right) = -\frac{4}{7}(17.5)$$
$$x = -10$$

check: $\frac{1}{4}x = 2x + 17.5$
$$\frac{1}{4}(-10) \overset{?}{=} 2(-10) + 17.5$$
$$-2.5 \overset{?}{=} -20 + 17.5$$
$$-2.5 = -2.5 ✔$$

16. Correct answer is b, since both withdrew money, both must involve subtraction.

17.

$$710 - 35w = 570 - 25w$$
$$710 - 35w + 35w = 570 - 25w + 35w$$
$$710 - 570 = 570 + 10w - 570$$
$$\frac{140}{10} = \frac{10w}{10}$$
$$14 = w$$

The accounts will be equal after 14 weeks.

18. $710 - 35(14) = 710 - 490 = \$220$

19.

$$27x + 11 = 7(3x + 5)$$
$$27x + 11 - 21x = 21x + 35 - 21x$$
$$6x + 11 - 11 = 35 - 11$$
$$\frac{6x}{6} = \frac{24}{6}$$
$$x = 4$$

length = $3x + 5 = 3(4) + 5 = 17$; 17 mi;
area = $27x + 11 = 27(4) + 11 = 119$; 119 mi^2

20. Answers will vary. Possible answer for 4:

$$9b + 8 = 10b$$
$$9b - 9b + 8 = 10b - 9b \qquad \text{Subtract } 9b \text{ from both sides to isolate the variable.}$$
$$8 = b \qquad \text{The solution is 8.}$$

PRACTICE

1.

$$m - 7 = -13 - m$$
$$m - 7 + m = -13 - m + m$$
$$2m - 7 + 7 = -13 + 7$$
$$\frac{2m}{2} = \frac{-6}{2}$$
$$m = -3$$

check: $m - 7 = -13 - m$
$$-3 - 7 \overset{?}{=} -13 - (-3)$$
$$-10 = -10 ✔$$

2.

$$2r = 35 - 3r$$
$$2r + 3r = 35 - 3r + 3r$$
$$\frac{5r}{5} = \frac{35}{5}$$
$$r = 7$$

check: $2r = 35 - 3r$
$$2(7) \overset{?}{=} 35 - 3(7)$$
$$14 = 14 ✔$$

3. $x + 6 = -6x + 13$ check: $x + 6 = -6x + 13$
    $x + 6 + 6x = -6x + 13 + 6x$ $1 + 6 \overset{?}{=} -6(1) + 13$
    $7x + 6 - 6 = 13 - 6$ $7 = 7$ ✔
    $\dfrac{7x}{7} = \dfrac{7}{7}$
    $x = 1$

4. $23 - 4e = -7e + 2$
    $23 - 4e + 7e = -7e + 2 + 7e$
    $23 + 3e - 23 = 2 - 23$
    $\dfrac{3e}{3} = \dfrac{-21}{3}$
    $e = -7$
    check: $23 - 4e = -7e + 2$
    $23 - 4(-7) \overset{?}{=} -7(-7) + 2$
    $23 + 28 \overset{?}{=} 49 + 2$
    $51 = 51$ ✔

5. $5y = 32 - 3y$ check: $5y = 32 - 3y$
    $5y + 3y = 32 - 3y + 3y$ $5(4) \overset{?}{=} 32 - 3(4)$
    $\dfrac{8y}{8} = \dfrac{32}{8}$ $20 \overset{?}{=} 32 - 12$
    $y = 4$ $20 = 20$ ✔

6. $5x - 7 + 2x = 3x - 2 + 5x$
    $7x - 7 - 7x = 8x - 2 - 7x$
    $-7 + 2 = x - 2 + 2$
    $-5 = x$
    check: $5x - 7 + 2x = 3x - 2 + 5x$
    $5(-5) - 7 + 2(-5) \overset{?}{=} 3(-5) - 2 + 5(-5)$
    $-25 - 7 - 10 \overset{?}{=} -15 - 2 - 25$
    $-42 = -42$ ✔

7. $-3a + 1 = -4a + 8$
    $-3a + 1 + 4a = -4a + 8 + 4a$
    $a + 1 - 1 = 8 - 1$
    $a = 7$
    check: $-3a + 1 = -4a + 8$
    $-3(7) + 1 \overset{?}{=} -4(7) + 8$
    $-21 + 1 \overset{?}{=} -28 + 8$
    $-20 = -20$ ✔

8. $17 + 7n = 8 + 10n$ check: $17 + 7n = 8 + 10n$
    $17 + 7n - 7n = 8 + 10n - 7n$ $17 + 7(3) \overset{?}{=} 8 + 10(8)$
    $17 - 8 = 8 + 3n - 8$ $17 + 21 \overset{?}{=} 8 + 30$
    $\dfrac{9}{3} = \dfrac{3n}{3}$ $38 = 38$ ✔
    $3 = n$

9. $-8c + 3(c - 2) = -3c + 2$
    $-8c + 3c - 6 + 6 = -3c + 2 + 6$
    $-5c + 3c = -3c + 8 + 3c$
    $\dfrac{-2c}{-2} = \dfrac{8}{-2}$
    $c = -4$
    check: $-8c + 3(c - 2) = -3c + 2$
    $-8(-4) + 3(-4 - 2) \overset{?}{=} -3(-4) + 2$
    $32 + 3(-6) \overset{?}{=} 12 + 2$
    $32 - 18 \overset{?}{=} 12 + 2$
    $14 = 14$ ✔

10. $(3y + 4) - y = 10y$ check: $(3y + 4) - y = 10y$
    $2y + 4 - 2y = 10y - 2y$ $(3(\tfrac{1}{2}) + 4) - \tfrac{1}{2} \overset{?}{=} 10(\tfrac{1}{2})$
    $\dfrac{4}{8} = \dfrac{8y}{8}$ $\dfrac{11}{2} - \dfrac{1}{2} \overset{?}{=} 5$
    $\dfrac{1}{2} = y$ $5 = 5$ ✔

11. $2(-3h + 5) = -9h - 17$
    $-6h + 10 + 9h = -9h - 17 + 9h$
    $3h + 10 - 10 = -17 - 10$
    $\dfrac{3h}{3} = \dfrac{-27}{3}$
    $h = -9$
    check: $2(-3h + 5) = -9h - 17$
    $2(-3(-9) + 5) \overset{?}{=} -9(-9) - 17$
    $2(27 + 5) \overset{?}{=} 81 - 17$
    $64 = 64$ ✔

12. $-17w - 4(w - 1) = -20w + 1$
    $-17w - 4w + 4 = -20w + 1$
    $-21w + 4 + 21w = -20w + 1 + 21w$
    $4 - 1 = w + 1 - 1$
    $3 = w$
    check: $-17w - 4(w - 1) = -20w + 1$
    $-17(3) - 4(3 - 1) \overset{?}{=} -20(3) + 1$
    $-51 - 8 \overset{?}{=} -60 + 1$
    $-59 = -59$ ✔

13. Answers will vary by student: example might be you have your car washed once a week and washed and waxed every fifth week. If you spend $36 more for the wash and wax than the washes alone, how much does each service cost? Solution: $4x = x + 36$, $3x = 36$, $x = 12$; $12 wash; $45 wash and wax

14. $6(p + 5) - 3(p - 2) = 12p + 18$

$6p + 30 - 3p + 6 = 12p + 18$

$3p + 36 - 3p = 12p + 18 - 3p$

$36 - 18 = 9p + 18 - 18$

$\dfrac{18}{9} = \dfrac{9p}{9}$

$2 = p$

check: $6(p + 5) - 3(p - 2) = 12p + 18$

$6(2 + 5) - 3(2 - 2) = 12(2) + 18$

$6(7) - 3(0) = 24 + 18$

$42 = 42$ ✔

15. $12(x - 7) + 3(2x + 2) = 50x - 62$

$12x - 84 + 6x + 6 = 50x - 62$

$18x - 78 - 18x = 50x - 62 - 18x$

$\dfrac{-16}{32} = \dfrac{32x}{32}$

$-\dfrac{1}{2} = x$

check: $12(x - 7) + 3(2x + 2) = 50x - 62$

$12\left(-\dfrac{1}{2} - 7\right) + 3\left(2\left(-\dfrac{1}{2}\right) + 2\right) \overset{?}{=} 50\left(-\dfrac{1}{2}\right) - 62$

$12\left(\dfrac{-15}{2}\right) + 3(1) \overset{?}{=} -25 - 62$

$-90 + 3 \overset{?}{=} -87$

$-87 = -87$ ✔

16. $\dfrac{1}{2}(a + 15) = 3a - 1.25$

$2\left(\dfrac{1}{2}(a + 15)\right) = 2(3a - 1.25)$

$a + 15 - a = 6a - 2.5 - a$

$15 + 2.5 = 5a - 2.5 + 2.5$

$\dfrac{17.5}{5} = \dfrac{5a}{5}$

$3.5 = a$

check: $\dfrac{1}{2}(a + 5) = 3a - 1.25$

$\dfrac{1}{2}(3.5 + 15) \overset{?}{=} 3(3.5) - 1.25$

$\dfrac{1}{2}(18.5) \overset{?}{=} 10.5 - 1.25$

$9.25 = 9.25$ ✔

17. $\dfrac{2}{3}(6d + 3) + (d - 8) = -4d + 12$

$4d + 2 + d - 8 = -4d = 12$

$5d - 6 + 4d = -4d + 12 + 4d$

$9d - 6 + 6 = 12 + 6$

$\dfrac{9d}{9} = \dfrac{18}{9}$

$d = 2$

check: $\dfrac{2}{3}(6d + 3) + (d - 8) = -4d + 12$

$\dfrac{2}{3}(6(2) + 3) + (2 - 8) \overset{?}{=} -4(2) + 12$

$\dfrac{2}{3}(15) + (-6) \overset{?}{=} -8 + 12$

$10 - 6 \overset{?}{=} -8 + 12$

$4 = 4$ ✔

18. $-3 + 2(4k - 13) = -40k + 7$

$-3 + 8k - 26 = -40k + 7$

$8k - 29 + 40k = -40k + 7 + 40k$

$48k - 29 + 29 = 7 + 29$

$\dfrac{48k}{48} = \dfrac{36}{48}$

$k = \dfrac{3}{4}$

check: $-3 + 2(4k - 13) = -40k + 7$

$-3 + 2\left(4\left(\dfrac{3}{4}\right) - 13\right) \overset{?}{=} -40\left(\dfrac{3}{4}\right) + 7$

$-3 + 2(-10) \overset{?}{=} -30 + 7$

$-23 = -23$ ✔

19. $-5(6 - w) = 9(w + 2) - 16$

$-30 + 5w = 9w + 18 - 16$

$-30 + 5w - 5w = 9w + 2 - 5w$

$-30 - 2 = 4w + 2 - 2$

$\dfrac{-32}{4} = \dfrac{4w}{4}$

$-8 = w$

check: $-5(6 - w) = 9(w + 2) - 16$

$-5(6 - (-8)) \overset{?}{=} 9(-8 + 2) - 16$

$-5(14) \overset{?}{=} 9(-6) - 16$

$-70 = -70$ ✔

20. $\dfrac{3}{5}(15c + 10) = 12c - 9$

$9c + 6 - 9c = 12c - 9 - 9c$

$6 + 9 = 3c - 9 + 9$

$\dfrac{15}{3} = \dfrac{3c}{3}$

$5 = c$

check: $\dfrac{3}{5}(15c + 10) = 12c - 9$

$\dfrac{3}{5}[15(5) + 10] \overset{?}{=} 12(5) - 9$

$\dfrac{3}{5}(85) \overset{?}{=} 60 - 9$

$51 = 51$ ✔

21.
$$2(-3m + 4) = 4(-2m + 6)$$
$$-6m + 8 + 8m = -8m + 24 + 8m$$
$$2m + 8 - 8 = 24 - 8$$
$$\frac{2m}{2} = \frac{16}{2}$$
$$m = 8$$
check: $2(-3m + 4) = 4(-2m + 6)$
$$2(-3(8) + 4) \stackrel{?}{=} 4(-2(8) + 6)$$
$$2(-20) \stackrel{?}{=} 4(-10)$$
$$-40 = -40 ✔$$

22. profit per \$7.50 shirt: $48 \div 12 = 4$; $7.50 - 4 = 3.50$;
   $3.50x = 700$, $x = 200$; 200 shirts
profit per \$9.50 shirt: $54 \div 12 = 4.50$; $9.50 - 4.50 = 5.00$;
   $5.00x = 700$, $x = 140$; 140 shirts
profit per \$10.00 shirt: $60 \div 12 = 5$; $10.00 - 5.00 = 5.00$;
   $5.00x = 700$, $x = 140$; 140 shirts
profit per \$12.00 shirt: $66 \div 12 = 5.50$; $12.00 - 5.50 = 6.50$;
   $6.50x = 700$, $x \approx 108$; 108 shirts
The \$12.00 shirt will make \$700 profit by selling the fewest number of shirts.

23.
$$90 + 9x = 50 + 13x$$
$$90 + 9x - 9x = 50 + 13x - 9x$$
$$90 - 50 = 50 + 4x - 50$$
$$\frac{40}{4} = \frac{4x}{4}$$
$$10 = x$$
10 yd^2 of carpet would cost the same at either store.

24. $90 + 9x = 90 + 9(10) = 90 + 90 = \$180$

25. Equation b is the correct model.

26.
$$6x + 3 = x - 17$$
$$6x + 3 - x = x - 17 - x$$
$$5x + 3 - 3 = -17 - 3$$
$$\frac{5x}{5} = \frac{-20}{5}$$
$$x = -4$$

EXTEND

27.
$$\frac{x + 6}{3} = \frac{5x}{9}$$
$$9(x + 6) = 3(5x)$$
$$9x + 54 - 9x = 15x - 9x$$
$$\frac{54}{6} = \frac{6x}{6}$$
$$9 = x$$

28.
$$\frac{2m + 1}{9} = \frac{12 - 7m}{6}$$
$$6(2m + 1) = 9(12 - 7m)$$
$$12m + 6 + 63m = 108 - 63m + 63m$$
$$75m + 6 - 6 = 108 - 6$$
$$\frac{75m}{75} = \frac{102}{75}$$
$$m = 1.36$$

29.
$$\frac{2(3a - 4)}{7} = \frac{-4(10 - a)}{-4}$$
$$-4 \cdot 2(3a - 4) = 7 \cdot (-4)(10 - a)$$
$$-24a + 32 + 24a = -280 + 28a + 24a$$
$$32 + 280 = -280 + 52a + 280$$
$$\frac{312}{52} = \frac{52a}{52}$$
$$6 = a$$

30.
$$2x + 3 - 4n = x + 7$$
$$2(-4) + 3 - 4n = -4 + 7$$
$$-5 - 4n + 5 = 3 + 5$$
$$\frac{-4n}{-4} = \frac{8}{-4}$$
$$n = -2$$

31.
$$2n + w - 10 = 3w + 1 - 5n$$
$$2n + (-3) - 10 = 3(-3) + 1 - 5n$$
$$2n - 13 + 5n = -8 - 5n + 5n$$
$$7n - 13 + 13 = -8 + 13$$
$$\frac{7n}{7} = \frac{5}{7}$$
$$n = \frac{5}{7}$$

32. Let n represent the first number, then $4n - 12$ represents the second number, and $\frac{3}{4}(4n - 12) + 25$ represents the third.
$$4n - 12 = \frac{3}{4}(4n - 12) + 25$$
$$4n - 12 = 4n - 9 + 25$$
$$4n - 12 - 3n = 3n + 16 - 3n$$
$$n - 12 + 12 = 16 + 12$$
$$n = 28$$
$$4(28) - 12 = 112 - 12 = 100$$
The numbers are 28, 100, and 100.

33. Let x represent the length of each board.

$$19x - 8 = 18x + 4$$
$$18x - 8 - 18x = 18x + 4 - 18x$$
$$x - 8 + 8 = 4 + 8$$
$$x = 12$$
$$19x - 8 = 19(12) - 8 = 220$$

Each board is 12 ft long and he needs 220 ft.

34.
$$\frac{x-3}{20} + \frac{x+4}{16} = 1$$
$$80\left(\frac{x-3}{20} + \frac{x+4}{16}\right) = (1)80$$
$$4x - 12 + 5x + 20 = 80$$
$$9x + 8 - 8 = 80 - 8$$
$$\frac{9x}{9} = \frac{72}{9}$$
$$x = 8$$

$P(\text{red}) = \frac{8-3}{20} = \frac{1}{4} = 0.25;\ P(\text{green}) = \frac{8+4}{16} = \frac{3}{4} = 0.75$

THINK CRITICALLY

35. The coefficients of the variable terms differ by 1.

36. The variable can be eliminated from one side using only the addition property.

37. a. $3y + 2 = 3y + 7,\ 0 = 5$
b. $6 - 8m = -12m + 9 + 4m,\ 0 = 15$
c. $3(5x - 2) - 6x = 7 + 9x,\ 15x - 6 - 6x = 7 + 9x,\ 0 = 13$
Each equation has no solution, solving produces a false statement.

PROJECT CONNECTION

1. Answers will vary. If mixing 10% fat and 20% fat, the yield would be 14% fat content.

2. Answers will vary. **3.** Answers will vary.

4. If there is much less shrinkage when cooked, the low-fat meat may be a better buy.

ALGEBRAWORKS

1. Let A represent the person's age.
Minimum $= 0.60(220 - A)$; maximum $= 0.75(220 - A)$

2. Answers will vary by age.

3. He is placing too much stress on his heart by exceeding the maximum rate for his age.

4. Possible answers: Use the age 35 in the expression $0.75(220 - 35)$ to determine the safe maximum: 138.75 or 139; or solve for the maximum age for which 150 beats per minute is appropriate using $(220 - A)0.75 = 150$, in which case $A = 20$; 150 beats per minute is an appropriate maximum for a 20-year-old.

Lesson 3.8, pages 152–155

EXPLORE

1. 8 boundary; 4 interior

2. Counting squares yields 7 units^2.

3. $A = \frac{8}{2} + 4 - 1 = 4 + 4 - 1 = 7$

4. Answers will vary by student.

TRY THESE

1. no; only one variable

2. no; only one variable

3. yes; two variables

4. yes; two variables

5. Subtract 7 from both sides.

6. Subtract x from both sides.

7. Add m to both sides.

8. Multiply both sides by 9.

9. Divide both sides by 9.

10. Subtract 12 from both sides.

11.
$$P = 2(l + w)$$
$$P - 2l = 2l + 2w - 2l$$
$$\frac{P - 2l}{2} = \frac{2w}{2}$$
$$\frac{p - 2l}{2} = w$$

12.
$$I = \frac{100P}{p}$$
$$pI = 100P$$
$$\frac{pI}{I} = \frac{100P}{I}$$
$$p = \frac{100P}{I}$$

13.
$$H = \frac{15N}{60}$$
$$\frac{60H}{15} = \frac{15N}{15}$$
$$4H = N$$

14. Fiberglass blanket: $3.25 \cdot 6.5 = 21.125$; yes
Loosefill fiberglass: $2.2 \cdot 6.5 = 14.3$; no
Loosefill cellulite: $3.7 \cdot 6.5 = 24.05$; yes

15. Possible answer: Both use the same properties and operations; literal equations may be solved for different variables, which cannot be done for an equation with only one variable.

1. $g = \dfrac{Gm}{s^2}$

$gs^2 = Gm$

$\dfrac{gs^2}{G} = m$

2. $m = \dfrac{2E}{v^2}$

$mv^2 = 2E$

$\dfrac{mv^2}{2} = E$

3. $V = \dfrac{1}{3}\pi r^2 h$

$3V = \pi r^2 h$

$\dfrac{3V}{\pi r^2} = h$

4. $I = Prt$

$\dfrac{I}{Pr} = t$

5. $C = c + rc$

$C = c(1 + r)$

$\dfrac{C}{1 + r} = c$

6. $W = VIt$

$\dfrac{W}{VI} = t$

7. $E = mc^2$

$\dfrac{E}{c^2} = m$

8. $P = IV$

$\dfrac{P}{V} = I$

9. $3x + 7 = y$

$3x = y - 7$

$x = \dfrac{y - 7}{3}$

$x = \dfrac{19 - 7}{3} = \dfrac{12}{3} = 4$

10. $9 - 4m + b = 15$

$b = 15 - (9 - 4m)$

$b = 15 - (9 - 4(2))$

$b = 15 - (9 - 8); \quad b = 15 - 1; \quad b = 14$

11. $V = lwh$

$\dfrac{V}{lw} = h$

$h = \dfrac{48}{2 \cdot 8} = \dfrac{48}{16} = 3$

12. $C = 2\pi r$

$\dfrac{C}{2\pi} = r$

$r = \dfrac{94.2}{2(3.14)} = 15$

13. $VI = \dfrac{E}{t}$

$V = \dfrac{E}{It}$

$V = \dfrac{900}{0.5(300)} = 6$

14. $P = \dfrac{kw}{H}$

$PH = kw$

$\dfrac{PH}{k} = w$

$w = \dfrac{18.8 \cdot 3}{1.2} = 47$

15. $d = rt$, so $t = \dfrac{d}{r}$

$t = \dfrac{715}{55} = 13$ hr

16. $d = rt$, so $r = \dfrac{d}{t}$

$r = \dfrac{988}{19} = 52$ mph

17. $d = rt = 61 \cdot 11.5 = 701.5$ miles

18. $A = \dfrac{1}{2}h(b_1 + b_2)$

$2A = h(b_1 + b_2)$

$\dfrac{2A}{h} = b_1 + b_2$

$\dfrac{2A}{h} - b_1 = b_2$

$\dfrac{2 \cdot 112}{8} - 12 = b_2$

$\dfrac{224}{8} - 12 = b_2$

$28 - 12 = b_2$

$16 = b_2$

19. $A = \dfrac{1}{2}h(b_1 + b_2)$

$2A = h(b_1 + b_2)$

$\dfrac{2A}{(b_1 + b_2)} = h$

$\dfrac{2 \cdot 39}{9 + 17} = h$

$\dfrac{78}{26} = h$

$3 = h$

20. $A = P + Prt$

$A = P(1 + rt)$

$\dfrac{A}{1 + rt} = P$

$\dfrac{434}{1 + 0.06 \cdot 4} = P$

$\dfrac{434}{1.24} = P$

$\$350 = P$

21. $A = P + Prt$

$A - P = Prt$

$\dfrac{A - P}{Pt} = r$

$\dfrac{12{,}200 - 8000}{8{,}000 \cdot 5} = r$

$\dfrac{4200}{40{,}000} = r$

$0.105 = r$

$10.5\% = r$

22. Answers will vary, e.g., the area of a circular lawn is 300 ft². What is the longest rope that could be used to keep a dog tied upon the lawn?

EXTEND

23. $Q(A + B + C) = 100A + 10B + C$

$Q = \dfrac{100A + 10B + C}{A + B + C}$

24. $Q = \dfrac{100A + 10B + C}{A + B + C} = \dfrac{100(3) + 10(9) + 6}{3 + 9 + 6} = \dfrac{396}{18} = 22$

25. $Q = \dfrac{100A + 10B + C}{A + B + C} = \dfrac{100(5) + 10(0) + 4}{5 + 0 + 4} = \dfrac{504}{9} = 56$

THINK CRITICALLY

26. Final cost is equal to cost before tax.

27. If $t = 0$, no interest is paid since no time elapsed.

28. Value of index is zero.

29. $3a - 4b = 8$

$-4b = 8 - 3a$

$b = \dfrac{8 - 3a}{-4}$

Incorrect solution was divided by 4, not -4.

30. $9c + 6d = 12$

$$6d = 12 - 9c$$

$$d = \frac{12 - 9c}{6}$$

Incorrect solution divided only part of the right hand side by 6.

MIXED REVIEW

31. $-9 - 7m = 12$

$$-7m = 21$$

$$m = -3$$

32. $15 + 5x = -25$

$$5x = -40$$

$$x = -8$$

33. $-13r + 28 = -89$

$$-13r = -117$$

$$r = 9$$

34. $4b + 8 = 6b$

$$8 = 2b$$

$$4 = b$$

35. $0.19 \cdot 80 = 15.2$, D

Lesson 3.9, pages 156–159

EXPLORE THE PROBLEM

1. percent of pure juice • amount of mixture = $0.60 \cdot 20$

2. percent of pure juice • amount of mixture = $0.10n$

3. n gal + 20 gal = $(n + 20)$ gal;
gal of pure juice = $0.50(n + 20)$

4. $0.10n + 0.60(20) = 0.50(n + 20)$

5. $0.10n + 0.60(20) = 0.50(n + 20)$

$$10n + 60(20) = 50(n + 20)$$

$$10n + 1200 = 50n + 1000$$

$$200 = 40n$$

$$5 = n$$

5 gal of 10% juice are needed.

6. $20 + 5 = 25$ gal of 50% juice

7.

Mixture	Pure juice in mixture	Amount of mixture, gal	Amount of pure juice
10%	0.10	n	$0.10n$
60%	0.60	20	$0.60(20)$
50%	0.50	$n + 20$	$0.50(n + 20)$

INVESTIGATE FURTHER

8. total − amount invested at 6% = $2000 - a$

9. $0.06a$

10.

Investment	Rate	Amount	Interest
6%	0.06	a	$0.06a$
8%	0.08	$2000 - a$	$0.08(2000 - a)$

11. $0.06a + 0.08(2000 - a) = 132$

12. $6a + 8(2000 - a) = 13,200$

$$6a + 16,000 - 8a = 13,200$$

$$-2a = -2800$$

$$a = 1400 @ 6\%;$$

$$2000 - a = 2000 - 1400 = 600$$

$1400 invested at 6%; $600 invested at 8%

APPLY THE STRATEGY

13. Let x represent the number of milliliters of 80% solution.

$$0.35(40) + 0.80(x) = 0.60(40 + x)$$

$$35(40) + 80x = 60(40 + x)$$

$$1400 + 80x = 2400 + 60x$$

$$20x = 1000$$

$$x = 50$$

She should add 50 mL of 80% solution.

14. Let b represent the amount of money invested in bonds.

$$0.06(b) + 0.04(11,000 - b) = 530$$

$$6b + 4(11,000 - b) = 53,000$$

$$6b + 44,000 - 4b = 53,000$$

$$2b = 9000$$

$$b = 4500$$

He invested $4500 in bonds.

15. Let x represent the amount of ground sirloin.

$$0.80(30) + 0.95(x) = 0.90(30 + x)$$

$$80(30) + 95(x) = 90(30 + x)$$

$$2400 + 95x = 2700 + 90x$$

$$5x = 300$$

$$x = 60$$

60 lbs of ground sirloin should be added.

16. Let x represent the amount of pure alcohol.

$$0.45(120) + 1.00(x) = 0.60(120 + x)$$

$$45(120) + 100x = 60(120 + x)$$

$$5400 + 100x = 7200 + 60x$$

$$40x = 1800$$

$$x = 45$$

45 mL of pure alcohol should be added.

17. Let x represent the amount of 33% silver alloy.

$$0.33(x) + 0.75(300 - x) = 0.58(300)$$

$$33x + 75(300 - x) = 58(300)$$

$$33x + 22,500 - 75x = 17400$$

$$-42x = -5100$$

$$x \approx 121$$

121 grams of 33% silver alloy should be added.

18. Let x represent the price of the blended oil per liter.

$$10.99(1200) + 4.99(800) = x(1200 + 800)$$
$$13,188 + 3992 = 2000x$$
$$17,180 = 2000x$$
$$8.59 = x$$

The blended oil is worth $8.59 per liter.

19a. Let x represent the amount of pure antifreeze.

$$0.20(8 - x) + 1.00(x) = 0.60(8)$$
$$20(8 - x) + 100(x) = 60(8)$$
$$160 - 20x + 100x = 480$$
$$80x = 320$$
$$x = 4$$

4 qts should be replaced with pure antifreeze.

b. $0.20(8) = 1.6$ qts

c. $0.20(4) = 0.8$ qts

d. $0.60(8) = 4.8$ qts

20. Let x represent the amount of water.

$$0.12(500) + 0.00(x) = 0.07(500 + x)$$
$$12(500) + 0x = 7(500 + x)$$
$$6000 = 3500 + 7x$$
$$2500 = 7x$$
$$357 \approx x$$

She should add 357 mL to the solution.

REVIEW PROBLEM SOLVING STRATEGIES

1. Quilt Questions

B	Y	P
W	P	W
Y	Y	B

2. Triangle Count

a. 3

b. 36

c. 11

3. Strategic Substitution

a. 1

b. Subtract 7, leaving 11. If the other player subtracts 9, you subtract 1; if the other player subtracts 9, you subtract 1; if the other player subtracts 8, you subtract 2, and so on.

c. The winning strategy is to leave your opponent having to subtract from a number that has 1 as the ones digit. The player who goes first can always win by subtracting 9 to leave 91. Thereafter, whatever number Player 2 subtracts, Player 1 subtracts so as to get the next number with 1 as the ones digit.

Chapter Review, pages 160–161

1. c

2. d

3. e

4. a

5. b

6. 18

7. 15

8. -6

9.
$$x + 3 = 8$$
$$x + 3 - 3 = 8 - 3$$
$$x = 5$$

10.
$$x - 3 = 4$$
$$x - 3 + 3 = 4 + 3$$
$$x = 7$$

11.
$$x - 2 = -5$$
$$x - 2 + 2 = -5 + 2$$
$$x = -3$$

12.
$$x - 4 = 5 \qquad \text{check: } x - 4 = 5$$
$$x - 4 + 4 = 5 + 4 \qquad\qquad 9 - 4 \overset{?}{=} 5$$
$$x = 9 \qquad\qquad\qquad 5 = 5 ✔$$

13.
$$x + 12 = 4 \qquad \text{check: } x + 12 = 4$$
$$x + 12 - 12 = 4 - 12 \qquad\qquad -8 + 12 \overset{?}{=} 4$$
$$x = -8 \qquad\qquad\qquad 4 = 4 ✔$$

14.
$$-5 + x = 14 \qquad \text{check: } -5 + x = 14$$
$$-5 + x + 5 = 14 + 5 \qquad\qquad -5 + 19 \overset{?}{=} 14$$
$$x = 19 \qquad\qquad\qquad 14 = 14 ✔$$

15.
$$-3x = 15 \qquad \text{check: } -3x = 15$$
$$\frac{-3x}{-3} = \frac{15}{-3} \qquad\qquad -3(-5) \overset{?}{=} 15$$
$$x = -5 \qquad\qquad\qquad 15 = 15 ✔$$

16.
$$10x = 40 \qquad \text{check: } 10x = 40$$
$$\frac{10x}{10} = \frac{40}{10} \qquad\qquad 10(4) \overset{?}{=} 40$$
$$x = 4 \qquad\qquad\qquad 40 = 40 ✔$$

17.
$$\frac{n}{3} = 6 \qquad \text{check: } \frac{n}{3} = 6$$
$$3\left(\frac{n}{3}\right) = 3(6) \qquad\qquad \frac{18}{3} \overset{?}{=} 6$$
$$n = 18 \qquad\qquad\qquad 6 = 6 ✔$$

18.
$$4y = 56 - 12 \qquad \text{check: } 4y = 56 - 12$$
$$4y = 44 \qquad\qquad 4(11) \overset{?}{=} 56 - 12$$
$$\frac{4y}{4} = \frac{44}{4} \qquad\qquad 44 = 44 ✔$$
$$y = 11$$

19. $w - \dfrac{2}{3} = 3$ check: $w - \dfrac{2}{3} = 3$

$w - \dfrac{2}{3} + \dfrac{2}{3} = 3 + \dfrac{2}{3}$ $3\dfrac{2}{3} - \dfrac{2}{3} \overset{?}{=} 3$

$w = 3\dfrac{2}{3}$ $3 = 3$ ✔

20. $5.2 = p - 1.3$ check: $5.2 = p - 1.3$

$5.2 + 1.3 = p - 1.3 + 1.3$ $5.2 \overset{?}{=} 6.5 - 1.3$

$6.5 = p$ $5.2 = 5.2$ ✔

21. $\dfrac{x}{75} = 0.46$ check: $\dfrac{x}{75} = 0.46$

$x = 75(0.46)$ $\dfrac{3.45}{75} \overset{?}{=} 0.46$

$x = 34.5$ $0.46 = 0.46$ ✔

22. $x(8) = 36$ check: $x(8) = 36$

$x = \dfrac{36}{8}$ $4.5(8) \overset{?}{=} 36$

$x = 4.5 = 450\%$ $36 = 36$ ✔

23. $x(50) = 12$ check: $x(50) = 12$

$x = \dfrac{12}{50}$ $0.24(50) \overset{?}{=} 12$

$x = 0.24 = 24\%$ $12 = 12$ ✔

24. $3x - 6 = 12$ check: $3x - 6 = 12$

$3x = 18$ $3(6) - 6 \overset{?}{=} 12$

$x = 6$ $12 = 12$ ✔

25. $\dfrac{x}{4} + 2 = -3$ check: $\dfrac{x}{4} + 2 = -3$

$\dfrac{x}{4} = -5$ $\dfrac{-20}{4} + 2 \overset{?}{=} -3$

$x = -20$ $-3 = -3$ ✔

26. $\dfrac{2}{3}y - 3 = -13$ check: $\dfrac{2}{3}y - 3 = -13$

$\dfrac{2}{3}y = -10$ $\dfrac{2}{3}(-15) - 3 \overset{?}{=} -13$

$y = -15$ $-13 = -13$ ✔

27. $2(2y + 3) = 18$ check: $2(2y + 3) = 18$

$4y + 6 - 6 = 18 - 6$ $2(2(3) + 3) \overset{?}{=} 18$

$4y = 12$ $2(9) \overset{?}{=} 18$

$y = 3$ $18 = 18$ ✔

28. $12(2 + p) - 4p = 96$ check: $12(2 + p) - 4p = 96$

$24 + 12p - 4p = 96$ $12(2 + 9) - 4(9) \overset{?}{=} 96$

$8p = 72$ $12(11) - 36 \overset{?}{=} 96$

$p = 9$ $96 = 96$ ✔

29. $1.3m + 4 = -48$ check: $1.3m + 4 = -48$

$1.3m = -52$ $1.3(-40) + 4 \overset{?}{=} -48$

$m = -40$ $-4.8 = -4.8$ ✔

30. $3x + 12 = -2x + 42$ check: $3x + 12 = -2x + 42$

$5x + 12 = 42$ $3(6) + 12 \overset{?}{=} -2(6) + 42$

$5x = 30$ $18 + 12 \overset{?}{=} -12 + 42$

$x = 6$ $30 = 30$ ✔

31. $15 - n = 2n$ check: $15 - n = 2n$

$15 = 3n$ $15 - 5 \overset{?}{=} 2(5)$

$5 = n$ $10 = 10$ ✔

32. $4(3 - y) = 2y$ check: $4(3 - y) = 2y$

$12 - 4y = 2y$ $4(3 - 2) \overset{?}{=} 2(2)$

$12 = 6y$ $4 = 4$ ✔

$2 = y$

33. $2(-5y + 3) = 3(-2y + 10)$

$-10y + 6 = -6y + 30$

$6 = 4y + 30$

$-24 = 4y$

$-6 = y$

check: $2(-5y + 3) = 3(-2y + 10)$

$2(-5(-6) + 3) \overset{?}{=} 3(-2(-6) + 10)$

$2(30 + 3) \overset{?}{=} 3(12 + 10)$

$66 = 66$ ✔

34. $\dfrac{1}{2}(3c - 1) = c$ check: $\dfrac{1}{2}(3c - 1) = c$

$3c - 1 = 2c$ $\dfrac{1}{2}(3(1) - 1) \overset{?}{=} 1$

$-1 = -c$ $\dfrac{3}{2} - \dfrac{1}{2} \overset{?}{=} 1$

$1 = c$ $1 = 1$ ✔

35. $\dfrac{1}{2}x + 2 = \dfrac{5}{2} - x$ check: $\dfrac{1}{2}x + 2 = \dfrac{5}{2} - x$

$x + 4 = 5 - 2x$ $\dfrac{1}{2}\left(\dfrac{1}{3}\right) + 2 \overset{?}{=} \dfrac{5}{2} - \dfrac{1}{3}$

$3x + 4 = 5$ $\dfrac{1}{6} + 2 \overset{?}{=} \dfrac{13}{6}$

$3x = 1$ $\dfrac{13}{6} = \dfrac{13}{6}$ ✔

$x = \dfrac{1}{3}$

36. $a - 4 = 2b - 7$ **37.** $8r + 7s = 12$

$a + 3 = 2b$ $8r = 12 - 7s$

$\dfrac{a + 3}{2} = b$ $r = \dfrac{12 - 7s}{8}$

38. $P = 2l + 2w$

$P - 2w = 2l$

$\dfrac{P - 2w}{2} = l$

39.

Investment	Rate	Amount	Interest
5%	0.05	x	$0.05x$
7%	0.07	$2100 - x$	$0.07(2100 - x)$

$$0.05x + 0.07(2100 - x) = 123$$
$$5x + 7(2100 - x) = 100(123)$$
$$5x + 14,700 - 7x = 12,300$$
$$-2x = -2400$$
$$x = 1200$$
$$2100 - x = 2100 - 1200 = 900$$

$1200 invested at 5%; $900 invested at 7%

Chapter Assessment, pages 162–163

1. Explanations will vary. Both sides of the equation can be multiplied by $\frac{1}{6}$, or divided by 6.

2. $x + 14 = 2$
$x = -12$

3. $26 = t - (-4)$
$22 = t$

4. $9 + b = 0$
$b = -9$

5. $w - 3.2 = 17.6$
$w = 20.8$

6. $13a = 39$
$a = 3$

7. $-6x = -72$
$x = 12$

8. $\frac{c}{2} = 2.5$
$c = 5$

9. $-\frac{2}{3}b = -6$
$b = 9$

10. $97 + 4j = 17$
$4j = -80$
$j = -20$

11. $\frac{1}{2}x + 5 = -9$
$\frac{1}{2}x = -14$
$x = -28$

12. $2(x + 2) = 46$
$x + 2 = 23$
$x = 21$

13. $8m - (8 - m) = 28$
$9m - 8 = 28$
$9m = 36$
$m = 4$

14. $6x - 7 = 13 - x$
$7x = 20$
$x = \frac{20}{7}$

15. $3x - 21 = -3x + 15$
$6x = 36$
$x = 6$

16. $\frac{5}{21} = \frac{h}{84}$
$420 = 21h$
$20 = h$

17. $\frac{16}{64} = \frac{a}{12}$
$192 = 64h$
$3 = a$

18. $2(x - 4) - 3(x + 4) = -50$
$2x - 8 - 3x - 12 = -50$
$-x - 20 = -50$
$-x = -30$
$x = 30$

19. Explanations should include the equation $x \cdot 40 = 35$ and the proportion $\frac{x}{100} = \frac{35}{40}$.

20. $x(50) = 80$
$x = \frac{80}{50} = 160\%$

21. $0.15(24) = 3.6$

22. $0.12(x) = 30$
$x = \frac{30}{0.12} = 250$

23. $3v - 16 = -6v + 2$
$9v = 18$
$v = 2$

B has a negative solution, so it cannot be equivalent to the equation.

24a. $6(-30) = -180$ so approximate solution is -30.

b. $75 - 25 = 50$ so approximate solution is 75.

c. $-20 + 4 = -16$ so approximate solution is -20.

d. $-12 = \frac{36}{-3}$ so approximate solution is -3.

25. Answers will vary but should include adding 5 to both sides and then dividing both sides by 3.

26. $d = rt$
$\frac{d}{r} = t$

27. $A = \frac{1}{2}h(b_1 + b_2)$
$2A = h(b_1 + b_2)$
$\frac{2A}{h} = b_1 + b_2$
$\frac{2A}{h} - b_2 = b_1$

28. $A = P + Prt$
$A - P = Prt$
$\frac{A - P}{Pt} = r$

29. $wb = \frac{nm}{a}$
$awb = nm$
$\frac{awb}{n} = m$

30. I: $1.5 - (-2) = 1.5 + 2 = 3.5$; yes
II: $\frac{-2}{4} = -\frac{1}{2} \neq 8$; no
III: $-2 + 4 = 2$; yes
Correct answer is B.

31. Let x represent Flavia's hourly rate.
$2(35x) = 1575$
$70x = 1575$
$x = 22.50$
Flavia's hourly rate is $22.50.

32. $P = 2l + 2w$
$400 = 2l + 2(80)$
$240 = 2l$
$120 = l$

33. Let x represent the number of kilograms of apricots.

$$6x + 4(14 - x) = 5.50(14)$$
$$6x + 56 - 4x = 77$$
$$2x = 21$$
$$x = 10.5$$

There are 10.5 kilograms of apricots in the mixture.

34. Let x represent the amount invested at 5%.

$$0.05(x) + 0.07(x + 1500) = 465$$
$$5x + 7(x + 1500) = 100(465)$$
$$5x + 7x + 10,500 = 46,500$$
$$12x = 36,000$$
$$x = 3000$$

$x + 1500 = 3000 + 1500 = 4500$; $3000 invested at 5%; $4500 invested at 7%

Cumulative Review, page 164

1. $1000 - 400 = 600$
2. fourth to fifth
3. C

4. Since the last 3 games increased sharply, the means > median, answer A.

5. $x - 6 = -2$
$x = 4$

6. $n + 21 = 15$
$n = -6$

7. $5y = -20$
$y = -4$

8. $\frac{2}{3}x = 12$
$x = 18$

9. $4(c - 2) - c = -11$
$4c - 8 - c = -11$
$3c = -3$
$c = -1$

10. $2(5k + 3) = k - 12$
$10k + 6 = k - 12$
$9k = -18$
$k = -2$

11. $\left(\left(\frac{2}{3}\right)2 - \frac{2}{3}\right) \cdot (-9) = \left(\frac{4}{3} - \frac{2}{3}\right) \cdot (-9) = \frac{2}{3} \cdot (-9) = -6$

12. $-7 \cdot 3 - 6 \div 3 + 12 \cdot (-2) = -21 - 2 - 24 = -47$

13. There is a separate key on the calculator to change a sign to a negative number rather than the subtraction key. Juan used the subtraction key twice.

14. $24 = 0.75x$
$32 = x$

15. $x \cdot 5 = 7$
$x = \frac{7}{5} = 1.4 = 140\%$

16. $0.28(150) = 42$

17. Let n represent the 1st integer; then $n + 1$ and $n + 2$ represent the next two consecutive integers.

$$n + (n + 1) + (n + 2) = -24$$
$$3n + 3 = -24$$
$$3n = -24$$
$$n = -9,$$

$n + 1 = -9 + 1 = -8$; $n + 2 = -9 + 2 = -7$; The largest integer is -7.

18. Let x represent the amount invested at 7%.

$$0.05(1200 - x) + 0.07(x) = 74.40$$
$$5(1200) - x) + 7x = 100(74.40)$$
$$6000 - 5x + 7x = 7440$$
$$2x = 1440$$
$$x = 720$$

Martina invested $720 at 7%.

19. $0.88 \cdot x = 22$
$x = \frac{22}{0.88} = 25$

20. Answers will vary, but Ralph's method avoids negative numbers.

Standardized Test, page 165

1. $20 - x \cdot 20 = 14$
$-20x = -6$
$x = 0.30$ or 30%

The correct answer is D.

2. $\sqrt{8}$ is approximately 3, but less, so correct answer is C.

3. C

4. Mean is most affected by outliers, correct answer is A.

5. Highest power of x is 4, correct answer is D.

6. $P = 2l + 2w$
$60 = 2l + 2 \cdot 10$
$40 = 2l$
$20 = l$

The correct answer is B.

7. C

8. $\frac{8}{20} = \frac{4}{10}$ or 40%
The correct answer is C.

9. E

10. $(-1)^{23} \cdot (-2)^3 = -1 \cdot (-8) = 8$;
The correct answer is B.

11. Lowest number is 32, since range is 26 the highest number is 58.
I is true, the mode could be 58, e.g. {32, 58, 58}
II is also true, the median could be 32, e.g. {32, 32, 58}
III is false, since the highest value is 58 and there is at least one point less than 58, the mean must be less than 58.
The correct answer is D.

Data Activity, pages 166–167

1. Answers will vary, but should be less than 120,000. Actual total should be less than estimate since numbers were probably rounded up.

2. $\frac{5}{14} \approx 0.357$ or 36%

   $\frac{10}{14} \approx 0.714$ or 71%

3. sum of capacities = 698,179

   mean = $\frac{731,361}{14} \approx 52,240$

   range = $63,000 - 38,765 = 24,235$

4. Answers will vary. Make a graph with years along the horizontal axis and seating capacity along the vertical axis. Graph an ordered pair for each stadium and see if there is a pattern to the points.

5.

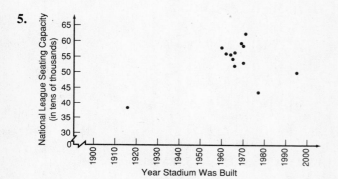

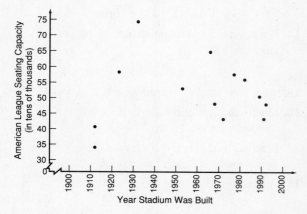

Answers will vary. In both leagues, seating capacity seems to have peaked and is presently declining. The American League has built stadiums fairly regularly while the National League built the majority of their stadiums in the 60's.

Lesson 4.1, pages 169–170

THINK BACK

1.

Fastballs	Curveballs	Fastballs	Curveballs
0	15	8	7
1	14	9	6
2	13	10	5
3	12	11	4
4	11	12	3
5	10	13	2
6	9	14	1
7	8	15	0

EXPLORE

2. Answers will vary. Possible response: Use a pair of numbered axes; for each ordered pair in the table, count the first number on the horizontal axis, the second number on the vertical axis, and make a point.

3. See graphing calculator screen on page 169.

4. number of fastballs; number of curveballs

5. No solution, this is an activity.

6. Number of curveballs decreases. Answers will vary, accept any answers students can justify.

7. Points are on a line.

MAKE CONNECTIONS

8. $c = 15 - f$

9. $c + f = 15$

10. One possible equation is $c = 15 - f$. Check:
    $(f, c) = (0, 15); 0 \overset{?}{=} 15 - 15, 0 = 0$ ✔;
    $(f, c) = (5, 10); 5 \overset{?}{=} 15 - 10, 5 = 5$ ✔;
    $(f, c) = (9, 6); 9 \overset{?}{=} 15 - 6, 9 = 9$ ✔

SUMMARIZE

11. Students should include tables, graphs, and ordered pairs. Answers for their respective advantages and disadvantages will vary.

12. No; only whole numbers of fastballs and curveballs make sense. Other points on such a line would represent numbers other than whole numbers.

13. The paired data for fastballs and curveballs would be included in the paired data for this relation. However, this relation is not limited to whole numbers. The graph would be a line through the points graphed for the fastball and curveball relation. It would be impossible to make a table of all possible values or to write all the ordered pairs because there is an infinite number of combinations of two real numbers with a sum of 15.

Lesson 4.2, pages 171–176

EXPLORE

1. Possible answers: Same—both have two columns and five rows; the numbers in the second column are less than the numbers in the first column; the numbers in both columns are related to each other; some numbers in the second column of both tables are the same. Different—all the numbers in the first column of Table 1 are the same, and all the numbers in the first column of Table 2 are different.

2. You couldn't name the corresponding number in the second column for Table 1 because the different second column numbers correspond to the same first column number. You could do this for Table 2, because each first-column number corresponds to only one second-column number.

TRY THESE

1. function; domain = $\{-1, -2, -3, -4, -5\}$; range = $\{2\}$

2. not a function; $\frac{1}{2}$ is paired with both $\frac{1}{3}$ and $\frac{1}{4}$.

3. function; domain = $\{3, 4, 5, 6\}$; range = $\{3, 4, 5, 6\}$

4. function; domain = $\{-5, -4, -3, -2, -1\}$; range = $\{1, 4, 9, 16, 25\}$

5. $f(-4) = \frac{-4}{4} = -1$

   $f(0) = \frac{0}{4} = 0$

   $f(8) = \frac{8}{4} = 2$

6. $f\left(\frac{1}{3}\right) = 3\left(\frac{1}{3}\right) + 1 = 1 + 1 = 2$

   $f\left(\frac{1}{2}\right) = 3\left(\frac{1}{2}\right) + 1 = \frac{3}{2} + 1 = 2\frac{1}{2}$

   $f\left(\frac{3}{4}\right) = 3\left(\frac{3}{4}\right) + 1 = \frac{9}{4} + 1 = 3\frac{1}{4}$

7. $f(x) = 2x - 2$

   $f(3) = 2(3) - 2 = 6 - 2 = 4$

8. $f(10) = 16(10)^2 = 16 \cdot 100 = 1600$

PRACTICE

1. function; domain = $\{\$0.50, \$0.60, \$0.70, \$0.80, \$0.90\}$; range = $\{\$0.02\}$

2. not a function; 2 is paired with -4, 0, and 4.

3. function; domain = $\{2, 4, 6, 8, 10\}$; range = $\{0\}$

4. function; domain = $-7, -5, -3, -1\}$; range = $\{-7, -5, -3, -1\}$

5. not a function; 3 is paired with $1.50, $1.80, and $2.49.

6. not a function; 55 is paired with all 5 output values.

7. function; domain = $\{1, 2, 3, 4, 5\}$; range = $\{50, 100, 150, 200, 250\}$

8. Possible answer: First look to see if there are two or more input values that are the same. If so, look at the corresponding output values. If they are the same, it is a function. If they are different, it is not a function.

9. $f(-4) = 0.5(-4) = -2$

   $f(1) = 0.5(1) = 0.5$

   $f(20) = 0.5(20) = 10$

10. $f(0) = 3(0 + 4) = 3(4) = 12$

    $f(3) = 3(3 + 4) = 3(7) = 21$

    $f(10) = 3(10 + 4) = 3(14) = 42$

11. $f(-3) = (-3)^2 - 2 = 9 - 2 = 7$

    $f(-1) = (-1)^2 - 2 = 1 - 2 = -1$

    $f(3) = (3)^2 - 2 = 9 - 2 = 7$

12. $f\left(\frac{1}{4}\right) = 2\left(\frac{1}{4}\right) - \frac{1}{2} = \frac{1}{2} - \frac{1}{2} = 0$

    $f\left(\frac{1}{2}\right) = 2\left(\frac{1}{2}\right) - \frac{1}{2} = 1 - \frac{1}{2} = \frac{1}{2}$

    $f\left(\frac{3}{4}\right) = 2\left(\frac{3}{4}\right) - \frac{1}{2} = \frac{3}{2} - \frac{1}{2} = 1$

13a. $f(0.5) = 9(0.5) = 4.5$; 4.5 calories

   b. $f(2) = 9(2) = 18$; 18 calories

   c. $f(17) = 9(17) = 153$; 153 calories

14. Both domain and range include all rational numbers ≥ 0.

EXTEND

15. not a function; a is paired with 1, 2, 3, and 4.

16. function if a, b, and c are all different; domain = $\{a, b, c\}$; range = $\{-5, -3, -1\}$

17. function if a, b, c, and d are all different; domain = $\{a, b, c, d\}$; range = $\{d, e, f, g\}$

18. change $x^2 - 2 \Rightarrow A$ to $x^2 + 3 \Rightarrow A$
    $f(-2) = 7$, $f(4) = 19$, $f(15) = 228$

19a. $f(\$3,500) = \$100 + 0.1(\$3,500) = \$450$

   b. $f(\$2,000) = \$100 + 0.1(\$2,000) = \$300$

   c. $f(\$1,900) = \$200 + 0.05(\$1,900) = \$295$

20. domain = $\{x \mid 0 \leq x < 2,000\}$
    range = $\{y \mid 200 \leq 300\}$
    domain = $\{x \mid x \geq 2,000\}$
    range = $\{y \mid y > 300\}$

21. Possible answer: Parent's age is represented by the function $f(x) = x + 25$. The range is 25 to 43.

22. Output is 3 less than input, so $f(x) = x - 3$.

23. Output is 1 more than twice the input, so $f(x) = 2x + 1$.

24. possible answer: perimeter of a square, where x = length of a side of the square.

25. possible answer: cost of a pizza, where x = number of toppings

26. $f(x) = \dfrac{x}{3} = 12$
$x = 36$

27. $f(x) = x^2 + 2 = 18$
$x^2 = 16$
$x = 4 \text{ or } -4$

MIXED REVIEW

28. Mean: $\dfrac{35 + 27 + 56 + 72 + 41}{5} = \dfrac{231}{5} = 46.2$

29. Median: 27, 35, **41**, 56, 72; median = 41

30. Mode: none

31. Range: $72 - 27 = 45$

32. $\dfrac{d}{7} = 42$
$d = 7(42)$
$d = 294$

33. $3t + 2 = -13$
$3t = -15$
$t = -5$

34. $f(-3) = 4(-3)^2 - 0.2(-3)$;
$4(9) - 0.2(-3)$; $36 - (-0.6)$; 36.6; C

ALGEBRAWORKS

1. Answers will vary depending on student's drawing.

2. Answers will vary depending on student's drawing.

3. Answers will vary depending on student's drawing.

4. Answers will vary depending on student's drawing.

5. $r = \dfrac{r_1 + r_2 + r_3 + \cdots + r_n}{n}$

$= \dfrac{12 + 10 + 14 + 15 + 11 + 9 + 11 + 12 + 13 + 15 + 11 + 14}{12}$

$= \dfrac{147}{12} = 12.25$

Area $= \pi r^2 = \pi (12.25)^2 = \pi (152.0625) \approx 471 \text{ m}^2$

Lesson 4.3, pages 177–182

EXPLORE

1. The center of the logo. She can match this to the center of the area on her wall in which she wants to paint the logo.

2. Possible response: Use an ordered pair of integers to describe how far from the center each point is. Let the first integer in the pair tell how many squares to the left or right of the center each point is. Let the second integer tell how many squares above or below the center each point is.

1. $(-4, -2)$

2. $(3, 0)$

3. $\left(-2\dfrac{1}{2}, 1\right)$

4. $\left(3\dfrac{1}{2}, -1\dfrac{1}{2}\right)$

5.
6.
7.
&
8.

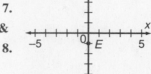

9. function; domain = $\{x \mid x \in \mathcal{R}\}$; range = $\{y \mid y \geq 0\}$

10. function; domain = $\{-5, -4, -3, -2, -1, 0, 1, 2, 3, 4, 5\}$; range = $\{-3\}$

11. Does not pass vertical line test, so not a function.

PRACTICE

1. $f(x)$ coordinate is zero; point is on the x-axis

2. x coordinate is < 0; $f(x)$ coordinate is > 0; point is in Quadrant II

3. x coordinate is > 0; $f(x)$ coordinate is < 0; point is in Quadrant IV

4. x coordinate is > 0; $f(x)$ coordinate is > 0; point is in Quadrant I

5. $(-2, -1)$

6. $(-3, 3)$

7. $\left(-3\dfrac{1}{2}, 0\right)$

8. $\left(2, \dfrac{1}{2}\right)$

9. $\left(0, -2\dfrac{1}{2}\right)$

10. $\left(\dfrac{1}{2}, -\dfrac{1}{2}\right)$

11.
12.
13.
&
14.

15. function; domain = $\{x \mid 0 \leq x \leq 120\}$; range = $\{x \mid 0 \leq x \leq 240\}$

16. function; domain: 0–6 oz; range: \$0.32, \$0.55, \$0.78, \$1.01, \$1.24, \$1.47 (By looking at the graph, students will only be able to give approximations of the values in the range.)

17. Possible answer: Look to see whether more than one point lies on an imaginary vertical line through each point in the graph.

18.

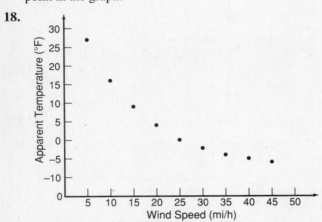

function; The graph passes the vertical line test and each input has a unique output.

19.

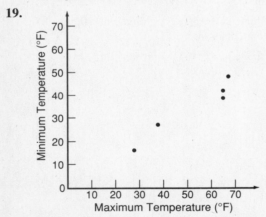

not a function; The graph fails vertical line test. When $x = 65$, $f(x)$ is both 39 and 42.

EXTEND

20. Tables will vary based on input values used. Input values should be the time remaining in the period before intermission. When added, each should be 20.

21. Each period is 20 minutes, so $f(x) = 20 - x$.

22.

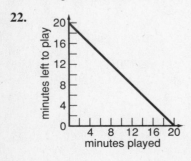

23. $f(8) = 20 - 8 = 12$. Could also refer to table or graph.

24. Possible story: Cindy started from a standstill and gradually picked up speed. At 5 min, she had to cross a busy street, so she slowed down. At 6 min, she finished crossing the street and again picked up speed until achieving a steady speed. Later on, Cindy cooled down by gradually decreasing her speed.

25. Answers will vary.

THINK CRITICALLY

26. In Quadrant III both coordinates must be < 0; $(-a, -b)$

27. In Quadrant I both soordinates must be > 0; (a, b)

28. On the x-axis $f(x)$ coordinate = 0; $(a, 0)$ or $(-a, 0)$

29. In Quadrant IV the x-coordinate is > 0 and $f(x)$ coordinate is < 0; $(a, -b)$

30. On the $f(x)$ axis, the x-coordinate is 0; $(0, b)$ or $(0, -b)$

31. In Quadrant II the x-coordinate is < 0 and $f(x)$ coordinate is > 0; $(-a, b)$

32. a; There are two output values that are the same next to two consecutive input values. There is a corresponding flat place on graph a.

33. b; There is a greater difference among output values than in Exercise 32. Graph b rises more steeply than graph a.

Lesson 4.4, pages 183–188

EXPLORE

1. $y = 2x - 1$

2.

Input (x)	Output (y)
–2	–5
0	–1
1	1
2	3

3. Possible answer: Yes; it appears that there will be a different output value for each input value substituted in the equation.

TRY THESE

1. $y = -x + 1$

x	y
–2	3
0	1
2	–1
4	–3

2.

x	y
−4	−1
−2	1
0	3
2	5

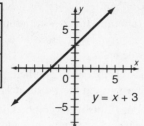

$y = x + 3$

3.

x	y
−2	−5
0	−4
2	−3
4	−2

$y = \dfrac{x}{2} - 4$

4.

x	y
−3	−2
−1	2
1	6
3	10

$y = 2x + 4$

5. No; the point is not on the line.

6. Yes; the point is on the line.

7. No; the point is not on the line.

8. Yes; the point is on the line.

9. No; the point is not on the line.

10. No; the point is not on the line.

11. not a linear function; each input value increases by 15 min; output values sometimes increase by different numbers of minutes.

PRACTICE

1.

x	y
−1	−3
1	−1
3	1

(−2, −4)
(0, −2)
(2, 0)

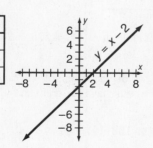

$y = x - 2$

2.

x	y
−2	−3
0	−3
2	−3

(−4, −3)
(−1, −3)
(3, −3)

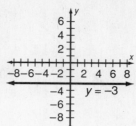

$y = -3$

3.

x	y
−1	−4
0	0
1	4

(2, 8)
(−2, −8)
$\left(\dfrac{1}{2}, 2\right)$

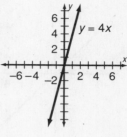

$y = 4x$

4.

x	y
−3	6
0	4
3	2

(−6, 8)
(6, 0)
$\left(1, 3\dfrac{1}{3}\right)$

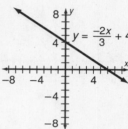

$y = \dfrac{-2x}{3} + 4$

5. Answers will vary. Possible answers: Use values close to zero so the graph is not too large. Use positive, 0, and negative numbers. Let $x = 0$ since it is usually easiest to solve. Only two ordered pairs are needed to graph a linear function; however, a third pair will be helpful in pointing out any errors.

6. Yes; the point is on the line.

7. No; the point is not on the line.

8. Yes; the point is on the line.

9. Yes; the point is on the line.

10. No; the point is not on the line.

11. No; the point is not on the line.

12. Linear; as each map distance increases by $\dfrac{1}{2}$ in., the actual distance increases by 50 mi.

13. Nonlinear; as each income amount increases by $15,000, the tax rate increases by a different amount (13%, 0%, 3%, 0%).

14. Linear; as the pieces of music increase by 1, the minutes of practice increase by zero.

EXTEND

15. $y - x = 5$
$y = x + 5$

x	y
−3	2
0	5
3	8

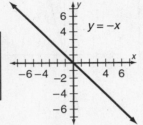

$y = x + 5$

16. $x + y = 0$
$y = -x$

x	y
−2	2
0	0
2	−2

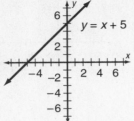

$y = -x$

17. $y - \frac{3x}{4} = 3$
$y = \frac{3x}{4} + 3$

x	y
−4	0
0	3
4	6

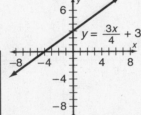

$y = \frac{3x}{4} + 3$

18. $2y + x = 4$
$2y = -x + 4$
$y = -\frac{1}{2}x + 2$

x	y
−2	3
0	2
2	1

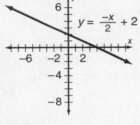

$y = \frac{-x}{2} + 2$

19.

sand (lb)	gravel (lb)
3	4
6	8
9	12
12	16

$y = \frac{4}{3}x$

20.

air time (min)	total charge
10	$26.40
20	$28.80
30	$31.20
40	$33.60

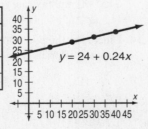

$y = 24 + 0.24x$

THINK CRITICALLY

21. not a linear function; If this were graphed, it would be a vertical line through the x-axis at 1. This graph fails the vertical line test.

22. linear function; If this were graphed, it would be a horizontal line through the y-axis at 1.

23. linear function; If this were graphed, it would be a diagonal line bisecting Quadrant I and Quadrant III.

24. not a linear function; This is not in the form $y = ax + b$.

25. Possible answer: Although x and y values may change by different amounts, for each equal change in x-values, there is an equal change in y-values.

26. Possible answer: It is in the form or can be written in the form $y = ax + b$, where a and b are constants.

27. Possible answer: The graph is a straight line.

PROJECT CONNECTION

1. Answers will vary depending on student's data.

2. Answers will vary depending on student's data.

3. Answers will vary depending on student's data.

4. Answers will vary depending on student's data.

ALGEBRAWORKS

1. Strike zone is rectangular so area = width • height; $A = 17h$

2. Answers will vary. Possible answer:

h	A
15	255
20	340
25	425
30	510

3.

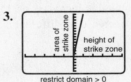

restrict domain > 0

4. 20 in., from table in **2**; $340 = 17 \cdot h$; $20 = h$.

5. $A = 17 \cdot 1\frac{1}{2} = 25.5$ in^2

6. Two people of the same height will be crouched differently so will have different strike zone heights.

Lesson 4.5, pages 189–191

THINK BACK

1. operator's error; Possible response: an estimate of 34×21 is 600. It appears that the operator entered one or both numbers incorrectly.

2. reasonable result; Possible response: an estimate of 25×60 is 1,500. So 1,575 is reasonable.

3. No solution, this is an activity.

4.
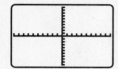

5. Screen shows only coordinate axes.

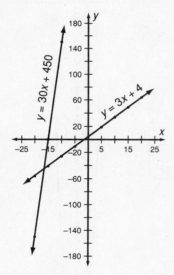

6. Answers will vary. In this case the points on the graph appear outside of the standard window defined in **3**.

7.

$y = 30x + 450$		$y = 3x + 4$	
x	y	x	y
−20	−150	−20	−56
−15	0	−15	−41
−10	150	−10	−26
−5	300	−5	−11
0	450	0	4
5	600	5	19
10	750	10	34
15	900	15	49
20	1,050	20	64

8. Both graphs are now visible.

MAKE CONNECTIONS

9. They are the corresponding y values when the minimum and maximum x values are entered into the equation.

Possible range values for Problems **10–13**:

	minimum x-value	maximum x-value	x-scale	minimum y-value	maximum y-value	y-scale
10.	−10	15	1	−800	100	100
11.	−10	10	1	0	85	5
12.	−60	10	5	−2	15	1
13.	−10	10	1	−1	1.5	0.5

SUMMARIZE

14. Answers will vary but should include that the graph may not fit on the part of the coordinate plane that first appears in the window. Explanation should go on to describe how to select and enter new window range values.

15. Always; if the smaller window included the graph, certainly the larger one will, too.

16. Sometimes; shrinking the window could result in the graph being outside its range.

17. The minimum and maximum values for the horizontal axis are the same.

18. They will be able to determine the effects on the graph of just changing the scale of the vertical axis.

19. Answers will vary. Students should try four sets of range values where the range for the vertical axis is kept the same and the range for the horizontal axis varies. In this way, they isolate the effect of changing the horizontal scale.

20. No; the square of any real number is greater than or equal to zero, the range of this function does not extend below 100.

Lesson 4.6, pages 192–196

EXPLORE

1. Both have x and y variables and a constant of 1 is added, but one equation has x squared and the other does not.

2.
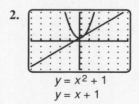

$y = x^2 + 1$
$y = x + 1$

3. Both are functions, both cross the y-axis at $(0, 1)$. One is a line and one is a curve.

1.

x	y
−2	12
−1	3
0	0
1	3
2	12

$y = 3x^2$

2.

x	y
−2	−4
−1	−1
0	0
1	−1
2	−4

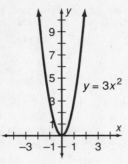

$y = -x^2$

3.

x	y
−2	4
−1	−2
0	−4
1	−2
2	4

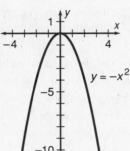

$y = 2x^2 - 4$

4.

x	y
−2	7
−1	3
0	1
1	1
2	3

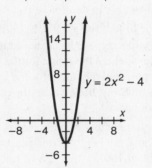

$y = x^2 - x + 1$

5.

x	y
−2	6
−1	−3
0	−6
1	−3
2	6

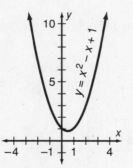

$y = 3x^2 - 6$

6. No; the point is not on the curve.

7. Yes; the point is on the curve.

8. Yes; the point is on the curve.

9. Yes; the point is on the curve.

10. 50 m

11. 55–58 sec

1.

x	y
−2	−8
−1	−2
0	0
1	−2
2	−8

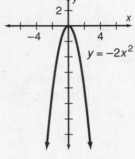

$y = -2x^2$

$(−3, −18)$
$(3, −18)$
$(4, −32)$

2.

x	y
−2	2
−1	$\frac{1}{2}$
0	0
1	$\frac{1}{2}$
2	2

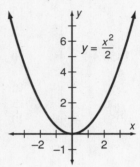

$y = \frac{x^2}{2}$

$\left(−3, \frac{9}{2}\right)$
$\left(3, \frac{9}{2}\right)$
$(4, 8)$

3.

x	y
−2	17
−1	8
0	5
1	8
2	17

$(−3, 22)$
$(3, 22)$
$(4, 53)$

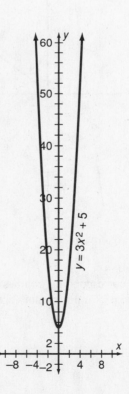

$y = 3x^2 + 5$

4.

x	y
−2	−2
−1	−5
0	−4
1	1
2	10

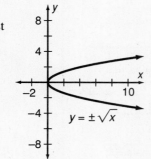

$y = 2x^2 + 3x - 4$

(−3, 5)
(2, 23)
(4, 40)

5. Possible answers: Use five values: two negative values, two positive values, and zero. It's easy to find y when zero is substituted for x. If a high or low point was not found when graphing, then substitute some more values for x in the table between the ordered pairs that are near where the high or low point should be.

6. 700

7. Either 92 or 93, too close to get exactly from graph.

8. Increase from 1970 to about 1983, decrease from about 1983 to 1995. Answers will vary but should be close to 1983 as the turn point.

9. non-linear

x	y
−2	−8
−1	−1
0	0
1	1
2	8

$y = x^3$

10. neither, fails vertical line test

x	y
0	0
4	±2
9	±3

$y = \pm\sqrt{x}$

11. non-linear

x	y
−2	16
−1	1
0	0
1	1
2	16

$y = x^4$

12. Either \$83 or \$84, too close to get exactly from graph.

13. Between 700 and 800 ft^2; too close to get exactly from graph

14. To cover the cost of employees, equipment, and supplies, the cleaner must charge a minimum amount regardless of how little carpet is cleaned.

15. The charge per square foot decreases as the square footage increases.

EXTEND

16. $y = 20x - 5x^2$

17.

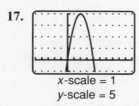

x-scale = 1
y-scale = 5

18. Max height is 20 m after 2 seconds.

19. Returns to ground in 4 seconds.

20. Regraph function; max is at (2.5, 31.25) so it went 11.25 m higher and stayed up 1 second longer.

THINK CRITICALLY

21. Answers will vary. Explanation should include that the graph is a curve because the output values do not change at an equal rate when the input values are squared. The graph of $y = x^2$ is symmetric to the y-axis because the same output results from squaring positive and negative input values. For example, when $x = 2$ or $x = -2$, $y = 4$. The graph of $y = x^3$ is not symmetric to the y-axis because the same output does not result from cubing positive and negative input values. For example, when $x = 2$, $y = 8$, but when $x = -2$, $y = -8$.

22. They have the same shape.

23. One opens up, the other opens down.

24. Only the point (0, 0).

MIXED REVIEW

25. $89,000 = \dfrac{80,000 + 97,000 + 42,000 + 150,000 + p}{5}$

$445,000 = 369,000 + p$

$76,000 = p$

26. $7a + 9b - 3ab + 2(a - b)$
$7a + 9b - 3ab + 2a - 2b$
$9a + 7b - 3ab$

27. $4r^2 - 2r + 7r^2 - 3 + 5r$
$11r^2 + 3r - 3$

28. $x \cdot 750 = 90$

$x = \dfrac{90}{750} = 0.12$ or 12%

Correct answer is B.

29. $x < 0$, $y > 0$, so point is in Quadrant II.

30. $x > 0$, $y < 0$, so point is in Quadrant IV.

31. $x > 0$, $y = 0$, so point is on x-axis.

32. $x < 0$, $y < 0$, so point is in Quadrant III.

Lesson 4.7, pages 197–201

EXPLORE

1. 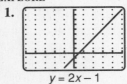 This equation has an infinite number of solutions.

$$y = 2x - 1$$

2. Locate 7 on the y-axis and move horizontally to the point on the graph. Now move vertically down to the x-axis and read the value of x at that point.

3. 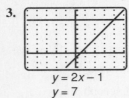 It is a linear equation with an infinite number of solutions.

$$y = 2x - 1$$
$$y = 7$$

4. Point of intersection is (4, 7). This has the value of x that solves the equation $7 = 2x - 1$.

5. 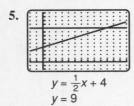 Graph $y = \frac{1}{2}x + 4$ and $y = 9$ and find the x- and y-values where the two lines intersect. The solution is (10, 9).

$$y = \frac{1}{2}x + 4$$
$$y = 9$$

Check: $\frac{1}{2}(10) + 4 = 5 + 4 = 9$

TRY THESE

1. $y = 9x - 5$ and $y = 22$

2. $y = 5x + 7$ and $y = 2x + 30$

3. $y = 3(x - 5)$ and $y = x + 7$

4. 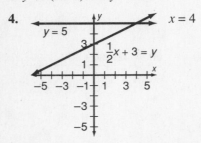 $x = 4$

5. 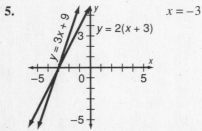 $x = -3$

6. 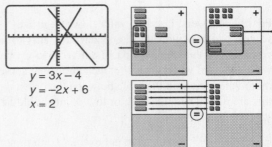 Both graphs are the same line, so $x = $ all real numbers.

7. $3x - 4 = -2x + 6$

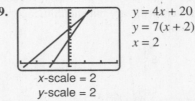

$$y = 3x - 4$$
$$y = -2x + 6$$
$$x = 2$$

8. $4x + 20 = 7(x + 2)$

9.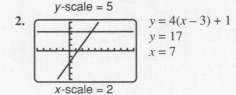

$$y = 4x + 20$$
$$y = 7(x + 2)$$
$$x = 2$$

x-scale = 2
y-scale = 2

10. Graph the function $y = 2x - 5$. Then graph the function $y = 11$. Determine that the graphs intersect at (8, 11). This means that when $x = 8$, $y = 11$ so $x = 8$ is a solution of $11 = 2x - 5$.

PRACTICE

1.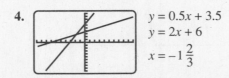

$$y = -3x + 5$$
$$y = -10$$
$$x = 5$$

x-scale = 1
y-scale = 5

2.

$$y = 4(x - 3) + 1$$
$$y = 17$$
$$x = 7$$

x-scale = 2
y-scale = 5

3.

$$y = 2x - 4$$
$$y = 2(x - 2)$$

Both graphs are the same line, so $x = $ all real numbers.

4.

$$y = 0.5x + 3.5$$
$$y = 2x + 6$$
$$x = -1\frac{2}{3}$$

5. Answers will vary. Advantage of graphing: it's easier to see if there is more than one solution. Advantage of using algebra: it takes less time and is more accurate.

6. Let x represent the hours the second plane flies.
$375(x + 1) = 500x$

7.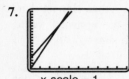
$y = 375(x + 1)$
$y = 500x$
$x = 3$

x-scale = 1
y-scale = 100

8. $500(3) = 375(3+1) = 1{,}500$ miles

9. Let x represent number of videos rented.
$1.99x = 20 + 0.89x$

10.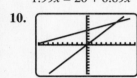
$y = 1.99x$
$y = 20 + 0.89x$
$18 < x < 19$

Plan B is more economical only after renting 19 or more videos.

11. Plan A: $1.99(19) = \$37.81$
Plan B: $20 + 0.89(19) = \$36.91$

EXTEND

12.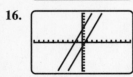
$y = x^2 - 5$
$y = -x + 1$
$x = -3$ or $x = 2$

13.
$y = x^2 + 1$
$y = -x^2 + 1$
$x = 0$

14.
$y = \dfrac{x}{2}$
$y = x - 4$
$x = 8$

15.
$y = -4(x + 4) + 5$
$y = 2(-2x + 3)$
Lines are parallel, so no solution.

16.
$y = 3x + 4$
$y = 3x - 2$
Lines are parallel, so no solution.

17.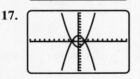
$y = x^2 - 2$
$y = -x^2 + 2$
$x = \pm\sqrt{2}$

18. Let $x =$ the number of minutes the hot water runs.
Let $y =$ the number of minutes the cold water runs.
$2.5x = 3(x - 1);$

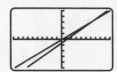
$y = 2.5x$
$y = 3(x - 1)$
$x = 6$
Hot water is on 6 minutes and cold water is on 5 minutes.

19. The graphs intersect at $(6, 15)$, meaning that 6 min after the hot water was turned on, 15 gal of hot water and 15 gal of cold water, or 30 gal altogether, have run. This is 10 gal more than the sink will hold. Vinje guessed corrctly.

THINK CRITICALLY

20. The two lines intersect in a single point.

21. Not possible for linear functions.

22. The two lines are parallel; they do not intersect.

23. Graphs coincide to form one line.

24. A curve and line that intersect in a single point.

25. A curve and line that intersect in two points.

MIXED REVIEW

26. $8450 \cdot \dfrac{f}{100} = \dfrac{8452f}{100}$; D

27. $(3 \cdot 4)^2 = 12^2 = 144$

28. $(3)(4)^2 = (3)(16) = 48$

29. $(-c)(-d)(-e) = (cd)(-e) = -cde$

30. $\dfrac{-m}{n} \cdot \dfrac{o}{-p} = \dfrac{-mo}{-np} = \dfrac{mo}{np}$

31. $f(-2) = 3(-2) - 5 = -6 - 5 = -11$

32. $f(0) = 3(0) - 5 = -5$

33. $f(3) = 3(3) - 5 = 9 - 5 = 4$

Lesson 4.8, pages 202–205

EXPLORE THE PROBLEM

1. No solution, this is an activity.

2. No solution, this is an activity.

3. No solution, this is an activity.

4. No solution, this is an activity.

5. No solution, this is an activity.

INVESTIGATE FURTHER

6. Olivis; The horizontal segment WX indicates that her distance from the starting point did not change.

7. Jorge; The line segment for the first part of his trip rises more steeply, showing that he moved more rapidly and was able to travel a greater distance from the starting point by t_1.

8. He turned sharply and walked back to the starting point.

9. No, her graph does not show $d = 0$ at the end of the trip.

10a. Karen sped up between **A** and **B**.

b. Karen changed directions twice.

APPLY THE STRATEGY

11. No; Answers will vary but may include that the vertical gap indicates he was in two places at once, which is impossible.

12. Possible Graph:

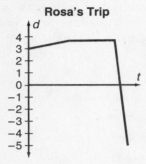

Rosa's Trip

13. Answers will vary. They do the same things except Phil is always 1 ft behind Jill.

14.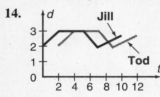

15. a; The container is wider at the bottom and narrower at the top, the height rises gently at first, then more steeply later.

16. Answers will vary depending on the bottle shape but should be similar in logic to **15** above.

17. The graph shows that over time the population increases.

18. The graphs have a similar shape.

19. The graph shows over time the population decreases.

20. The graphs have a similar shape.

REVIEW PROBLEM SOLVING STRATEGIES

1. Monkey Business

Answers will vary. 1 capuchin eats $\frac{1}{4}$ lb in 1 day, so 12 capuchins eat 3 lb in 1 day and 12 capuchins eat 36 lb in 12 days. Similarly 1 spider eats $\frac{1}{3}$ lb in 1 day, so 12 spiders eat 4 lb in 1 day and 12 spiders eat 48 lb in 12 days; 1 howler eats $\frac{1}{2}$ lb in 1 day, so 12 howlers eat 6 lb in 1 day and 12 howlers eat 72 lb in 12 days.
Add: $36 + 48 + 72 = 156$ lb

2. Crypto-Digit

SUB	138
+ TUB	+ 938
SINK	1076

a. S must stand for 1 because the greatest possible sum of two three-digit numbers is less than 2000.

b. No; if T were 7 or less, there would not be a digit in the thousands place of the sum.

c. T cannot be less than or equal to 7 so it stands for 8 or 9. If T stands for 8, then U stands for a number that is 5 or greater. Otherwise, there would be no thousands digit. If T stands for 9, then U must stand for a number less than 5. If not the number 1 would be represented by two different letters, S and I.

3. Line-Segments

a. Methods will vary. Students may count 2-dot, 3-dot, and 4-dot segments.

b. Answers will vary.

c. Answers will vary.

d. 35 segments

Lesson 4.9, pages 206–211

EXPLORE/WORKING TOGETHER

1. Data will vary.

2. Graphs will vary with data collected in Question **1**.

3. points; The graph displays a particular number of ordered pairs of data.

4. Possible answer: in general, as height increases, foot length incteases.

TRY THESE

1.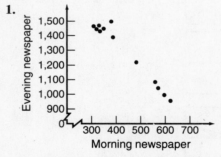

2. As the number of morning papers increases, the number of evening papers decreases.

PRACTICE

1.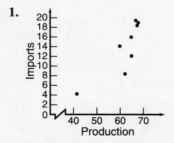

2. As energy production increased, so did energy imports.

3.

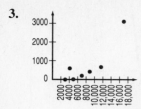

4. With the exception of Europe, as area increases, so does population. Because of Europe, one might say no relationship exists.

5. Possible answer: For Problems **1** and **2**, the increase in both production and imports can be seen equally as well in both the table and scatter plot. Because neither list of data in the table for Problems **3** and **4** is in increasing or decreasing order, it is easier to determine that no relationship exists by looking at the scatter plot.

EXTEND

6.

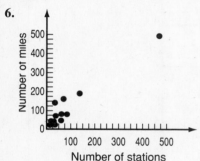

7. As the number of miles increase, do so the number of stations.

8. Accept 150–300; Possible answer: The point for (300, 200) would fall in the general pattern of points already on the scatter plot.

THINK CRITICALLY

9. d; The points for these ordered pairs of data form a straight line because the perimeter of a square increases 4 units for every unit in length that the side of a square increases.

10. e; The sales of soft drinks would probably increase as the temperature increased, but not necessarily at a steady rate.

11. b; People's heights should not affect their Science grade, and their Science grade should not affect their height.

12. c; Home heating costs would probably increase as temperature decreased, but not necessarily at a steady rate.

13. a; The points for these ordered pairs of data form a straight line because the number of hours left in a day decreases as much as the number of hours that have passed increases.

MIXED REVIEW

14. $\begin{bmatrix} 3+(-1) & 5+(-4) \\ -4+4 & 2+7 \\ 2+0 & 8+1 \end{bmatrix} = \begin{bmatrix} 2 & 1 \\ 0 & 9 \\ 2 & 9 \end{bmatrix}$

15. $\begin{bmatrix} -1-3 & -4-5 \\ 4-(-4) & 7-2 \\ 0-2 & 1-8 \end{bmatrix} = \begin{bmatrix} -4 & -9 \\ 8 & 5 \\ -2 & -7 \end{bmatrix}$

16. Range $= 19,000 = S - 1,000$, so $S = 20,000$

17. $|-15| = 15$

18. $-|15| = -15$

19. $-|-15| = -15$

20. $y = 6^2 + 2 \cdot 6 - 3 = 36 + 2 \cdot 6 - 3 = 36 + 12 - 3 = 45$

21. $y = 0^2 + 2 \cdot 0 - 3 = -3$

22. $y = (-2)^2 + 2(-2) - 3 = 4 + 2(-2) - 3$
$= 4 + (-4) - 3 = -3$

23. $4(x - 3) = 4(x) - 4(3) = 4x - 12$

24. $-3(a - 7) = -3(a) - (-3)(7) = -3a + 21$

25. $4(y + 4y^2) = 4(y) + 4(4y^2) = 4y + 16y^2$

26. $7(2x - 2) = 7(2x) - 7(2) = 14x - 14$

27. $\quad 6(x + 2) = 4(x + 12)$
$\quad\quad 6x + 12 = 4x + 48$
$\quad\quad\quad\quad 2x = 36$
$\quad\quad\quad\quad\ x = 18,\ \text{C}$

28. $\quad 4(x + 1) = 6 - 3(1 - 2x)$
$\quad\quad 4x + 4 = 6 - 3 + 6x$
$\quad\quad\quad\ 1 = 2x$
$\quad\quad\quad\ \frac{1}{2} = x,\ \text{A}$

29. $\quad 3x - 4 = 5$
$\quad\quad\ 3x = 9$
$\quad\quad\quad x = 3$

30. $\quad 4(x - 7) = 12$
$\quad\quad 4x - 28 = 12$
$\quad\quad\quad\ 4x = 40$
$\quad\quad\quad\quad x = 10$

31. $\quad \frac{x}{3} - 8 = 16$
$\quad\quad\ \frac{x}{3} = 24$
$\quad\quad\ x = 72$

32. $\quad \frac{x}{5} - 16 = -1$
$\quad\quad\ \frac{x}{5} = 15$
$\quad\quad\ x = 75$

1. Lists, data, and graphs will vary.

2. Lists, data, and graphs will vary.

Chapter Review, pages 212-213

1. b   **2.** e   **3.** d   **4.** c   **5.** a

6. function; domain = {3, 5, 7, 9, 11}; range = {2, 4}

7. not a function; -5 is paired with both $\frac{1}{5}$ and $-\frac{1}{5}$.;
-4 is paired with both $\frac{1}{4}$ and $-\frac{1}{4}$.

8. function; domain = {5, 10, 15, 20, 25};
range = {5, 10, 15, 20, 25}

9. $(-4, -2)$ **10.** $(-2, 3)$ **11.** $(2, 1)$

12. $(5, 0)$ **13.** $(1, -2)$

14. Yes; It passes the vertical line test and for each x there is a unique y.

For 15 to 20, tables and additional solutions will vary.

15.

x	y
-2	5
0	3
2	1
4	-1

$(3, 0)$

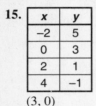
$y = -x + 3$

16.

x	y
-1	-5
0	-2
1	1
2	4

$(-2, -8)$

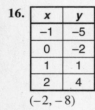
$y = 3x - 2$

17.

x	y
-4	-3
0	0
4	3
8	6

$\left(-1, -\frac{3}{4}\right)$

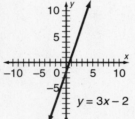
$y = \frac{3}{4}x$

18.

x	y
-1	-2
0	0
1	-2
2	-8

$(-2, -8)$

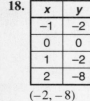
$y = -2x^2$

19.

x	y
-3	0
0	-9
1	-8
3	0

$(-1, -8)$

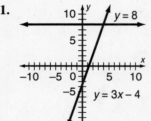
$y = x^2 - 9$

20.

x	y
-1	0
0	-3
1	-4
2	-3

$(-2, 5)$

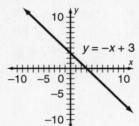
$y = x^2 - 2x - 3$

21.

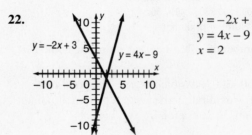
$y = 8$
$y = 3x - 4$

$y = 3x - 4$
$y = 8$
$x = 4$

22.

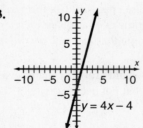
$y = -2x + 3$
$y = 4x - 9$

$y = -2x + 3$
$y = 4x - 9$
$x = 2$

23.

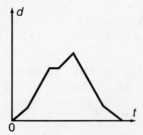
$y = 4x - 4$

$y = 4(x - 1)$
$y = 4x - 4$

Both graphs are the same line so $x =$ all real numbers.

24. Possible graph:

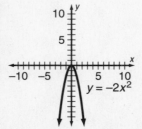

25.

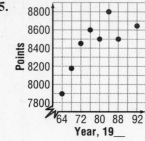

26. As the years increase, the number of points tend to increase.

Chapter Assessment, pages 214-215

1. Look to see if any input value occurs more than once. -8 and -4 both occur twice, so look to see if each is paired with the same output value both times. If so, the set is a function. If the output values for one or both are different each time, then the set is not a function.

2. $f(12) = -\dfrac{12}{3} = -4$

3. $f(3) = (3 - 5)^2 = (-2)^2 = 4$

4. K

5. F

6. E

7. B

8.

x	y
-2	2
0	4
2	6
4	8

linear

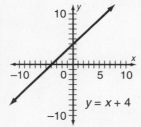

$y = x + 4$

9.

x	y
-2	1
-1	-2
0	-3
2	1

nonlinear

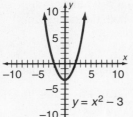

$y = x^2 - 3$

10.

x	y
-4	11
-2	5
0	-1
4	-13

linear

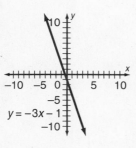

$y = -3x - 1$

11.

x	y
-4	9
0	6
4	3
8	0

linear

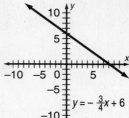

$y = -\dfrac{3}{4}x + 6$

12. C

13. D

14. C

15. Write an equation in the form $y = f(x)$ for each side of the original equation. Then graph both equations. The x-coordinate of the point where the two graphs intersect is the solution of the original equation.

16. False; Vicki began walking faster than Mickey.

17. True

18. In general, as the number of days since payday increased, the amount of available money decreased. This is shown by the points on the graph which tend to form a line that falls from left to right.

19. Let x represent the hours Karla drove.
$55(x + 1) = 65x$

20.

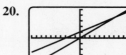

$y = 55(x + 1)$
$y = 65x$
$x = 5.5$

Cumulative Review, page 216

1. $(2, -3)$ **2.** $(0, 1)$

3. $(-3, -4)$ **4.** $(3, 0)$

5. $f(3) = 4(3) - 5 = 7$
$f(10) = 4(10) - 5 = 35$
$f(-5) = 4(-5) - 5 = -25$

6. $f(12) = \dfrac{12}{3} + 2 = 6$
$f(-24) = \dfrac{-24}{3} + 2 = -6$
$f(0) = \dfrac{0}{3} + 2 = 2$

7. $7x - 3 = 25$
$7x = 28$
$x = 4$

8. $\dfrac{2}{3}x = -10$
$2x = -30$
$x = -15$

9.
$$6x - 2(x + 4) = 3(4x - 3) - 15$$
$$6x - 2x - 8 = 12x - 9 - 15$$
$$4x - 8 = 12x - 24$$
$$16 = 8x$$
$$2 = x$$

10.
$$\begin{bmatrix} -4 + (-7) & 0 + (-6) & \frac{3}{4} + \frac{5}{8} \\ 2 + (-5) & -3 + 9 & 13 + (-13) \end{bmatrix}$$

$$= \begin{bmatrix} -11 & -6 & \frac{11}{8} \\ -3 & 6 & 0 \end{bmatrix}$$

11.
$$\begin{bmatrix} -7 - (-4) & -6 - 0 & \frac{5}{8} - \frac{3}{4} \\ -5 - 2 & 9 - (-3) & -13 - 13 \end{bmatrix}$$

$$= \begin{bmatrix} -3 & -6 & -\frac{1}{8} \\ -7 & 12 & -26 \end{bmatrix}$$

12. B

13.

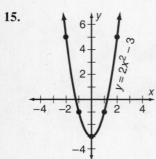

14.

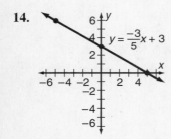

15.

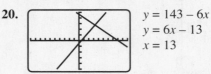

16. If x and y are both in Quadrant III, their sum would have to be less than zero and $7 \not< 0$.

17. $0.15(120) = 18$

18. $27\% + 5\% = 32\%$

19.

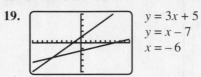

$y = 3x + 5$
$y = x - 7$
$x = -6$

20.

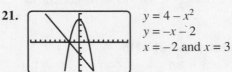

$y = 143 - 6x$
$y = 6x - 13$
$x = 13$

21.

$y = 4 - x^2$
$y = -x - 2$
$x = -2$ and $x = 3$

22. $0.76(75) = 57$

23.
$$A = P + Prt$$
$$A - P = Prt$$
$$\frac{A - P}{Pr} = t$$

24. $5(2x - 3) + 7x - 4(x + 3) - 10$
$10x - 15 + 7x - 4x - 12 - 10$
$13x - 37$

Standardized Test, page 217

	Column 1	Column 2

1. A;
$4 + x(4) = 5$	$5 - 5(x) = 4$
$4x = 1$	$-5x = -1$
$x = \frac{1}{4} = 25\%$	$x = \frac{1}{5} = 20\%$

2. B; $\qquad x + 4 \qquad\qquad x + 7$
$x + 7$ is 3 more than $x + 4$.

3. C; $\qquad |a - b| \qquad\qquad |b - a|$
$a - b$ and $b - a$ only differ in sign and the absolute value makes both the same sign.

4. D; $\qquad xy \qquad\qquad \frac{x}{y}$
It depends on what x and y equal.

5. C;
$f(3) = -2(3)^2 + 5$	$f(-3) = -2(-3)^2 + 5$
$= -2(9) + 5$	$= -2(9) + 5$
$= -18 + 5$	$= -18 + 5$
$= -13$	$= -13$

6. A; $\qquad 3^4 = 81 \qquad\qquad 4^3 = 64$

7. C; $\qquad a + 0 = a \qquad\qquad a \cdot 1 = a$

8. A;
$-2 - (-5)$	$-5 - (-2)$
$-2 + 5$	$-5 + 2$
3	-3

	Column 1	Column 2		Column 1	Column 2

9. B;   0 $\dfrac{x}{-9} = -6$

$x = 54$

10. B; $-2 + 3 + (-5)$ $4 + (-1) + (-6)$

-4 -3

11. B; since $x < 0$ and $y > 0$.

12. D; since it depends on what x equals.

13. A; $\dfrac{3}{4} \div \dfrac{2}{3} = \dfrac{3}{4} \cdot \dfrac{3}{2} = \dfrac{9}{8}$ $\dfrac{3}{4} \cdot \dfrac{2}{3} = \dfrac{6}{12} = \dfrac{1}{2}$

14. C; $\dfrac{2+3+6+6+6+9+10}{7} = \dfrac{42}{7} = 6$ 2, 3, 6, **6**, 6, 9, 10

15. B; $\dfrac{10}{20} = \dfrac{1}{2}$ $\dfrac{20}{30} = \dfrac{2}{3}$

16. A; since both coordinates in Quadrant I are greater than zero and both coordinates in Quadrant III are less than zero.

17. C; $(0.50)(100) \div 4$ $(0.25)(50)$

$50 \div 4$ 12.5

12.5

18. C; since both y-coordinates are 1.

19. B; $3x - 2 = 4x + 5$

$-7 = x$

So $2x = -14$ and $x - 1 = -8$

20. B; $xy < 0$ $x - y > 0$

21. B; $\left(\dfrac{1}{4}x\right)^2 = \dfrac{1}{16}x^2$ $\dfrac{1}{4}x^2$

22. C; $4 \div 6 = \dfrac{4}{6} = \dfrac{2}{3}$ $\dfrac{2}{3}$

Chapter 5   Linear Inequalities

Data Activity, page 218

1. The new advertisements are being introduced to the consumer. More people are becoming interested in the product, and sales are increasing slowly.

2. Maturity; sales are leveling off. Answers will vary. This may be because the ads are losing their power to attract new customers.

3. Decline; $t > C$ and $t < D$; the advertising campaign is no longer effective.

4. Answers will vary. The introductory stage would be much longer than the other stages. Even during the growth stage, the increase would not be as great as for other (low learning) products, such as snack foods.

Lesson 5.1, pages 221–226

EXPLORE

1. $1,000,000

2. Any amount smaller than $1,000,000 is acceptable.

3. Any amount greater than $1,000,000 is acceptable.

4. Yes; any amount between $0 and $1,000,000 is acceptable; to the left of $1,000,000 on the number line

5. Yes, infinite number; to the right of $1,000,000

6. This is not possible because of the trichotomy property.

TRY THESE

1. $3.5 > -3.5$ so $c > -3.5$

2. $-3 > -3.5$ so $c > -3.5$

3. $-3.8 < -3.5$ so $c < -3.5$

4.

5.

6.

7.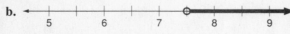

8a. $r > \$7.50$

b.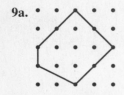

9a. 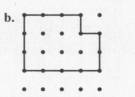 b.

PRACTICE

1. Yes, because $-6 \le -6$ is true.

2. Yes, because $-0.2 \ne 0.2$ is true.

3. No, because $100 < 100$ is not true.

4. No, because $4 \ge 4\frac{1}{2}$ is not true.

5. Possible answers: Substitute the given number of the variable in the inequality and check whether the inequality is true or not true; graph the inequality on a number line and check whether the given number is included in the graph; use a graphing utility to test the inequality.

6.

7.

8.

9.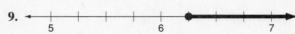

10.

11.

12.

13.

14a. $a \ge 18$

b.

15. Answers will vary. Possible answer: "The amount of gas it took to fill my car this time is greater than the amount it took last time."

16a. $e < 29,028$

b.

EXTEND

17. No, because $7.3 > 7.3$ is not true; possible answers: $g \ge 7.3$ or $g \le 7.3$

18. Yes, because $-3 \ne -2$, so -3 is true.

19. No, because $11 \le 10$, is not true; possible answers: $v > 10$ or $v \ge 10$

20. No, because $-4 \ge -3\frac{1}{4}$ is not true; possible answers: $z < -3\frac{1}{4}$ or $z \le -3\frac{1}{4}$.

21. Yes, because $3 < 5$, is true.

22. No, because $-4 \geq 0$ is not true; possible answers: $-4 < c$ or $-4 \leq c$

23. $x \geq 6$

24. $x < -2$

25. $x \neq \dfrac{1}{2}$

26. $x \leq 96$

27. $x \leq 37.4$

28. $x < -3\dfrac{1}{2}$

29a. $s \leq 55$

   b. No, numbers less than zero are impossible in the situation.

30a. $8.2 \leq p \leq 10$

   b. Disagree; points are bounded both above by 10 and below by 8.2. An arrow is inappropriate since it indicates all solutions in the direction of that arrow.

THINK CRITICALLY

31. Always true, since $a + 1$ is always 1 number to the right of a on the number line

32. Always true, since $b - 1$ is always 1 number to the left of b on the number line

33. Never true, since $a < b$ falls to the left of b on the number line and cannot also be to the right of b

34. Sometimes true, if $a = b$ both inequalities can be satisfied simultaneously

35. Sometimes true. It is possible for $a > b$ and satisfy hypothesis, e.g. $a = b$ and $b = -5$, but $a \not< b$.

36. Sometimes true. It is possible for $a < b$ and still satisfy hypothesis, e.g. $a = 1$ and $b = 0$, but $a \not< b$

ALGEBRAWORKS

1. Possible answers: $A < D + G$; $B < C$; $B + E \leq H$; $F < H \leq G$.

2.-5. Answers will vary.

Lesson 5.2, pages 227–229

THINK BACK

1.

2. Possible answer: the solution of the equation is a single point, whereas the inequality has an infinite solution set.

EXPLORE

3. Possible inequalities (if numbers selected are 2 and -2):

   a. $9 < 10$; **b.** $5 < 6$; **c.** $5 < 6$; **d.** $9 < 10$; **e.** $14 < 16$;

   f. $-14 < -16$; **g.** $3\dfrac{1}{2} < 4$; **h.** $-3\dfrac{1}{2} < -4$

4.

New inequality resulting from	Partner A	Partner B
	True or not true?	
Adding positive integer	true	true
Adding negative integer	true	true
Subtracting positive integer	true	true
Subtracting negative integer	true	true
Multiplying by positive integer	true	true
Multiplying by negative integer	not true	not true
Dividing by positive integer	true	true
Dividing by negative integer	not true	not true

5. multiplying and dividing both sides by a negative number

6. Reverse the inequality symbol in the above cases.

7. & 8. Results should be similar to results in Activities 3-6.

MAKE CONNECTIONS

9. True, subtracting a positive number does not change the inequality symbol.

10. False, dividing by a positive number does not change the inequality symbol.

11. False, subtracting a negative number does not change the inequality symbol.

12. True, multiplying by a negative number reverses the inequality symbol.

13. False, multiplying by a positive number does not change the inequality symbol.

14. $<$ **15.** $<$ **16.** $\leq$ **17.** $<$

18. $\geq$ **19.** $\leq$ **20.** $>$

SUMMARIZE

21. Adding the same real number to each side of an inequality and subtracting the same real number from each side of an inequality, multiplying each side of an inequality by the same positive real number, or dividing each side of an inequality by the same positive real number results in an inequality that is true. Multiplying each side of an inequality by the same negative real number or dividing each side of an inequality by the same negative real number results in an inequality that is untrue. To make the untrue inequality true, you must reverse the inequality symbol.

22. Adding zero to or subtracting zero from each side of an inequality results in a true inequality. Multiplying each side of a "less than" inequality by zero results in an untrue inequality because both sides are then equal to zero. It isn't appropriate to consider the effect of dividing both sides of an inequality by zero because division by zero is undefined.

23. Some values; each inequality will only be true when the number that is on the right of the original inequality (-4) is substituted for the variable.

EXPLORE/WORKING TOGETHER

1. $x + 3 > 5$

2. $x > 2$

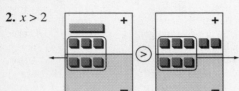

3. x is any real number greater than 2.

4. As with equations, use Algeblocks to form zero pairs to isolate the variable. Unlike solutions of equations, however, the solution is not only what is seen on the mat, but can include numbers greater than or less than what is seen on the mat.

TRY THESE

1. $a + 2 \leq -5$

   $a \leq -7$ Subtract 2.

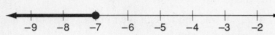

2. $\frac{-b}{3} > -2$

   $b < 6$ Multiply both sides by -3 and reverse the inequality symbol.

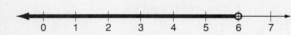

3. $c - 0.3 \geq 4.5$

   $c \geq 4.8$ Add 0.3.

4. $\frac{2}{5}d < 10$

   $d < 25$ Multiply by $\frac{5}{2}$.

5. $\geq m + 3$

   m or $m \leq 4$ Subtract 3.

6. $-2 < \frac{x}{4}$

   $-8 < x$ or $x > -8$ Multiply by 4.

7. $w - \frac{5}{6} \geq \frac{1}{6}$

   $w \geq 1$ Add $\frac{1}{6}$.

8. $3.1 < t - 1.8$

   $4.9 < t$ or $t > 4.9$ Add 1.8.

9. $-z + 3 \geq 0$

   $3 \geq z$ or $z \leq 3$ Add z.

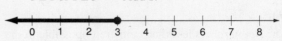

10. Parts + Labor $\leq$ \$300
   \$59.40 + c $\leq$ \$300
             c $\leq$ \$240.60

11. $2x \leq 6$

   $x \leq 3$

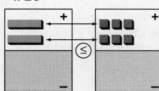

PRACTICE

1. $\frac{3}{4}k \leq 12$

   $k \leq 16$ Multiply by $\frac{4}{3}$.

2. $d - \frac{1}{2} < 2$

   $d < 2\frac{1}{2}$ Add $\frac{1}{2}$.

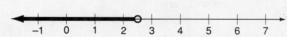

3. $p - 7 > 9$

   $p > 16$ Add 7.

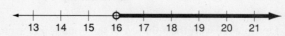

4. $-30 < -6n$

   $5 > n$ or $n < 5$ Divide by -6 and reverse the inequality symbol.

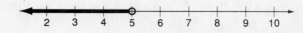

5. $\dfrac{e}{-1.2} > 6$

$e > -7.2$ Multiply by -1.2 and reverse the inequality symbol.

6. $-\dfrac{4}{5}b \le 20$

$b \ge -25$ Multiply by $-\dfrac{5}{4}$ and reverse the inequality symbol.

7. $c + \dfrac{2}{3} < 1\dfrac{1}{3}$

$c < \dfrac{2}{3}$ Subtract $\dfrac{2}{3}$.

8. $z - (-3) > -2.4$

$z + 3 > -2.4$ Simplify.

$z > -5.4$ Subtract 3.

9. $f - (-0.6) \ge 1.2$

$f + 0.6 \ge 1.2$ Simplify.

$f \ge 0.6$ Subtract 0.6.

10. $\dfrac{h}{-10} \ge -1$

$h \le 10$ Multiply by -10 and reverse the inequality symbol.

11. $0 \ge c - 4$

$4 \ge c$ or $c \le 4$ Add 4.

12. $\dfrac{m}{0.5} \ge -6$

$m \ge -3$ Multiply by 0.5.

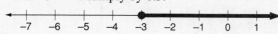

13. $d - 14 \le 11$

$d \le 25$ Add 14.

14. $0.25 < 5g$

$0.05 < g$ or $g > 0.05$ Divide by 5.

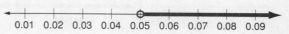

15. $\dfrac{j}{-8} \le 3$

$j \ge -24$ Multiply by -8 and reverse the inequality symbol.

16. $-3.2w < 9.6$

$w > -3$ Divide by -3.2 and reverse the inequality symbol.

17. $\dfrac{r}{4} \le -2$

$r \le -8$ Multiply by 4.

18. $-\dfrac{1}{3}e > -\dfrac{5}{6}$

$e < \dfrac{5}{2}$ Multiply by -3 and reverse the inequality symbol.

19. $10 \le q + 1.4$

$8.6 \le q$ or $q \ge 8.6$ Subtract 1.4.

20. $a - \left(-\dfrac{1}{3}\right) < 6$

$a + \dfrac{1}{3} < 6$ Simplify.

$a < 5\dfrac{2}{3}$ Subtract $\dfrac{1}{3}$.

21. $3.6 < \dfrac{z}{2}$

$7.2 < z$ or $z > 7.2$ Multiply by 2.

22. $s - 25 \le 7$

$s \le 18$ Add 25.

23. $9 < h + (-4.5)$

$\quad 13.5 < \text{ or } h > 13.5 \qquad$ Add 4.5.

24. $-10 > 2.5m$

$\quad -4 > m \text{ or } m < -4 \qquad$ Divide by 2.5.

25. $-3t > \dfrac{1}{2}$

$\quad t \le -\dfrac{1}{6} \qquad$ Divide by -3 and reverse the inequality symbol.

26. $3 + n - 7 \le 2$

$\quad n - 4 \le 2 \qquad$ Simplify.

$\quad\quad n \le 6 \qquad$ Add 4.

27. $2 - (3 - s) < 4$

$\quad -1 + s < 4 \qquad$ Simplify.

$\quad\quad s < 5 \qquad$ Add 1.

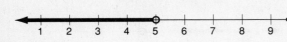

28. $3(4d) > -48$

$\quad 12d > -48 \qquad$ Simplify.

$\quad\quad d > -4 \qquad$ Divide by 12.

29. $A \ge 12(10)$

$\quad A \ge 120$

30. The term at least means "at or above." For example, the statement "You must be at least 18 years of age to vote." means you must be 18 or older than 18 to vote. The term at most means "at or below." For example, the statement "This elevator will hold at most 2040 lb." means the elevator will hold up to and including 2040 lb but not above that amount.

EXTEND

31. $(h - 3) \le 5$

$\quad\quad h \le 8 \qquad$ Add 3.

32. $-\dfrac{1}{2}p > -5$

$\quad\quad p < 10 \qquad$ Multiply by -2 and reverse the inequality symbol.

33. $-4 \le g - 7$

$\quad 3 \le g \text{ or } g \ge 3 \qquad$ Add 7.

34. $1.2s < 9$

$\quad s < 7.5 \qquad$ Divide by 12.

35. The graph of $y = x - 4$ is displayed only for values of x less than 3.

36. $624.32 + d - 279.38 \ge 600.00$

$\quad\quad 344.94 + d \ge 600.00 \qquad$ Divide by 12.

$\quad\quad\quad\quad d \ge 255.06 \qquad$ Subtract 44.94.

37a. $0.75w \le 25.00$

  b. $0.75w \le 25.00$

$\quad\quad w \le 33\dfrac{1}{3} \qquad$ Divide by 0.75.

Any integer from 1 to 33 solves the inequality.

THINK CRITICALLY

38. sometimes true; true if $b \ge 0$

39. always true

40. sometimes true; not true when three of the four numbers are negative

41. multiplication by zero always yields zero and then an equality is formed; division by zero is undefined

42. Answers will vary; can have infinite possibilities such as $x + 3 < 0$

MIXED REVIEW

43. Mean:

$$\dfrac{7.59 + 6.95 + 8.59 + 9.25 + 7.75 + 8.25}{6} = \dfrac{48.38}{6} \approx \$8.06$$

44. Median: 6.95, 7.59, 7.75, 8.25, 8.59, 9.25;

$$\dfrac{7.75 + 8.25}{2} = \dfrac{16}{2} = \$8.00$$

45. No mode

46. Range: $9.25 - 6.95 = \$2.30$

47. $\dfrac{e}{0.2} = 8$

$\quad\quad e = 1.6$

48. $\dfrac{3}{4}k = -15$

$\quad\quad k = -20$

49. $\dfrac{1}{3}n = -12$

$\quad\quad n = -36$

50. x-coordinate is zero, so R is on the y-axis

51. x-coordinate < 0, y-coordinate > 0,
so S is in Quadrant II

52. x-coordinate > 0, y-coordinate < 0,
so T is in Quadrant IV

53. $x + \geq 9$, subtract 4 to get $x \geq 5$; C

Lesson 5.4, pages 235–240

EXPLORE/WORKING TOGETHER

1. Power Print $= 0.05x$; Quality Print $= 50.00 + 0.04x$

2. $0.05x \leq 200.00$ and $50.00 + 0.04x \leq 200$

3. Possible answer: Same—you need to isolate the variable on one side. Different—you can solve the first inequality using one operation, but you need to use more than one operation to solve the second inequality.

TRY THESE

1. $7c - 4 \geq 24$

$\quad 7c \geq 28 \qquad$ Add 4.

$\quad c \geq 4 \qquad$ Divide by 7.

2. $2(4 - d) - 5d > 8$

$\quad 8 - 7d > 8 \qquad$ Distribute and simplify.

$\quad -7d > 0 \qquad$ Subtract 8.

$\quad d < 0 \qquad$ Divide by -7 and reverse inequality sign.

3. $\frac{2}{3}(s + 6) < -20$

$\quad \frac{2}{3}s + 4 < -20 \qquad$ Distribute.

$\quad \frac{2}{3}s < -24 \qquad$ Subtract 4.

$\quad s < -36 \qquad$ Multiply by $\frac{3}{2}$.

4. $6z + \frac{9}{2} < 5z - 7$

$\quad z < -11\frac{1}{2} \qquad$ Subtract $5z$ and $\frac{9}{2}$.

5. $8 - 4x \leq 6x - 2$

$\quad 10 \leq 10x \qquad$ Add $4x$ and 2.

$\quad 1 \leq x$ or $x \geq 1 \qquad$ Divide by 10.

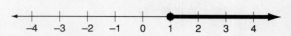

6. $5 > \frac{1}{3}b + 14$

$\quad -9 > \frac{1}{3}b \qquad$ Subtract 14.

$\quad -27 > b$ or $b < -27 \qquad$ Multiply by 3.

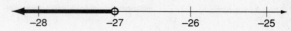

7. $\frac{2}{5}x + 3 > \frac{1}{5}x + 1$

$\quad \frac{1}{5}x > -2 \qquad$ Subtract $\frac{1}{5}x$ and 3.

$\quad x > -10 \qquad$ Multiply by 5.

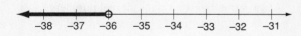

8. $8y + 9 \leq 4 + 8y$

$\quad 5 \leq 0 \qquad$ Subtract $8y$ and 4.

Untrue, so no solution.

9. $40 + 25h \geq 150$

$\quad 25h \geq 110 \qquad$ Subtract 40.

$\quad h \geq 4.4 \qquad$ Divide by 25.

10. $2x - 5 > 4 - x$

$\quad x > 3 \qquad$ Add x and 5 to both sides.

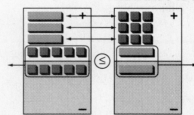

Algeblocks only show integer solutions. However, the solution is all real numbers > 3.

PRACTICE

1. $7 \geq 2 - d$

$\quad d \geq -5 \qquad$ Add d and Subtract 7.

2. $\frac{m}{5}+3>9$

$\qquad \frac{m}{5}>6 \qquad$ Subtract 3.

$\qquad m>30 \qquad$ Multiply by 5.

3. $3c-7\ge 29$

$\qquad 3c\ge 36 \qquad$ Add 7.

$\qquad c\ge 12 \qquad$ Divide by 3.

4. $\quad 72<-3h+4-5h$

$\qquad 68<-8h \qquad$ Subtract 4 and simplify.

$\qquad -8\frac{1}{2}>h$ or $h<-8\frac{1}{2} \qquad$ Divide by -8 and reverse inequality symbol.

5. $106\ge 19p+6+6p$

$\qquad 100\ge 25p \qquad$ Subtract 6 and simplify.

$\qquad 4\ge p$ or $p\le 4 \qquad$ Divide by 25.

6. $\frac{2}{3}f-4<12$

$\qquad \frac{2}{3}f<16 \qquad$ Add 4.

$\qquad f<24 \qquad$ Multiply by $\frac{3}{2}$.

7. $\frac{5}{8}e-3-\frac{3}{8e}>-5$

$\qquad \frac{1}{4}e>-2 \qquad$ Add 3 and simplify.

$\qquad e>-8 \qquad$ Multiply by 4.

8. $\frac{2}{5}(g-3)\ge -4$

$\qquad g-3\ge -10 \qquad$ Multiply by $\frac{5}{2}$.

$\qquad g\ge -7 \qquad$ Add 3.

9. $7x+4<39+2x$

$\qquad 5x<35 \qquad$ Subtract 4 and $2x$.

$\qquad x<7 \qquad$ Divide by 5.

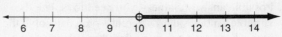

10. $13+5n\ge 25+5n$

$\qquad 13\ge 25 \qquad$ Subtract $5n$.

Untrue, so no solution.

11. $6(k-2)>48$

$\qquad k-2>8 \qquad$ Divide by 6.

$\qquad k>10 \qquad$ Add 2.

12. $2x-5(x+3)\ge -20$

$\qquad 2x-5x-15\ge -20 \qquad$ Distribute.

$\qquad -3x\ge -5 \qquad$ Add 15 and simplify.

$\qquad x\le \frac{5}{3} \qquad$ Divide by -3 and reverse the inequality symbol.

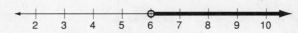

13. $4a+6\ge 7+4a$

$\qquad 6\ge 7 \qquad$ Subtract $4a$.

Untrue, so no solution.

14. $3(q+4)-5(q-1)<5$

$\qquad 3q+12-5q+5<5 \qquad$ Distribute.

$\qquad -2q+17<5 \qquad$ Simplify.

$\qquad -2q<-12 \qquad$ Subtract 17.

$\qquad q>6 \qquad$ Divide by -2 and reverse the inequality symbol.

15. $3y-7+5y<2+8y$

$\qquad 8y-7<2+8y \qquad$ Simplify.

$\qquad -7<2 \qquad$ Subtract $8y$.

Always true, so solution is all real numbers.

16. $5(7+r)>12r$

$\qquad 35+5r>12r \qquad$ Distribute.

$\qquad 35>7r \qquad$ Subtract $5r$.

$\qquad 5>r$ or $r<5 \qquad$ Divide by 7.

17. $7(2-e) \geq 3(e+8)$

    $14 - 7e \geq 3e + 24$ Distribute.

    $-10 \geq 10e$ Add $7e$ and subtract 24.

    $-1 \geq e$ or $e \leq -1$ Divide by 10.

18. $4(p-3) < 4(p-4)$

    $p - 3 < p - 4$ Divide by 4.

    $-3 < -4$ Subtract p.

Untrue, so no solution.

19. $3(3b+1) - (b-1) \leq 6(b+8)$

    $9b + 3 - b + 1 \leq 6b + 48$ Distribute.

    $8b + 4 \leq 6b + 48$ Simplify.

    $2b \leq 44$ Subtract $6b$ and 4.

    $b \leq 22$ Divide by 2.

20. $6(d+4) - (d-5) > 5d - 1$

    $6d + 24 - d + 5 > 5d - 1$ Distribute.

    $5d + 29 > 5d - 1$ Simplify.

    $29 > 1$ Subtract $5d$.

Always true, so solution is all real numbers.

21. In $x > 7$, all x-values which are to the right of 7 on the number line make the inequality true. In $7 < x$, 7 is less than all values that make the statement true. That is, all x-values are to the right of 7 on the number line. Therefore the two statements are equivalent.

22a. not possible; $\dfrac{582+p}{8} \geq 90$; $p \geq 138$

  b. $\dfrac{582+p}{8} \geq 80$; $p \geq 58$

23. $2x - 3 \leq 3x + 2$

    $-5 \leq x$ or $x \geq -5$ Subtract $2x$ and 2.

The solution is $x \geq -5$. The graph is a straight horizontal line beginning at (5, 1) and extending to the right.

EXTEND

24. didn't reverse the inequality sign when dividing both sides by -4; $a > -4$

25. distributed 2 over h but not over 4; $h \geq 2$

26. combined like terms incorrectly; $x \leq -2$

27. incorrectly thought that if all variables were eliminated, the solution is all real numbers; no solution

28. $90 - 3n > 48$

    $-3n > -42$ Subtract 90.

    $n < 14$ Divide by -3 and reverse the inequality symbol.

29. $z + 27 < 4z - 6$

    $33 < 3z$ Subtract z and add 6.

    $11 < z$ or $z > 11$ Divide by 3.

30. $40(8) + 1.5(8)h \geq 360$

    $320 + 12h \geq 360$ Simplify.

    $12h \geq 40$ Subtract 320.

    $h \geq 3\frac{1}{3}$ hours   Divide by 12.

31. $100 + 0.06s \geq 360$

    $0.06s \geq 260$ Subtract 100.

    $s \geq \$4333.34$ Divide by 0.06.

32. $100 + 0.06s > 40(8) + 1.5(5)(8)$

    $100 + 0.06s > 380$ Simplify.

    $0.06s > 280$ Subtract 100.

    $s > \$4666.67$ Divide by 0.06.

THINK CRITICALLY

33. Answers will vary.

34. Possible problem: If Laura received a \$5 rebate for buying two identical pairs of jeans, she would still pay at least \$35. What is the original price of the jeans?

    $2c - 5 \geq 35 \;\rightarrow\; 2c \geq 40 \;\rightarrow\; c \geq 20$

35. Possible problem: One more person was added to each of 3 groups of equal size. Altogether, there were fewer than 18 people. How many people were in each group to start?

    $3(m+1) < 18 \;\rightarrow\; m + 1 < 6 \;\rightarrow\; m < 5$

36. Must consider 3 cases: (I) $a > 0$; (II) $a < 0$; (III) $a = 0$

  I. $ax + b < c$

    $ax < c - b$ Subtract b.

    $x < \dfrac{c-b}{a}$ Divide by a.

  II. $ax + b < c$

    $ax < c - b$ Subtract b.

    $x > \dfrac{c-b}{a}$ Divide by a.

                           Reverse inequality symbol since $a < 0$.

  III. $ax + b < c$

    $0x + b < c$ Substitute 0 for a.

  true for all real number values for x

37. $1144 = \dfrac{995 + 1600 + 875 + 1250 + t}{5}$

    $5720 = 4720 + t$ Multiply by 5.

    $1000 = t$ Subtract 4720.

38. $|-8| = 8$ 39. $-|-8| = -8$ 40. $-|8| = -8$

41. Substitute and solve; B

42. $\dfrac{1}{3}x - 4 > 8$

    $\dfrac{1}{3}x > 12$ Add 4.

    $x > 36$ Multiply by 3.

43. $y + 3 - 3(y - 1) < 4$

    $y + 3 - 3y + 3 < 4$ Distribute.

    $-2y + 6 < 4$ Simplify.

    $-2y < -2$ Subtract 6.

    $y > 1$ Divide by -2 and reverse the inequality symbol.

ALGEBRA WORKS

1. $x + \left(x - \dfrac{1}{8}\right) + \left(x - \dfrac{3}{16}\right) \le 28\dfrac{3}{4}$

2. $x + \left(x - \dfrac{1}{8}\right) + \left(x - \dfrac{3}{16}\right) \le 28\dfrac{3}{4}$

    $3x - \dfrac{5}{16} \le 28\dfrac{3}{4}$ Simplify.

    $3x \le 29\dfrac{1}{16}$ Add $\dfrac{5}{16}$.

    $x \le 9\dfrac{11}{16}$ Divide by 3.

Width of A $\le 9\dfrac{11}{16}$

Width of B $\le 9\dfrac{9}{16}$ $\left(9\dfrac{11}{16} - \dfrac{1}{8}\right)$

Width of C $\le 9\dfrac{1}{2}$ $\left(9\dfrac{11}{16} - \dfrac{3}{16}\right)$

3. Answers will vary depending on magazine.

Lesson 5.5, pages 241–246

EXPLORE/WORKING TOGETHER

1. $c \ge 7$ 2. $c \le 20$

3. no; no; To qualify, both inequalities must be satisfied at the same time.

TRY THESE

1.

2.

3.

4. Impossible, no solution

5.

6. $0 < 2z - 4 < 18$

    $4 < 2z < 22$ Add 4.

    $2 < z < 11$ Divide by 2.

7. $x < 2$ or $x > 2$; could also use $x \ne 2$

8. $x < -1$ or $x > 2$ 9. $0 < x < 4$

10. $x \le -6$ or $x \ge 6$

11. $4 < x + 6 < 10$

    $-2 < x < 4$ Subtract 6.

12. $7 \le 2x - 3 \le 9$

    $10 \le 2x \le 12$ Add 3.

    $5 \le x \le 6$ Divide by 2.

13. $6 + 2x < -12$

    $2x < -18$ Subtract 12.

    $x < -9$ Divide by 2.

    or

    $3x + 8 > 20$

    $3x > 12$ Subtract 8.

    $x > 4$ Divide by 3.

14. $4 - 2x < 6$

    $-2x < 2$ Subtract 4.

    $x > -1$ Divide by -2 and reverse the inequality symbol.

    or

    $3x + 7 < -2$

    $3x < -9$ Subtract 7.

    $x < -3$ Divide by 3.

15.

15a. The graphs have no common points.

   b. $s \le 0$ or $s > 6$

   c. All real numbers less than or equal to zero or greater than six.

16a. $20 \le w \le 22$

   b.

   all basketball weights greater than or equal to 20 oz and less than or equal to 22 oz

1. No, since $9 < 9$ is not true. **2.** Yes, since $h \geq 5$ is true.

3. Yes, since -2 is between -3 and 0.

4. Yes, since $d > -1\frac{1}{2}$ is true.

5. No, since $0 > 2$ and $0 \leq -3$ are both untrue.

6. No, since $0 > 6$ and $0 < 4$ are both untrue.

7. $x < 0$ or $x > 0$; could also use $x \neq 0$

8. $x \leq -2$ or $x \geq 4$ **9.** $0 \leq x \leq 5$ **10.** $x < -3$ or $x > 4$

11.

12.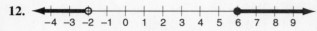

13.

14.

15.

16. Impossible, so no graph

17. $3a + 2 < -13$ or $3a + 5 \geq 2$

   $3a < -15$ or $3a \geq -3$

   $a < -5$ or $a \geq -1$

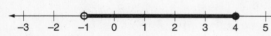

18. $u + 3 > 2$ and $u + 3 \leq 7$

   $u > -1$ and $u \leq 4$

19. $3j + 5 < 20$ or $2j - 1 > 13$

   $3j < 15$ or $2j > 14$

   $j < 5$ or $j > 7$

20. $-3 \leq 2c + 9 \leq 7$

   $-12 \leq 2c \leq -2$

   $-6 \leq c \leq -1$

21. $5 < \frac{1}{2}b - 7 < 9$

   $12 < \frac{1}{2}b < 16$

   $24 < b < 32$

22. $h - 3 < 6$ and $h - 3 > 12$

   $h < 9$ and $h > 15$
Impossible, no graph

23. $50 \leq 7(n + 12) \leq 68$

   $50 \leq 7n + 84 \leq 68$ Distribute.

   $-34 \leq 7n \leq -16$ Subtract 84.

   $-4 \leq n \leq -3$ Divide by 7.

(Note: n represents an integer so, the nearest integer that works is expressed in solution)

24. $-8 < k + 3 < 5$

   $-11 < k < 2$ Subtract 3.

25. $2 - x > -3$

   $-x > -5$ Subtract 2.

   $x < 5$ Multiply by -1 and reverse the inequality symbol.
   or
$3x + 1 > 22$

   $3x > 21$ Subtract 1.

   $x > 7$ Divide by 3.

26. $-15 \leq 2b + 5 \leq 3$

   $-20 \leq -2b \leq -2$ Subtract 5.

   $10 \geq b \geq 1$ or $1 \leq b \leq 10$ Divide by -2 and reverse the inequality symbol.

27. $3 - x > -5$

   $-x > -8$ Subtract 3.

   $x < 8$ Multiply by -1 and reverse the inequality symbol.

   or
$2x + 2 > 28$

   $2x > 26$ Subtract 2.

   $x > 13$ Divide by 2.

28. $2 - 3x > -6$

   $-3x > -8$ Subtract 2.

   $x < \frac{8}{3}$ Divide by -3 and reverse the inequality symbol.

   or
$-x + 8 < 16$

   $-x < 8$ Subtract 8.

   $x > -8$ Multiply by -1 and reverse the inequality symbol.

29. $2 - x < -3$

$\quad -x < -5 \qquad$ Subtract 2.

$\quad x > 5 \qquad$ Multiply by -1 and reverse the inequality symbol.

$\quad$ and

$4 - 3x > -17$

$\quad -3x > -21 \qquad$ Subtract 4.

$\quad x < 7 \qquad$ Divide by -3 and reverse the inequality symbol.

30. Possible answers: If the graph is between two values, the inequality includes *and*; if the graph is to the left of one value and the right of another value, the inequality includes *or*; open and closed circles indicate whether particular values are included or not.

31a. $t < 55°$ or $t > 80°$

b.

temperatures less than 55° or greater than 80°

EXTEND

32. $-5 \le \dfrac{4 - 3m}{2} < 1$

$\quad -10 \le 4 - 3m < 2 \qquad$ Multiply by 2.

$\quad -14 \le -3m < -2 \qquad$ Subtract 4.

$\quad \dfrac{14}{3} \ge m > \dfrac{2}{3}$ or $\dfrac{2}{3} < m \le \dfrac{14}{3} \qquad$ Divide by -3 and reverse the inequality symbol.

33. $0 < 2 - \dfrac{3}{4}d \le \dfrac{1}{2}$

$\quad -2 < -\dfrac{3}{4}d \le -1\dfrac{1}{2} \qquad$ Subract 2.

$\quad \dfrac{8}{3} > d \ge 2$ or $2 \le d < \dfrac{8}{3} \qquad$ Multiply by $-\dfrac{4}{3}$ and reverse the inequality symbol.

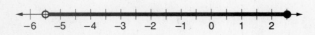

34. $-2 \le \dfrac{3 + 2y}{-4} < 2$

$\quad 8 \ge 3 + 2y > -8 \qquad$ Multiply by -4 and reverse the inequality symbol.

$\quad -8 < 3 + 2y \le 8 \qquad$ Rewrite.

$\quad -11 < 2y \le 5 \qquad$ Subtract 3.

$\quad -\dfrac{11}{2} < y \le \dfrac{5}{2} \qquad$ Divide by 2.

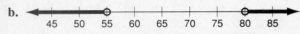

35. $3 - x > -5$

$\quad -x > -8 \qquad$ Subtract 3.

$\quad x < 8 \qquad$ Multiply by -1 and reverse the inequality symbol.

$\quad$ or

$2x - 2 > 28$

$\quad 2x > 30 \qquad$ Add 2.

$\quad x > 15 \qquad$ Divide by 2.

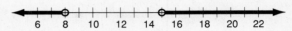

36. $-20 < 5(t + 3) < 5$

$\quad -4 < t + 3 < 1 \qquad$ Divide by 5.

$\quad -7 < t < -2 \qquad$ Subtract 3.

37. $-20 - 2b < -4b < -4 - 2b$

$\quad -20 < -2b < -4 \qquad$ Add $2b$.

$\quad 10 > b > 2$ or $2 < b < 10 \qquad$ Divide by -2 and reverse the inequality symbol.

38a. $15 \le 9c \le 50$

b. $2 \le c \le 5 \qquad$ Divide by 9 (remember: c must be an integer value)

39a. $155 \le w - 30 \le 199$

b. $185 \le w \le 229 \qquad$ Add 30. any weight between 185 and 229

THINK CRITICALLY

40a. Accept any inequality with $\le$ or $\ge$.

b. Accept any compound inequality with *or* and with $\le$ or $\ge$.

c. Accept any compound inequality with *and* and with $\le$ or $\ge$.

d. Accept any compound inequality with *and* with $\le$ and $\ge$, and with the same number such as $x \ge 5$ and $x \le 5$.

e. Accept any compound inequality with *or*, with $\le$ and $\ge$, and with the same number such as $x \ge 5$ and $x \le 5$.

41a. $2x > 5$ and $x + 7 > 6$ and $3x + 2 < 20$
$$3x < 18$$
$$x > \frac{5}{2} \text{ and } x > -1 \text{ and } x < 6$$

b. $6 - x < 9$ or $-3x > 18$ or $-2 > x + 6$
$$-x < 3$$
$$x > -3 \text{ or } x > -6 \text{ or } -8 > x$$
So, $x > -6$ or $x < -8$.

42. There are no solutions to the disjunction under these conditions.

43. all real values of x

MIXED REVIEW
44. Range = greatest value − least value
$$1:25 = s - 3:75 \text{ or } 1:25 = 5:10 - s$$
$$5:40 = s \qquad \text{or} \qquad s = 3:45$$

45. $5b + 8 = -2b - 6$ **46.** $3(4m - 2) = 0.5(8m + 4)$
$$7b = -14 \qquad\qquad\qquad 12m - 6 = 4m + 2$$
$$b = -2 \qquad\qquad\qquad\quad 8m = 8$$
$$m = 1$$

47.

x	y
−4	5
0	−3
2	−7

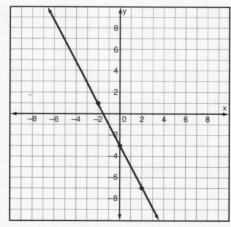

48.

x	y
−3	3
0	4
3	5

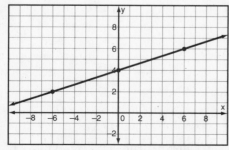

49. None of the values make the inequalities true; D.

ALGEBRAWORKS
1. Possible answer: Full-size color illustrations of paper money could be passed for the real thing or used to make counterfeit printing plates; nobody could pass a paper coin for the real thing.

2, 3. to the nearest tenth of an inch, $y < 4.6$ or $y > 9.2$, $w < 2.0$ or $w > 3.9$; to the nearest tenth of a centimeter, $y < 11.8$ or $y > 23.6$, $w < 5.0$ or $w > 9.9$

4, 5. to the nearest tenth of an inch, $4.6 \le y \le 9.2$, $2.0 \le w \le 3.9$; to the nearest tenth of a centimeter, $11.8 \le y \le 23.6$, $5.0 \le w \le 9.9$

6. Answers will vary.

Lesson 5.6, pages 247-251

EXPLORE/WORKING TOGETHER
1. Ordered data: 21, 21, 22, 22, 23, 25, 25, 26, 30, 32, 34, 35, 37, 40, 44, 52, 54, 55, 58, 60,
$$\text{Median} = \frac{32 + 34}{2} = \frac{66}{2} = 33$$

2. Half the buyers are younger than 33, and half are older.

3. possible answer: No, the manufacturer would have no way to know how much younger or older than the median most buyers are.

TRY THESE
1. Ordered data: 400, 480, 500, 525, 550, 575, 650, 675, 700, 950

Q_1 median: $500

Q_2 median: $\frac{550 + 575}{2} = \frac{1125}{2} = \$562.50$

Q_3 median: $675

2.

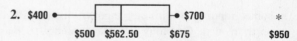

$950 is an outlier. The range is $700 − $400 = $300.

3. $400 \le c < 500$

PRACTICE
1. Ordered data: 23, 23, 24, 24, 24, 25, 25, 26, 26, 27, 29, 33, 33, 33, 42, 48, 48, 48, 50, 65

Q_1 median: $\frac{24 + 25}{2} = 24.5$

Q_2 median: $\frac{27 + 29}{2} = 28$

Q_3 median: $\frac{42 + 48}{2} = 45$

2.

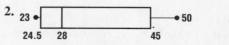

65 is an outlier. The range is $50 - 23 = 27$

3. $45 < p \le 65$

4. Ordered data: 49, 51, 51, 59, 60, 61, 63, 63, 66, **67**, 69, 70, 71, 72, 73, 76, 76, 76, 76

Q_1 median: 60

Q_2 median: 67

Q_3 median: 73

5.

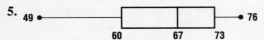

No outliers; range: $76 - 49 = 27$

6. Look for statements such as: The life expectancy in the 25% of the regions with the longest life expectancy is from 73 to 76. This range (73-76) is much less than the range of life expectancies in the 25% of the regions with the shortest life expectancy (49-60).

7. Team A had the highest score of 198.
Team A had the lowest score of 130.

8. Team B with a median of 183

9. Answers will vary. Team B scores were higher in the middle 50% of the data and also higher in the bottom 25%.

EXTEND

10.

11. Ordered data: 3, 3, 3, 3, 3, 3, 3, 3, 4, 4, 4, 4, 4, 4, 5, 5, 5, 5, 6, 6, 7, 7, 7, 8, 8, **8**, 8, 8, 8, 9, 9, 10, 10, 11, 11, 11, 11, 12, 12, 13, 13, 14, 15, 18, 21, 22, 23, 25, 32, 33, 54

Q_1 median: 4

Q_2 median: 8

Q_3 median: 12

12. On some graphing utilities the range will include all values in the data. Graphing the data on others, such as a statistics software program, may show outliers. Graphing by hand would show 54 as an outlier.

13a. Lower 75%: $3 \le v \le 12$
Upper 75%: $4 \le v \le 54$

b. Lower 50%: $3 \le v \le 8$
Middle 50%: $4 \le v \le 12$
Upper 50%: $8 \le v \le 54$

c. $3 \le v \le 54$

14. Possible answer: Concentrate on the 25% of states having greatest number of electoral votes because they have so many more than the other states do.

THINK CRITICALLY

15. d; Temperatures in San Francisco are more consistent and moderate than other places.

16. a; Minneapolis is coldest and temperatures vary widely.

17. c; Temperatures vary, but not as much as Minneapolis.

18. b; Phoenix is hottest of the four choices.

19. Answers will vary. Possible answer: The following set of math scores could be represented by the boxplot: 63, 65, 68, 70, 72, 74, 74, 76, 76, 80, 80, 82, 83, 84, 85, 89, 91, 93, 94, 95.

Lesson 5.7, pages 252–255

EXPLORE THE PROBLEM

1. $100 - a$.

2. Translate "has at least" into $\ge$; $100 - a \ge \frac{2}{3}a$

3. $100 \ge \frac{5}{3}a$
$60 \ge a$ or $a \le 60$, but a is also ≥ 0.
So $0 \le a \le 60$ and is an integer.

4. Fran has at most 60; Stan has at least 40.

5. It is easier to work with $\frac{2}{3}a$ rather than $\frac{2}{3}(100 - a)$ which occurs when a represents the autographs Stan has.

6. Answers will vary.

INVESTIGATE FURTHER

7a. $F > 98.6$

b. $\frac{9}{5}C + 32$; $\frac{9}{5}C + 32 > 98.6$

c. $\frac{9}{5}C > 66.6$
$C > 37$
temperatures greater than 37° C

8a. $C < 15°$ or $C > 24°$

b. $\frac{5}{9}(F - 32°) < 15°$ or $\frac{5}{9}(F - 32°) > 24°$

c. $F - 32° < 27°$ or $F - 32° > 43.2°$
$F < 59°$ or $\quad\quad$ F > 75.2 is unsafe
$59° \le F \le 75.2°$ is safe

APPLY THE STRATEGY

9. $87t + 78t \ge 858$
$165t \ge 858$
$t \ge 5.2$ hours

10. Let $\quad\quad\quad\quad w = $ width
$2w + 25 = $ length
$2w + 2(2w + 25) = $ perimeter

$2w + 2(2w + 25) \le 335$
$2w + 4w + 50 \le 335$
$6w \le 285$
$w \le 47.5$

Width is 47; length is $2(47.5) + 25 = 120$

11. Let r = Red and $360 - r$ = Blue

$$r \le 1\tfrac{1}{2}(360 - r)$$

$$r \le 540 - \tfrac{3}{2}r$$

$$\tfrac{5}{2}r \le 540$$

$$r \le 216,$$

Maxumum of red is 216; Minimum of blue is 144.

12. $86° \le \tfrac{9}{5}C + 32° \le 104°$

$$54° \le \tfrac{9}{5}C \le 72°$$

$$30° \le C \le 40°$$

13. $3700° \le \tfrac{5}{9}(t - 32°) \le 4200°$

$$6660° \le t - 32° \le 7560°$$

$$6692° \le t \le 7592°$$

14. $22a \ge 330$

$$a \ge 15 \text{ gallons}$$

However, the tank is limited to 20 gallons. So, $15 \le a \le 20$.

15. $15 + 17.50m < 20m$

$$15 < 2.50m$$

$$6 < m$$

Since m is an integer, $m \ge 7$.

16a. $1000 - 5p \ge 500$

$$-5p \ge -500$$

$$p \le 100$$

However, the minimum price is 40 cents. So $40 \le p \le 100$.

b. Midpoint $= \dfrac{40 + 100}{2} = 70$ cents

Sales: $1000 - 5(70) = 1000 - 350 = 650$ bagels

REVIEW PROBLEM SOLVING STRATEGIES

1a. Answers will vary. Solving a simpler problem and making a diagram are effective.

b. B; Since A came forward as the shortest student in the column, A is shorter than B.

c. B; B came forward as the tallest student in the row so B is taller than A.

d. Pick a student C in A's column and B's row. A is shorter than C or C would have come forward: B is taller than C or C would have come forward; $A < C$ and $C < B$ means $A < B$.

e. No; since the arrangement is analogous to the original problem, the same reasoning applies, so $B > A$.

2a. Keep one strip intact and cut each of the other strips in a different place (total of six cuts). After the cuts have been made, many arrangements are possible; one example is shown. If each number appears exactly once in each row, column, and diagonal, the row, column and diagonal sums will all be equal.

1	2	3	4	5	6	7
3	4	5	6	7	1	2
5	6	7	1	2	3	4
7	1	2	3	4	5	6
2	3	4	5	6	7	1
4	5	6	7	1	2	3
6	7	1	2	3	4	5

b. The magic sum will be
$1 + 2 + 3 + 4 + 5 + 6 + 7 = 28$.

3a. $31 + 36 - 58 = 9$

b. 31, the smaller of the two

c. $58 - 36 = 22$

Chapter Review, pages 256–257

1. e   **2.** d   **3.** a   **4.** c   **5.** b

6.

7.

8.

9.

10. $<$ **11.** $\le$ **12.** $<$ **13.** $>$

14. $a - (-3) > 8$

$$a + 3 > 8 \qquad \text{Simplify.}$$

$$a > 5 \qquad \text{Subtract 3.}$$

15. $\tfrac{5}{6}x < 5$

$$x < 6 \qquad \text{Multiply by } \tfrac{6}{5}.$$

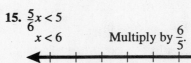

16. $-8 \le \tfrac{4}{5}y$

$$-10 \le y \qquad \text{Multiply by } \tfrac{5}{4}.$$

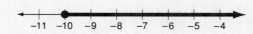

17. $-4.2 > -1.4w$

$$3 < w \qquad \text{Divide by } -1.4 \text{ and reverse the inequality symbol.}$$

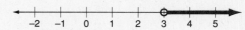

18. $\frac{x}{4} + 2 > 7$

$\frac{x}{4} > 5$ Subtract 2.

$x > 20$ Multiply by 4.

19. $3 - 2b < 15$

$-2b < 12$ Subtract 3.

$b > -6$ Divide by -2 and reverse the inequality symbol.

20. $4x + 2 \leq 47 + x$

$3x \leq 45$ Subtract x and 2.

$x \leq 15$ Divide by 3.

21. $3(r+) > 4(r+1)$

$3r + 6 > 4r + 4$ Distribute.

$2 > r$ or $r < 2$ Subtract $3r$ and 4.

22. $\frac{x}{2} + 3 > 9 + \frac{x}{3}$

$\frac{x}{6} > 6$ Subtract $\frac{x}{3}$ and 3.

$x > 36$ Multiply by 6.

23. $3x + 4 < 4x - 8$

$4 < x - 8$ Subtract $3x$.

$12 < x$ or $x > 12$ Add 8.

24. $2(x+3) > 5x + 5$

$2x + 6 > 5x + 5$ Distribute.

$1 > 3x$ Subtract $2x$ and 5.

$\frac{1}{3} > x$ or $x < \frac{1}{3}$ Divide by 3.

25. $2(y+2) \geq 4(y+3)$

$2y + 4 \geq 4y + 12$ Distribute.

$-8 \geq 2y$ Subtract $2y$ and 12.

$-4 \geq y$ or $y \leq -4$ Divide by 2.

26. $x + 8 > 6$ and $x + 2 < 10$

$x > -2$ and $x < 8$

$-2 < x < 8$

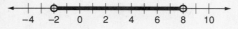

27. $2x - 6 > 8$ or $3x - 4 < 5$

$2x > 14$ or $3x < 9$

$x > 7$ or $x < 3$

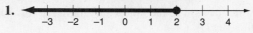

28. $4 + y \leq 1$ or $37 \leq 12 + 5y$

$y \leq -3$ or $25 \leq 5y$

$5 \leq y$

29. $x - 4 > 1$ and $x + 4 < -8$

$x > 5$ and $x < -12$

impossible, so no solution

30. $4x - 7 \geq 13$ or $2x + 8 \leq 6$

$4x \geq 20$ or $2x \leq -2$

$x \geq 5$ or $x \leq -1$

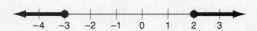

31. $13 + 4x \leq 1$ or $28 \leq 14 + 7x$

$4x \leq -12$ or $14 \leq 7x$

$x \leq -3$ or $2 \leq x$

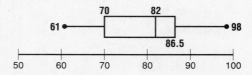

32. $Q_1 = 70$, $Q_2 = 82$, $Q_3 = 86.5$

33. $70 \leq s \leq 86.5$

34. $2(47) + 2w \geq 108$

$94 + 2w \geq 108$

$2w \geq 14$

$w \geq 7$

35. $2w + 2(3w + 15) \leq 94$

$2w + 6w + 30 \leq 94$

$8w \leq 64$

$w \leq 8$

width $= 8$; length $= 3(8) + 15 = 24 + 15 = 39$

Chapter Assessment, pages 258–259

1.

2.

3.

4. $x - 4 < 7$

$\quad x < 11 \qquad$ Add 4.

5. $x + 7 \geq -1$

$\quad x \geq -8 \qquad$ Subtract 7.

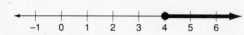

6. $1.3n \geq 5.2$

$\quad n \geq 4 \qquad$ Divide by 1.3.

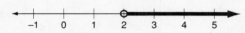

7. $-4n < -8$

$\quad n > 2 \qquad$ Divide by -4 and reverse the inequality symbol.

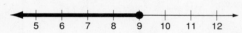

8. $\frac{1}{3}y \leq 3$

$\quad y \leq 9 \qquad$ Multiply by 3.

9. $-\frac{3}{4}x < 6$

$\quad x > -8 \qquad$ Multiply by $-\frac{4}{3}$ and reverse the inequality symbol.

10. $4 - n \leq -2$

$\quad -n \leq -6 \qquad$ Subtract 4.

$\quad n \geq 6 \qquad$ Multiply by -1 and reverse the inequality symbol.

11. $2y + 3 \leq 9$

$\quad 2y \leq 6 \qquad$ Subtract 3.

$\quad y \leq 3 \qquad$ Divide by 2.

12. $9 - 4x < 7$

$\quad -4x < -2 \qquad$ Subtract 9.

$\quad x > \frac{1}{2} \qquad$ Divide by -4 and reverse the inequality symbol.

13. $4(x + 2) < 6(x - 1)$

$\quad 4x + 8 < 6x - 6 \qquad$ Distribute.

$\quad 8 < 2x - 6 \qquad$ Subtract $4x$.

$\quad 14 < 2x \qquad$ Add 6.

$\quad 7 < x$ or $x > 7 \qquad$ Divide by 2.

14. $0.37x + 0.17 > 3.5$

$\quad 0.37x > 3.33 \qquad$ Subtract 0.17.

$\quad x > 9 \qquad$ Divide by 0.37.

15. $4x - 2 < 13 - x$

$\quad 5x < 15 \qquad$ Add x and 2.

$\quad x < 3 \qquad$ Divide by 5.

16. $2x - 2 > 3x + 6$

$\quad -8 > x$ or $x < -8 \qquad$ Subtract $2x$ and 6.

17. $\frac{2}{3}x - 4 > 2$

$\quad \frac{2}{3}x > 6 \qquad$ Add 4.

$\quad x > 9 \qquad$ Multiply by $\frac{3}{2}$.

18. $x + 5 \geq 6(x - 4) + 9$

$\quad x + 5 \geq 6x - 15 \qquad$ Distribute and simplify.

$\quad 20 \geq 5x \qquad$ Subtract x and add 15.

$\quad 4 \geq x$ or $x \leq 4 \qquad$ Divide by 5.

19. $2x + 17 \leq 5(x - 4) - 8$

$\quad 2x + 17 \leq 5x - 28 \qquad$ Distribute and simplify.

$\quad 45 \leq 3x \qquad$ Subtract $2x$ and add 28.

$\quad 15 \leq x$ or $x \geq 15 \qquad$ Divide by 3.

20. Answers will vary. Example: To solve $-3x + 4 > 13$, first subtract 4 from both sides using the subtraction property of inequality. Then divide both sides by -3 using the division property of inequality. Dividing by a negative number changes the direction of the inequality symbol. This leaves the solution $x < -3$.

21. $x + 5 > 4 \quad$ and $\quad x + 7 < 12$

$\quad x > -1 \quad$ and $\qquad x < 5$

$\qquad -1 < x < 5$

22. $3x - 1 > 5 \quad$ or $\quad 2x - 7 < -11$

$\quad 3x > 6 \quad$ or $\qquad 2x < -4$

$\quad x > 2 \quad$ or $\qquad x < -2$

$\qquad x > 2$ or $x < -2$

23. $3 - y \geq 4 \quad$ or $\quad 22 \leq 3y + 4$

$\quad -y \geq 1 \quad$ or $\quad 18 \leq 3y$

$\quad y \leq -1 \quad$ or $\quad 6 \leq y$

$\qquad y \leq -1$ or $6 \leq y$

24. $2x+3>7$ and $x+9<16$

$\qquad 2x>4$ and $\qquad x<7$

$\qquad x>2$ and $\qquad x<7$

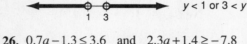

$2<x<7$

25. $5y-2<3$ or $-3<3(y-4)$

$\qquad 5y<5$ or $-3<3y-12$

$\qquad y<1$ or $9<3y$

$\qquad y<1$ or $3<y$

$y<1$ or $3<y$

26. $0.7a-1.3\le3.6$ and $2.3a+1.4\ge-7.8$

$\qquad 0.7a\le4.9$ and $\qquad 2.3a\ge-9.2$

$\qquad a\le7$ and $\qquad a\ge-4$

$-4\le a\le7$

27. $y\ge4$ **28.** $n\le24$

29. $2x+3\le13$ **30.** $11.95\le c\le17.95$

31. $k+(k+5)\ge37$

$\qquad 2k\ge32$

$\qquad k\ge16$

$l=k+5$, so $k=5\ge16+5\rightarrow l\ge21$

Keith worked at least 16 hours and Lisa worked at least 21 hours.

32. $r+(r+4)+[(r+4)+6]\le38$

$\qquad 3r+14\le38$ Simplify.

$\qquad 3r\le24$ Subtract 14.

$\qquad r\le8$ Divide by 3.

33. $n+(n+2)<50$

$\qquad 2n+2<50$ Simplify.

$\qquad 2n<48$ Subtract 2.

$\qquad n<24$ Divide by 2.

Since the greatest even integer less than 24 is 22, the two integers are $n=22$ and $n+2=24$.

34. $97-52=45$ **35.** 77 **36.** Between 60 and 86

37. 60 **38.** 86 **39.** 25% **40.** 25%

PERFORMANCE ASSESSMENT
Use Algeblocks

Students should check their solutions by choosing appropriate values for the variable and substituting unit blocks in the original inequality.

Use Number Lines

a. Answers will vary. Students should indicate that adding the same number shifts the points but does not change order.

b. Students' diagrams should show that the order after subtraction does not change. Sample diagram:

$1<3$

subtract 3

$-2<0$

Inequalities in Advertising

$Chronicle>Times$	$Chronicle<Times$
$155d>585+90d$	$155d<585+90d$
$65d>585$	$65d<585$
$d>9$	$d<9$

For $d<9$, *The Chronicle* is less expensive. After 9 days, *The Chronicle* is more expensive. Select the more costly newspaper if its readership matches your desired market or if it has a larger circulation.

Magazine Analysis

Answers will vary.

Cumulative Review, page 260

1.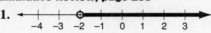

2.

3. $y+5>8$ **4.** $2x\le-12$

$\qquad y>3$ Subtract 5. $\qquad x\le-6$ Divide by 2.

5. $4(3-c)\le-4$

$\qquad 12-4c\le-4$ Distribute.

$\qquad -4c\le-16$ Subtract 12.

$\qquad c\ge4$ Divide by -4 and reverse the inequality symbol.

6. total of 4, 2 miles going and 2 miles returning

7. about 15 minutes **8.** about 40 minutes to get home; B

9. $6*\dfrac{4}{5}=\dfrac{24}{5}=4.8$ **10.** $5*6-4=30-4=26$

11.

12. $4(x-12)-3(5x+3)+30$

$4x-48-15x-9+30$ Distribute.

$-11x-27$ Simplify.

13. $-3c+\dfrac{2}{3}(6c-5)-\dfrac{1}{4}(12c-2)$

$-3c+4c-\dfrac{10}{3}-3c+\dfrac{1}{2}$ Distribute.

$-2c-\dfrac{17}{6}$ Simplify.

14. $3x-7<5$ and $5x+11>1$

     $3x<12$ and $5x>-10$

      $x<4$ and $x>-2$

          $-2<x<4$

15. $7-2n\le 13$ or $4(n+6)\le 8$

    $-2n\le 6$ or $4n+24\le 8$

     $n\ge -3$ or $4n\le -16$

     $n\ge -3$ or $n\le -4$

16. $-9\cdot 8=-17$ **17.** $-9-(-8)=72$

18. $-38-(-145)=-38+145=107$

19. $(-4)(-5)-6=20-6=14$

20. $24.95+0.10x\le 50.00$

        $0.10x\le 25.05$

          $x\le 250.5$ miles

21. $6n-2=-32$

        $6n=-30$

         $n=-5$

22. $x+(x+2)>72$

      $2x+2>72$

      $2x>70$

       $x>35$

Since first odd integer greater than 35 is 37, the two integers are 37 and 39.

23. $p-0.20p=27$

      $0.80p=27$

        $p=33.75$

Standardized Test, 261

1. C; $-4x>-12$

     $x<3$ Divide by -4 and reverse the inequality symbol.

2. E   **3.** B

4. I; not true, because the median would be a data point which must be a whole number.

II; true, since you average the two middle points if number of data values is even.

III; There is not enough information to know mean.

IV; true, since the mode must be a data point and 7.5 is not a whole number; D

5. D   **6.** $(-2)^4-2^4=16-16=0$; A

7. Since the y-coordinate is 0, it is on x-axis; D

8. A; by substitution   **9.** C

10. $\dfrac{3}{8}=0.375$ or 37.5%; B

Chapter 6   Linear Functions and Graphs

Data Activity, pages 262-264

1. $[(114.6 - 100.3) + (126.0 - 114.6) + (128.7 - 126.0) +$

   $(133.7 - 128.7)] \div 4 = \dfrac{33.4}{4} = 8.35$

   $8.35 billion; $183.8 billion. Answers will vary. The prediction may be high because revenue increased more slowly. The prediction may be low because revenue increased more rapidly.

2. $[(967 - 857) + (1047 - 967) + (1052 - 1047) +$

   $(1073 - 1052)] \div 4 = \dfrac{216}{4} = 54$

   54 thousand; 1505 thousand. Answers will vary. The prediction may be too high because employment increased more slowly. The prediction may be low because employment increased more rapidly.

3. $\dfrac{13.3}{126.0} = 0.10\overline{5}$ or 10.6%

4. 133.7 billion $= 1.337 \times 10^{11}$

   1073 thousand $= 1.073 \times 10^{6}$

5. Answers will vary. Since revenue is in billions and employment is in thousands, it would not be possible to construct a useful vertical scale so both sets of data could be graphed.

Lesson 6.1, pages 265-267

THINK BACK

1. Graphs of b, c, and d will pass through (2, 3). Answers will vary, but may include solving graphically or substituting the values 2 and 3 for x and y, respectively, to see whether those values satisfy the equation.

2. Points a and b lie on the graph. Possible answers: Substitute the x- and y-values into the equation to determine whether they satisfy it.

EXPLORE

3.

4. The graphs should look the same.

5. All equations have the same graph.

6. Solve each equation for y in terms of x to see they are equivalent.

7. Possible answers: The graphs are all parallel; they all slant from lower left to upper right; the lines are not equidistant from one another; each graph intersects the y-axis at the point named by the last term of its equation.

8. Possible answer: This graph will be parallel to and between $y = 2x + 1$ and $y = 2x + 5$. Since the other graphs are all parallel, and the form of this equation, including the coefficient of the x-terms, is the same as the others, it appears that this line will be parallel to the others. The last constant term, which shows where the graph crosses the y-axis, seems to determine the position of the graph in relation to the others.

9. The y-coordinate increases by 2 for every 1 unit increase in x.

10. not parallel since y changes by 5 for each 1 unit change in x

MAKE CONNECTIONS

11. Answers will vary but should include the fact that two lines are parallel when both graphs have the same change in y per unit change in x. Students may also note that when the coefficient of y in both equations is 1, the x-coefficients are equal.

12. The lines appear to be parallel, and in each equation, the x-coefficient is 3 and the y-coefficient is 1.

13. The lines are not parallel. Students should note that although the y-coefficient in both equations is 1, the x-coefficient in the first question is 4 and in the second is 2; thus the changes in y per unit change in x are not equal.

14. Graph a shows a decreasing graph from left to right.

15. In $y = 4x$ the y-value increases 4 units for each 1-unit increase in x; in $y = -4x$ the y-value decreases 4 units for each 1-unit increase in x. The x-coefficients are not the same, so the lines cannot be parallel but must intersect.

16a. & b. $y = 3x + 1$

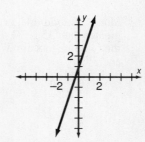

x	y	y + 4
0	1	5
1	4	8
−1	−2	2

   c. Each point $(x, y + 4)$ is 4 units directly above (x, y).

   d. parallel with one graph 4 units above the other

17. The measurements can be used to determine that staircase B is steeper, since for every unit change in the horizontal, staircase B has a greater vertical change.

18. Design *a* is the steepest. It has the greatest change of *y* for a 1-unit change in *x*.

SUMMARIZE

19 . Use integral values of *x* to compute the corresponding *y*-values. If the difference between *y*-values for each equation is the same, the graphs will be parallel.

20. Graph *b* is the steepest increase, *a* has equal changes in *y* as *x* changes, and *c* decreases from left to right.

21. $y = 10x - 1$ since the coefficient of *x* is greater

Lesson 6.2, pages 268-275

EXPLORE/WORKING TOGETHER

1. The order of steepness is A, C, B, D, F, E.

2. Compare the number of units a slope falls with the number of units it moves horizontally.

3. Answers will vary.

TRY THESE

Problems 1-6 are done by counting rise units and dividing by run units.

1. *k*   **2.** *l*   **3.** *j*   **4.** *t*   **5.** *s*   **6.** *r*

7. $\dfrac{3-0}{6-0} = \dfrac{3}{6} = \dfrac{1}{2}$ **8.** $\dfrac{4-1}{5-2} = \dfrac{3}{3} = 1$

9. $\dfrac{6-6}{2-(-8)} = \dfrac{0}{10} = 0$ **10.** $\dfrac{10-(-5)}{4-1} = \dfrac{15}{3} = 5$

11. $\dfrac{7-(-1)}{4-4} = \dfrac{8}{0}$ undefined   **12.** $\dfrac{-6-(-1)}{3-(-2)} = \dfrac{-5}{5} = -1$

13. Line *a*: $\dfrac{8-1}{6-3} = \dfrac{7}{3}$ Line *b*: $\dfrac{7-2}{0-(-5)} = \dfrac{5}{5} = 1$

Line *a* is steeper since $\left|\dfrac{7}{3}\right| > |1|$.

14. Line *c*: $\dfrac{9-2}{-1-2} = \dfrac{7}{-3}$ Line *d*: $\dfrac{8-1}{3-7} = \dfrac{7}{-4}$

Line *c* is steeper since $\left|-\dfrac{7}{3}\right| > \left|-\dfrac{7}{4}\right|$.

15. Line *e*: $\dfrac{12-4}{4-12} = \dfrac{8}{-8} = -1$

Line *f*: $\dfrac{-2-(-8)}{-3-5} = \dfrac{6}{-8} = -\dfrac{3}{4}$

Line *e* is steeper since $|-1| > \left|-\dfrac{3}{4}\right|$.

16. $\dfrac{14,100,000 - 12,400,000}{1993 - 1983} = 170,000$ vehicles per year

17. Answers will vary, but should include a statement that one of the planes climbs more quickly than the other, and students should give the altitude of each plane at a particular time following takeoff.

PRACTICE

1. $\dfrac{4-2}{5-1} = \dfrac{2}{4} = \dfrac{1}{2}$ **2.** $\dfrac{5-2}{4-1} = \dfrac{3}{3} = 1$

3. $\dfrac{9-1}{2-5} = -\dfrac{8}{3}$ **4.** $\dfrac{0-(-3)}{0-(-7)} = \dfrac{3}{7}$

5. $\dfrac{-2-1}{1-2} = \dfrac{-3}{-1} = 3$ **6.** $\dfrac{6-(-3)}{1.6-1.6} = \dfrac{9}{0}$ undefined

7. $\dfrac{14-(-4)}{-7-(-1)} = \dfrac{18}{-6} = -3$ **8.** $\dfrac{2-2}{4-(-2)} = \dfrac{0}{6} = 0$

9. $\dfrac{13-10}{-3-(-3)} = \dfrac{3}{0}$ undefined

10. Line *a*: $\dfrac{16-2}{12-6} = \dfrac{14}{6} = \dfrac{7}{3}$ Line *b*: $\dfrac{6-0}{1-(-3)} = \dfrac{6}{4} = \dfrac{3}{2}$

Line *a* is steeper since $\left|\dfrac{7}{3}\right| > \left|\dfrac{3}{2}\right|$.

11. Line *c*: $\dfrac{7-3}{-1-3} = \dfrac{4}{-4} = -1$ Line *d*: $\dfrac{-1-(-5)}{8-0} = \dfrac{4}{8} = \dfrac{1}{2}$

Line *c* is steeper since $|-1| > \left|\dfrac{1}{2}\right|$.

12. Line *k*: $\dfrac{7-2}{5-1} = \dfrac{5}{4}$ Line *l*: $\dfrac{6-1}{0-(-4)} = \dfrac{5}{4}$

Lines *k* and *l* have equal steepness since $\left|\dfrac{5}{4}\right| = \left|\dfrac{5}{4}\right|$.

13. Line *p*: $\dfrac{8-0}{-2-1} = \dfrac{8}{-3}$ Line *q*: $\dfrac{8-0}{2-6} = \dfrac{8}{-4} = -2$

Line *p* is steeper since $\left|-\dfrac{8}{3}\right| > |-2|$.

14. $(1,4)$ and $(4,2)$; slope $= \dfrac{4-2}{1-4} = -\dfrac{2}{3}$

15. Points will vary; slope $= \dfrac{3-1}{2-0} = \dfrac{2}{2} = 1$

16. $(2,4)$ and $(-1,-1)$; slope $= \dfrac{4-(-1)}{2-(-1)} = \dfrac{5}{3}$

17. $\dfrac{45,000,000 - 4,500,000}{1988 - 1970} = \dfrac{40,500,000}{18} =$ 2,250,000 homes per year

18. $\dfrac{30,400,000 - 26,100,000}{1988 - 1980} = \dfrac{4,300,000}{8} =$ 537,500 licenses per year

19. $\dfrac{3.5 - 2.9}{1988 - 1960} = \dfrac{0.6}{28} = 0.02$ lb/yr

20. $\dfrac{139,000,000 - 61,700,000}{1987 - 1960} = \dfrac{77,300,000}{27} \approx$ 2,862,963 cars per year

21. positive, since as *x* goes from -2 to 7, *y* increases

22. negative, since as *x* goes from -4 to 0, *y* decreases

23. positive, since as *x* goes from -2 to 12, *y* increases

24. negative, since as *x* goes from -9 to 2, *y* decreases

25. zero, since *y* does not change

26. positive, since as x goes from -14 to -5, y increases

27.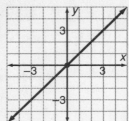

28.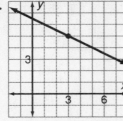

29.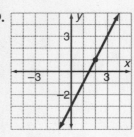

30.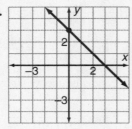

31. $\dfrac{1}{3} = \dfrac{435}{x}$

$x = 3 \cdot 435 = 1305 \text{ ft}$

32. $\dfrac{2}{5} = \dfrac{x}{1305}$

$\dfrac{2 \cdot 1305}{5} = x$

$522 = x$

33. $0.25 = \dfrac{177}{x}$

$x = \dfrac{177}{0.25} = 708$

$x = 708' \text{ from center to edge}$

So, the length of each base is $2x$ or 1416 ft.

34. Following each turn, slope will still be 2.

wheel   half-turn   full turn

EXTEND

35. Except for 1987 to 1988, the slope for each year is positive. The negative slope for 1987 to 1988 indicates that the number of flights decreased.

36. Increase was greatest from 1982-84; its slope is the steepest positive slope

37. $\dfrac{14,400,000 - 9,000,000}{1992 - 1982} = \dfrac{5,400,000}{10} =$

540,000 flights per year

38. Possible answer: A straight line connecting the endpoints of the first half (1982-87) would be steeper than one connecting the endpoints of the second half. Therefore, the rate of change was greater during the first half.

39. $0.75 \text{ miles} = 0.75(5280 \text{ ft}) = 3960 \text{ ft}$

$\text{Slope} = \dfrac{\text{Rise}}{\text{Run}} = 0.07 = \dfrac{7}{100}$

$\dfrac{7}{100} = \dfrac{x}{3960} \rightarrow x = \dfrac{7.3960}{100} = 277.2$

So, the elevation at the beginning of the road is $2500 - 277.2$ or 2222.8 ft

THINK CRITICALLY

40a. Square $= 8 \cdot 8 = 64$ square units

Rectangle $= 5 \cdot 13 = 65$ square units

b. The rectangle cannot exist as drawn. Two parts of the diagonal have different slopes; one part has a slope of $\dfrac{3}{8}$. The other has a slope of $\dfrac{2}{5}$. The "diagonal" is not a straight line.

MIXED REVIEW

41. $(-3)^2 - 5(-3)(-5) = 9 - 5(-3)(-5) = 9 - 75 = -66$

42. $\dfrac{(-3)(-5) - 4^2}{7} = \dfrac{(-3)(-5) - 16}{7} = \dfrac{15 - 16}{7} = \dfrac{-1}{7}$

43. $-\dfrac{5}{5}\left[(-3)^2(4)\right] = -\dfrac{5}{5}[9(4)] = -(-1)(36) = 36$

44. linear, as x increases by 1, y always increases by 2

45. nonlinear, as x increases by 1, y increases at a nonconstant rate

46. nonlinear, as x increases by 2, y increases at a nonconstant rate

47. B; $3.5 + 2.5x \le 23$

$\quad\quad 2.5x \le 19.5$

$\quad\quad\quad x \le 7.8$

ALGEBRA WORKS

1. No; possible answer: the bill both increases and decreases from month to month.

2. High heating costs in winter and cooling costs in summer

3. Mean is $246.83, rounded to $247 per month

4. a horizontal line with equation $y = 247$

5. The slope would be zero indicating no change from month to month.

EXPLORE

1.

Equation	In "$y =$" form	Slope	$y =$ intercept
$y = 3x + 4$	$y = 3x + 4$	$\frac{4-7}{0-1} = 3$	4
$2y = 8x - 10$	$y = 4x - 5$	$\frac{-5-(-1)}{0-1} = 4$	-5
$x + y = 6$	$y = -x + 6$	$\frac{6-5}{0-1} = -1$	6
$y = x$	$y = x$	$\frac{0-1}{0-1} = 1$	0

2. When the equation is in "$y =$" form the slope is the coefficient of x and the y-intercept is the constant.

TRY THESE

1. $3; -2$ **2.** $-2; 1$ **3.** $3; 3$ **4.** $-3; -1$

5.

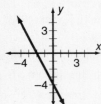

6.

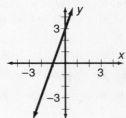

7.

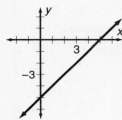

8.

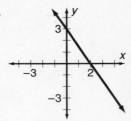

9. $m = 4, b = 8$ **10.** $m = -\frac{3}{2}, b = 3$

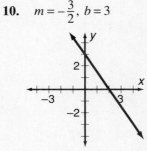

11. $m = \frac{5}{2}, b = -5$ **12.** $m = -\frac{1}{2}, b = 4$

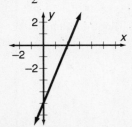

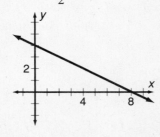

13. $\frac{1}{20}$ **14.** $\frac{-5}{3}$ **15.** $-\frac{1}{1.25} = -0.8$ **16.** 6

17. $y = \frac{2}{3}x - 6$ Slope is $\frac{2}{3}$.

$2x - 4 = 3y \rightarrow \frac{2}{3}x - \frac{4}{3} = y$ Slope is $\frac{2}{3}$.

Slopes are equal, so the lines are parallel.

18. $y = \frac{1}{2}x + 4$ Slope is $\frac{1}{2}$.

$4x - 3y = 12 \rightarrow y = \frac{4}{3}x - 4$ Slope is $\frac{4}{3}$.

Slopes are not equal and are not negative reciprocals, so the lines are neither parallel nor perpendicular.

19. $2y = 10x - \frac{2}{5} \rightarrow y = 5x - \frac{1}{5}$ Slope is 5.

$\frac{1}{5}x + y = 3 \rightarrow y = -\frac{1}{5}x + 3$ Slope is $-\frac{1}{5}$.

The slopes are negative reciprocals, so the lines are perpendicular.

20. $m = 4, b = -3, y = 4x - 3$

21. $m = \frac{1}{3}, \; b = 1, \; y = \frac{1}{3}x + 1$

22. $m = -2, \; b = -\frac{1}{2}, \; y = -2x - \frac{1}{2}$

23. $m = \frac{4}{5}, \; b = 0, \; y = \frac{4}{5}x$

24. The slope is 1.175; it represents the approximate annual growth in the number of tons recycled (1.175 million tons); the T-intercept is 8; it represents the number of tons that were recycled in 1960 (8 million tons).

25. Possible answer: Find the slope-intercept equation of the line: $y = -2x + 6$. Then substitute 20 for x into the equation to find $y = -34$.

PRACTICE

1.

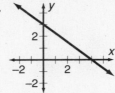

2.

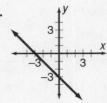

3.

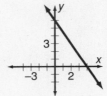

4.

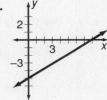

5. $y - 7 = -x$

$\quad y = -x + 7$

$\quad m = -1,\ b = 7$

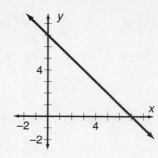

6. $5y = -2x + 10$

$\quad y = -\dfrac{2}{5} + 2$

$\quad m = -\dfrac{2}{5},\ b = 2$

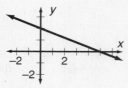

7. $x + 2y = 3$

$\quad 2y = -x + 3$

$\quad y = -\dfrac{1}{2}x + \dfrac{3}{2}$

$\quad m = -\dfrac{1}{2},\ b = \dfrac{3}{2}$

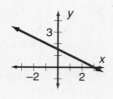

8. $5x - y = 15$

$\quad -y = -5x + 15$

$\quad y = 5x - 15$

$\quad m = 5,\ b = -15$

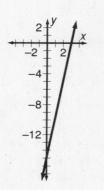

9. $y = x - 1,\ m = 1,\ b = -1$; c

10. $y = x + 1,\ m = 1,\ b = 1$; d

11. $y = -x + 1,\ m = -1,\ b = 1$; b

12. $y = -x - 1,\ m = -1,\ b = -1$; a

13. $m = -2.5,\ b = 0,\ y = -2.5x$

14. $m = 3.1,\ b = -0.5,\ y = 3.1x - 0.5$

15. $m = -\dfrac{7}{8},\ b = \dfrac{3}{5},\ y = -\dfrac{7}{8}x + \dfrac{3}{5}$

16. $m = 9.1,\ b = -1.7,\ y = 9.1x - 1.7$

17. $m = 3$, so the slope of parallel line is 3.

18. $7y = 5x - 3 \rightarrow y = \dfrac{5}{7}x - \dfrac{3}{7}$

$\quad m = \dfrac{5}{7}$ so, the slope of the parallel line is $\dfrac{5}{7}$.

19. $x - 3y = 9 \rightarrow y = \dfrac{1}{3}x - 3$

$\quad m = \dfrac{1}{3}$ so, the slope of the parallel line is is $\dfrac{1}{3}$.

20. $4 - 2x = 6y \rightarrow y = -\dfrac{1}{3} + \dfrac{2}{3}$

$\quad m = -\dfrac{1}{3}$ so, the slope of the parallel line is $-\dfrac{1}{3}$.

21. $m = 1$ so, the slope of the perpendicular line is -1.

22. $m = -\dfrac{1}{2}$ so, the slope of the perpendicular line is 2.

23. $2y - x = 5 \rightarrow y = \dfrac{1}{2}x + \dfrac{5}{2}$

$\quad m = \dfrac{1}{2}$ so, the slope of perpendicular line is -2.

24. $3x + 2y = 4 \rightarrow y = -\dfrac{3}{2}x + 2$

$\quad m = -\dfrac{3}{2}$ so, the slope of perpendicular line is $\dfrac{2}{3}$.

25a. Slope describes average annual change in sales; s-intercept indicates sales for the first year, 1983.

b. In 1986, $x = 3$ so $s = 0.5(3) + 1.8 = 1.5 + 1.8 = 3.3$ differs by $3.3 - 2.93 = 0.37$ billion or $370 million

26a. Both x-coordinates are zero, so the line is on the y-axis.

b. Both y-coordinates are zero, so the line is on the x-axis.

EXTEND

27. $m = 3$, so $y = 3x + b$

$\quad 0 = 3(0) + b$ Substitute $(0, 0)$ in equation.

$\quad b = 0$

$\quad y = 3x$

28. $x - 3y = 9 \rightarrow y = \dfrac{1}{3}x - 3$

$\quad m = \dfrac{1}{3}$, so $y = \dfrac{1}{3}x + b$

$\quad 3 = \dfrac{1}{3}(6) + b$ Substitute $(6, 3)$ in the equation.

$\quad b = 1$

$\quad y = \dfrac{1}{3}x + 1$

29. $2y = -4x + 7 \rightarrow y = -2x + \dfrac{7}{2}$

$\quad m = -2$, so $y = -2x + b$

$\quad -5 = -2(1) + b$ Substitute $(1, -5)$ in the equation.

$\quad b = -3$

$\quad y = -2x - 3$

30. $m = 1$, perpendicular slope is -1, so $y = -x + b$

$\quad 3 = -1(1) + b$ Substitute $(1, 3)$ in the equation.

$\quad b = 4$

$\quad y = -x + 4$

31. $2y = 3x + 5 \rightarrow y = \frac{3}{2}x + \frac{5}{2}$

$m = \frac{3}{2}$, perpendicular slope is $-\frac{2}{3}$,

so $y = -\frac{2}{3}x + b$

$0 = -\frac{2}{3}(-3) + b$ Substitute $(-3,0)$ in the equation.

$b = -2$

$y = -\frac{2}{3}x - 2$

32. $\frac{1}{2}x = 1 - y \rightarrow y = -\frac{1}{2}x + 1$

$m = -\frac{1}{2}$, perpendicular slope is 2, so $y = 2x + b$

$-5 = 2(1) + b$ Substitute $(1,-5)$ in the equation.

$b = -7$

$y = 2x - 7$

33a. Slopes must show two distinct pairs of parallel sides.

b. No, no two lines have the same slope.

34. In 1900 there were approximately 83 accidental deaths per 100,000 and the rate has been dropping at an average of 0.5 deaths per year.

35. the sub's depth the moment it begins surfacing

36. length of time required to surface

37. Yes; possible answer: the difference between y_2 and y_1 will be negative, while the difference between x_2 and x_1 will be positive.

THINK CRITICALLY

38. Counting rise over run yields $m = 2$

39. $(-4, 2)$

40. $-\frac{1}{2}$

41. Slopes are negative reciprocals; product is -1

42a. The lines are distinct and parallel.

b. The lines are not parallel but share the same point on the y-axis.

43. Possible answer: Write the first equation, then write a second equation whose slope is the negative reciprocal of the first. Write a third by taking the negative reciprocal of the second or by using the same slope as the first. Complete the rectangle by writing a fourth equation whose slope is equal to that of the second equation. The equations will have at least two different y-intercepts.

MIXED REVIEW

44. Edgar, since his ratio 7:4 or $\frac{7}{4}$ was greatest

45. 4 out of 7 which is $\frac{4}{7}$ or about 57%

46. $\frac{2}{2+3} = \frac{2}{5}$ or 40%

47. She watches television $\frac{3}{1+3} = \frac{3}{4}$ of the time, so

$\frac{3}{4} \cdot 6 = 4\frac{1}{2}$ hours

48. distributive property of multiplication over addition

49. associative property of addition

50. commutative property of multiplication

51. additive inverse property

52. $\frac{1}{2}x - 7 = 14 \rightarrow \frac{1}{2}x = 21 \rightarrow x = 42$; D

ALGEBRAWORKS

1. $\text{BTU} = \frac{\text{WHILE}}{60}$

2. $\text{BTU} = (20 \cdot 9 \cdot 18 \cdot 18 \cdot 18) \div 60 = \frac{1,049,760}{60} =$

$17,496 \approx 17,500$ BTU's

3. $L_1 = R - 0.05R \rightarrow L_1 = 0.95R$

4. $L_2 = R + 0.05R \rightarrow L_2 = 1.05R$

5. no, because the slopes are not equal

6. Answers will vary.

Lesson 6.4, pages 285-292

EXPLORE/WORKING TOGETHER

1. exactly one line   **2.** infinite number

3. infinite number   **4.** infinite number

5. exactly one line

6. Points are not colinear, so no line passes through those 3 points.

7. infinite number   **8.** exactly one line

9. No lines, since to be both parallel and perpendicular, slopes would have to be equal and negative reciprocals

10. exactly one line   **11.** Question 1, $y = 8x + 3$

TRY THESE

1. $y - 8 = 0.5(x - 9) \rightarrow y = 0.5x + 3.5$

2. $y - 1 = 2[x - (-4)] \rightarrow y = 2x + 9$

3. $y - 5 = -2(x - 1) \rightarrow y = -2x + 7$

4. $y - 3 = \frac{1}{3}(x - 3) \rightarrow y = \frac{1}{3}x + 2$

5. $y - (-6) = 0[x - (-2)] \rightarrow y = -6$

6. $y - 2 = 1.5(x - 4) \rightarrow y = 1.5x - 4$

7. $m = \frac{4 - 1}{5 - 3} = \frac{3}{2}$

$y - 1 = \frac{3}{2}(x - 3) \rightarrow y = \frac{3}{2}x - \frac{7}{2}$

8. $m = \dfrac{0-(-6)}{3-(-3)} = \dfrac{6}{6} = 1$

$y - 0 = 1(x-3) \rightarrow y = x - 3$

9. $m = \dfrac{-1-(-4)}{-2-(-4)} = \dfrac{3}{2}$

$y - (-1) = \dfrac{3}{2}[x-(-2)] \rightarrow y = \dfrac{3}{2}x + 2$

10. $m = \dfrac{4-(-2)}{1-4} = \dfrac{6}{-3} = -2$

$y - 4 = -2(x-1) \rightarrow y = -2x + 6$

11. $y - 4 = -3(x-1)$

$\quad y - 4 = -3x + 3$ Distribute.

$\quad\quad y = -3x + 7$ Add 4.

12. $y - (-2) = 4[x-(-2)]$

$\quad y + 2 = 4(x+2)$ Simplify.

$\quad y + 2 = 4x + 8$ Distribute.

$\quad\quad y = 4x + 6$ Subtract 2.

13. $y - 1 = \dfrac{1}{2}[x-(-2)]$

$\quad y - 1 = \dfrac{1}{2}(x+2)$ Simplify.

$\quad y - 1 = \dfrac{1}{2}x + 1$ Distribute.

$\quad\quad y = \dfrac{1}{2}x + 2$ Add 1.

14. $y - (-3) = \dfrac{2}{3}[x-(-2)]$

$\quad y + 3 = \dfrac{2}{3}(x+2)$ Simplify.

$\quad y + 3 = \dfrac{2}{3}x + \dfrac{4}{3}$ Distribute.

$\quad\quad y = \dfrac{2}{3}x - \dfrac{5}{3}$ Subtract 3.

15. $m = -0.5$, so, $y = -0.5x + b$

$17 = -0.5(40) + b$ Substitute (40, 17) in the equation.

$37 = b$

$y = -0.5x + 37$

16. Answers will vary. The rate of decrease may have changed. As the percent continued to decrease, it is not likely that the relation would have remained linear.

PRACTICE

1. $-0.5 = -2(-1) + b$

$\quad -0.5 = 2 + b$

$\quad -2.5 = b$

2. $7 = 3(-2) + b$

$\quad 7 = -6 + b$

$\quad 13 = b$

3. $4.5 = 2(-3) + b$

$\quad 4.5 = -6 + b$

$\quad 10.5 = b$

4. $0 = -7(-1) + b$

$\quad 0 = 7 + b$

$\quad -7 = b$

5. $y - 2 = -4(x-0) \rightarrow y = -4x + 2$

6. $y - (-4) = \dfrac{1}{2}(x-1) \rightarrow y = \dfrac{1}{2}x - 4\dfrac{1}{2}$

7. $y - 5 = -1.5[x-(-5)] \rightarrow y = -1.5x - 2.5$

8. $y - 2 = \dfrac{5}{2}[x-(-4)] \rightarrow y = \dfrac{5}{2}x + 12$

9. $y - (-0.5) = 6(x-3) \rightarrow y = 6x - 18.5$

10. $y - 3 = -0.1(x-2) \rightarrow y = -0.1x + 3.2$

11. $m = \dfrac{2-1}{-3-2} = -\dfrac{1}{5}$

$y - 2 = -\dfrac{1}{5}[x-(-3)]$

$y - 2 = -\dfrac{1}{5}x - \dfrac{3}{5}$ Simplify and distribute.

$\quad\quad y = -\dfrac{1}{5}x + \dfrac{7}{5}$ Add 2.

12. $m = \dfrac{7-1}{5-(-1)} = \dfrac{6}{6} = 1$

$y - 7 = 1(x-5)$

$y - 7 = x - 5$ Distribute.

$\quad\quad y = x + 2$ Add 7.

13. $m = \dfrac{4-(-2)}{-6-(-4)} = \dfrac{6}{-2} = -3$

$y - 4 = -3[x-(-6)]$

$y - 4 = -3x - 18$ Simplify and distribute.

$\quad\quad y = -3x - 14$ Add 4.

14. $m = \dfrac{3-(-1)}{0-(-1)} = \dfrac{4}{1} = 4$

$y - 3 = 4(x-0)$

$y - 3 = 4x$ Distribute.

$\quad\quad y = 4x + 3$ Add 3.

15. $m = \dfrac{70-40}{90-30} = \dfrac{30}{60} = \dfrac{1}{2}$

$y - 70 = \dfrac{1}{2}(x-90) \rightarrow y = \dfrac{1}{2}x + 25$

16. $m = -30$

The 1988 point is $(8, 1141)$.

$y - 1141 = -30(x-8) \rightarrow y = -30x + 1381$

17. $m = \dfrac{-31-(-3)}{5-25} = \dfrac{-28}{-20} = 1.4$

$y - (-31) = 1.4(x-5) \rightarrow y = 1.4x - 38$

18. Yes; if the wind chill is $-46°F$, the actual temperature is approximately $-6°F$; for a temperature of $-20°F$, the wind chill would be approximately $-66°F$.

19. Answers will vary.

20. $m = \dfrac{8-8}{7-(-2)} = \dfrac{0}{9} = 0$

$y - 8 = 0[x - (-2)] \rightarrow y = 8$

21. $m = \dfrac{1-(-2)}{-6-6} = \dfrac{3}{-12} = -\dfrac{1}{4}$

$y - 1 = -\dfrac{1}{4}[x - (-6)] \rightarrow y = -\dfrac{1}{4}x - \dfrac{1}{2}$

22. $m = \dfrac{2-(-4)}{-9-3} = \dfrac{6}{-12} = -\dfrac{1}{2}$

$y - 2 = -\dfrac{1}{2}[x - (-9)] \rightarrow y = -\dfrac{1}{2}x - \dfrac{5}{2}$

23. $m = \dfrac{0-(-8)}{2-1} = \dfrac{8}{1} = 8$

$y - 0 = 8(x - 2) \rightarrow y = 8x - 16$

24. $m = \dfrac{1-(-1)}{0.5-2.5} = \dfrac{2}{-2} = -1$

$y - 1 = -1(x - 0.5) \rightarrow y = -x + 1.5$

25. $m = \dfrac{0-(-2.5)}{5-3} = \dfrac{2.5}{2} = 1.25$

$y - 0 = 1.25(x - 5) \rightarrow y = 1.25x - 6.25$

26. $m = \dfrac{1.5-(-5.5)}{1.5-(-2)} = \dfrac{7}{3.5} = 2$

$y - 1.5 = 2(x - 1.5) \rightarrow y = 2x - 1.5$

27. $m = \dfrac{-1-(-2.5)}{6-4} = \dfrac{1.5}{2} = 0.75$

$y - (-2.5) = 0.75(x - 4) \rightarrow y = 0.75x - 5.5$

28. $m = \dfrac{6-(-4)}{-2.2-(-4.2)} = \dfrac{10}{2} = 5$

$y - 6 = 5[x - (-2.2)] \rightarrow y = 5x + 17$

29. $y - 2 = -\dfrac{7}{2}(x + 1)$ **30.** $y = -\dfrac{7}{2}x - \dfrac{3}{2}$

31a. $m = 10$ December 94 is point (11, 185) on the graph.

$y - 185 = 10(x - 11) \rightarrow y = 10x + 75$

b. April 95 is 15 months after January 93.

$y = 10(15) + 75 = 150 + 75 = 225$ pairs

THINK CRITICALLY

32. (2, 1) and (−4, −5)

33. Write both in slope-intercept form: $y = x - 1$

34. Slope is 1 so perpendicular slope is −1

$y - 1 = -1(x - 2)$ and $y + 5 = -1(x + 4)$

35. $y - 1 = -1(a - 2) \rightarrow y = -a + 3$

$y + 5 = -1(a + 4) \rightarrow y = -a - 9$

MIXED REVIEW

36. $\sqrt{21} - \sqrt{8} \approx 4.58 - 2.83 = 1.75$

37. $\sqrt{21 - 8} = \sqrt{13} \approx 3.61$

38. $\sqrt{12} + \sqrt{8} \approx 3.46 + 2.83 = 6.29$

39. $\sqrt{12 + 8} = \sqrt{20} \approx 4.47$

40. $|4| = 4$ **41.** $\left|-\dfrac{1}{3}\right| = \dfrac{1}{3}$ **42.** $|0| = 0$

43. $|-9| = 9$ **44.** $-|-2.4| = -2.4$

45. $-\dfrac{3}{8} + 2\dfrac{4}{5} = -\dfrac{3}{8} + \dfrac{14}{5} = -\dfrac{15}{40} + \dfrac{112}{40} = \dfrac{97}{40} = 2\dfrac{17}{40}$

46. $-3\dfrac{1}{4} - \left(-8\dfrac{2}{5}\right) = -\dfrac{13}{4} + \dfrac{42}{5} = -\dfrac{65}{20} + \dfrac{168}{20} = \dfrac{103}{20} = 5\dfrac{3}{20}$

47. $6\dfrac{8}{10} - \left(-2\dfrac{1}{6}\right) = \dfrac{68}{10} + \dfrac{13}{6} = \dfrac{204}{30} + \dfrac{65}{30} = \dfrac{269}{30} = 8\dfrac{29}{30}$

48. $x + 18 \geq 3x + 6$, $12 \geq 2x$, $6 \geq x$ or $x \leq 6$;   D

ALGEBRAWORKS

1. Possible answer: Using the points (20, 3.65) and (28, 4.12), the equation $y = 0.06x + 2.45$ results.

2.–5. In all cases, answers will vary slightly depending on which (x, y) pairs students used to determine slope and y-intercept. In Question 5, students should remember to divide by 2000 to convert pounds to tons, once they have found a total.

Lesson 6.5, pages 293-298

EXPLORE

1. Activity, no answer.   **2.** Activity, no answer.

3.

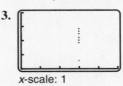

x-scale: 1
y-scale: 10

4. 7 points; Numerous data points are repeated. There are only 7 distinct points.

5. $\dfrac{87 - 85}{3 - 3} = \dfrac{2}{0}$, undefined

6.

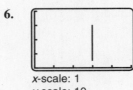

x-scale: 1
y-scale: 10

7. yes, at the point (3, 67)

8. No, it would be (4, 87).

9. Undefined slope cannot be written in $y = mx + b$ form.

TRY THESE

1. $y = 8x + 5$

$-8x + y = 5$ Subtract $8x$.

2.
$$2y = 5x - 1$$
$$-5x + 2y = -1 \qquad \text{Subtract } 5x.$$

3.
$$3x = 2y$$
$$3x - 2y = 0 \qquad \text{Subtract } 2y.$$

4.
$$y = \frac{1}{2}x + 4$$
$$2y = x + 8 \qquad \text{Multiply by 2.}$$
$$-x + 2y = 8 \qquad \text{Subtract } x.$$

5. $y - 9 = 0$
$$y = 9 \qquad \text{Add 9.}$$

6. $y - 2 = 3(x - 4)$
$$y - 2 = 3x - 12 \qquad \text{Distribute.}$$
$$-3x + y = -10 \qquad \text{Subtract } 3x \text{ and add 2.}$$

7. $x + 7 = 0$
$$x = -7 \qquad \text{Subtract 7.}$$

8.
$$y = -\frac{2}{3}x + \frac{1}{2}$$
$$6y = -4x + 3 \qquad \text{Multiply by 6.}$$
$$4x + 6y = 3 \qquad \text{Add } 4x.$$

9.
$$3y = \frac{1}{2}x - 2$$
$$6y = x - 4 \qquad \text{Multiply by 2.}$$
$$-x + 6y = -4 \qquad \text{Subtract } x.$$

10. $m = \dfrac{5-3}{3-(-1)} = \dfrac{2}{4} = \dfrac{1}{2}$
$$y - 5 = \frac{1}{2}(x - 3)$$
$$2y - 10 = x - 3 \qquad \text{Multiply by 2.}$$
$$-x + 2y = 7 \qquad \text{Subtract } x \text{ and add 10.}$$

11. $m = \dfrac{9-7}{2-4} = \dfrac{2}{-2} = -1$
$$y - 9 = -1(x - 2)$$
$$y - 9 = -x + 2 \qquad \text{Distribute.}$$
$$x + y = 11 \qquad \text{Add } x \text{ and 9.}$$

12. $m = \dfrac{1-(-2)}{-1-(-4)} = \dfrac{3}{3} = 1$
$$y - 1 = 1[x - (-1)]$$
$$y - 1 = x + 1$$
$$-x + y = 2 \qquad \text{Subtract } x \text{ and add 1.}$$

13. $m = \dfrac{7-(-1)}{-3-(-3)} = \dfrac{8}{0}$, undefined slope
$$x = -3$$

14. $m = \dfrac{-2-(-2)}{6-(-2)} = \dfrac{0}{8} = 0$
$$y - (-2) = 0(x - 6)$$
$$y = -2 \qquad \text{Distribute and subtract 2.}$$

15. $m = \dfrac{3-(-3)}{3-(-3)} = \dfrac{6}{6} = 1$
$$y - 3 = 1(x - 3)$$
$$-x + y = 0 \qquad \text{Subtract } x \text{ and add 3.}$$

16. $x = 3$

17. $y = -3$

18. $x = -3$

19a. $60x + 90y = 1260$

b.

60-minute tapes, x	0	6	12	15	21
90-minute tapes, y	14	10	6	4	0

20. Sample answers: Slope-intercept form may be easier because the y-intercept can be determined immediately; standard form may be easier because both intercepts can be determined from the given equation without transposing the variable x from one side of the equation to the other.

PRACTICE

1. $y - 2 = -1[x - (-2)]$
$$y - 2 = -x - 2 \qquad \text{Simplify and distribute.}$$
$$x + y = 0 \qquad \text{Add } x \text{ and 2.}$$

2. $y - 1 = \dfrac{1}{2}(x - 4)$
$$2y - 2 = x - 4 \qquad \text{Multiply by 2.}$$
$$-x + 2y = -2 \qquad \text{Subtract } x \text{ and add 2.}$$

3. $m = \dfrac{3-(-3)}{3-(-1)} = \dfrac{6}{4} = \dfrac{3}{2}$
$$y - 3 = \frac{3}{2}(x - 3)$$
$$2y - 6 = 3x - 9 \qquad \text{Multiply by 2 and distribute.}$$
$$-3x + 2y = -3 \qquad \text{Subtract } 3x \text{ and add 6.}$$

4. $m = \dfrac{7-2}{5-0} = \dfrac{5}{5} = 1$
$$y - 7 = 1(x - 5)$$
$$-x + y = 2 \qquad \text{Subtract } x \text{ and add 7.}$$

5. $y - 2 = -2(x - 8)$
$$y - 2 = -2x + 16 \qquad \text{Distribute.}$$
$$2x + y = 18 \qquad \text{Add } 2x \text{ and 2.}$$

6. $y - 1 = -\dfrac{1}{2}[x - (-2)]$
$$2y - 2 = -x - 2 \qquad \text{Multiply by 2 and distribute.}$$
$$x + 2y = 0 \qquad \text{Add } x \text{ and 2.}$$

7. $m = \dfrac{-3-(-3)}{3-0} = \dfrac{0}{3} = 0 \rightarrow y = -3$

8. $m = 0 \rightarrow y = 2$

9. $m = \frac{-1-(-3)}{-1-(-2)} = \frac{2}{1} = 2$

$y-(-1) = 2[x-(-1)]$

$y+1 = 2x+2$ Simplify and distribute.

$-2x+y = 1$ Subtract $2x$ and 1.

10. $m = \frac{7-(-5)}{4-4} = \frac{12}{0}$, undefined slope

$x = 4$

11. Slope is $m = \frac{0-0}{3-(-7)} = \frac{0}{10} = 0 \rightarrow y = 0$

12. $m = \frac{11-1}{0-0} = \frac{10}{0}$, undefined slope

$x = 0$

13a. $x + 2y = 18$

b.

x	0	2	4	6	12	18
y	9	8	7	6	3	0

14. Answers will vary; possible answer: Let a equal number of avocados and m equal number of mangos.

Equation: $1a + 2m = 24$

EXTEND

15. $y = -0.2x + 1.5$

$10y = -2x + 15$ Multiply by 10.

$2x + 10y = 15$ Add $2x$.

16. $y = 1.4x - 2.8$

$5y = 7x - 14$ Multiply by 5.

$-7x + 5y = -14$ Subtract $7x$.

17. $y = 3.6x + 8$

$5y = 18x + 40$ Multiply by 5.

$-18x + 5y = 40$ Subtract $18x$.

18. $y = 0.25x - 1.25$

$4y = x - 5$ Multiply by 4.

$-x + 4y = -5$ Subtract x.

19. $y = -1.35x + 0.15$

$20y = -27x + 3$ Multiply by 20.

$27x + 20y = 3$ Add $27x$.

20. $y = -2.44x - 1$

$25y = -61x - 25$ Multiply by 25.

$61x + 25y = -25$ Add $61x$.

21. $3y = 4.5x - 1.75$

$12y = 18x - 7$ Multiply by 4.

$-18x + 12y = -7$ Subtract $18x$.

22. $2.5y = -1.5x + 2$

$5y = -3x + 4$ Multiply by 2.

$3x + 5y = 4$ Add $3x$.

23. $0.2y = 3x - 15$

$y = 15x - 75$ Multiply by 5.

$-15x + y = -75$ Subtract $15x$.

24. $x = 2.5y - 3$

$2x = 5y - 6$ Multiply by 2.

$2x - 5y = -6$ Subtract $5y$.

25. $2x = 15 - 3.2y$

$10x = 75 - 16y$ Multiply by 5.

$10x + 16y = 75$ Add $16y$.

26. $-x = 2.4 + 3y$

$5x = -12 - 15y$ Multiply by -5.

$5x + 15y = -12$ Add $15y$.

27. Using average rate of change, slope is approximately 0.03 or 0; the corresponding y-intercept is approximately 15.8; equation in standard form is $y = 15.8$.

28. $m = \frac{1.00 - 0.25}{1960 - 1938} = \frac{0.75}{22} \approx \$0.03$ per year

29. $m = \frac{4.25 - 1.00}{1993 - 1960} = \frac{3.25}{33} \approx \$0.10$ per year

Change was $\frac{0.10}{0.03}$ or about 3 times greater.

30. $w = 1.60$ since the line is horizontal

31. $w = 3.35$ from 1980 to 1990

32. 6 amounts; each horizontal segment represents a different wage

THINK CRITICALLY

33. horizontal: $y = -5$;   vertical: $x = 2$

34. $y = -3$

35. $x = 1$ and $x = -1$ are both 1 unit from y-axis

36. $x = 4$

37. $y = -8$

38. horizontal: $y = 0$ vertical: $x = 0$

MIXED REVIEW

39. 12, 16, 17, 19, 21, 24, 28, 29; median $= \frac{19 + 21}{2} = 20$; B

40. $0.35(125) = 43.8$

41. $x(120) = 250 \rightarrow x = \frac{250}{120} = 2.08\overline{3}$ or 208.3%

42. $221 = 0.24x \rightarrow \frac{221}{0.24} = x \rightarrow x \approx 920.8$

43. $0.12\,(18) = 2.2$

44. $-2x - 18 = -44$

$-2x = -26$ Add 18.

$x = 13$ Divide by -2.

45. $45 - 3x = -75$

$\quad\quad -3x = -120$ \quad\quad Subtract 45.

$\quad\quad\quad\quad x = 40$ \quad\quad\quad Divide by -3.

46. $8x - 3(2x + 1) = -9$

$\quad 8x - 6x - 3 = -9$ \quad\quad Distribute.

$\quad\quad\quad\quad 2x = -6$ \quad\quad\quad Add 3 and simplify.

$\quad\quad\quad\quad\quad x = -3$ \quad\quad\quad Divide by 2.

47. $\dfrac{1.35 - 0.10}{0.10} = \dfrac{1.25}{0.10} = 12.5$ or 1250%

Lesson 6.6, pages 299-306

EXPLORE

1. In general, the more TV watched, the lower the score.

2. Independent: hours of TV watched
Dependent: test score

3. They decreased.

4. $\bar{x} =$

$\dfrac{1+3+0+1+2+2+3+4+2+2+1+3}{12} = \dfrac{24}{12} = 2.0$

$\bar{y} =$

$\dfrac{88+71+87+80+78+76+70+68+84+90+93+74}{12} =$

$\dfrac{959}{12} \approx 79.9$

5. Activity, so no solution

6. Answers will vary, but there should be about the same number of points above and below the line.

7. an approximation of the graph of a linear equation that describes this set of data

TRY THESE

1. More cars mean more emissions, so positive correlation

2. Higher weight generally means less mileage, so negative correlation

3. Taller generally means heavier, so positive correlation

4. Salary does not depend on hand size, so no correlation

5. As temperature increases snow ski sales decrease, so negative correlation

6. Possible answer: Using calculator for linear regression, equation is

$y = 823x + 5917$. In 1998, $x = 23$ so income should be

$y = 823(23) + 5917 = 18,929 + 5917 = \$24,846$

7. $30,000 < 823x + 5917$

$24,083 < 823x$

$29.26 < x$ so income will pass $30,000 in 2005
(1975 + 30)

8. Answers will vary. Possible choice is to let x represent the number of years past 1987.

PRACTICE

1. As x increases, y decreases, so negative correlation

2. As x increases, y increases, so positive correlation

3. Graph is nonlinear, so correlation is not appropriate

4. No apparent pattern, so zero correlation

5. As x increases, y increases, so positive correlation

6. As x increases, y decreases, so negative correlation

7. $m = \dfrac{33-14}{6.1-2.1} = \dfrac{19}{4} = 4.75$ \quad **8.** $y - 8 = 4.1(x-10)$

$\quad y - 14 = 4.75(x - 2.1)$ \quad\quad\quad\quad $y - 8 = 4.1x - 41$

$\quad y - 14 = 4.75x - 9.975$ \quad\quad\quad\quad $y = 4.1x - 33$

$\quad\quad\quad y = 4.75x + 4.025$

9. Points are very close to linear with a positive slope, so choose a; 0.99

10. Answers will vary.

EXTEND

11. negative; as the number of cans redeemed increases, the number in trash bin decreases.

12. $y = -x + 601.19$

13. $r = -0.997$. Points are negatively correlated and lie very close to the line.

14. For each can redeemed, one less is found in trash.

15. $y = -(298) + 601.19 \approx 303$ cans

16. Predicted is $y = -210 + 601.19 \approx 391$.
Differs by $400 - 391 = 9$ cans

17.

Week, x	1	2	3	4	5	6	7	8
Redeemed, y	324	309	310	288	523	509	210	251

Using this data set and a calculator, $r = -0.044$, almost no relationship.

18. The equation for the line of best fit is $y = -28.7x + 1368.6$ with variables rounded to the nearest tenth. With $x = 0$ representing 1983 and 2015 representing $x = 32$, the equation gives a value of 450 for the number of evening newspapers that will be published in 2015. In the earlier exercise, using slope and one point, the equation is $y = -30x + 1381$. Using $x = 0$ to represent 1980, the equation gives a value of 331 for the year 2015.

THINK CRITICALLY

19. Predicted from equation is

$y = 2.07(70) + 4.26 = 149.16$

Difference is $149.16 - 149 = 0.16$

% difference is $\dfrac{0.16}{149} \approx 0.001$ or 0.1%

20. Predicted from equation is
$y = 2.07(98) + 4.26 = 207.12$

Difference is $210 - 207.12 = 2.88$

% difference is $\frac{2.88}{210} \approx 0.014$ or 1.4%

This point is farther from the line of best fit than for 70°.

21. $\begin{bmatrix} 1+2 & 0+(-2) & 1+3 \\ 4+(-4) & 3+(-3) & -1+5 \\ 7+(-3) & -5+9 & -3+(-1) \end{bmatrix} = \begin{bmatrix} 3 & -2 & 4 \\ 0 & 0 & 4 \\ 4 & 4 & -4 \end{bmatrix}$

22. $\begin{bmatrix} 8-6 & -3-(-4) & 5-1 \\ 2-3 & -1-(-2) & 4-6 \\ -5-2 & 9-(-1) & -6-3 \end{bmatrix} = \begin{bmatrix} 2 & 1 & 4 \\ -1 & 1 & -2 \\ -7 & 10 & -9 \end{bmatrix}$

23. $(3^2) - 2(3)(-2) + (-2)^2 = 9 - 2(3)(-2) + 4 =$
$9 + 12 + 4 = 25$

24. $(12)^2 + 5(3) - (3)(-2)(12) + (3)(12) =$
$144 + 5(3) - (3)(-2)(12) + (3)(12) =$
$144 + 15 + 72 + 36 = 267$

25. $(3)^4 - (3)^2(-2) + 2(3)(-2)^2 + (-2)(12) =$
$81 - 9(-2) + 2(3)(4) + (-2)(12) = 81 + 18 + 24 - 24 = 99$

26. $4 + x > 12$
$\quad x > 8 \qquad$ Subtract 4.

27. $4x - 4 < x + 5$
$\quad 3x < 9 \qquad$ Subtract x and add 4.
$\quad x < 3 \qquad$ Divide by 3.

28. $\quad 2.2x \le 3x - 4$
$-0.8x \le -4 \qquad$ Subtract $3x$.
$\quad x \ge 5 \qquad$ Divide by -0.8 & reverse inequality.

29. A: $m = \frac{4-0}{2-(-5)} = \frac{4}{7}$

B: $m = \frac{2-2}{7-(-3)} = \frac{0}{10} = 0$

C: $m = \frac{2-(-4)}{-1-3} = \frac{6}{-4} = -\frac{3}{2}$

D: $m = \frac{6-(-2)}{-3-(-7)} = \frac{8}{4} = 2$

30. $m = 0.25, b = -25$

31. $m = 4.2, b = 0.3$

32. $m = 1, b = 0$

33. $m = -1, b = -1$

34. Parallel: $m = 1.2, b = 7 + 12 = 19$
$y = 1.2x + 19$

Perpendicular: $m = -\frac{1}{1.2} = -\frac{5}{6}$, point is $(-2, 4.6)$

$y - 4.6 = -\frac{5}{6}(x + 2)$

$y - 4.6 = -\frac{5}{6}x - \frac{5}{3}$

$y = -\frac{5}{6}x + 2\frac{14}{15}$

Lesson 6.7, pages 307-311

EXPLORE

1. Activity, so no solution

2.

$y = 2.4x$			$y = 2.4x + 2$		
x	y	$\frac{y}{x}$	x	y	$\frac{y}{x}$
1	2.4	2.4	1	4.4	4.4
3	7.2	2.4	3	9.2	3.0$\overline{6}$
5	12.0	2.4	5	14.0	2.8
7	16.8	2.4	7	18.8	2.7

3. In the first chart, the third column is constant. In the second chart, the third column varies.

4. Graphs will vary depending on the value chosen.

5. The third column in the new chart will also be different for each row.

6. The quotient $\frac{y}{x}$ will be a constant only when the equation is in the form $y = kx$ where k is a non zero real number.

TRY THESE

1. $y = kx$
$15 = 12k$
$k = \frac{15}{12} = \frac{5}{4}$
So, $y = \frac{5}{4}x$.
When $x = 40$, $y = \frac{5}{4}(40) = 50$.

2. $y = kx$
$10 = 25k$
$k = \frac{10}{25} = \frac{2}{5}$
So, $y = \frac{2}{5}x$.
When $x = 18$, $y = \frac{2}{5}(18) = \frac{36}{5} = 7.2$.

3. $y = kx$

$14 = 50k$

$k = \dfrac{14}{50} = \dfrac{7}{25}$

So, $y = \dfrac{7}{25}x$.

If $x = 85$, $y = \dfrac{7}{25}(85) = 23.8$.

4.

x	y	$\dfrac{y}{x}$
4	7	$\dfrac{7}{4}$
6	9	$\dfrac{3}{2}$
10	15	$\dfrac{3}{2}$
18	27	$\dfrac{3}{2}$

Not constant, so no direct variation.

5.

x	y	$\dfrac{y}{x}$
84	67.2	0.8
60	48.0	0.8
48	38.4	0.8
20	16.0	0.8

Constant, so $k = 0.8$ and $y = 0.8x$.

6.

x	y	$\dfrac{y}{x}$
0.3	1.2	4
0.7	2.8	4
12	48.0	4
21	84.0	4

Constant, so $k = 4$ and $y = 4x$.

7. no, since equation cannot be written in the form $y = kx$

8. yes, since equation is in the form $y = kx$

9. yes, since equation is equivalent to $y = \dfrac{9}{2}x$ which is of the form $y = kx$

10. no, since equation cannot be written in the form $y = kx$

11. yes, since equation is in the form $y = kx$

12. no, since equation cannot be written in the form $y = kx$

13. no, since equation cannot be written in the form $y = kx$

14. no, since equation cannot be written in the form $y = kx$

15. $y = kx$

$51 = 3k$

$k = \dfrac{51}{3} = 17$

So, $y = 17x$

If $x = 10$, $y = 17 \cdot 10 = 170$ words in 10 minutes

16. $y = 55x$, where x is hours, y is miles. Distance varies directly with time.

1. $14 = k \cdot 8$

$k = \dfrac{14}{8} = \dfrac{7}{4}$, so $y = \dfrac{7}{4}x$ or $1.75x$

2. $9 = -3k$

$k = -\dfrac{9}{3} = -3$, so $y = -3x$

3. $12 = 5k$

$k = \dfrac{12}{5}$, so $y = \dfrac{12}{5}x$ or $2.4x$

4. $-6 = -8k$

$k = \dfrac{-6}{-8} = \dfrac{3}{4}$, so $y = \dfrac{3}{4}x$ or $0.75x$

5. $6 = 1.5k$

$k = \dfrac{6}{1.5} = 4$, so $y = 4x$

6. $7 = 2.5k$

$k = \dfrac{7}{2.5} = 2.8$, so $y = 2.8x$

7. $-1.8 = 9k$

$k = -\dfrac{1.8}{9} = -0.2$, so $y = -0.2x$

8. $3 = -0.5k$

$k = -\dfrac{3}{0.5} = -6$, so $y = -6x$

9. $-3 = 4k$

$k = -\dfrac{3}{4}$, so $y = -\dfrac{3}{4}x$ or $-0.75x$

10. $1 = 1.5k$

$k = \dfrac{1}{1.5} = 0.\overline{6}$, so $y = 0.\overline{6}x$ or $\dfrac{2}{3}x$

When $y = 8$, $8 = \dfrac{2}{3}x$ or $12 = x$

11. $-5 = -8k$

$k = \dfrac{-5}{-8} = \dfrac{5}{8}$, so $y = \dfrac{5}{8}x$

When $x = 18$, $y = \dfrac{5}{8}(18) = \dfrac{90}{8} = 11.25$

12. $12 = 16k$

$k = \dfrac{12}{16} = \dfrac{3}{4}$, so $y = \dfrac{3}{4}x$

When $x = 48$, $y = \dfrac{3}{4}(48) = 36$

13. $4 = -10k$

$k = -\dfrac{4}{10} = -0.4$, so $y = -0.4x$

When $y = 14$, $14 = -0.4x$ and $x = -35$

14. $13.5 = -4.5k$

$k = -\dfrac{13.5}{4.5} = -3$, so $y = -3x$

When $y = 24$, $24 = -3(x)$ and $x = -8$

15a. $3.05 = 2.5k$

$k = \dfrac{3.05}{2.5} = 1.22$

b. $y = 1.22(3) = 3.66$

c. $11.34 = 6k$

$k = \dfrac{11.34}{6} = 1.89$, so $y = 1.89x$

When $x = 4$, $y = 1.89(4) = 7.56$

16. $C = \pi d$, the constant is π, approximately 3.14

17. 5-lb bag: $1.79 = 5k$

$k = \dfrac{1.79}{5} = 0.358$

8-lb bag: $2.76 = 8k$

$k = \dfrac{2.76}{8} = 0.345$

Constants are unequal, so no direct variation

18. $5 = 7k$

$k = \dfrac{5}{7}$, so $y = \dfrac{5}{7}k$

When $x = 11$, $y = \dfrac{5}{7}(11) \approx 7.9$ cm

19. Actual number = 149; estimated from equation = 149.16; percent difference is 0.1%

EXTEND

20. $550 = 40k$

$k = \dfrac{550}{40} = \dfrac{55}{4}$, so $V = \dfrac{55}{4}T$

When $T = 25°$, $V = \dfrac{55}{4}(25) = 343.75$ ml

21. $66.5 = 175k$

$k = \dfrac{66.5}{175} = 0.38$

22. $J = 2.54E$

So, if the weight on Jupiter is 381,

then $381 = 2.54E$.

$E = \dfrac{381}{2.54} = 150$ lbs on earth

23a. for A, $h = 3$ and $S = 24$, so $S/h = 8$;
for B, $h = 6$ and $S = 66$, so $S/h = 11$;
The height does not vary directly as surface area.

b. for A, $h = 3$ and $V = 6$, so $V/h = 2$;
for B, $h = 6$ and $V = 21$, so $V/h = 3.5$;
The height does not vary directly as volume.

THINK CRITICALLY

24. yes; only functions of the type $y = kx$ show direct variation and (0, 0) satisfies $y = kx$ for all k, so the graph passes through (0,0).

25. no, an undefined slope indicates a vertical line and means that k, the constant of variation, will be undefined.

26. no, if the slope is 0 then $k = 0$, in which case y must be 0 and cannot vary directly as x

Lesson 6.8, pages 312-315

EXPLORE THE PROBLEM

1. Activity so no solution   **2.** Activity so no solution

3. Answers will vary. The first graph makes it appear that admissions have been rising slowly; the second graph makes it appear that admissions have increased dramatically.

4. the second graph; answers will vary; someone who wants to show that the park isn't doing very well

5. $\dfrac{516,633 - 512,376}{30} = \dfrac{4257}{30} = 141.9$

Slopes are actually the same but the second graph appears steeper because of the choice of scale.

6. Answers will vary. The scale on the second graph has been altered so that a small range of numbers is spread much farther apart. This exaggerates the increase in admissions for each month.

7. Answers will vary. Check the range and interval size on the vertical scale; check that intervals are equal.

INVESTIGATE FURTHER

8. Height of tallest bar is approximately 4 times that of the shortest

9. 1,600,000; 1,400,000; 1,250,000; 1,300,000; 1,400,000; 1,200,000; 1,800,000; 1,600,000; 1,050,000; 1,500,000

10. Greatest is 1,800,000; least is 1,000,000; Greatest is about 1.8 times as much as the least.

11. Because a portion of the graph from 0 to 0.9 was omitted, the amount of vertical space for the remaining part of the bar was disproportionate to the amount of money represented.

12. Answers will vary. Students may choose a vertical scale from $0 to $2,000,000 with no breaks in increments of $200,000.

13a.

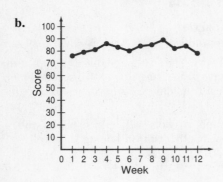

b.

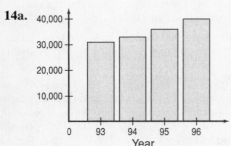

14a.

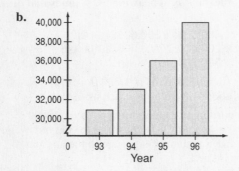

b.

15a. The height of each pizza is correct using the vertical scale but because the radius has been doubled, then tripled from 1994, the area of the 1995 pizza is 4 times the 1994 pizza, and the 1996 pizza is 9 times the area. To the casual observer, the picture may be misleading.

b. Answers will vary. One way is to draw pizzas with radii of about 1 cm, 1.41 cm, and 1.73 cm so the areas will be in the ratio 1:2:3. Omit the vertical scale and label each pizza with the sales figure.

16. Answers will vary.

REVIEW PROBLEM SOLVING STRATEGIES

1. Answers will vary.

a. $1 = \dfrac{44}{44}$

$2 = 4 + 4 - 4 + \sqrt{4}$ or $\dfrac{4}{4} + \dfrac{4}{4}$

$3 = (4 + 4 + 4) \div 4$ or $\dfrac{4}{\sqrt{4}} + \dfrac{4}{4}$

$4 = 4 - 4 + \sqrt{4} + \sqrt{4}$ or $\left(4 \cdot 4 \div \sqrt{4}\right) - 4$

$5 = \sqrt{4} + \sqrt{4} + \dfrac{4}{4}$

$6 = (4 + 4 + 4) \div \sqrt{4}$ or $4\sqrt{4} - 4 + \sqrt{4}$

   or $4 \cdot 4 \div \sqrt{4} - \sqrt{4}$

$7 = 4 + 4 - \dfrac{4}{4}$ or $4\sqrt{4} - \dfrac{4}{4}$ or $\dfrac{44}{4} - 4$

$8 = (4 + 4) \cdot 4 \div 4$ or $\dfrac{4}{4}\left(4\sqrt{4}\right)$

   or $\sqrt{4} + \sqrt{4} + \sqrt{4} + \sqrt{4}$

   or $4 \cdot 4 \div 4 + 4$

$9 = 4\sqrt{4} + \dfrac{4}{4}$ or $\dfrac{44}{4} - \sqrt{4}$ or $4 + \dfrac{4}{4} + 4$

$10 = 4 + 4 + 4 - \sqrt{4}$ or $\left(4 \cdot 4 \div \sqrt{4}\right) + \sqrt{4}$

   or $\dfrac{44 - 4}{4}$

$11 = \left(44 \div \sqrt{4}\right) \div \sqrt{4}$

$12 = 4\left[\sqrt{4} + (4 \div 4)\right]$ or $\left(\dfrac{4}{\sqrt{4}} \cdot 4\right) + 4$

   or $\left(\sqrt{4} \cdot 4\sqrt{4}\right) - 4$

$13 = \dfrac{44}{4} + \sqrt{4}$

$14 = 4 + 4 + 4 + \sqrt{4}$ or $(4 \cdot 4) - 4 + \sqrt{4}$

$15 = 4 \cdot 4 - \dfrac{4}{4}$ or $\dfrac{44}{4} + 4$

$16 = (4 \cdot 4) + 4 - 4$ or $4 + 4 + 4 + 4$

   or $4\sqrt{4} + 4\sqrt{4}$ or $4 \cdot 4 \cdot 4 \div 4$

$17 = (4 \cdot 4) + \dfrac{4}{4}$ or $\left(\sqrt{4}\right)^4 + \dfrac{4}{4}$

$18 = (4 \cdot 4) + 4 - \sqrt{4}$ or $\dfrac{44}{\sqrt{4}} - 4$

   or $\left(\sqrt{4}\right)^4 + 4 - \sqrt{4}$

b. $19 = 4! - 4 - \dfrac{4}{4}$

$20 = (4 \cdot 4) + \sqrt{4} + \sqrt{4}$ or $\left(4 + \dfrac{4}{4}\right) \cdot 4$

or $\dfrac{44}{\sqrt{4}} - \sqrt{4}$ or $\left(4\sqrt{4} \cdot \sqrt{4}\right) + 4$ or $\dfrac{44 - 4}{\sqrt{4}}$

$21 = \dfrac{44 - \sqrt{4}}{\sqrt{4}}$

$22 = \sqrt{4}\left(\dfrac{44}{4}\right)$ or $4 \cdot 4 + 4 + \sqrt{4}$

$23 = \dfrac{44 + \sqrt{4}}{\sqrt{4}}$

$24 = 4 \cdot 4 + 4 + 4$ or $\dfrac{44 + 4}{\sqrt{4}}$ or $\dfrac{44}{\sqrt{4}} + \sqrt{4}$

$25 = \dfrac{\sqrt{4} + \sqrt{4}}{0.4(0.4)}$ or $4! + \sqrt{4} - \dfrac{4}{4}$

$26 = \dfrac{44}{\sqrt{4}} + 4$

$27 = 4! + \sqrt{4} + \dfrac{4}{4}$

$28 = 4\left(4\sqrt{4}\right) - 4$

$29 = 4! + 4 + \dfrac{4}{4}$

$30 = 4\left(4\sqrt{4}\right) - \sqrt{4}$

2a. No, Nina must cover 17 km in 1 h and she walks at a rate of 5 km/h, so she walks less than 5 km.

b. She walks less than 3 km. Since 3 km is $\dfrac{3}{5}$ of the distance she can walk in an hour, it would take her $\dfrac{3}{5}$ of an hour, or 36 min, to walk 3 km. This would leave only 24 min for the bus ride. Even if the bus came as soon as she stopped walking, it could only cover 12 km in 24 min $\left(\dfrac{24}{60} = \dfrac{x}{30}\right)$. Since 12 km plus 3 km equals 15 km, 24 min will not be enough to finish the 17-km trip.

c. Since a bus comes along every 6 min, Nina should allow 6 min waiting time.

d. Nina should get on the bus after walking 2 km. Since 2 km is $\dfrac{2}{5}$ of the distance she can walk in 1 h, it will take her $\dfrac{2}{5}$ h, or 24 min, to walk 2 km. The 24 min she walks and 6 min waiting time leave 30 min for the bus ride. Since the bus travels 30 km/h, it will take 30 min for the bus to travel 15 km. Nina will get to her office in 1 h or less, depending on how long she has to wait for the bus.

3a. To find the true statement, assume for each statement in turn that it is true and that the other three statements are false. This is impossible in all cases except one: only Dan's statement can be true and the other three statements false. If Dan's statement is the only true statement, then Carol did it.

b. To find the false statement, assume for each statement in turn that it is false and that the other three statements are true. This is impossible in all cases except one: only Barbara's statement can be false and the other three statements true. If Barbara's statement is the only false statement, then Barbara is guilty.

Chapter Review, pages 316–317

1. b   **2.** c   **3.** a

4. c; The slope of $y = 4x + 3$ is 4.
The only line with slope of 4 is $y = 4x + 2$.

5. $m = \dfrac{4 - 2}{8 - 3} = \dfrac{2}{5}$ **6.** $m = \dfrac{9 - 3}{5 - 2} = \dfrac{6}{3} = 2$

7. $m = \dfrac{8 - 5}{3 - 4} = \dfrac{3}{-1} = -3$ **8.** $y = \dfrac{3}{4}x + 8$

9. $y = -2x + 3$ **10.** $y = -\dfrac{1}{3}x - 2$

11. $\left.\begin{array}{l} y = 3x + 2 \\ y = -\dfrac{1}{3}x + 5 \end{array}\right\}$ Slopes are negative reciprocals, so lines are perpendicular.

12. $\left.\begin{array}{l} y = 3x + 4 \\ y = 3x + 2 \end{array}\right\}$ Slopes are equal, so lines are parallel.

13. $\left.\begin{array}{l} y = \dfrac{1}{5}x - \dfrac{7}{5} \\ y = 3x + 2 \end{array}\right\}$ Slopes are unequal and not negative reciprocals, so the lines are neither parallel or perpendicular.

14. $y - 2 = -9(x - 6)$
$y - 2 = -9x + 54$ Distribute.
$y = -9x + 56$ Add 2.

15. $m = \dfrac{9 - 3}{4 - 3} = \dfrac{6}{1} = 6$
$y - 9 = 6(x - 4)$
$y - 9 = 6x - 24$ Distribute.
$y = 6x - 15$ Add 9.

16. $y - 2 = -\dfrac{1}{2}(x - 9)$
$y - 2 = -\dfrac{1}{2}x + \dfrac{9}{2}$ Distribute.
$y = -\dfrac{1}{2}x + \dfrac{13}{2}$ Add 2.

17. $y - 5 = 2(x - 2)$
$y - 5 = 2x - 4$ Distribute.
$-2x + y = 1$ Add 5 and subtract $2x$.

18. $m = \dfrac{7-3}{2-4} = \dfrac{4}{-2} = -2$

$y - 7 = -2(x - 2)$

$y - 7 = -2x + 4$ Distribute.

$2x + y = 11$ Add $2x$ and 7.

19. $m = \dfrac{4-1}{5-(-3)} = \dfrac{3}{8}$

$y - 4 = \dfrac{3}{8}(x - 5)$

$8y - 32 = 3x - 15$ Multiply by 8 and distribute.

$-3x + 8y = 17$ Add 32 and subtract $3x$.

20. $y - 7 = -\dfrac{3}{2}(x + 2)$

$2y - 14 = -3x - 6$ Multiply by 2 and distribute.

$3x + 2y = 8$ Add $3x$ and 14.

21. No relationship, so zero correlation

22. As x increases, y increases, so positive correlation

23. As x increases, y decreases, so negative correlation

24. $4 = k \cdot 6$

$k = \dfrac{4}{6} = \dfrac{2}{3}$,   so $y = \dfrac{2}{3}x$

25. $8 = 2.5k$

$k = \dfrac{8}{2.5} = 3.2$,   so $y = 3.2x$

26. $-7 = 4k$

$k = -\dfrac{7}{4}$,   so $y = -\dfrac{7}{4}k$

27.

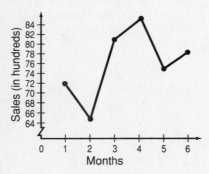

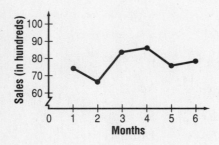

1. $y - 3 = -2(x - 2)$ Distribute.

$y - 3 = -2x + 4$ Add 3.

$y = -2x + 7$

2. $y - 2 = \dfrac{1}{2}(x + 2)$ Distribute.

$y - 2 = \dfrac{1}{2}x + 1$ Add 2.

$y = \dfrac{1}{2}x + 3$

3. $y - 0 = \dfrac{2}{3}(x - 0)$ Distribute.

$y = \dfrac{2}{3}x$

4. $y - (-3) = 3[x - (-3)]$ Simplify and distribute.

$y + 3 = 3x + 9$ Subtract 3.

$y = 3x + 6$

5. $y - (-1) = 2(x - 2)$ Distribute and simplify.

$y + 1 = 2x - 4$ Subtract 1.

$y = 2x - 5$

6. $y - 10 = 3(x - 2)$ Distribute.

$y - 10 = 3x - 6$ Add 10.

$y = 3x + 4$

7. $m = \dfrac{5-0}{0-(-5)} = \dfrac{5}{5} = 1$

$y - 5 = 1(x - 0)$

$y = x + 5; \; -x + y = 5$

8. $m = \dfrac{3-(-3)}{-5-1} = \dfrac{6}{-6} = -1$

$y - 3 = -1[x - (-5)]$

$y - 3 = -x - 5$

$y = -x - 2; \; x + y = -2$

9. $m = \dfrac{4-2}{2-(-4)} = \dfrac{2}{6} = \dfrac{1}{3}$

$y - 4 = \dfrac{1}{3}(x - 2)$

$y - 4 = \dfrac{1}{3}x - \dfrac{2}{3}$

$y = \dfrac{1}{3}x + 3\dfrac{1}{3}; \; -x + 3y = 10$

10. $m = \dfrac{7-4}{0-(-2)} = \dfrac{3}{2}$

$y - 7 = \dfrac{3}{2}(x - 0)$

$y = \dfrac{3}{2}x + 7; \; -3x + 2y = 14$

11. $m = \dfrac{4-(-2)}{-6-6} = \dfrac{6}{-12} = -\dfrac{1}{2}$

$y - 4 = -\dfrac{1}{2}[x-(-6)]$

$y - 4 = -\dfrac{1}{2}x - 3$

$y = -\dfrac{1}{2}x + 1; \; x + 2y = 2$

12. $m = \dfrac{6-(-3)}{4-1} = \dfrac{9}{3} = 3$

$y - 6 = 3(x-4)$

$y - 6 = 3x - 12$

$y = 3x - 6; \; -3x + y = -6$

13. $m = \dfrac{2-(-2)}{6-(-6)} = \dfrac{4}{12} = \dfrac{1}{3}$

$y - 2 = \dfrac{1}{3}(x-6)$

$y - 2 = \dfrac{1}{3}x - 2$

$y = \dfrac{1}{3}x; \; -x + 3y = 0$

14. $m = \dfrac{4-(-2)}{-2-2} = \dfrac{6}{-4} = -\dfrac{3}{2}$

$y - 4 = -\dfrac{3}{2}[x-(-2)]$

$y - 4 = -\dfrac{3}{2}x - 3$

$y = -\dfrac{3}{2}x + 1; \; 3x + 2y = 2$

15. A. $3(1) + 2(1) = 5$

  B. $3(4) + 2(-1) = 10$

  C. $3(-1.2) + 2(2) = 0.4$

  D. $3(-3) + 2(4) = -1$

16. Answers will vary. First determine the value of the slope m for both lines. Parallel lines have identical slopes, so the value of m in each equation should be the same. The slopes of perpendicular lines are negative reciprocals of each other.

17. Both slopes are -2, so lines are parallel.

18. First line has slope -2, second has slope $\dfrac{1}{2}$, so lines are perpendicular.

19. First line has slope 3, second has slope -3, so lines are neither parallel or perpendicular.

20. Both lines have slope of 2, so lines are parallel.

21. A. $3 \neq 2 - 1$

  B. $3 + 2 = 5$

  C. $3 - 2 = 1$

  D. $2(2) + 3(3) = 13$

22. Slope is 2, so perpendicular slope is $-\dfrac{1}{2}$.

$y - 1 = -\dfrac{1}{2}(x-0)$

$y = -\dfrac{1}{2}x + 1$

23. Slope is $\dfrac{2}{3}$, so perpendicular slope is $-\dfrac{3}{2}$.

$y - 0 = -\dfrac{3}{2}(x-0)$

$y = -\dfrac{3}{2}x$

24. Slope is $\dfrac{1}{3}$, so perpendicualr slope is -3.

$y - 3 = -3(x-3)$

$y - 3 = -3x + 9$

$y = -3x + 12$

25. Slope is 2, so perpendicular slope is $-\dfrac{1}{2}$.

$y - 1 = -\dfrac{1}{2}(x-3)$

$y - 1 = -\dfrac{1}{2}x + \dfrac{3}{2}$

$y = -\dfrac{1}{2}x + \dfrac{5}{2}$

26a. $2x + 4y = 60$

b.

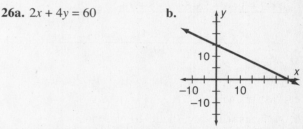

b.

x	0	2	4	6	12	16	20	30
y	15	14	13	12	9	7	5	0

27.

x	y	$\dfrac{y}{x}$
3.24	0.36	9
4.23	0.47	9
4.77	0.53	9
6.66	0.74	9
12.33	1.37	9

So, $k = 9$ and $y = 9x$.

28. Depending on the scale used, the relationship between items of data in the graph can be exaggerated by increasing or decreasing the vertical distance between the items.

Slope Models

a. It rises 1 and runs 2, so the slope is $\frac{1}{2}$.

b. Answers will vary.

Drawing Directions

a.–e. Possible graph:

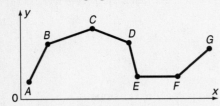

f. AB, BC, and (possibly) FG show an increase; CD, DE show a decrease; EF shows no change

A Good Plot

a.

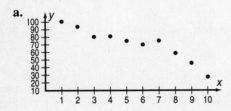

The data is negatively correlated.

b. $\frac{1+2+3+4+5+6+7+8+9+10}{10} = \frac{55}{10} = 5.5 = \bar{x}$

$\frac{100+95+80+80+75+70+75+55+45+30}{10} =$

$\frac{705}{10} = 70.5 = \bar{y}$

c. Answers will vary.

d. Answers will vary; should be close to:
$y = 108 - 6.82x$

Student Survey

a.–e. Answers will vary.

Cumulative Review, page 320

1. $m = \frac{5-3}{6-1} = \frac{2}{5}$ **2.** $m = \frac{7-(-5)}{-2-1} = \frac{12}{-3} = -4$

3. $m = \frac{3}{4}$ **4.** $m = \frac{1}{3}$, the negative reciprocal of -3

5. undefined slope   **6.** $97 - 36 = 61$

7. median $= Q_2 = 78$ **8.** $84 - 65 = 19$

9. A. between 36 and 65 is 25% of data

    B. between 65 and 78 is 25% of data

    C. between 78 and 84 is 25% of data

    D. between 84 and 97 is 25% of data

    E. All quartiles are the same.

10. $\begin{bmatrix} 5-(-1) & -3-(-9) \\ -2-6 & 8-3 \end{bmatrix} = \begin{bmatrix} 6 & 6 \\ -8 & 5 \end{bmatrix}$

11. $\begin{bmatrix} 5.8+(-4.6) & 1.6+(-1.6) \\ -7.1+(-0.7) & -1.6+(-1.6) \\ 4+2.7 & -9.5+4.8 \end{bmatrix} = \begin{bmatrix} 1.2 & 0 \\ -7.8 & -3.2 \\ 6.7 & -4.7 \end{bmatrix}$

12. $\frac{x}{7} = -8$

$x = -56$ Multiply by 7.

13. $8(2n - 3) = -3(4 - 5n)$

$16n - 24 = -12 + 15n$ Distribute.

$n = 12$ Subtract $15n$ and add 24.

14. Let n represent the number.

$n - 11 = 13$

$n = 24$ Add 11.

15. Graph $y = 2x - 4$ and $y = x + 3$;
$x = 7$ at point of intersection

16. Graph $y = 60 - 3x$ and $y = 3x + 24$;
$x = 6$ at point of intersection

17. Graph $y = x^2 - 4$ and $y = x + 2$;
$x = 3$ and -2 at points of intersection

18. $f(-2) = 2(-2 - 6) = 2(-8) = -16$

$f(5) = 2(5 - 6) = 2(-1) = -2$

$f\left(\frac{1}{2}\right) = 2\left(\frac{1}{2} - 6\right) = 2\left(-5\frac{1}{2}\right) = -11$

19. $f(1) = \frac{1}{5} - 4 = -\frac{19}{5}$

$f(20) = \frac{20}{5} - 4 = 4 - 4 = 0$

$f(-5) = \frac{-5}{5} - 4 = -1 - 4 = -5$

20. $m = \frac{5-(-3)}{4-(-2)} = \frac{8}{6} = \frac{4}{3}$

$y - 5 = \frac{4}{3}(x - 4)$

$y - 5 = \frac{4}{3}x - \frac{16}{3}$ Distribute.

$y = \frac{4}{3}x - \frac{1}{3}$ Add 5.

21. Parallel to x-axis is horizontal line $y = -1$

22. Slope-intercept form is the only way to enter an equation into a graphing utility.

23. $2a - 3b = 7$

$-3b = 7 - 2a$

$b = \frac{7-2a}{-3}$ or $\frac{2a-7}{3}$

24. $27 = 18k$

$k = \dfrac{27}{18} = \dfrac{3}{2}$, so $y = \dfrac{3}{2}x$

When $x = 60, y = \dfrac{3}{2}(60) = 90$

25. $550 = 6k$

$k = \dfrac{550}{6} = \dfrac{275}{3}$, so $y = \dfrac{275}{3}x$

When $y = 2475$, then $2475 = \dfrac{275}{3}x$

$27 = x$

26. In an ordered pair, the first coordinate is always the horizontal movement and the second coordinate is always the vertical movement. (3, 5) is 3 units right and 5 units up; (5, 3) is 5 units right and 3 units up.

Standardized Test, page 321

1. $3x - 4y = 12$

$-4y = -3x + 12$

$y = \dfrac{3}{4}x - 3$ So, slope $= \dfrac{3}{4}$.

2. $f(-3) = 2(-3)^2 - 3(-3) - 12 =$

$2(9) - 3(-3) - 12 = 18 + 9 - 12 = 15$

3. $0.08x = 24$

$x = \dfrac{24}{0.08} = 300$

4. $72 \div 2 - 4 \cdot 3^2 = 72 \div 2 - 4 \cdot 9 = 36 - 36 = 0$

5. Length $= 8 - (-3) = 11$

Width $= 2 - (-5) = 7$

$P = 2(7) + 2(11) = 14 + 22 = 36$

$A = (7)(11) = 77$

$P + A = 36 + 77 = 113$

6. $3x - 8(kx + 2) = 19x - 16$

$3x - 8kx - 16 = 19x - 16$ Distribute.

$(3 - 8k)x = 19x$ Add 16; factor out G.C.F.

$3 - 8k = 19$ Divide by x.

$-8k = 16k$ Subtract 3.

$k = -2$ Divide by -8.

So, $k = -2$ when $x \neq 0$.

7. $\dfrac{12}{12 + 3 + 1 + 2 + 62} = \dfrac{12}{80} = 0.15$ or 15%

8. $\dfrac{62}{80} = 0.775$ or 77.5%

9. $\dfrac{3 + 5 + 4 + 4 + 3 + 4 + 5 + 4 + 5}{9} = \dfrac{37}{9} \approx 4.11$

10. $4x + 2.5 < 6x + 8.2$

$-5.7 < 2x$ Subtract $4x$ and 8.2.

$-2.85 < x$ So least integer is -2.

11. Let x represent the amount invested at 8%.

$0.08x + 0.06(1500 - x) = 111.60$

$8x + 9000 - 6x = 11160$ Multiply by 10

$2x = 2160$ Subtract 9000.

$x = 1080$ Divide by 2.

12. $4(t) + 2(6) = b$ and $4(2) + 2(t) = b$

$4(t) + 2(6) = 4(2) + 2(t)$

$4t + 12 = 8 + 2t$

$2t = -4$ Subtract $2t$ and 12.

$t = -2$ Divide by 2.

13. Since $y = x^2$, x can be either positive or negative and return the same value for y.

So, the quotient of x values will be $\dfrac{x}{-x}$ or $\dfrac{-x}{x} = -1$.

14. $7 - 6x > 9$

$-6x > 2$ Subtract 7.

$x < -\dfrac{1}{3}$ Divide by -6 and reverse inequality.

So, the first integer is -1.

15. x - intercept is where $y = 0$, so $5x = 15$ and $x = 3$

y - intercept is where $x = 0$, so $2y = 15$ and $y = 7\dfrac{1}{2}$

The product of these coordinates is $(3)\left(7\dfrac{1}{2}\right) = 22\dfrac{1}{2}$.

16. $\dfrac{78 + 85 + 82 + 81 + x}{5} \geq 80$

$326 + x \geq 400$ Multiply by 5; simplify.

$x \geq 74$ Subtract 326.

17. $d = $ number of dimes; $3d = $ number of quarters

$d + 3d = 24$ Simplify.

$4d = 24$ Divide by 4.

$d = 6;\ 3d = 3(6) = 18$

6 dimes is \$0.60, 18 quarters is \$4.50,

total money is \$5.10

18. $5x - 7y = 35$

$-7y = -5x + 35$

$y = \dfrac{5}{7}x - 5$

Slope is $\dfrac{5}{7}$; the slope of perpendicular line is $-\dfrac{7}{5}$.

19. $y - (-2) = 5$,   so $y = 3$

$3z - 5 = 10$,   so $z = 5$

$5 - z = x$,   so $5 - (y + z) = x$

Since $y + z = 3 + 5 = 8$,

then $x = 5 - (y + z) = 5 - 8 = -3$

Chapter 7   Systems of Linear Equations

Data Activity, pages 322–324

1. Drop the 6 from 546. $54 \div 2 + 18 = 27 + 18 = $ 45th St

2. $80 \div 2 + 22 = $ 62nd St.; 62nd St. to 42nd St. is 20 blocks south.

3. No, she added 31 to 82 when she should have subtracted to get 51st Street.

4. $(57 - 12)2 = (45)2 = 90$; 900 Seventh Avenue

Lesson 7.1, pages 325–327

THINK BACK

1a. 0   **b.** 10   **c.** 2

2a. −7   **b.** −3   **c.** 4

3. If the *x*- and *y*-coordinates of a point on the line are input into the equation, they satisfy the equation.

4. $m = 1$, $b = 4$ so $y = x + 4$

EXPLORE

5.
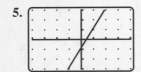

6. Ordered pairs may vary but should satisfy the equation $y = 2x - 0.8$.

Possible pairs:

x	−4	−3	−2	−1	0	1	2	3	4
y	−8.8	−6.8	−4.8	−2.8	−0.8	1.2	3.2	5.2	7.2

7.
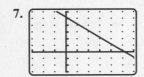

8. Ordered pairs may vary but should satisfy the equation $y = -x + 5.2$.

Possible pairs:

x	−4	−3	−2	−1	0	1	2	3	4
y	9.2	8.2	7.2	6.2	5.2	4.2	3.2	2.2	1.2

9. Point (2, 3.2) or close is in both tables.

10. Lines have different slopes so they are not parallel.

11.
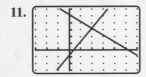

Point is (2, 3.2).

12.
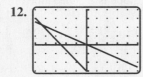

Point is (− 4.5, 2.4).

13.
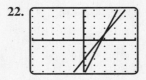

The lines are parallel, so there is no intersection.

14. Answers will vary, but equations must have different slopes.

MAKE CONNECTIONS

15. Substitute (2, 3) into each equation, and verify that a true equation results.

16. A point whose coordinates satisfy the equation is on the graph.

17. intersection

18. Intersecting lines are lines with exactly one point in common.

19. If a point is on the graph of an equation, then the coordinates of the point are a solution of the equation.

20. are a solution of both equations

SUMMARIZE

21. Answers will vary. Graph both equations. The point of intersection of the lines, if there is one, is the ordered pair you are looking for. Substitute the coordinates in each equation to see whether a true equation results.

22.

23.

Intersection is at (2, 1).   Intersection is at (−1, 3).

24. Graphs coincide, equations describe the same line.

25. They must be parallel.

26. Using each ranger's position and the direction of the fire, a line can be drawn on a map connecting the ranger and the fire. The point of intersection of the lines is the position of the fire.

27. Using the ZOOM feature, students should be able to approximate the two points of intersection, (1, 4) and (−1, 2).

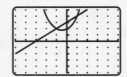

Lesson 7.2, page 328–334

EXPLORE

1. **Population Change of Baltimore and San Diego 1970–1990**

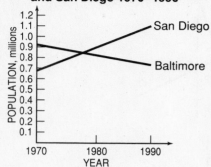

2. that the changes in population were constant

3. the moment in time when the populations were equal

4. Estimate from the graph: 1977; about 850,000

5. Possibly, but not necessarily. They will only intersect if the populations were the same at some time between 1970 and 1990.

TRY THESE

1. $5 \overset{?}{=} 4(3) - 7$ $5 \overset{?}{=} -3 + 8$

$5 \overset{?}{=} 12 - 7$ $5 = 5 ✔$

$5 = 5 ✔$

Yes, $(3, 5)$ is a solution.

2. $2(2) + (-3) \overset{?}{=} 1$ $3(2) - (-3) \overset{?}{=} 3$

$4 + (-3) \overset{?}{=} 1$ $6 + 3 \overset{?}{=} 3$

$1 = 1 ✔$ $9 \neq 3$

No, $(2, -3)$ is not a solution.

3. $2 \overset{?}{=} -0 + 2$ $3(2) \overset{?}{=} -3(0) + 6$

$2 = 2 ✔$ $6 \overset{?}{=} 0 + 6$

$6 = 6 ✔$

Yes, $(0, 2)$ is a solution.

4. $2 \overset{?}{=} -3 + 5$ $2 \overset{?}{=} -3 + 3$

$2 = 2 ✔$ $2 \neq 0$

No, $(-3, 2)$ is not a solution.

5.

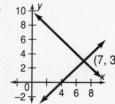

6.

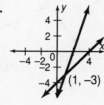

7.

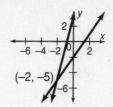

8. $y = 3x + 8$ and $y = 4x + 5$

9.

They each charge \$17 for 3 hours.

10. Answers will vary. If the job takes less than 3 hours, the sander should be rented from Premier Rentals. If the job takes more than 3 hours, the sander should be rented from Thrifty Rentals.

11. Answers will vary. Possible advantage for paper and pencil: You don't have to write the equations in slope-intercept form. Possible advantage for graphing utility: It is faster, especially when the equations are already in slope-intercept from. Newer calculators give the point of intersection directly. Possible disadvantage for paper and pencil: It takes more time, and graph paper is needed to be precise. Possible disadvantage for graphing utility: The point of intersection is sometimes difficult to identify.

PRACTICE

1. $2 \overset{?}{=} (3)(1) - 1$ $2 \overset{?}{=} -1 + 3$

$2 \overset{?}{=} 3 - 1$ $2 = 2 ✔$

$2 = 2 ✔$

Yes, $(1, 2)$ is a solution.

2. $5(0) - 4 \overset{?}{=} 1$ $3(0) + 1 \overset{?}{=} 4$

$-4 \neq 1$ $1 \neq 4$

No, $(0, 4)$ is not a solution.

3. $3 \overset{?}{=} 3(-2) + 3$ $3 \overset{?}{=} -(-2) + 1$

$3 \overset{?}{=} -6 + 3$ $3 \overset{?}{=} 2 + 1$

$3 \neq -3$ $3 = 3 ✔$

No, $(-2, 3)$ is not a solution.

4. $2(0.5) + 3(1) \overset{?}{=} 4$ $-8(0.5) - 2(1) \overset{?}{=} -6$

$1 + 3 \overset{?}{=} 4$ $-4 - 2 \overset{?}{=} -6$

$4 = 4 ✔$ $-6 = -6 ✔$

Yes, $(0.5, 1)$ is a solution.

5. $3x + 2y = 28$ and $2x + y = 17$

6.

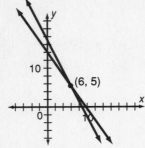

(6, 5)

7. $6 /yd for cotton, $5/yd for rayon

8.

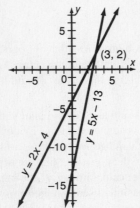

(3, 2)

$y = 5x - 13$

$y = 2x - 4$

9.

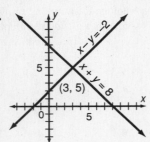

$x - y = -2$

$x + y = 8$

(3, 5)

10.

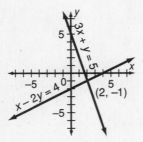

$3x + y = 5$

$x - 2y = 4$

(2, −1)

11.

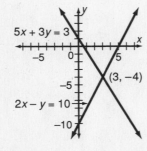

$5x + 3y = 3$

(3, −4)

$2x - y = 10$

12.

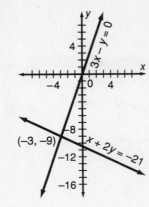

$3x - y = 0$

(−3, −9)

$x + 2y = -21$

13.

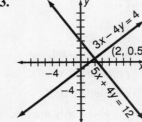

$3x - 4y = 4$

(2, 0.5)

$5x + 4y = 12$

14.

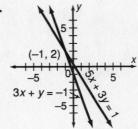

(−1, 2)

$3x + y = -1$

$5x + 3y = 1$

15.

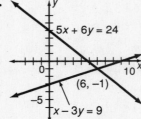

$5x + 6y = 24$

(6, −1)

$x - 3y = 9$

16.

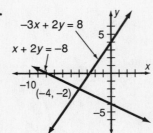

$-3x + 2y = 8$

$x + 2y = -8$

(−4, −2)

17. $y = 0.06x + 44;\ y = 0.09x + 35$

18.

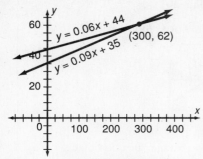

Each charges $62 for 300 miles.

19. Answers will vary. Answers should include a statement indicating expected miles to be either more than or less than 300 mi. For less than 300 mi, Getaway Auto Rentals offers the better deal. For more than 300 mi, Prestige Car Rentals offers the better deal.

20. Explanations will vary, but students should mention writing each equation in a form convenient for graphing, using either paper and pencil or a graphing utility, then graphing each equation, finding the point of intersection, and relating the x- and y-values of the point of intersection to the variables in the system of equations.

21. $y = 15x + 250$ and $y = 20x + 200$

22.

Intersection is at $(10, 400)$.

23. 10 weeks; $400 in each account

EXTEND

24. $y = x - 1$ and $x + y = 7$

25.

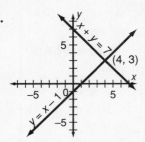

So, $(4, 3)$ solves the system.

26.

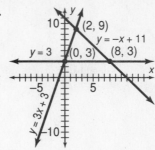

27. Area $= \frac{1}{2} bh$

b is the distance from $(0,3)$ to $(8,3)$, so $b = 8$.
h is the distance from $(2,9)$ to line $y = 3$, so $h = 6$.

Area $= \frac{1}{2}(8)(6) = 24$

28.

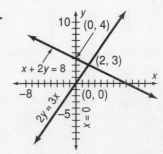

Vertices are $(0, 0)$, $(0, 4)$, and $(2, 3)$.

29. Graph the system $y = 3x - 2$ and $y = x - 4$.

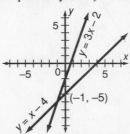

At the point of intersection $(-1, -5)$.

THINK CRITICALLY

30. Answers will vary. Possible solution:
$x - y = 6, x + y = -2$

31. No, because two distinct lines may intersect in at most one point. There are an infinite number of systems with $(-3, -5)$ as solution.

32. First diagonal from $(4, 9)$ to $(8, 1)$:

slope is $m = \frac{9-1}{4-8} = \frac{8}{-4} = -2$

equation is $y - 1 = -2(x - 8)$ or $y = -2x + 17$
Second diagonal from $(8, 9)$ to $(4, 1)$:

slope is $m = \frac{9-1}{8-4} = \frac{8}{4} = 2$,

equation is $y - 1 = 2(x - 4)$ or $y = 2x - 7$
Graph the system.

(Solution continues on next page.)

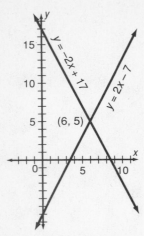

So, the intersection point is (6, 5).

33. $(0, b)$ is the only point of intersection.

ALGEBRA WORKS

1. revenues; expenses; the revenues graph is a straight line with a constant slope; the expenses graph is a curved line with a gradually increasing slope

2. just prior to 1993; The graphs intersect at a point with a horizontal coordinate between 1992 and 1993.

3. individual: $y = 1.48n + 2.16$
central:　　$y = 1.95n + 0.71$

4.

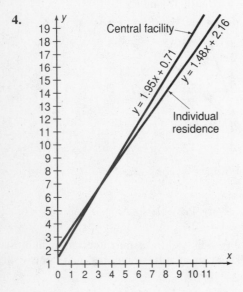

5. Upon examining the graph, during year 4.

6. Answers will vary. Students should note that the total costs of the central facility system are less than those of the individual residence system for 4 years but that thereafter the individual system total costs are less. The manager's decision on which system to choose will be based on whether the city wants immediate savings, in which case the central facility system is best, or whether it should look for long-term savings, in which case the individual residence system is best.

Lesson 7.3, pages 335–341

EXPLORE / WORKING TOGETHER

1.

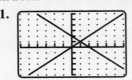

2. Graphs intersect at $(1, 1)$.

3. You have to approximate the coordinates at $(3.43, 3.43)$.

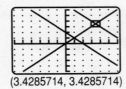

(3.4285714, 3.4285714)

4. The intersection points of systems with solutions that are integers or simple fractions can easily be pinpointed, making graphical solutions ideal. With many solutions that are fractions or decimals, however, the points of intersection, and therefore the solutions, are difficult to locate.

TRY THESE

1. 　$x - 3y = 6$　　Equation 2
　　$5y - 3y = 6$　　Substitute $5y$ for x.
　　　　$2y = 6$　　Divide by 2.
　　　　　$y = 3$

Substitute 3 for y in Equation 1 and solve for x.

$x = 5y = 5(3) = 15$

So $(15, 3)$ solves the system.

2. 　　　$3x - 2y = 10$　　Equation 2
　　$3(-2y - 2) - 2y = 10$　　Substitute $-2y - 2$ for x.
　　　$-6y - 6 - 2y = 10$　　Distribute.
　　　　　　$-8y = 16$　　Add 6 and simplify.
　　　　　　　$y = -2$　　Divide by -8.

Substitute -2 for y in Equation 1 and solve for x.

$x = -2y - 2 = -2(-2) - 2 = 4 - 2 = 2$

So $(2, -2)$ solves the system.

3. 　　　$x - y = 1$　　Equation 1
　　$x - (-x + 5) = 1$　　Substitute $-x + 5$ for y.
　　　$x + x - 5 = 1$　　Distribute.
　　　　　$2x = 6$　　Simplify and add 5.
　　　　　　$x = 3$　　Divide by 2.

Substitute 3 for x in Equation 2 and solve for y.

$y = -x + 5 = -3 + 5 = 2$

So $(3, 2)$ solves the system.

4.

$$y = 3x - 6 \qquad \text{Solve equation 1 for } y.$$
$$9x - 2y = 3 \qquad \text{Equation 2}$$
$$9x - 2(3x - 6) = 3 \qquad \text{Substitute } 3x - 6 \text{ for } y.$$
$$9x - 6x + 12 = 3 \qquad \text{Distribute.}$$
$$3x = -9 \qquad \text{Subtract 12 and simplify.}$$
$$x = -3 \qquad \text{Divide by 3.}$$

Substitute -3 for x in Equation 1 and solve for y.
$$y + 6 = 3(-3) \rightarrow y + 6 = -9 \rightarrow y = -15$$
So, $(-3, -15)$ solves the system.

5.

$$y = -3x + 3 \qquad \text{Solve equation 1 for } y.$$
$$3x - 2y = -3 \qquad \text{Equation 2}$$
$$3x - 2(-3x + 3) = -3 \qquad \text{Substitute } -3x + 3 \text{ for } y.$$
$$3x + 6x - 6 = -3 \qquad \text{Distribute.}$$
$$9x = 3 \qquad \text{Add 6 and simplify.}$$
$$x = \frac{3}{9} = \frac{1}{3} \qquad \text{Divide by 9.}$$

Substitute $\frac{1}{3}$ for x in Equation 2 and solve for y.

$$3\left(\frac{1}{3}\right) + y = 3 \rightarrow 1 + y = 3 \rightarrow y = 2$$

So, $\left(\frac{1}{3}, 2\right)$ solves the system.

6.

$$4x + y = 17 \qquad \text{Equation 1}$$
$$y = -4x + 17 \qquad \text{Solve equation 1 for } y.$$
$$3x - 4y = 8 \qquad \text{Equation 2}$$
$$3x - 4(-4x + 17) = 8 \qquad \text{Substitute } -4x + 17 \text{ for } y.$$
$$3x + 16x - 68 = 8 \qquad \text{Distribute.}$$
$$19x = 76 \qquad \text{Simplify and add 68.}$$
$$x = 4 \qquad \text{Divide by 19.}$$

Substitute 4 for x in Equation 2 and solve.
$$4(4) + y = 17 \rightarrow 16 + y = 17 \rightarrow y = 1$$
So, $(4, 1)$ solves the system.

7. $x + y = 8000$ and $0.06x + 0.08y = 530$

8.

$$x + y = 8000 \qquad \text{Equation 1}$$
$$y = -x + 8000 \qquad \text{Solve for } y.$$
$$0.06x + 0.08y = 530 \qquad \text{Equation 2}$$
$$0.06x + 0.08(-x + 8000) = 530 \qquad \text{Substitute } -x + 8000 \text{ for } y.$$
$$0.06x - 0.08x + 640 = 530 \qquad \text{Distribute.}$$
$$-0.02x = -110 \qquad \text{Simplify and subtract 640.}$$
$$x = 5500 \qquad \text{Divide by } -0.02.$$

Substitute 5500 for x in Equation 1 and solve.
$$5500 + y = 8000 \rightarrow y = 2500$$
So, Ricardo invested \$5500 at 6% and \$2500 at 8%.

9. Answers will vary but students should state that they will begin by looking for a variable with a coefficient of 1 or -1. In the second equation the coefficient of y is 1, so they can solve for y in the second equation and substitute that expression in the first to find x.

PRACTICE

1.

$$x = 2y + 16 \qquad \text{Solve Equation 1 for } x.$$
$$4x + y = 1 \qquad \text{Equation 2}$$
$$4(2y + 16) + y = 1 \qquad \text{Substitute } 2y + 16 \text{ for } x.$$
$$8y + 64 + y = 1 \qquad \text{Distribute.}$$
$$9y = -63 \qquad \text{Subtract 63 and simplify.}$$
$$y = -7 \qquad \text{Divide by 9.}$$

Substitute -7 for y in Equation 1 and solve.
$$x - 2(-7) = 16 \rightarrow x + 14 = 16 \rightarrow x = 2$$
So, $(2, -7)$ solves the system.

2.

$$x = -2y + 6 \qquad \text{Solve Equation 1 for } x.$$
$$4x + 3y = 4 \qquad \text{Equation 2}$$
$$4(-2y + 6) + 3y = 4 \qquad \text{Substitute } -2y + 6 \text{ for } x.$$
$$-8y + 24 + 3y = 4 \qquad \text{Distribute.}$$
$$-5y = -20 \qquad \text{Subtract 24.}$$
$$y = 4 \qquad \text{Divide by 5.}$$

Substitute 4 for y in Equation 1 and solve.
$$x + 2(4) = 6 \rightarrow x + 8 = 6 \rightarrow x = -2$$
So, $(-2, 4)$ solves the system.

3.

$$2x - 3y = 13 \qquad \text{Equation 2}$$
$$2x - 3(-4x + 5) = 13 \qquad \text{Substitute } -4x + 5 \text{ for } y.$$
$$2x + 12x - 15 = 13 \qquad \text{Distribute.}$$
$$14x = 28 \qquad \text{Add 15 and simplify.}$$
$$x = 2 \qquad \text{Divide by 14.}$$

Substitute 2 for x in Equation 1 and solve.
$$y = -4(2) + 5 = -8 + 5 = -3$$
So, $(2, -3)$ solves the system.

4.

$$y = 5x + 23 \qquad \text{Solve Equation 1 for } y.$$
$$3x - y = -15 \qquad \text{Equation 2}$$
$$3x - (5x + 23) = -15 \qquad \text{Substitute } 5x + 23 \text{ for } y.$$
$$3x - 5x - 23 = -15 \qquad \text{Distribute.}$$
$$-2x = 8 \qquad \text{Add 23 and simplify.}$$
$$x = -4 \qquad \text{Divide by } -2.$$

Substitute -4 for x in Equation 2 and solve.
$$3(-4) - y = -15 \rightarrow -12 - y = -15 \rightarrow -y = -3 \rightarrow y = 3$$
So, $(-4, 3)$ solves the system.

5.

$y = 2x + 2$	Solve Equation 1 for y.
$2x + 3y = 6$	Equation 2
$2x + 3(2x + 2) = 6$	Substitute $2x + 2$ for y.
$2x + 6x + 6 = 6$	Distribute.
$8x = 0$	Subtract 6 and simplify.
$x = 0$	Divide by 8.

Substitute 0 for x in Equation 1 and solve.

$-2(0) + y = 2 \rightarrow y = 2$

So, $(0, 2)$ solves the system.

6.

$x = -5y + 11$	Solve Equation 1 for x.
$4x - y = 2$	Equation 2
$4(-5y + 11)y = 2$	Substitute $5y + 11$ for x.
$-20y + 44y = 2$	Distribute.
$-21y = -42$	Subtract 44 and simplify.
$y = 2$	Divide by -21.

Substitute 2 for y in Equation 1 and solve.

$x + 5(2) = 11 \rightarrow x + 10 = 11 \rightarrow x = 1$

So $(1, 2)$ solves the system.

7.

$-\frac{2}{3}x + \frac{11}{3} = y$	Solve Equation 1 for y.
$3x + 3y = 18$	Equation 2
$3x + 3\left(-\frac{2}{3}x + \frac{11}{3}\right) = 18$	Substitute $\frac{2}{3}x + \frac{11}{3}$ for y.
$3x - 2x + 11 = 18$	Distribute.
$x = 7$	Subtract 11 and simplify.

Substitute 7 for x in Equation 1 and solve.

$2(7) + 3y = 11 \rightarrow 14 + 3y = 11 \rightarrow 3y = -3 \rightarrow y = -1$

So, $(7, -1)$ solves the system.

8.

$y = 2x - \frac{1}{2}$	Solve Equation 2 for y.
$5x + 3y = 4$	Equation 1
$5x + 3\left(2x - \frac{1}{2}\right) = 4$	Substitute $2x - \frac{1}{2}$ for y.
$5x + 6x - \frac{3}{2} = 4$	Distribute.
$11x = \frac{11}{2}$	Add $\frac{3}{2}$ and simplify.
$x = \frac{1}{2}$	Divide by 11.

Substitute $\frac{1}{2}$ for x in Equation 2 and solve.

$4\left(\frac{1}{2}\right) - 2y = 1 \rightarrow 2 - 2y = 1 \rightarrow -2y = -1 \rightarrow y = \frac{1}{2}$

So, $\left(\frac{1}{2}, \frac{1}{2}\right)$ solves the system.

9.

$x = -7y + 1$	Solve Equation 2 for x.
$5x + 7y = -3$	Equation 1
$5(-7y + 1) + 7y = -3$	Substitute $-7y + 1$ for x.
$-35y + 5 + 7y = -3$	Distribute.
$-28y = -8$	Subtract 5 and simplify.
$y = \frac{2}{7}$	Divide by -28.

Substitute $\frac{2}{7}$ for y in Equation 1 and solve.

$5x + 7\left(\frac{2}{7}\right) = -3 \rightarrow 5x + 2 = -3 \rightarrow 5x = -5 \rightarrow x = -1$

So, $\left(-1, \frac{2}{7}\right)$ solves the system.

10.

$x = y$	Solve Equation 2 for x.
$3x + 6y = 6$	Equation 1
$3x + 6x = 6$	Substitute x for y.
$9x = 6$	Simplify.
$x = \frac{2}{3}$	Divide by 9.

Substitute $\frac{2}{3}$ for x in Equation 2 and solve.

$2\left(\frac{2}{3}\right) = 2y \rightarrow y = \frac{2}{3}$

So, $\left(\frac{2}{3}, \frac{2}{3}\right)$ solves the system.

11.

$\frac{1}{2}x = 3y + 2$	Equation 2
$\frac{1}{2}(10y) = 3y + 2$	Substitute $10y$ for x.
$5y = 3y + 2$	Simplify.
$2y = 2$	Subtract $3y$.
$y = 1$	Divide by 2.

Substitute 1 for y in Equation 1 and solve.

$x = 10(1) = 10$

So, $(10, 1)$ solves the system.

12.

$x - 7 = y$	Equation 1
$x - 7 = \frac{1}{4}x - 1$	Substitute $\frac{1}{4}x - 1$ for y.
$\frac{3}{4}x = 6$	Add 7 and subtract $\frac{1}{4}x$.
$x = 8$	Multiply by $\frac{4}{3}$.

Substitute 8 for x in Equation 1 and solve.

$8 - 7 = y \rightarrow y = 1$

So, $(8, 1)$ solves the system.

13. $x + y = 90$ and $x = 5y$

14.

$x + y = 90$	Equation 1
$5y + y = 90$	Substitute $5y$ for x.
$6y = 90$	Simplify.
$y = 15$	Divide by 6.

Substitute 15 for y in Equation 2 and solve.
$x = 5(15) = 75$
So, the two angles are 75° and 15°.

15. $y = 40x + 50$ and $y = 45x + 30$

16.

$y = 40x + 50$	Equation 1
$45x + 30 = 40x + 50$	Substitute $45x + 30$ for y.
$5x = 20$	Subtract 30 and $40x$.
$x = 4$	Divide by 5.

Substitute 4 for x in Equation 1 and solve.
$y = 40(4) + 50 = 160 + 50 = 210$
4 hours; $210

17. For a job lasting less than 4 h, Paula's is cheaper. For a job lasting more than 4 h, Reliable would be cheaper.

18. Answers may vary. Students should mention that the graphical method is useful when solutions are integers because the intersections can be pinpointed easily. In addition, a graph gives a pictorial representation of changes in the variables that may be useful in a real world problem. When solutions are fractions or decimals, however, points of intersection are usually difficult to determine exactly on a graph. In such cases, the substitution method is more precise.

EXTEND

19.

$2(x + 1) + 3(y - 3) - 4y + 4 = -5$	Equation 1
$2x + 2 + 3y - 9 - 4y + 4 = 5$	Distribute.
$2x - y - 3 = 5$	Simplify.
$2x + 2 = y$	Solve for y.
$4(2x + y) - 3(x + y) - x = 5$	Equation 2
$8x + 4y - 3x - 3y - x = 5$	Distribute.
$4x + y = 5$	Simplify.
$y = -4x + 5$	Solve for y.
$2x + 2 = -4x + 5$	Substitute $2x + 2$ for y.
$x = \dfrac{1}{2}$	Solve for x.

Use Equation 1 to solve for y.
$$y = 2x + 2 = 2\left(\frac{1}{2}\right) + 2 = 1 + 2 = 3$$
So, $\left(\frac{1}{2}, 3\right)$ solves the system.

20.

$4(2x + 2y - 1) - 3(x + 2y + 4) - 2(x + \frac{1}{2}y) = 0$	Equation 1
$8x + 8y - 4 - 3x - 6y - 12 - 2x - y = 0$	Distribute.
$3x + y - 16 = 0$	Simplify.
$y = -3x + 16$	Solve for y.
$2(x + y + 1) + 3(x - y - 2) + 2(3 - 2x) = -2$	Equation 2
$2x + 2y + 2 + 3x - 3y - 6 + 6 - 4x = -2$	Distribute.
$x - y + 2 = -2$	Simplify.
$y = x + 4$	Solve for y.
$-3x + 16 = x + 4$	Substitute $-3x + 16$ for y.
$3 = x$	Solve for x.

Use Equation 2 to solve for y.
$y = x + 4 = 3 + 4 = 7$
So, $(3, 7)$ solves the system.

21. Answers will vary. Example: $y = x - 8$ and $y = -2x + 1$

22. $\begin{cases} x + y = 395 \\ 28x + 22y = 10,130 \end{cases}$ $\rightarrow y = 395 - x.$

$28x + 22(395 - x) = 10,130$ Substitute $395 - x$ for y.

$28x + 8690 - 22x = 10,130$ Distribute.

$6x = 1440$ Subtract 8690 and simplify.

$x = 240$ Divide by 6.

Substitute 240 for x in Equation 1 and solve.

$240 + y = 395 \rightarrow y = 155$

So: 240 tickets sold at \$28 and 155 tickets at \$22

23. $\begin{cases} x + y = 8 \\ 10x + y = 7y \end{cases}$ $\rightarrow x = 8 - y.$

$10(8 - y) + y = 7y$ Substitute $8 - y$ for x.

$80 - 10y + y = 7y$ Distribute.

$80 - 9y = 7y$ Simplify.

$80 = 16y$ Add $9y$.

$5 = y$ Divide by 16.

Substitute 5 for y in Equation 1 and solve.

$x + 5 = 8 \rightarrow x = 3$

So, the number is 35.

24. $\begin{cases} x + y = 200 \\ 0.10x + 0.35y = 0.25(200) \end{cases}$ So $y = 200 - x.$

$0.10x + 0.35(200 - x) = 0.25(200)$ Substitute $200 - x$ for y.

$10x + 7000 - 35x = 5000$ Multiply by 100 and distribute.

$x = 80$ Solve for x.

Substitute 80 for x in Equation 1 and solve.

$80 + y = 200 \rightarrow y = 120$

Mix 80 cm^3 of solution A and 120 cm^3 of solution B.

25. $\begin{cases} 4 = (r - c)2 \\ 4 = (r + c)1 \end{cases}$ $\rightarrow 4 = 2r - 2c$ $\rightarrow r = 4 - c$

$4 = 2(4 - c) - 2c$ Substitute $4 - c$ for r.

$4 = 8 - 2r - 2c$ Distribute.

$1 = c$ Solve for c.

Substitute 1 for c in Equation 1 and solve.

$4 = (r + 1)1 \rightarrow r = 3$

The rate of the boat in still water is 3 mph and the rate of the current is 1 mph.

26. $\begin{cases} 10 = (r + c)\dfrac{5}{6} \\ 12 = (r - c)\dfrac{3}{2} \end{cases}$ $\rightarrow 60 = 5r + 5c$ $\rightarrow r = 8 + c$

$60 = 5(8 + c) + 5c$ Substitute $c + 8$ for r.

$60 = 40 + 5c + 5c$ Distribute.

$2 = c$ Solve for c.

Substitute 2 for c in Equation 2 and solve.

$12 = (r - 2)\dfrac{3}{2} \rightarrow 8 = r - 2 \rightarrow r = 10$

The rate of the current is 2 mph and the rate of the boat in still water is 10 mph.

27. $\begin{cases} 2x - 3y - z = -1 \\ 5y - 3z = 7 \\ 5z = 5 \end{cases}$ $\rightarrow z = 1$

$5y - 3(1) = 7$ Substitute 1 for z in equation 2.

$y = 2$ Solve for y.

$2x - 3(2) - 1 = -1$ Substitute 2 for y and 1 for z in Equation 1.

$x = 3$ Solve for x.

So, $(3, 2, 1)$ solves the system.

28. $\begin{cases} 4x - y + 3z = 4 \\ -3y + 2z = 1 \\ -2z = -10 \end{cases}$ $\rightarrow z = 5$

$-3y + 2(5) = 1$ Substitute 5 for z in Equation 2.

$y = 3$ Solve for y.

$4x - 3 + 3(5) = 5$ Substitute 3 for y in Equation 1.

$x = -2$ Solve for x.

So, $(-2, 3, 5)$ solves the system.

29. In applying the distributive property to $2x - (4x - 10) = 4$, she forgot to multiply -10 by -1.

30. Substitute $(-1, 3)$ for (x, y) and solve.

$A(-1) + 3(3) = 7$ Equation 1

$A = 2$

$2(-1) - B(3) = 13$ Equation 2

$B = -5$

THINK CRITICALLY

31. $\begin{cases} m+n=\frac{3}{4} \\ 3m-n=\frac{1}{4} \end{cases}$ $\rightarrow n=\frac{3}{4}-m$

$3m-\left(\frac{3}{4}-m\right)=\frac{1}{4}$ Substitute $\frac{3}{4}-m$ for n in Equation 2.

$3m-\frac{3}{4}+m=\frac{1}{4}$ Distribute.

$4m=1$ Add $\frac{3}{4}$ and simplify.

$m=\frac{1}{4}$ Divide by 4.

$m=\frac{1}{4}$ and $n=\frac{3}{4}-m=\frac{3}{4}-\frac{1}{4}=\frac{1}{2}$

$m=\frac{1}{x}=\frac{1}{4}\rightarrow x=4;\ n=\frac{1}{y}=\frac{1}{2}\rightarrow y=2$

So, $(4,2)$ solves original system.

32. $\begin{cases} 2m-2n=\frac{1}{2} \\ m+5n=\frac{3}{4} \end{cases}$ $\rightarrow m=\frac{3}{4}-5n$

$2\left(\frac{3}{4}-5n\right)-2n=\frac{1}{2}$ Substitute $\frac{3}{4}-5n$ for m in Equation 1.

$\frac{3}{2}-10n-2n=\frac{1}{2}$ Distribute.

$-12n=-1$ Subtract $\frac{3}{2}$ and simplify.

$n=\frac{1}{12}$ Divide by -12.

$n=\frac{1}{12}$ and $m=\frac{3}{4}-5n=\frac{3}{4}-5\left(\frac{1}{12}\right)=\frac{9}{12}-\frac{5}{12}=\frac{1}{3}$

$m=\frac{1}{x}=\frac{1}{3}\rightarrow x=3;\ n=\frac{1}{y}=\frac{1}{12}\rightarrow y=12$

So, $(3,12)$ solves the original system.

33. $\begin{cases} 2A+B=9 \\ -3A+3B=9 \end{cases}$ $\rightarrow B=9-2A$

$-3A+3(9-2A)=9$ Substitute $9-2A$ for B in Equation 2.

$-3A+27-6A=9$ Distribute.

$A=2$ Solve for A.

Substitute 2 for A in Equation 1 and solve.
$2(2)+B=9\rightarrow B=5$

34. $\begin{cases} -A+2B+3=0 \\ 3A+0B+3=0 \end{cases}$ $\rightarrow A=-1$

Substitute -1 for A in Equation 1 and solve.
$-(-1)+2B+3=0\Rightarrow 2B+4=0\Rightarrow B=-2$

35. The variable x is eliminated, leaving a false statement (for example, $9=-8$). A check of both equations shows that they have the same slope, meaning that they are parallel and do not intersect.

ALGEBRAWORKS

1. $y=2.145x+2.99$

2. $r=0.999961967167$; Yes it is linear.

3. $35=2.145x+2.99\rightarrow x\approx 15$

4. $\begin{cases} y=2.1x+3.7 \\ y=2.145x+2.99 \end{cases}$

$2.1x+3.7=2.145x+2.99$ Substitute $2.1x+3.7$ for y.

$0.71=0.045x$ Subtract $2.1x$ and 2.99.

$15.78\approx x$ Divide by 0.045.

Substitute 15.78 for x in Equation 1 and solve.
$y=2.1(15.78)+3.7=36.84$

5. Green and red each for about 37 seconds

6. About 16 cars will move through each cycle.

Lesson 7.4, page 342–347

EXPLORE/WORKING TOGETHER

1a.

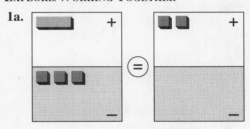

b.

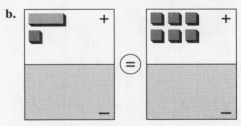

2. $x-3=2$ and $x+1=6$

     $x=5$ and $x=5$

3.

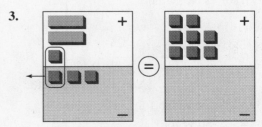

The equation is $2x-2=8$.

4. $2x - 2 = 8$

$\quad\quad 2x = 10$

$\quad\quad\;\; x = 5$

5. The sum of the two equations has the same solution as the two original equations. Adding doesn't affect the solution.

6. Equations will vary, but the solution of the "sum" equation is the same as the solution of the original equations.

TRY THESE

1. Multiply Equation 1 by 7 and Equation 2 by 4, or multiply Equation 1 by 3 and Equation 2 by 5.

2. Multiply Equation 1 by 3 and Equation 2 by 2, or multiply Equation 1 by 9 and Equation 2 by 8.

3. $x - y = 5$

$\underline{2x - y = 13}$

$-x = -8 \quad$ Subtract the equations.

$\quad\; x = 8 \quad$ Solve for x.

Substitute 8 for x in Equation 1 and solve.

$8 - y = 5 \rightarrow y = 3$

So, $(8, 3)$ solves the system.

4. $3x + 2y = 41$

$\underline{5x - 2y = 15}$

$8x = 56 \quad$ Add the equations.

$\quad x = 7 \quad$ Divide by 8.

$3(7) + 2y = 41 \quad$ Substitute 7 for x in Equation 1.

$\quad\quad y = 10 \quad$ Solve for y.

So, $(7, 10)$ solves the system.

5. $5x + 4y = 10 \quad\quad \rightarrow \quad 5x + 4y = 10$

$(6x - 2y = -22) \cdot 2 \rightarrow \underline{12x - 4y = -44}$

$\quad\quad\quad\quad\quad\quad\quad\quad 17x = -34 \quad$ Add.

$\quad\quad\quad\quad\quad\quad\quad\quad\;\; x = -2 \quad$ Divide by 17.

$5(-2) + 4y = 10 \quad$ Substitute -2 for x in Equation 1.

$\quad\quad\quad y = 5 \quad$ Solve for y.

So, $(-2, 5)$ solves the system.

6. $(2x - 3y = 8) \cdot 2 \rightarrow 4x - 6y = 16$

$(3x + 2y = -1) \cdot 3 \rightarrow \underline{9x + 6y = -3}$

$\quad\quad\quad\quad\quad\quad\quad\quad 13x = 13 \quad$ Add.

$\quad\quad\quad\quad\quad\quad\quad\quad\; x = 1 \quad$ Divide by 13.

$2(1) - 3y = 8 \quad$ Substitute 1 for x in Equation 1.

$\quad\quad y = -2 \quad$ Solve for y.

So, $(1, -2)$ solves the system.

7. $-8x + 4y = 0 \quad\quad \rightarrow -8x + 4y = 0$

$(3x - 2y = 6) \cdot 2 \rightarrow \underline{6x - 4y = 12}$

$\quad\quad\quad\quad\quad\quad\quad\quad -2x = 12 \quad$ Add.

$\quad\quad\quad\quad\quad\quad\quad\quad\;\; x = -6 \quad$ Divide by -2.

$-8(-6) + 4y = 0 \quad$ Substitute -6 for x in Equation 1.

$\quad\quad y = -12 \quad$ Solve for y.

So, $(-6, -12)$ solves the system.

8. $(3x + 2y = 1) \cdot 3 \quad \rightarrow 9x + 6y = 3$

$(4x + 3y = -1) \cdot 2 \rightarrow \underline{8x + 6y = -2}$

$\quad\quad\quad\quad\quad\quad\quad\quad x = 5 \quad$ Subtract.

$2y = 1 - 3(5) \quad$ Substitute 5 for x in Equation 1.

$\;\; y = -7 \quad\quad$ Solve for y.

So, $(5, -7)$ solves the system.

9. $200 \cdot (x + y = 16) \quad \rightarrow 200x + 200y = 3200$

$200x + 165y = 2745 \rightarrow \underline{200x + 165y = 2745}$

$\quad\quad\quad\quad\quad\quad\quad\quad\quad\quad 35y = 455 \quad$ Subtract.

$\quad\quad\quad\quad\quad\quad\quad\quad\quad\quad\; y = 13 \quad$ Divide by 35.

$x + 13 = 16 \quad$ Substitute 13 for y in Equation 1.

$\quad\; x = 3 \quad$ Solve for x.

There are 3 workers at \$200 per day and 13 at \$165 per day.

10. Problems will vary. Example: Mr. Martinez is four times as old as his son. The sum of his age and his son's age is 45. How old are he and his son?

PRACTICE

1. Multiply Equation 1 by 3 and Equation 2 by 5 or multiply Equation 1 by 7 and Equation 2 by 6.

2. Multiply Equation 1 by 9 and Equation 2 by 8 or multiply Equation 1 by 7 and Equation 2 by 3.

3. $2x + 2y = -2$

$\underline{5x - 2y = 9}$

$7x = 7 \quad$ Add the equations.

$\; x = 1 \quad$ Divide by 7.

$2(1) + 2y = -2 \quad$ Substitute 1 for x in Equation 1.

$\quad\quad y = -2 \quad$ Solve for y.

So, $(1, -2)$ solves the system.

4. $2x + 2y = 8$

$\underline{2x - y = 5}$

$3y = 3 \quad$ Subtract the equations.

$\; y = 1 \quad$ Divide by 3.

$2x - 1 = 5 \quad$ Substitute 1 for y in Equation 2.

$\quad x = 3 \quad$ Solve for y.

So, $(3, 1)$ solves the system.

5. $3x + 3y = 9$

$\underline{4x - 3y = -16}$

$7x = -7 \qquad$ Add the equations.

$x = -1 \qquad$ Divide by 7.

$3(-1) + 3y = 9 \qquad$ Substitute -1 for x in Equation 1.

$y = 4 \qquad$ Solve for y.

So, $(-1, 4)$ solves the system.

6. $(8x - 3y = 17) \cdot 2 \;\rightarrow\; 16x - 6y = 34$

$-7x + 6y = 2 \rightarrow\; \underline{-7x + 6y = 2}$

$9x = 36 \qquad$ Add.

$x = 4 \qquad$ Divide by 9.

$8(4) - 3y = 17 \qquad$ Substitute 4 for x in Equation 1.

$y = 5 \qquad$ Solve for y.

So, $(4, 5)$ solves the system.

7. $7x - 10y = -1 \rightarrow\; 7x - 10y = -1$

$(3x + 2y = -13) \cdot 5 \rightarrow \underline{15x + 10y = -65}$

$22x = -66 \qquad$ Add.

$x = -3 \qquad$ Divide by 22.

$3(-3) + 2y = -13 \qquad$ Substitute -3 for x in Equation 1.

$y = -2 \qquad$ Solve for y.

So, $(-3, -2)$ solves the system.

8. $(4x - 3y = 15) \cdot 2 \;\rightarrow\; 8x - 6y = 30$

$8x + 2y = -10 \rightarrow\; \underline{8x + 2y = -10}$

$-8y = 40 \qquad$ Subtract.

$y = -5 \qquad$ Divide by -8

$4x - 3(-5) = 15 \qquad$ Substitute -5 for x in Equation 1.

$x = 0 \qquad$ Solve for x.

So, $(0, -5)$ solves the system.

9. $2x + 8y = -1 \rightarrow\; 2x + 8y = -1$

$(-10x + 4y = 16) \cdot 2 \rightarrow \underline{-20x + 8y = 32}$

$22x = -33 \qquad$ Subtract.

$x = -\dfrac{3}{2} \qquad$ Divide by 22.

$2\left(-\dfrac{3}{2}\right) + 8y = -1 \qquad$ Substitute $-\dfrac{3}{2}$ for x in Equation 1.

$-3 + 8y = -1 \qquad$ Simplify.

$y = \dfrac{1}{4} \qquad$ Divide by 8.

So, $\left(-\dfrac{3}{2}, \dfrac{1}{4}\right)$ solves the system.

10. Write the equations in standard form.

$(5x + 3y = -9) \cdot 4 \;\rightarrow\; 20x + 12y = -36$

$(3x - 4y = -17) \cdot 3 \rightarrow \underline{9x - 12y = -51}$

$29x = -87 \qquad$ Add.

$x = -3 \qquad$ Divide by 29.

$5(-3) + 3y + 9 = 0 \qquad$ Substitute -3 for x in Equation 1.

$y = 2 \qquad$ Solve for y.

So, $(-3, 2)$ solves the system.

11. Write the equations in standard form.

$4x + 7y = 5 \rightarrow\; 4x + 7y = 5$

$(2x - 5y = -6) \cdot 2 \;\rightarrow\; \underline{4x - 10y = -12}$

$17y = 17 \qquad$ Subtract.

$y = 1 \qquad$ Divide by 17.

$2x = 5(1) - 6 \qquad$ Substitute 1 for y in Equation 2.

$x = -\dfrac{1}{2} \qquad$ Solve for x.

So, $\left(-\dfrac{1}{2}, 1\right)$ solves the system.

12. $(4x - 3y = 9) \cdot 3 \;\rightarrow\; 12x - 9y = 27$

$(-3x + 5y = 7) \cdot 4 \;\rightarrow\; \underline{-12x + 20y = 28}$

$11y = 55 \qquad$ Add.

$y = 5 \qquad$ Divide by 11.

$4x - 3(5) = 9 \qquad$ Substitute 5 for y in Equation 1.

$x = 6 \qquad$ Solve for x.

So, $(6, 5)$ solves the system.

13. $(3x + 4y = -1) \cdot 7 \;\rightarrow\; 21x + 28y = -7$

$(7x + 9y = 0) \cdot 3 \rightarrow\; \underline{21x + 27y = 0}$

$y = -7 \qquad$ Subtract.

$3x + 4(-7) = -1 \qquad$ Substitute -7 for y in Equation 1.

$x = 9 \qquad$ Solve for x.

So, $(9, -7)$ solves the system.

14. $(4x - 2y = -19) \cdot 3 \;\rightarrow\; 12x - 6y = -57$

$(-6x - 3y = 1.5) \cdot 2 \;\rightarrow\; \underline{-12x - 6y = 3}$

$-12y = -54 \qquad$ Add.

$y = 4.5 \qquad$ Divide.

$4x - 2(4.5) = -19 \qquad$ Substitute 4.5 for y in Equation 1.

$x = -2.5 \qquad$ Solve for x.

So, $(-2.5, 4.5)$ solves the system.

15. $79 \cdot (m + w = 89) \rightarrow 79m + 79w = 7031$

$79m + 65w = 6289 \rightarrow \underline{79m + 65w = 6289}$

$\hspace{5.5cm} 14w = 742 \quad$ Subtract.

$\hspace{5.7cm} w = 53 \quad$ Divide.

$m + 53 = 89 \hspace{1cm}$ Substitute 53 for w in Equation 1.

$\hspace{0.9cm} m = 36 \hspace{1.3cm}$ Solve for m.

So, it sold 36 pairs of Mercury
and 53 pairs of Whirlwind.

16. $7c + 8s = 50 \hspace{1cm} \rightarrow \hspace{0.5cm} 7c + 8s = \hspace{0.3cm} 50$

$(5c + 4s = 31) \cdot 2 \rightarrow \underline{10c + 8s = \hspace{0.3cm} 62}$

$\hspace{3.3cm} -3c \hspace{1.2cm} = -12 \quad$ Subtract.

$\hspace{4.5cm} c = 4 \hspace{1cm}$ Divide.

$5(4) + 4s = 31 \hspace{1cm}$ Substitute 4 for x in Equation 1.

$\hspace{1.2cm} s = 2.75 \hspace{0.7cm}$ Solve for s.

Chedder sold at $4.00 per pound.
Swiss sold at $2.75 per pound.

17. The solution is $x = 5$. Explanations will vary. Students
may point out that the elimination method includes
applying the multiplication property of equality. Both
the original and the equivalent equations have the same
solution. That is, applying the property does not
change the solution.

EXTEND

18. Write the equations in standard form.

$8x - 10y = 90$

$\underline{8x - \hspace{0.2cm} 2y = 66}$

$\hspace{0.7cm} -8y = 24 \hspace{1cm}$ Subtract.

$\hspace{1.1cm} y = -3 \hspace{1cm}$ Divide by -8.

$3[x - (-3)] - \frac{1}{3}[x + (-3)] = 30 \hspace{0.5cm}$ Substitute -3 for y in Equation 1.

$\hspace{1.5cm} 3x + 9 - \frac{1}{3}x + 1 = 30 \hspace{0.7cm}$ Distribute.

$\hspace{3cm} \frac{8}{3}x = 20 \hspace{0.7cm}$ Subtract 10 and simplify.

$\hspace{3cm} x = 7\frac{1}{2} \hspace{0.7cm}$ Multiply by $\frac{3}{8}$.

So, $\left(7\frac{1}{2}, -3\right)$ solves the system.

19. Write the equations in standard form.

$5x - 2y = 10 \hspace{1.3cm} \rightarrow \hspace{0.7cm} 5x - \hspace{0.3cm} 2y = \hspace{0.3cm} 10$

$(-x + 3y = 24) \cdot 5 \rightarrow \underline{-5x + 15y = 120}$

$\hspace{5.3cm} 13y = 130 \quad$ Add.

$\hspace{5.7cm} y = 10 \hspace{0.3cm}$ Divide.

$\frac{x}{2} - \frac{10}{5} = 1 \hspace{1cm}$ Substitute 10 for y in Equation 1.

$5x - 20 = 10 \hspace{0.5cm}$ Multiply by 10.

$\hspace{0.9cm} x = 6 \hspace{1cm}$ Solve for x.

So, $(6, 10)$ solves the system.

20. Write the equations in standard form.

$2x - 2y = 9$

$\underline{2x - \hspace{0.3cm} y = 7}$

$\hspace{0.7cm} -y = 2 \hspace{1cm}$ Subtract.

$\hspace{1.1cm} y = -2$

$7x + 3(-2 - 3) = 5[x + (-2)] \hspace{0.5cm}$ Substitute -2 for y in Equation 1.

$\hspace{1.6cm} 7x - 15 = 5x - 10 \hspace{0.7cm}$ Distribute.

$\hspace{3.2cm} x = \frac{5}{2} \hspace{1cm}$ Solve for x.

So, $\left(\frac{5}{2}, -2\right)$ solves the system.

21. Write the equations in standard form.

$3x - 2y = 8$

$\underline{3x + 7y = 1}$

$\hspace{0.5cm} -9y = 7 \hspace{1cm}$ Subtract.

$\hspace{0.9cm} y = -\frac{7}{9} \hspace{0.7cm}$ Divide.

$3x + 7\left(-\frac{7}{9}\right) = 1 \hspace{0.7cm}$ Substitute $-\frac{7}{9}$ for y in Equation 2.

$\hspace{1.8cm} 3x - \frac{49}{9} = 1 \hspace{0.7cm}$ Simplify.

$\hspace{1.6cm} 27x - 49 = 9 \hspace{0.7cm}$ Multiply by 9.

$\hspace{3cm} x = \frac{58}{27} \hspace{0.7cm}$ Solve for x.

So, $\left(\frac{58}{27}, -\frac{7}{9}\right)$ solves the system.

22. $2l + 2w = 200 \hspace{1cm} \rightarrow \hspace{0.5cm} 2l + 2w = 200$

$\hspace{0.8cm} l = w + 16 \hspace{0.7cm} \rightarrow \underline{\hspace{0.3cm} 2l - 2w = \hspace{0.3cm} 32}$

$\hspace{4.8cm} 4l = 232 \quad$ Add.

$\hspace{5.1cm} l = 58 \hspace{0.5cm}$ Divide.

$58 = w + 16 \hspace{1cm}$ Substitute 58 for l in Equation 2.

$42 = w \hspace{1.5cm}$ Solve for w.

The dimensions are 58 yd by 42 yd.

23. $y = 3.2x + 124$ where $x =$ years past 1970

$y = 6.5x + 70$

$\overline{0 = -3.3x + 54}$ Subtract the equations.

$x = 16.\overline{36}$ Solve for x.

The redwood surpassed the fir in 1986.

24. $x + y = 500$ $\rightarrow$ $20x + 20y = 10,000$

$0.20x + 0.30y = 0.24(500)$ $\rightarrow$ $\underline{20x + 30y = 12,000}$

$-10y = -2000$

$y = 200$

$x + 200 = 500$ Substitute 200 for y in Equation 1.

$x = 300$ Solve for x.

300 mL of 20% solution, 200 mL of 30% solution

25. the first equation, $N = 1200 - 10t$

26. If the length of time needed to complete the route increases, the bus slows down, and some riders will find another, faster way to travel to their destination.

27. $N = 1200 - 10t$ $\rightarrow$ $10t + N = 1200$

$t = 18 + 0.001N$ $\rightarrow$ $\underline{10t - 0.01N = 180}$

$1.01N = 1020$ Subtract.

$N \approx 1010$

$t \approx 18 + 0.001(1010) = 18 + 1.01 \approx 19$

The equilibrium point of about 1010 passengers is reached when the bus completes its route in about 19 minutes.

28. The length of time needed to complete the route has been reduced. The faster speed of the trip is likely to increase the number of riders.

THINK CRITICALLY

29. $5x + 4y = 9a + b$ $\rightarrow$ $15x + 12y = 27a + 3b$

$7x - 6y = a + 13b$ $\rightarrow$ $\underline{14x - 12y = 2a + 26b}$

$29x = 29a + 29b$ Add.

$x = a + b$

$5(a + b) + 4y = 9a + b$ Substitute $a + b$ for x in Equation 1.

$y = a - b$ Solve for y.

So, $(a + b, a - b)$ solves the system.

30. $4ax - 3cy = 3b + a$ $\rightarrow$ $4ax - 3cy = 3b + a$

$ax - cy = b$ $\rightarrow$ $\underline{3ax - 3cy = 3b}$

$ax = a$ Subtract.

$x = 1$

$a(1) - cy = b$ Substitute 1 for x in Equation 2.

$y = \dfrac{a - b}{c}$ Solve for y.

So, $\left(1, \dfrac{a - b}{c}\right)$ solves the system.

31. $A = \frac{1}{2}bh$; $y = 3x$ crosses x-axis at $x = 0$, $y = -\frac{1}{2}x + 7$ crosses x-axis at $x = 14$, so $b = 14 - 0 = 14$.

$h = y$-coordinate of the intersection of $y = -\frac{1}{2}x + 7$ and $y = 3x$.

$\begin{cases} y = 3x \\ y = -\frac{1}{2}x + 7 \end{cases}$

$0 = 3\frac{1}{2}x - 7$ Subtract.

$x = 2$ Solve for x.

$y = 3(2)$ Substitute 2 for x in Equation 1.

$y = 6$

Area $= \frac{1}{2}(14)(6) = 42$

32.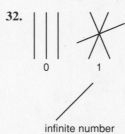

0   1

infinite number

Answers will vary but should include the following: The three lines may intersect at one point, two of the lines may be parallel, the three lines may be parallel, or all three may coincide. The solution to a system of three linear equations must satisfy each equation.

So, there will be one solution, no solution, or infinitely many solutions if all three lines are collinear.

33. $\begin{cases} x - 2y = -6 & \text{Equation 1} \\ x + 6y = -6 & \text{Equation 2} \end{cases}$

$-8y = 0$ Subtract.

$y = 0$ Solve for y.

$x - 2(0) = -6$ Substitute 0 for y in Equation 1.

$x = -6$ Solve for x.

So, $(-6, 0)$ is one vertex.

$\begin{cases} x - 2y = -6 & \text{Equation 1} \\ 3x + 2y = 14 & \text{Equation 3} \end{cases}$

$4x = 8$ Add.

$x = 2$ Solve for x.

$2 - 2y = -6$ Substitute 2 for x in Equation 1.

$y = 4$ Solve for y.

So, $(2, 4)$ is the second vertex.

$x + 6y = -6$ $\rightarrow$ $3x + 18y = -18$

$3x + 2y = 14$ $\rightarrow$ $\underline{3x + 2y = 14}$

$16y = -32$ Subtract.

$y = -2$

$x + 6(-2) = -6$ Substitute -2 for y in Equation 2.

$x = 6$ Solve for x.

So, $(6, -2)$ is the third vertex.

34. She added $-3y$ and $8y$ instead of subtracting. She should have gotten

$$-11y = 11$$
$$y = -1$$
$$x + 4(-1) = 1 \rightarrow x = 5$$

So, the solution is $(5,1)$.

MIXED REVIEW

35. $5(a - 7) = 5(a) + 5(-7) = 5a - 35$

36. $-2(x - 6) = -2(x) - 2(-6) = -2x + 12$

37. $2(b + 3b^2) = 2(b) + 2(3b^2) = 2b + 6b^2$

38. $4(2m - 1) = 4(2m) + 4(-1) = 8m - 4$

39. $d = kt \rightarrow 93,500 = k \cdot 5 \rightarrow k = 18,700$

Solve for 4 hours:

$$d = 18,700t$$
$$d = 18,700(4) = 74,800$$

40. $-x - 5 < 4$

$$-x < 9 \qquad \text{Add 5.}$$
$$x > -9 \qquad \text{Multiply by 1 and reverse inequality symbol.}$$

So, the correct answer is D.

41. $3x - 2y = -4 \quad \rightarrow \quad 9x - 6y = -12$

$7x - 3y = 9 \quad \rightarrow \quad \underline{14x - 6y = 18}$

$$-5x = -30 \qquad \text{Subtract.}$$
$$x = 6 \qquad \text{Divide by 4.}$$

$3(6) - 2y = -4 \qquad$ Substitute 6 for x in Equation 1.

$$y = 11 \qquad \text{Solve for } y.$$

So, $(6,11)$ solves the system.

Lesson 7.5 , pages 348–352

EXPLORE

1. Go north 5 blocks.

2. 59th Street is Central Park South. Any point on Central Park South is also on 59th Street. The driver needs further instructions concerning where on the street the tourist wishes to go.

3. Fifth Avenue and the Avenue of the Americas are parallel and therefore do not intersect. The driver cannot follow the instructions

4. The lines might intersect at one point (as do Seventh Avenue and 56th Street), at an infinite number of points (Sixth Avenue and the Avenue of the Americas), or at no points (48th Street and 49th Street).

TRY THESE

1. $x + y = 13$

$\underline{x - y = 5}$

$2x = 18 \qquad$ Add the equations.

$x = 9 \qquad$ Solve for x.

$9 + y = 13 \qquad$ Substitute 9 for x in Equation 1.

$y = 4 \qquad$ Solve for y.

So, $(9,4)$ solves the system.

2. $x - 2y = 4 \quad \rightarrow \quad 3x - 6y = 12$

$3x - 6y = 12 \quad \rightarrow \quad \underline{3x - 6y = 12}$

$0 = 0 \qquad$ Subtract.

True, so the system is dependent.

3. $-2x + y = 0 \quad \rightarrow \quad -4x + 2y = 0$

$-4x + 2y = 2 \quad \rightarrow \quad \underline{-4x + 2y = 2}$

$0 = 2 \qquad$ Subtract.

False, so the system is inconsistent.

4. Two identical lines, so the system is dependent.

5. One point of intersection, so the system is independent.

6. Parallel lines, so the system is inconsistent.

7. Consolidated: $y = 5.60x - 35$

Co-op: $y = 5.60x - 25$

8. Both lines have slope of 5.60 so they are parallel and will never intersect. Co-op's payment will always exceed consolidated.

9. Charts may vary. Here is one possible chart.

Number of solutions	Picture	Description	Characterization
0		parallel lines	inconsistent
1		intersecting lines	consistent and independent
infinite		same line	consistent and dependent

PRACTICE

1. Both lines have the same slope so they are parallel and the system is inconsistent.

2. $4x - y = 3$

$\underline{2x - y = -1}$

$2x = 4 \qquad$ Subtract the equations.

$x = 2 \qquad$ Solve for x.

$4(2) - y = 3 \qquad$ Substitute 2 for x in Equation 1.

$y = 5 \qquad$ Solve for y.

So, $(2,5)$ solves the system.

3. $x - 2y = 4 \quad \rightarrow \quad x - 2y = 4$

$3y - x = y - 8 \quad \rightarrow \quad \underline{-x + 2y = -8}$

$0 = -4 \quad$ Add.

False, so the system is inconsistent.

4. $y = -x + 4 \quad \rightarrow \quad 3y = -3x + 12$

$3y = -3x + 12 \quad \rightarrow \quad \underline{3y = -3x + 12}$

$0 = 0 \qquad$ Subtract.

True, so the system is dependent.

5. $2(6x + 10y) + 8 = 0 \quad \rightarrow \quad 12x + 20y = -8$

$-2 = 3x + 5y \qquad \rightarrow \quad \underline{12x + 20y = -8}$

$0 = 0 \qquad$ Subtract.

True, so the system is dependent.

6. $y = -4x + 2 \rightarrow 4x + y = 2$

$x + 2y = 11 \quad \rightarrow \quad \underline{4x + 8y = 44}$

$-7y = -42 \quad$ Subtract.

$y = 6 \qquad$ Solve for y.

$6 = -4x + 2 \qquad$ Substitute 6 for y in Equation 1.

$-1 = x \qquad\qquad$ Solve for x.

So, $(-1, 6)$ solves the system.

7. Lines are parallel, so the system is inconsistent

8. Two identical lines, so the system is dependent.

9. One point of intersection, so the system is independent.

10. $2x + 3y = 96 \quad \rightarrow \quad 2x + 3y = 96$

$x + 2y = 42 \quad \rightarrow \quad \underline{2x + 4y = 84}$

$-y = 12 \quad$ Subtract.

$y = -12 \quad$ Solve for y.

$x + 2(-12) = 42 \quad$ Substitute -12 for y in Equation 2.

$x = 66 \quad$ Solve for x.

The freighters meet at 66° E, 12° S.

11. Answers will vary.

EXTEND

12. $12x + 10y = 20 \rightarrow 12x + 10y = 20$

$ax + 2y = b \quad \rightarrow \quad \underline{5ax + 10y = 5b}$

$12x - 5ax = 20 - 5b \quad$ Subtract.

If the system is dependent, then:

$12x - 5ax = 0 \qquad$ and $\qquad 20 - 5b = 0$

$12x = 5ax \qquad$ and $\qquad 20 = 5b$

$2.4 = a \qquad$ and $\qquad 4 = b$

13. $y = 3x - 5$

$5y = ax + b \rightarrow y = \dfrac{a}{5}x + \dfrac{b}{5}$

To be inconsistent, slopes must be equal. So $3 = \dfrac{a}{5}$ or

$15 = a$. If $\dfrac{b}{5} = -5$ or $b = -25$, the system would be

dependent, so $b \neq 25$.

14. First table:

Slope is $m = \dfrac{-4 - (-6)}{1 - 2} = \dfrac{2}{-1} = -2$ and

$y - (-4) = -2(x - 1)$ or $y = -2x - 2$.

Second table:

Slope is $m = \dfrac{-9 - (-12)}{1 - 2} = \dfrac{3}{-1} = -3$ and

$y - (-9) = -3(x - 1)$ or $y = -3x - 6$.

$-2x - 2 = -3x - 6 \qquad$ Substitute $-2x - 2$ for y.

$x = -4 \qquad\qquad$ Solve for x.

$y = -2(-4) - 2 = 6 \qquad$ Solve for y.

So, $(-4, 6)$ solves the system.

15. $a(3) + 4(2) = -1 \qquad$ Substitute $(3, 2)$ in Equation 2.

$a = -3 \qquad\qquad$ Solve for a.

THINK CRITICALLY

16. The slopes are unequal. The y-intercepts may or may not be equal.

17. The slopes are equal. The y-intercepts are unequal.

18. The slopes and the y-intercepts are equal.

19. $A + B + C = 11{,}900 \quad \rightarrow \quad A + B + C = 11{,}900$

$\left.\begin{matrix} A + B = 7{,}700 \\ B + C = 8{,}300 \end{matrix}\right\} \rightarrow \underline{A + 2B + C = 16{,}000}$

$-B = -4{,}100$

$B = 4{,}100$

$A + 4{,}100 = 7{,}700$, so $A = 3{,}600$

$4{,}100 + C = 8{,}300$, so $C = 4{,}200$

Lesson 7.6, pages 353–359

EXPLORE / WORKING TOGETHER

1. Graphing:

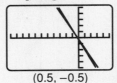

$(0.5, -0.5)$

(Solution continues on next page.)

Substitution:

$$4x + 6y = -1 \rightarrow 4x = -6y - 1 \rightarrow x = -\frac{3}{2}y - \frac{1}{4}$$

$$3\left(-\frac{3}{2}y - \frac{1}{4}\right) + 5y = -1 \quad \text{Equation 2}$$

$$-\frac{9}{2}y - \frac{3}{4} + 5y = -1$$

$$\frac{1}{2}y = -\frac{1}{4}$$

$$y = -\frac{1}{2} \quad \text{Solve for } x.$$

$$4x + 6\left(-\frac{1}{2}\right) = -1 \quad \text{Equation 1}$$

$$4x - 3 = -1$$

$$x = \frac{1}{2} \quad \text{Solve for } y.$$

So, $\left(\frac{1}{2}, -\frac{1}{2}\right)$ is the solution.

Elimination:

$$4x + 6y = -1 \rightarrow \quad 12x + 18y = -3$$
$$3x + 5y = -1 \rightarrow \quad \underline{12x + 20y = -4}$$
$$-2y = 1 \quad \text{Subtract.}$$
$$y = -\frac{1}{2} \quad \text{Solve for } y.$$

$$4x + 6\left(-\frac{1}{2}\right) = -1 \quad \text{Substitute} -\frac{1}{2} \text{ for } y \text{ in Equation 1.}$$

$$x = \frac{1}{2} \quad \text{Solve for } x.$$

So, $\left(\frac{1}{2}, -\frac{1}{2}\right)$ is the solution.

2. Answers may vary. Because none of the coefficients is 1 or -1, students will have to solve both equations for y before solving the system graphically or by substitution. For this reason, they may find the elimination method easiest for this example. Solved for y, both equations contain fractions that are difficult to substitute. For this reason, students may find substitution harder. Graphing is also hard. The slopes and intercept of the two lines are very close, so it is difficult to plot them accurately and to distinguish them as two separate lines.

3. Graphing provides a visual representation of the equations, often an advantage in interpreting real world problems. When one coefficient is 1 or -1, substitution can often be accomplished easily. Otherwise, elimination is usually the simplest method to use.

TRY THESE

1. $\begin{vmatrix} 1 & 2 \\ 3 & 4 \end{vmatrix} = 1(4) - 2(3) = 4 - 6 = -2$

2. $\begin{vmatrix} 0 & 1 \\ 0 & 1 \end{vmatrix} = 0(1) - 1(0) = 0 - 0 = 0$

3. $\begin{vmatrix} 2 & 5 \\ 4 & -1 \end{vmatrix} = 2(-1) - 5(4) = -2 - 20 = -22$

4. $\begin{vmatrix} -6 & 2 \\ 7 & -3 \end{vmatrix} = -6(-3) - 2(7) = 18 - 14 = 4$

5. $\begin{vmatrix} 5 & -2 \\ -4 & 8 \end{vmatrix} = 5(8) - (-2)(-4) = 40 - 8 = 32$

6. $\begin{vmatrix} 3 & -3 \\ -3 & 3 \end{vmatrix} = 3(3) - (-3)(-3) = 9 - 9 = 0$

7. $D = \begin{vmatrix} 5 & 6 \\ 2 & 1 \end{vmatrix} = 5(1) - 6(2) = 5 - 12 = -7$

$D_x = \begin{vmatrix} 1 & 6 \\ 6 & 1 \end{vmatrix} = 1(1) - 6(6) = 1 - 36 = -35$

$D_y = \begin{vmatrix} 5 & 1 \\ 2 & 6 \end{vmatrix} = 5(6) - 1(2) = 30 - 2 = 28$

$x = \dfrac{D_x}{D} = \dfrac{-35}{-7} = 5$ and $y = \dfrac{D_y}{D} = \dfrac{28}{-7} = -4$

Solution is $(5, -4)$.

8. $D = \begin{vmatrix} -4 & 3 \\ 3 & 4 \end{vmatrix} = -4(4) - 3(3) = -16 - 9 = -25$

$D_x = \begin{vmatrix} 6 & 3 \\ 8 & 4 \end{vmatrix} = 6(4) - 3(8) = 24 - 24 = 0$

$D_y = \begin{vmatrix} -4 & 6 \\ 3 & 8 \end{vmatrix} = -4(8) - 6(3) = -32 - 18 = -50$

$x = \dfrac{D_x}{D} = \dfrac{0}{-25} = 0$ and $y = \dfrac{D_y}{D} = \dfrac{-50}{-25} = 2$

Solution is $(0, 2)$.

9. Rewrite equations in standard form.

$$x - 2y = 4$$
$$2x + 3y = 1$$

$D = \begin{vmatrix} 1 & -2 \\ 2 & 3 \end{vmatrix} = 1(3) - (-2)(2) = 3 + 4 = 7$

$D_x = \begin{vmatrix} 4 & -2 \\ 1 & 3 \end{vmatrix} = 4(3) - (-2)(1) = 12 + 2 = 14$

$D_y = \begin{vmatrix} 1 & 4 \\ 2 & 1 \end{vmatrix} = 1(1) - 4(2) = 1 - 8 = -7$

$x = \dfrac{D_x}{D} = \dfrac{14}{7} = 2$ and $y = \dfrac{D_y}{D} = \dfrac{-7}{7} = -1$

Solution is $(2, -1)$

10. CDs, $8; cassettes, $6

11. Descriptions will vary. Students should mention writing both equations in standard form before beginning and then using the coefficients and constants from the equations to write three determinants. They should specify how the three determinants are used to find x and y. Advantages include that the method is easy to remember

and apply; some graphing utilities evaluate determinants. Disadvantages include that the method does not work when the determinant of coefficients, D, is zero and that opportunities for errors in arithmetic are many.

PRACTICE

1. $\begin{vmatrix} 0 & 1 \\ 3 & 2 \end{vmatrix} = 0(2) - 1(3) = 0 - 3 = -3$

2. $\begin{vmatrix} 5 & 2 \\ 1 & 4 \end{vmatrix} = 5(4) - 2(1) = 20 - 2 = 18$

3. $\begin{vmatrix} 4 & 4 \\ 4 & 4 \end{vmatrix} = 4(4) - 4(4) = 16 - 16 = 0$

4. $\begin{vmatrix} 2 & 2 \\ 5 & 5 \end{vmatrix} = 2(5) - 2(5) = 10 - 10 = 0$

5. $\begin{vmatrix} -1 & 3 \\ 4 & -8 \end{vmatrix} = -1(-8) - 3(4) = 8 - 12 = -4$

6. $\begin{vmatrix} 6 & 4 \\ 3 & 2 \end{vmatrix} = 6(2) - 4(3) = 12 - 12 = 0$

7. $\begin{vmatrix} -6 & 8 \\ 5 & -7 \end{vmatrix} = -6(-7) - 8(5) = 42 - 40 = 2$

8. $\begin{vmatrix} -10 & 9 \\ -8 & 6 \end{vmatrix} = -10(6) - 9(-8) = -60 + 72 = 12$

9. $\begin{vmatrix} 12 & 11 \\ 9 & -6 \end{vmatrix} = 12(-6) - 11(9) = -72 - 99 = -171$

10. $\begin{vmatrix} x & y \\ 2 & 3 \end{vmatrix} = x(3) - y(2) = 3x - 2y$

11. $\begin{vmatrix} 2k & 4 \\ 3k & 2 \end{vmatrix} = 2k(2) - 4(3k) = 4k - 12k = -8k$

12. $\begin{vmatrix} 3m & 2m \\ 7 & 5 \end{vmatrix} = 3m(5) - 2m(7) = 15m - 14m = m$

13. Answers will vary but need to include that the value of the determinant is the difference of the products of the diagonals.

14. $D = \begin{vmatrix} 4 & -1 \\ 1 & -3 \end{vmatrix} = -4(-3) - (-1)(1) = -11$

$D_x = \begin{vmatrix} 9 & -1 \\ 16 & -3 \end{vmatrix} = 9(-3) - (-1)(16) = -11$

$D_y = \begin{vmatrix} 4 & 9 \\ 1 & 16 \end{vmatrix} = 4(16) - 9(1) = 55$

$x = \dfrac{D_x}{D} = \dfrac{-11}{-11} = 1$ and $y = \dfrac{D_y}{D} = \dfrac{55}{-11} = -5$

Solution is $(1, -5)$.

15. $D = \begin{vmatrix} 3 & -5 \\ 5 & 4 \end{vmatrix} = 3(4) - (-5)(5) = 37$

$D_x = \begin{vmatrix} -23 & -5 \\ 11 & 4 \end{vmatrix} = -23(4) - (-5)(11) = -37$

$D_y = \begin{vmatrix} 3 & -23 \\ 5 & 11 \end{vmatrix} = 3(11) - (-23)(5) = 148$

$x = \dfrac{D_x}{D} = \dfrac{-37}{37} = -1$ and $y = \dfrac{D_y}{D} = \dfrac{148}{37} = 4$

Solution is $(-1, 4)$.

16. $D = \begin{vmatrix} 1 & 4 \\ 5 & -7 \end{vmatrix} = 1(-7) - 4(5) = -27$

$D_x = \begin{vmatrix} 13 & 4 \\ -16 & -7 \end{vmatrix} = 13(-7) - 4(-16) = -27$

$D_y = \begin{vmatrix} 1 & 13 \\ 5 & -16 \end{vmatrix} = 1(-16) - 13(5) = -81$

$x = \dfrac{D_x}{D} = \dfrac{-27}{-27} = 1$ and $y = \dfrac{D_y}{D} = \dfrac{-81}{-27} = 3$

Solution is $(1, 3)$.

17. $D = \begin{vmatrix} 3 & 1 \\ 2 & -3 \end{vmatrix} = 3(-3) - 1(2) = -11$

$D_x = \begin{vmatrix} -1 & 1 \\ -8 & -3 \end{vmatrix} = -1(-3) - 1(-8) = 11$

$D_y = \begin{vmatrix} 3 & -1 \\ 2 & -8 \end{vmatrix} = 3(-8) - (-1)(2) = -22$

$x = \dfrac{D_x}{D} = \dfrac{11}{-11} = -1$ and $y = \dfrac{D_y}{D} = \dfrac{-22}{-11} = 2$

Solution is $(-1, 2)$.

18. Rewrite equations in standard form.

$9x - 2y = 3$

$3x - y = 6$

$D = \begin{vmatrix} 9 & -2 \\ 3 & -1 \end{vmatrix} = 9(-1) - (-2)(3) = -3$

$D_x = \begin{vmatrix} -3 & 2 \\ 6 & -1 \end{vmatrix} = 3(-1) - (-2)(6) = 9$

$D_y = \begin{vmatrix} 9 & 3 \\ 3 & 6 \end{vmatrix} = 9(6) - 3(3) = 45$

$x = \dfrac{D_x}{D} = \dfrac{9}{-3} = -3$ and $y = \dfrac{D_y}{D} = \dfrac{45}{-3} = -15$

Solution is $(-3, -15)$.

19. Rewrite equations in standard form.

$x - 3y = 7$

$5x - 6y = -14$

$D = \begin{vmatrix} 1 & -3 \\ 5 & -6 \end{vmatrix} = 1(-6) - (-3)(5) = 9$

$D_x = \begin{vmatrix} 7 & -3 \\ -14 & -6 \end{vmatrix} = \begin{matrix} 7(-6) - (-3)(-14) \\ = -84 \end{matrix}$

$D_y = \begin{vmatrix} 1 & 7 \\ 5 & -14 \end{vmatrix} = 1(-14) - 7(5) = -49$

$x = \dfrac{D_x}{D} = \dfrac{-84}{9} = -\dfrac{28}{3}$ and $y = \dfrac{D_y}{D} = \dfrac{-49}{9}$

Solution is $\left(\dfrac{-28}{3}, \dfrac{-49}{9}\right)$

20. Rewrite equations in standard form.

$3x - 7y = 2$

$6x - 13y = 4$

$D = \begin{vmatrix} 3 & -7 \\ 6 & -13 \end{vmatrix} = 3(-13) - (-7)(6) = 3$

$D_x = \begin{vmatrix} 2 & -7 \\ 4 & -13 \end{vmatrix} = 2(-13) - (-7)(4) = 2$

$D_y = \begin{vmatrix} 3 & 2 \\ 6 & 4 \end{vmatrix} = 3(4) - 2(6) = 0$

$x = \dfrac{D_x}{D} = \dfrac{2}{3}$ and $y = \dfrac{D_y}{D} = \dfrac{0}{3} = 0$

Solution is $\left(\dfrac{2}{3}, 0\right)$.

21. Rewrite equations in standard form.

$9x - 6y = 0$

$3x - 4y = -18$

$D = \begin{vmatrix} 9 & -6 \\ 3 & -4 \end{vmatrix} = 9(-4) - (-6)(3) = -18$

$D_x = \begin{vmatrix} 0 & -6 \\ -18 & -4 \end{vmatrix} = 0(-4) - (-6)(-18) = -108$

$D_y = \begin{vmatrix} 9 & 0 \\ 3 & -18 \end{vmatrix} = 9(-18) - 0(3) = -162$

$x = \dfrac{D_x}{D} = \dfrac{-108}{-18} = 6$ and $y = \dfrac{D_y}{D} = \dfrac{-162}{-18} = 9$

Solution is $(6, 9)$.

22. Rewrite equations in standard form.

$4x + 6y = 16$

$x - 2y = 1.2$

$D = \begin{vmatrix} 4 & 6 \\ 1 & -2 \end{vmatrix} = 4(-2) - 6(1) = -14$

$D_x = \begin{vmatrix} 16 & 6 \\ 1.2 & -2 \end{vmatrix} = 16(-2) - (6)(1.2) = -39.2$

$D_y = \begin{vmatrix} 4 & 16 \\ 1 & 1.2 \end{vmatrix} = 4(1.2) - 16(1) = -11.2$

$x = \dfrac{D_x}{D} = \dfrac{-39.2}{-14} = 2.8$ and $y = \dfrac{D_y}{D} = \dfrac{-11.2}{-14} = 0.8$

Solution is $(2.8, 0.8)$.

23. $D = \begin{vmatrix} 3 & 2 \\ 15 & -2 \end{vmatrix} = 3(-2) - 2(15) = -36$

$D_x = \begin{vmatrix} 24 & 2 \\ 48 & -2 \end{vmatrix} = 24(-2) - 2(48) = -144$

$D_y = \begin{vmatrix} 3 & 24 \\ 15 & 48 \end{vmatrix} = 3(48) - 24(15) = -216$

$x = \dfrac{D_x}{D} = \dfrac{-144}{-36} = 4$ and $y = \dfrac{D_y}{D} = \dfrac{-216}{-36} = 6$

Solution is $(4, 6)$.

24. $D = \begin{vmatrix} 1 & -6 \\ 1 & 2 \end{vmatrix} = 1(2) - (-6)(1) = 8$

$D_x = \begin{vmatrix} 3 & -6 \\ 5 & 2 \end{vmatrix} = 3(2) - (-6)(5) = 36$

$D_y = \begin{vmatrix} 1 & 3 \\ 1 & 5 \end{vmatrix} = 1(5) - 3(1) = 2$

$x = \dfrac{D_x}{D} = \dfrac{36}{8} = \dfrac{9}{2}$ and $y = \dfrac{D_y}{D} = \dfrac{2}{8} = \dfrac{1}{4}$

Solution is $\left(\dfrac{9}{2}, \dfrac{1}{4}\right)$.

25. $D = \begin{vmatrix} 5 & -2 \\ 4 & 5 \end{vmatrix} = 5(5) - (-2)(4) = 33$

$D_x = \begin{vmatrix} 1 & -2 \\ 47 & 5 \end{vmatrix} = 1(5) - (-2)(47) = 99$

$D_y = \begin{vmatrix} 5 & 1 \\ 4 & 47 \end{vmatrix} = 5(47) - 1(4) = 231$

$x = \dfrac{D_x}{D} = \dfrac{99}{33} = 3$ and $y = \dfrac{D_y}{D} = \dfrac{231}{33} = 7$

Solution is $(3, 7)$.

26. $9l + 8s = 220$

$\quad 7l + 6s = 170$

$D = \begin{vmatrix} 9 & 8 \\ 7 & 6 \end{vmatrix} = 9(6) - (8)(7) = -2$

$D_l = \begin{vmatrix} 220 & 8 \\ 170 & 6 \end{vmatrix} = 220(6) - 8(170) = -40$

$D_s = \begin{vmatrix} 9 & 220 \\ 7 & 170 \end{vmatrix} = 9(170) - 220(7) = -10$

$l = \dfrac{D_l}{D} = \dfrac{-40}{-2} = 20$ and $s = \dfrac{D_s}{D} = \dfrac{-10}{-2} = 5$

He used $20 bills and $5 bills.

27. $\quad w + c = 40$

$\quad 30w + 22c = 1000$

$D = \begin{vmatrix} 1 & 1 \\ 30 & 22 \end{vmatrix} = 1(22) - 1(30) = -8$

$D_w = \begin{vmatrix} 40 & 1 \\ 1000 & 22 \end{vmatrix} = 40(22) - 1(1000) = -120$

$D_c = \begin{vmatrix} 1 & 40 \\ 30 & 1000 \end{vmatrix} = 9(1000) - 40(30) = -200$

$w = \dfrac{D_w}{D} = \dfrac{-120}{-8} = 15$ and $c = \dfrac{D_c}{D} = \dfrac{-200}{-8} = 25$

He sold 15 wool blend slacks and 25 chinos.

28. $\quad x + y = 30{,}000$

$\quad 15x + 25y = 530{,}000$

$D = \begin{vmatrix} 1 & 1 \\ 15 & 25 \end{vmatrix} = 1(25) - 1(15) = 10$

$D_x = \begin{vmatrix} 30{,}000 & 1 \\ 530{,}000 & 25 \end{vmatrix} = \begin{aligned} &30{,}000(25) - 530{,}000(1) \\ &= 220{,}000 \end{aligned}$

$D_y = \begin{vmatrix} 1 & 30{,}000 \\ 15 & 530{,}000 \end{vmatrix} = \begin{aligned} &1(530{,}000) - 30{,}000(15) \\ &= 80{,}000 \end{aligned}$

$x = \dfrac{D_x}{D} = \dfrac{220{,}000}{10} = 22{,}000$ and

$y = \dfrac{D_y}{D} = \dfrac{80{,}000}{10} = 8{,}000$

So $22{,}000$ sold at $15 and $8{,}000$ sold at $25.

29. $\quad 8x + 11y = 76$

$\quad 15x + 40y = 220$

$D = \begin{vmatrix} 8 & 11 \\ 15 & 40 \end{vmatrix} = 8(40) - 11(15) = 155$

$D_x = \begin{vmatrix} 76 & 11 \\ 220 & 40 \end{vmatrix} = 76(40) - 11(220) = 620$

$D_y = \begin{vmatrix} 8 & 76 \\ 15 & 220 \end{vmatrix} = 8(220) - 76(15) = 620$

$x = \dfrac{D_x}{D} = \dfrac{620}{155} = 4$ and $y = \dfrac{D_y}{D} = \dfrac{620}{155} = 4$

So 4 g each.

30. Answers may vary.

31. $2[0(2) - 4(6)] - 3[3(-5) - 5(-4)] = 2(-24) - 3(5) = -63$

32. $-5[(-1)(-4) - (-2)(-3)] + 4[6(-2) - 2(-3)] =$
$\quad -5(-2) + 4(-6) = -14$

33. $\quad a - 3 = f + 3 \qquad \rightarrow a - f = 6$

$\quad (a - 3) + 5 = 2[(f + 3) - 5] \quad \rightarrow a - 2f = -6$

$D = \begin{vmatrix} 1 & -1 \\ 1 & -2 \end{vmatrix} = 1(-2) - 1(-1) = -1$

$D_a = \begin{vmatrix} 6 & -1 \\ -6 & -2 \end{vmatrix} = 6(-2) - (-1)(-6) = -18$

$D_f = \begin{vmatrix} 1 & 6 \\ 1 & -6 \end{vmatrix} = 1(-6) - 6(1) = -12$

$a = \dfrac{D_a}{D} = \dfrac{-18}{-1} = 18$ and $f = \dfrac{D_f}{D} = \dfrac{-12}{-1} = 12$

So André had $18, and Franco had $12.

34. $6x - 4x = 10(2) - 6(2) \rightarrow 2x = 8 \rightarrow x = 4$

35. $-[5(x - 1) - 2(x + 1)] = 4(x + 1) - 3(x - 1) \rightarrow$
$\quad -(5x - 5 - 2x - 2) = 4x + 4 - 3x + 3 \rightarrow$
$\quad -3x + 7 = x + 7 \rightarrow -2x = 0 \rightarrow x = 0$

36. $\begin{bmatrix} 7 & 2 \\ 5 & 3 \end{bmatrix}^{-1} \begin{bmatrix} 4 \\ -5 \end{bmatrix} = \begin{bmatrix} 2 \\ -5 \end{bmatrix}$

37. $\begin{bmatrix} 3 & -2 \\ -6 & 5 \end{bmatrix}^{-1} \begin{bmatrix} -2 \\ 14 \end{bmatrix} = \begin{bmatrix} 6 \\ 10 \end{bmatrix}$

38. $\begin{bmatrix} -7 & -12 \\ 4 & 9 \end{bmatrix}^{-1} \begin{bmatrix} 13 \\ -1 \end{bmatrix} = \begin{bmatrix} -7 \\ 3 \end{bmatrix}$

39. $\begin{vmatrix} a & b \\ c & d \end{vmatrix} = ad - bc = 4$ The value of $\begin{vmatrix} c & d \\ a & b \end{vmatrix}$ is

$cb - ad$. And $cb - ad = bc - ad$, which

is the additive inverse of $ad - bc$.

So, $bc - ad = -(ad - bc) = -(4)$.

40. $\begin{vmatrix} b & a \\ d & c \end{vmatrix} = bc - da = bc - ad = -(-bc + ad) = -(4)$

41. $\begin{vmatrix} d & c \\ b & a \end{vmatrix} = da - bc = ad - bc = 4$

42. $D = \begin{vmatrix} 3 & -5 \\ 9 & -15 \end{vmatrix} = 3(-15) - (-5)(9) = 0$

Therefore $\dfrac{D_x}{D}$ and $\dfrac{D_x}{D}$ are undefined, so x and y are

undefined.

43. $\dfrac{78}{200} = 0.39$ or 39%

44. $0.04x = 112 \rightarrow x = \$2{,}800$

45–48.

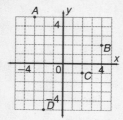

49. Correct answer is B because it is the only equation which passes through $(2, 3)$ and has slope $m = 4$.

50. $D = \begin{vmatrix} 3 & 7 \\ 5 & 9 \end{vmatrix} = 3(9) - 7(5) = -8$

$D_x = \begin{vmatrix} 1 & 7 \\ 7 & 9 \end{vmatrix} = 1(9) - 7(7) = -40$

$D_y = \begin{vmatrix} 3 & 1 \\ 5 & 7 \end{vmatrix} = 3(7) - 1(5) = 16$

$x = \dfrac{D_x}{D} = \dfrac{-40}{-8} = 5$ and $y = \dfrac{D_y}{D} = \dfrac{16}{-8} = -2$

The solution is $(5, -2)$.

ALGEBRAWORKS

1. $y = 8x$

2. $y = 2.25x + 2400 + 1280 = 2.25x + 3680$

3. Substitute $8x$ for y: $8x = 2.25x + 3680$

$\qquad\qquad\qquad 5.75 = 3680$

$\qquad\qquad\qquad\quad x = 640$

Substitute 640 for x: $y = 8(640) = 5120$

If 640 people attend, the cost of \$5120 will equal the revenue of \$5120.

4. Profit = Revenue – Cost, so

$\quad 15,000 = 8x - (2.25x + 3680)$

$\quad 18,680 = 5.75x$

$\quad\;\; 3249 \approx x$

5. Raising the price will increase the amount raised per ticket but may reduce the number of people who decide to attend the event. Lowering the price will lower the per-ticket profit but may increase the number of people who attend. Setting the ticket price that will result in maximum efforts will require the organization to balance these two competing factors.

Lesson 7.7, pages 360–363

EXPLORE

1. $35 - x$

2. Each dime is 10 cents, so $10x$

3. Each quarter is 25 cents, so $25(35 - x)$

4. $10x + 25(35 - x) = 515$

5. $10x + 875 - 25x = 515 \rightarrow 15x = 360 \rightarrow x = 24$;
$35 - x = 35 - 24 = 11$
24 dimes and 11 quarters

6. $d + q = 35$

7. $10d$; $25q$; $10d + 25q = 515$

8. $\begin{aligned} d + q &= 35 &\rightarrow\quad 25d + 25q &= 875 \\ 10d + 25q &= 515 &\rightarrow\quad 10d + 25q &= 515 \\ \hline & & 15d\qquad\quad &= 360 \quad \text{Subtract.} \\ & & d &= 24 \end{aligned}$

$24 + q = 35$ Substitute 24 for d in Equation 2.
$\quad\; q = 11$

9. Answers will vary.

10. Answers will vary. Possible answers are trial and error and using a computer spreadsheet.

INVESTIGATE FURTHER

11. x represents the greater number, y the lesser number

12. $\begin{aligned} x + y &= 50 &\rightarrow\quad 2x + 2y &= 100 \\ 3x - 2y &= 60 &\rightarrow\quad 3x - 2y &= 60 \\ \hline & & 5x\qquad\; &= 160 \quad \text{Add.} \\ & & x &= 32 \end{aligned}$

$32 + y = 50$ Substitute 32 for x in Equation 1.
$\quad\;\; y = 18$

So, $(32, 18)$ is the solution.

13. Let x = greater number, $50 - x$ = lesser number

14. $3x - 2(50 - x) = 60$

15. $3x - 100 + 2x = 60$

$\qquad\qquad\; 5x = 160$

$\qquad\qquad\;\; x = 32$

$32 + y = 50 \rightarrow y = 18$; same solution

16. **Two-variable solution:**
Let x = lesser number
Let y = greater number

$\qquad x + y = 87$

$\qquad\quad\; y = 2x + 3$

$x + (2x + 3) = 87$ $\qquad$ Substitute $2x + 3$ for y.

$\qquad\;\; x = 28$ $\qquad$ Solve for x.

$\qquad\; y = 2(28) + 3$ $\quad$ Substitute 28 for x in Equation 2.

$\qquad\; y = 59$ $\qquad$ Solve for y.

One-variable solution:
Let x = lesser number.
Then $2x + 3$ = greater number.

$x + (2x + 3) = 87$

$\qquad\;\; x = 28$ $\qquad$ Solve for x.

$2(28) + 3 = 59$ $\qquad$ Solve for $2x + 3$.

The lesser number is 28 and the greater number is 59.

17a. No. Answers will vary. The information in the problem does not make it easy to relate the price of one item to the price of the other.

b.
$$6b + 8n = 140 \quad \rightarrow \quad 18b + 24n = 420$$
$$9b + 6n = 132 \quad \rightarrow \quad \underline{18b + 12n = 264}$$
$$12n = 156 \quad \text{Subtract.}$$
$$n = 13 \quad \text{Divide by 12.}$$
$$6b + 8(13) = 140 \quad \text{Substitute 12 for } n \text{ in Equation 1.}$$
$$b = 6 \quad \text{Solve for } b.$$

So necklaces cost $13 and bracelets cost $6.

18. Could solve with a system of linear equations using one of several methods.
$$x + y = 7000 \quad \rightarrow \quad 55x + 55y = 385,000$$
$$0.04x + 0.055y = 337 \quad \rightarrow \quad \underline{40x + 55y = 337,000}$$
$$15x \quad = 48,000$$
$$x = 3200$$

$$3200 + y = 7000 \quad \text{Substitute 3200 for } x \text{ in Equation 1.}$$
$$y = 3800$$

$3200 at 4% and $3800 at 5.5%

APPLY THE STRATEGY

19.
$$10d + 25q = 1160$$
$$d = q + 32$$
$$10(q + 32) + 25q = 1160 \quad \text{Substitute } q + 32 \text{ for } d.$$
$$10q + 320 + 25q = 1160$$
$$q = 24$$
$$d = 24 + 32 = 56$$

He had 24 quarters and 56 dimes.

20.
$$n + q = 16 \quad \rightarrow n = 16q$$
$$5n + 25q = 220$$
$$5(16 - q) + 25q = 220 \quad \text{Substitute } 16 - q \text{ for } n.$$
$$805q + 25q = 220$$
$$q = 7$$
$$n = 16 - 7 = 9$$

9 nickels and 7 quarters

21.
$$x + y = 104$$
$$x = 2y - 1$$
$$2y - 1 + y = 104 \quad \text{Substitute } 2y - 1 \text{ for } x.$$
$$y = 35$$
$$x = 2(35) - 1 = 69$$

35 and 69

22.
$$2x - 5y = 16$$
$$x + 3y = 63 \quad \rightarrow x = 63 - 3y$$
$$2(63 - 3y) - 5y = 16 \quad \text{Substitute } 63 - 3y \text{ for } x.$$
$$y = 10$$
$$x = 63 - 3(10) = 33$$

10 and 33

23.
$$x = 15 + y$$
$$2x + 3y = 100$$
$$2(15 + y) + 3y = 100 \quad \text{Substitute } 15 + y \text{ for } x.$$
$$y = 14$$
$$x = 15 + 14 = 29$$

The product is $(14)(29) = 406$

24.
$$x + y = 80 \quad \rightarrow 9x + 9y = 720$$
$$6.00x + 9.00y = 6.60(80) \quad \rightarrow \underline{6x + 9y = 528}$$
$$3x \quad = 192$$
$$x = 64$$

$$64 + y = 80 \quad \text{Substitute 64 for } x \text{ in Equation 1.}$$
$$y = 16 \quad \text{Simplify.}$$
$$x + 16 = 80 \rightarrow x = 64$$

64 kg of almonds and 16 kg of cashews

25.
$$r + m = 57,000$$
$$r + 4000 = m$$
$$r + r + 4000 = 57,000 \quad \text{Substitute } r + 4000 \text{ for } m.$$
$$r = 26,500$$

$26,500

26.
$$x + y = 14,000 \quad \rightarrow 8x + 8y = 112,000$$
$$0.05x + 0.08y = 1000 \quad \rightarrow \underline{5x + 8y = 100,000}$$
$$3x \quad = 12,000$$
$$x = 4,000$$

$$4,000 + y = 14,000 \quad \text{Substitute 4,000 for } x \text{ in Equation 1.}$$
$$y = 10,000$$

$4,000 at 5% and $10,000 at 8%

27.
$$2l + 2w = 38$$
$$l = 3w - 1$$
$$2(3w - 1) + 2w = 38 \quad \text{Substitute } 3w - 1 \text{ for } l.$$
$$w = 5$$
$$l = 3(5) - 1 = 15 - 1 = 14$$

28.
$$7h + 3p = 19 \quad \rightarrow \quad 14h + 6p = 38$$
$$8h + 2p = 19.5 \quad \rightarrow \underline{24h + 6p = 58.5}$$
$$-10h \quad = -20.5 \quad \text{Subtract.}$$
$$h = 2.05$$

$$7(2.05) + 3p = 19 \quad \text{Substitute 2.05 for } h \text{ in Equation 1.}$$
$$p = 1.55$$

$$4h + 4p = 4(2.05) + 4(1.55) = \$14.40$$

You would get $20.00 − $14.40 = $5.60 change from $20.

29. $x =$ number of \$10 bills; $y =$ number of \$5 bills;
$z =$ number of \$1 bills

$$x + y + z = 94$$
$$10x + 5y + z = 446$$
$$y = x + 10$$

$$\left. \begin{array}{l} x + (x+10) + z = 94 \\ 10x + 5(x+10) + z = 446 \end{array} \right\} \text{ Substitute } x + 10 \text{ for } y.$$

$$2x + z = 84$$
$$\underline{15x + z = 396}$$
$$-13x \quad\; = -312 \quad \text{Subtract.}$$
$$x = 24$$
$$y = 24 + 10 = 34$$
$$2(24) + z = 84$$
$$z = 36$$

24 \$10 bills, 34 \$5 bills, and 36 \$1 bills

REVIEW PROBLEM SOLVING STRATEGIES
Pine Tree Puzzle

a. 19 years; $30 + 19 = 49$

b. 11 years ago because $30 - 19 = 11$; 8 years because $19 - 11 = 8$; no

c. tree is 28 years, ranger is 21 years.

d. Answers will vary, but most students will use two equations since a single expression would be complex. A possible system of equations is $p + r = 49$ and $p = 2[r - (p - r)]$. The system can be solved by substitution. As above, $p = 28$ and $r = 21$.

e. Answers will vary.

Red and Blue I

a. It should have been labeled Blue and Red.

b. Nothing, because a red sock could indicate either all red socks or a mixture of blue and red

c. The package cannot be a mixture or the label would be correct, nor can it have only red socks. So it must contain all blue socks.

d. The selection should be made from the package labeled Blue and Red. Since it cannot be a mixture, if you see a red sock the package should be labeled Red and if you see a blue sock the package should be labeled Blue. Then the other two packages can be determined: If the package labeled Blue and Red is blue, then the package labeled Red is neither red nor blue and must be mixed. If the package labeled Blue and Red is red, then the package labeled Blue is neither blue nor red and must be mixed.

Red and Blue II

1st move

2nd move

3rd move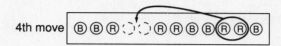

4th move

5th move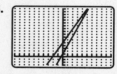

Chapter Review, pages 364–365

1. c **2.** a **3.** d **4.** b

5. $-2 = -2(4) + 6 = -8 + 6 = -2$
$-2 = -(4) + 2 = -2$
Yes, both are true.

6. $7(4) - (-2) = 28 + 2 = 30 \neq -8$
$-2 \neq -4(4) - 3 = -16 - 3 = -19$
No, $(4, -2)$ is not a solution.

7.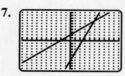
Intersection is $(5, 7)$.

8.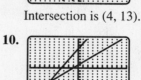
Intersection is $(4, 13)$.

9.
Intersection is $(1, 0)$.

10.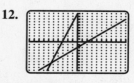
Intersection is $(-6, -5)$.

11.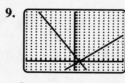
Intersection is $(1, -2)$.

12.
Intersection is $(-6, -8)$.

13. $x + 2y = -4 \rightarrow x = -2y - 4$

$3(-2y - 4) - 2y = 12$ Substitute $-2y - 4$ for x in Equation 2.

$\qquad\qquad y = -3$ Solve for y.

$\qquad x + 2(-3) = -4$ Substitute -3 for x in Equation 1.

$\qquad\qquad x = 2$ Solve for x.

So, $(2, -3)$ solves the system.

14. $x - 2y = 4 \rightarrow x = 2y + 4$

$2y + 4 + 4y = 7$ Substitute $2y + 4$ for x in Equation 2.

$\qquad\quad y = 0.5$ Solve for y.

$\quad x - 2(0.5) = 4$ Substitute 0.5 for y in Equation 1.

$\qquad\quad x = 5$ Solve for x.

So, $(5,\ 0.5)$ solves the system.

15. $x - y = 5 \rightarrow x = y + 5$

$y + 5 + 3y = -3$ Substitute $y + 5$ for x in Equation 2.

$\qquad\quad y = -2$ Solve for y.

$\quad x - (-2) = 5$ Substitute -2 for y in Equation 1.

$\qquad\quad x = 3$ Solve for x.

So, $(3, -2)$ solves the system.

16. $y - 2x = -3 \rightarrow y = 2x - 3$

$2x - 3 + 5 = 3x$ Substitute $2x - 3$ for y in Equation 2.

$\qquad\quad 2 = x$ Solve for x.

$\quad y - 2(2) = -3$ Substitute 2 for x in Equation 1.

$\qquad\quad y = 1$ Solve for y.

So, $(2, 1)$ solves the system.

17. $\qquad v = \frac{1}{2}m$

$\quad v + m = 54$

$\frac{1}{2}m + m = 54$ Substitute $\frac{1}{2}m$ for v in Equation 2.

$\qquad\quad m = 36$ Solve for m.

$\quad v = \frac{1}{2}(36)$ Substitute 36 for m in Equation 1.

$\qquad\quad v = 18$ Solve for v.

Victor is 18.

18. $3x + 2y = 18$

$\underline{\ x - 2y = -6\ }$

$4x \qquad = 12$ Add the equations.

$\quad x = 3$ Divide by 4.

$3 - 2y = -6$ Substitute 3 for x in Equation 2.

$\quad y = 4.5$ Solve for y.

So, $(3, 4.5)$ solves the system.

19. $\ x + y = 3 \ \rightarrow\ 3x + 3y = 9$

$2x + 3y = 1 \ \rightarrow\ \underline{2x + 3y = 1\ }$

$\qquad\qquad\qquad\quad x \quad\ = 8$ Subtract.

$8 + y = 3$ Substitute 8 for x in Equation 1.

$\quad y = -5$ Solve for y.

So, $(8, -5)$ solves the system.

20. $8x + 3y = -27 \ \rightarrow\ 24x + 9y = -81$

$6x - 9y = -9 \ \rightarrow\ \underline{\ 6x - 9y = -9\ }$

$\qquad\qquad\qquad\qquad 30x \qquad = -90$ Add.

$\qquad\qquad\qquad\qquad\quad x = -3$ Divide by 30.

$8(-3) + 3y = -27$ Substitute -3 for x in Equation 1.

$\qquad\quad y = -1$ Solve for y.

So, $(-3, -1)$ solves the system.

21. $\ x - y = -3 \ \rightarrow\ 4x - 4y = -12$

$2x - 4y = 22 \ \rightarrow\ \underline{\ 2x - 4y =\ \ 22\ }$

$\qquad\qquad\qquad\qquad 2x \qquad = -34$ Subtract.

$\qquad\qquad\qquad\qquad\ x = -17$ Divide by 2.

$-17x - y = -3$ Substitute -17 for x in Equation 1.

$\qquad\quad y = -14$ Solve for y.

So, $(-17, -14)$ solves the system.

22. $\quad c + l = 90 \qquad\qquad \rightarrow l = 90 - c$

$\quad c - 8 = l + 8$

$c - 8 = 90 - c + 8$ Substitute $90 - c$ for l in Equation 2.

$\qquad\quad c = 53$ Solve for c.

$\quad 53 + l = 90$ Substitute 53 for c in Equation 1.

$\qquad\quad l = 37$ Solve for l.

Chen has \$53 and Lisa has \$37.

23. $\quad x + y = 3$

$\quad \underline{x - y = 7\ }$

$\ 2x \quad = 10$ Add.

$\quad x = 5$

$5 + y = 3 \rightarrow y = -2$

So $(5, -2)$ solves the system.

24. $-3x + y = -6 \qquad\qquad \rightarrow y = 3x - 6$

$x + 1 = \frac{1}{3}(3x - 6)$ Substitute $3x - 6$ for y in Equation 2.

$x + 1 = x - 2$

$\quad 1 = -2$ False

The system is inconsistent.

25.
$$x + 3y = -4 \quad \rightarrow x = -3y - 4$$
$$2[3(-3y-4)+9y] = -24 \quad \text{Substitute} -3y-4 \text{ for } x \text{ in Equation 2.}$$
$$2(-12) = -24$$
$$-24 = -24 \quad \text{True}$$

The system is dependent.

26. $D = \begin{vmatrix} 3 & 1 \\ 2 & 4 \end{vmatrix} = 3(4) - 1(2) = 10$

$D_x = \begin{vmatrix} -4 & 1 \\ -6 & 4 \end{vmatrix} = -4(4) - 1(-6) = -10$

$D_y = \begin{vmatrix} 3 & -4 \\ 2 & -6 \end{vmatrix} = 3(-6) - (-4)(2) = -10$

$x = \dfrac{D_x}{D} = \dfrac{-10}{10} = -1$ and $y = \dfrac{D_y}{D} = \dfrac{-10}{10} = -1$

So, $(-1, -1)$ solves the system.

27. $D = \begin{vmatrix} 2 & 5 \\ 3 & -4 \end{vmatrix} = 2(-4) - 5(3) = -23$

$D_x = \begin{vmatrix} 9 & 5 \\ 2 & -4 \end{vmatrix} = 9(-4) - 5(2) = -46$

$D_y = \begin{vmatrix} 2 & 9 \\ 3 & 2 \end{vmatrix} = 2(2) - (9)(3) = -23$

$x = \dfrac{D_x}{D} = \dfrac{-46}{-23} = 2$ and $y = \dfrac{D_y}{D} = \dfrac{-23}{-23} = 1$

So, $(2, 1)$ solves the system.

28. $4x - 3y = 0$
$5x - 4y = -1$

$D = \begin{vmatrix} 4 & -3 \\ 5 & -4 \end{vmatrix} = 4(-4) - (-3)(5) = -1$

$D_x = \begin{vmatrix} 0 & -3 \\ -1 & -4 \end{vmatrix} = 0(-4) - (-3)(-1) = -3$

$D_y = \begin{vmatrix} 4 & 0 \\ 5 & -1 \end{vmatrix} = 4(-1) - 0(5) = -4$

$x = \dfrac{D_x}{D} = \dfrac{-3}{-1} = 3$ and $y = \dfrac{D_y}{D} = \dfrac{-4}{-1} = 4$

So, $(3, 4)$ solves the system.

29.
$$2l + 2w = 180$$
$$l = 4w$$
$$2(4w) + 2w = 180 \quad \text{Substitute } 4w \text{ for } l \text{ in Equation 1.}$$
$$w = 18$$
$$l = 4(18) = 72$$

So, the rectangle is 18 in. wide and 72 in. long.

30.
$$4s + 3g = 130 \quad \rightarrow \quad 4s + 3g = 130$$
$$g - 2s = -10 \quad \rightarrow \quad \underline{-4s + 2g = -20}$$
$$5g = 110 \quad \text{Add.}$$
$$g = 22 \quad \text{Divide by 5.}$$
$$22 - 2s = -10 \quad \text{Substitute 22 for } g \text{ in Equation 2.}$$
$$s = 16$$

So, the smaller number is 16 and the greater is 22.

Chapter Assessment, pages 366–367

1.
$$x + 2y = 4 \quad \rightarrow x = 4 - 2y$$
$$2(4 - 2y) + 3y = 7 \quad \text{Substitute } 4 - 2y \text{ for } x \text{ in Equation 2.}$$
$$8 - 4y + 3y = 7 \quad \text{Distribute.}$$
$$-y = -1 \quad \text{Subtract 8 and simplify.}$$
$$y = 1 \quad \text{Multiply by } -1.$$
$$x + 2(1) = 4 \quad \text{Substitute 1 for } y \text{ in Equation 1.}$$
$$x = 2 \quad \text{Solve for } x.$$

So, $(2, 1)$ solves the system.

2.
$$b + 4c = -5 \quad \rightarrow b = -4c - 5$$
$$2(-4c - 5) - c = 8 \quad \text{Substitute } -4c - 5 \text{ for } b \text{ in Equation 2.}$$
$$-8c - 10 - c = 8 \quad \text{Distribute.}$$
$$c = -2 \quad \text{Solve for } c.$$
$$b + 4(-2) = -5 \quad \text{Substitute } -2 \text{ for } c \text{ in Equation 1.}$$
$$b = 3 \quad \text{Solve for } b.$$

So, $(3, -2)$ solves the system.

3.
$$x - 2y = 0.5 \quad \rightarrow x = 2y + 0.5$$
$$2(2y + 0.5) - y = -0.5 \quad \text{Substitute } 2y + 0.5 \text{ for } x \text{ in Equation 2.}$$
$$4y + 1 - y = -0.5 \quad \text{Distribute.}$$
$$y = -0.5 \quad \text{Solve for } y.$$
$$x - 2(0.5) = 0.5 \quad \text{Substitute } -0.5 \text{ for } y \text{ in Equation 1.}$$
$$x = -0.5 \quad \text{Solve for } x.$$

So $(-0.5, -0.5)$ solves the system.

4.
$$3x + y = 1 \quad \rightarrow y = -3x + 1$$
$$6x + (-3x + 1) = 3 \quad \text{Substitute } -3x + 1 \text{ for } y \text{ in Equation 2.}$$
$$x = \frac{2}{3} \quad \text{Solve for } x.$$
$$3\left(\frac{2}{3}\right) + y = 1 \quad \text{Substitute } \frac{2}{3} \text{ for } y \text{ in Equation 1.}$$
$$y = -1 \quad \text{Solve for } y.$$

So, $\left(\dfrac{2}{3}, -1\right)$ solves the system.

5.
$a = 3l$
$a - l = 12$
$3l - l = 12$ Substitute $3l$ for a in Equation 2.
$l = 6$ Solve for l.
$a = 3(6)$ Substitute 6 for l in Equation 1.
$a = 18$ Solve for a.
Alana is 18 years old.

6. Answers will vary, but should include the following information. A consistent system has at least one solution. In a dependent system, the two equations have the same graph. The system has an infinite number of solutions, since any of the infinite number of ordered pairs that satisfy one equation also satisfy the other so it is consistent. Attempting to solve a consistent dependent system results in a true equation such as $0 = 0$. An inconsistent system has no solutions, since the graphs of the equations are parallel and do not intersect. Solving an inconsistent system results in a false equation like $0 = 7$.

7. $3n + 2t = 645 \quad \rightarrow \quad 6n + 4t = 1290$
$5n + 4t = 1145 \quad \rightarrow \quad \underline{5n + 4t = 1145}$
$n \quad = 155$ Subtract.
$3(155) + 2t = 645$ Substitute 155 for n in Equation 1.
$t = 90$ Solve for t.
The hotel is \$155 per night.
The tickets are \$90 per pair.

8. $10x - 6y = 42 \rightarrow 5x - 3y = 21$
$15x - 9y = 63 \rightarrow \underline{5x - 3y = 21}$
$0 = 0$ True, so dependent
The answer is C.

9. $2x - 3y = 1$
$\underline{2x + 3y = 7}$
$4x \quad = 8$ Add the equations.
$x = 2$ Divide by 4.
$2(2) - 3y = 1$ Substitute 2 for x in Equation 1.
$y = 1$ Solve for y.
So, $(2,1)$ solves the system.

10. $x + 2y = -6 \rightarrow 2x + 4y = -12$
$3x - 4y = 32 \rightarrow \underline{3x - 4y = 32}$
$5x \quad = 20$ Add.
$x = 4$ Divide by 5.
$4 + 2y = -6$ Substitute 4 for x in Equation 1.
$y = -5$ Solve for y.
So, $(4, -5)$ solves the system.

11. $x + y = 0.5 \rightarrow 4x + 4y = 2$
$2x + 4y = -5 \rightarrow \underline{2x + 4y = -5}$
$2x \quad = 7$ Subtract.
$x = 3.5$ Divide by 2.
$3.5 + y = 0.5$ Substitute 3.5 for x in Equation 1.
$y = -3$ Solve for y.
So, $(3.5, -3)$ solves the system.

12. $-4m + 3n = -25 \rightarrow -8m + 6n = -50$
$-3m - 2n = -6 \rightarrow \underline{-9m - 6n = -18}$
$-17m \quad = -68$ Add.
$m = 4$
$-4(4) + 3n = -25$ Substitute 4 for m in Equation 1.
$n = -3$ Solve for n.
So, $(4, -3)$ solves the system.

13. $3r + s = 0$
$\underline{3r + 4s = -9}$
$-3s = 9$ Subtract the equations.
$s = -3$ Divide by -3.
$-3 = -3r$ Substitute -3 for r in Equation 1.
$r = 1$ Solve for r.
So, $(1, -3)$ solves the system.

14. $3x + 2y = 0 \rightarrow 6x + 4y = 0$
$-2x + 3y = 13 \rightarrow \underline{-6x + 9y = 39}$
$13y = 39$ Add the equations.
$y = 3$ Divide by 13.
$3x = -2(3)$ Substitute 3 for y in Equation 1.
$x = -2$ Solve for x.
So, $(-2, 3)$ solves the system.

15. $2w + 2l = 158$
$l = w + 5$
$2w + 2(w + 5) = 158$ Substitute $w + 5$ for l in Equation 1.
$w = 37$ Solve for w.
$l = 37 + 5$ Substitute 5 for w in Equation 2.
$l = 42$ Solve for l.
width = 37 cm and length = 42 cm

16. $a + c = 240 \rightarrow 1.5a + 1.5c = 360$
$3a + 1.5c = 510 \rightarrow \underline{3a + 1.5c = 510}$
$-1.5a \quad = -150$ Subtract.
$a = 100$
$100 + c = 240$ Substitute 100 for a in Equation 1.
$c = 140$ Solve for c.
100 adult tickets; 140 children's tickets

17. By trial and error, the solution is B.

18.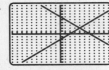

Intersection at $(4, 2)$

19.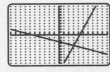

Intersection at $(3, -4)$

20. $\begin{vmatrix} 3 & -4 \\ 6 & 2 \end{vmatrix} = 3(2) - (-4)(6) = 30$

21. $\begin{vmatrix} 7 & 2 \\ -2 & 3 \end{vmatrix} = 7(3) - 2(-2) = 25$

22. $D = \begin{vmatrix} 6 & 5 \\ 1 & 2 \end{vmatrix} = 6(2) - 5(1) = 7$

$D_x = \begin{vmatrix} -1 & 5 \\ -6 & 2 \end{vmatrix} = -1(2) - 5(-6) = 28$

$D_y = \begin{vmatrix} 6 & -1 \\ 1 & -6 \end{vmatrix} = 6(-6) - (-1)(1) = -35$

$x = \dfrac{D_x}{D} = \dfrac{28}{7} = 4$ and $y = \dfrac{D_y}{D} = \dfrac{-35}{7} = -5$

So, $(4, -5)$ solves the system.

23. Write the equations in standard form.

$3x + y = -15$

$2x - 3y = 12$

$D = \begin{vmatrix} 3 & 1 \\ 2 & -3 \end{vmatrix} = 3(-3) - 1(2) = -11$

$D_x = \begin{vmatrix} -15 & 1 \\ 12 & -3 \end{vmatrix} = -15(-3) - 1(12) = 33$

$D_y = \begin{vmatrix} 3 & -15 \\ 2 & 12 \end{vmatrix} = 3(12) - (-15)(2) = 66$

$x = \dfrac{D_x}{D} = \dfrac{33}{-11} = -3$ and $y = \dfrac{D_y}{D} = \dfrac{66}{-11} = -6$

So, $(-3, -6)$ solves the system.

24. Write the equations in standard form.

$3x - 4y = -1$

$48x - 30y = 1$

$D = \begin{vmatrix} 3 & -4 \\ 48 & -30 \end{vmatrix} = 3(-30) - (-4)(48) = 102$

$D_x = \begin{vmatrix} -1 & -4 \\ 1 & -30 \end{vmatrix} = -1(-30) - (-4)(1) = 34$

$D_y = \begin{vmatrix} 3 & -1 \\ 48 & 1 \end{vmatrix} = 3(1) - (-1)(48) = 51$

$x = \dfrac{D_x}{D} = \dfrac{34}{102} = \dfrac{1}{3}$ and $y = \dfrac{D_y}{D} = \dfrac{51}{102} = \dfrac{1}{2}$

So, $\left(\dfrac{1}{3}, \dfrac{1}{2}\right)$ solves the system.

25. $D = \begin{vmatrix} 0.3 & 0.2 \\ 1 & -0.5 \end{vmatrix} = 0.3(-0.5) - 0.2(1) = -0.35$

$D_x = \begin{vmatrix} 0.7 & 0.2 \\ 0 & -0.5 \end{vmatrix} = 0.7(-0.5) - 0.2(0) = -0.35$

$D_y = \begin{vmatrix} 0.3 & 0.7 \\ 1 & 0 \end{vmatrix} = 0.3(0) - 0.7(1) = -0.7$

$x = \dfrac{D_x}{D} = \dfrac{-0.35}{-0.35} = 1$ and $y = \dfrac{D_y}{D} = \dfrac{-0.7}{-0.35} = 2$

So, $(1, 2)$ solves the system.

26. Answers should include the method for finding D using the coefficients of x and y, for finding D_x by substituting the constants for the x-column in D, and for finding D_y by substituting the constants for the y-column in D and then using those determinants to find x and y according to Cramer's formulas.

27.
$$\begin{array}{rl} x + y = 5000 & \rightarrow 8x + 8y = 40{,}000 \\ 0.08x + 0.06y = 360 & \rightarrow \underline{8x + 6y = 36{,}000} \\ & 2y = 4000 \quad \text{Subtract.} \\ & y = 2000 \end{array}$$

$x + 2000 = 5000$ Substitute 2000 for y in Equation 1.

$x = 3000$ Solve for x.

\$3000 at 8% and \$2000 at 6%

28.
$$\begin{array}{rl} x - y = 22 & \rightarrow 2x - 2y = 44 \\ 3x + 2y = 246 & \rightarrow \underline{3x + 2y = 246} \\ & 5x = 290 \quad \text{Add equations.} \\ & x = 58 \quad \text{Divide by 5.} \end{array}$$

$58 - y = 22$ Substitute 58 for x in Equation 1.

$y = 36$ Solve for y.

The numbers are 58 and 36.

Cumulative Review, page 368

1.
$$\begin{array}{rl} 4x + y = 11 & \rightarrow 12x + 3y = 33 \\ 5x - 3y = 1 & \rightarrow \underline{5x - 3y = 1} \\ & 17x = 34 \quad \text{Add the equations.} \\ & x = 2 \quad \text{Divide by 17.} \end{array}$$

$4(2) + y = 11$ Substitute 2 for x in Equation 1.

$y = 3$ Solve for y.

So, $(2, 3)$ solves the system.

2.
$$\begin{array}{rl} 3x - 2y = -4 & \rightarrow 3x - 2y = -4 \\ x - 5y = 3 & \rightarrow \underline{3x - 15y = 9} \\ & 13y = -13 \quad \text{Subtract.} \\ & y = -1 \quad \text{Divide by 13.} \end{array}$$

$x - 5(-1) = 3$ Substitute -1 for y in Equation 1.

$x = -2$ Solve for x.

So, $(-2, -1)$ solves the equation.

3. $\begin{vmatrix} 4 & 5 \\ 2 & 3 \end{vmatrix} = 4(3) - 5(2) = 2$

4. $\begin{vmatrix} -2 & 7 \\ -3 & 9 \end{vmatrix} = -2(9) - 7(-3) = 3$

5. C; No correlation exists.

6. $n - 8 > -6$

$\qquad n > 2$

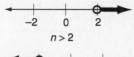

$n > 2$

7. $6c \le -12$

$\qquad c \le -2$

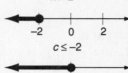

$c \le -2$

8. $15x - 4(3x - 1) \ge 5x + 6$

$\qquad 3x + 4 \ge 5x + 6$

$\qquad -2 \ge 2x$

$\qquad -1 \ge x \text{ or } x \le -1$

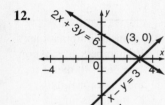

$x \le -1$

9. $\dfrac{6(3+2) - 3^2}{2^3 - 1} = \dfrac{6(5) - 9}{8 - 1} = \dfrac{30 - 9}{7} = \dfrac{21}{7} = 3$

10. $\dfrac{|-4 - (-6)|}{|-4| - |-6|} = \dfrac{|2|}{4 - 6} = \dfrac{2}{-2} = -1$

11. $3 + 4 \cdot 5 \;-\; 6 \div 3 \cdot 4 \;+\; 3^2 \;-\; 10$

$\;\; 3 + 20 \quad - \quad 2 \cdot 4 \;+\; 9 \;-\; 10$

$\;\; 23 \qquad - \qquad 8 \;+\; 9 \;-\; 10 = 14$

12. 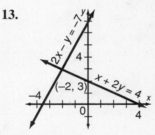 **13.**

14. $\quad n + \quad d = 25 \;\rightarrow\; 10n + 10d = 250$

$5n + 10d = 205 \;\rightarrow\; \underline{5n + 10d = 205}$

$\qquad\qquad\qquad\quad 5n \qquad = 45 \quad$ Subtract.

$\qquad\qquad\qquad\quad\; n = 9 \qquad$ Divide by 5.

$\;\; 9 + d = 25 \qquad$ Substitute 9 for n in Equation 1.

$\qquad d = 16 \qquad$ Solve for d.

She had 16 dimes.

15. $\quad x + y = 2$

$\quad \underline{x - y = 10}$

$\quad 2x \quad = 12 \quad$ Subtract.

$\qquad x = 6 \quad$ and $\quad y = -4$

16. $y = -2x + 2$ by counting rise over run and using the y-intercept of 2.

17. $y - 0 = 2(x + 2)$ is one of many possible answers.

18. They are perpendicular because slopes are -2 and $\frac{1}{2}$.

19. Domain $= \{1, 2, 3, 4\}$; Range $= \{5, 6, 7, 8\}$

20. Domain $= \{x: x \text{ is all real numbers}\}$

Range $= \{f(x): f(x) \text{ is all real numbers}\}$

21. Domain $= \{x: x \text{ is all real numbers}\}$

Range $= \{f(x): f(x) \text{ is all real numbers}\}$

22. $3\frac{1}{2}(20) = 70$ **23.** $\frac{90}{20} = 4\frac{1}{2}$ symbols

24. $80 + 70 + 50 + 120 + 90 + 90 = 500$

Standardized Test, Page 369

1. $\frac{21}{25} = 0.84$; Answer is E.   **2.** Answer is D since $-4 < -3$

3. $15 - (3x - 5) = 2(5 - 2x) - 4$

$\qquad 20 - 3x = 6 - 4x$

$\qquad\quad x = -14 \quad$ Answer is A.

4. $\quad \begin{vmatrix} 5 & 7 \\ 3 & -2 \end{vmatrix} - \begin{vmatrix} -4 & 2 \\ -5 & 6 \end{vmatrix}$

$[5(-2) - 7(3)] \;-\; [-4(6) - 2(-5)]$

$\qquad\quad -31 \;-\; (-14)$

$\qquad\qquad\quad -17 \quad$ Answer is B.

5. $4x - 7(0) = 28$

$\qquad\quad x = 7$

$x = 7$ when $y = 0$; Answer is A.

6. 18 respondents so median is mean of 9th and 10th, i.e.

$\frac{1+2}{2} = 1.51$; Answer is C.

7. $x + (x + 2) + (x + 4) \le 36$

$\qquad\qquad\qquad 3x \le 30$

$\qquad\qquad\qquad\; x \le 10$

So, value is ≤ 13; Answer is C.

8. $3x - 5y = 4 \quad \rightarrow \; 6x - 10y = \quad 8$

$2x + 7y = -18 \rightarrow \underline{6x + 21y = -54}$

$\qquad\qquad\qquad\qquad -31y = \; 62 \quad$ Subtract.

$\qquad\qquad\qquad\qquad\quad y = -2 \quad$ Answer is D.

9. $2x + 1 > 7$ **10.** $0.60x = 28,575$

$\quad 2x > 8$ $x = 47,625$

$\quad\; x < 4$ Answer is B.

Answer A is correct.

Data Activity, page 371

1. 15% of 200 = 0.15(200) = 30   **2.** $\frac{16}{100} = \frac{4}{25}$ or 16%

3. Answers will vary. One possible answer is a person between 35 and 44 with an income between $35,000 and $49,999.

4.

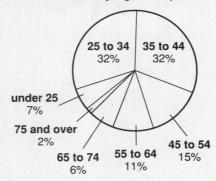

ATM Use by Age Group

35 to 44
32%

25 to 34
32%

under 25
7%

75 and over
2%

65 to 74
6%

55 to 64
11%

45 to 54
15%

ATM Use by Income

$75,000 and over
11%

under $15,000
14%

$50,000 to $74,999
18%

$15,000 to $24,999
16%

$35,000 to $49,999
23%

$25,000 to $34,999
18%

Lesson 8.1, pages 373–378

EXPLORE/WORKING TOGETHER

1a. Since marigolds cost $1.00. He can buy up to 10 and as few as 0, so (10, 0) and (0, 0).

b. Since geraniums cost $2.00, he can buy up to 5 and as few as 0, so (0, 5) and (0, 0).

c. If he buys 3 geraniums, he spends $6.00 and has $4.00 left. He can buy up to 4 marigolds or as few as 0, so (4, 3) and (0, 3).

2.

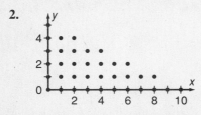

3. Answers will vary. Possible answer: Points form the interior of a right triangle.

4. $x + 2y = 10$

5.

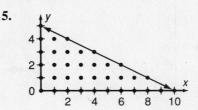

TRY THESE

1. $16 > -4(-4) \rightarrow 16 > 16$

Not a solution

2. $3(2) - 2(-3) \geq 12 \rightarrow 6 + 6 \geq 12 \rightarrow 12 \geq 12$

Yes, it is a solution.

3. $6(0) + 3 < 5(0) \rightarrow 3 < 0$

Not a solution

4.

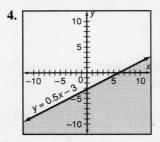

$y = 0.5x - 3$

5.

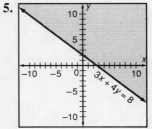

$3x + 4y = 8$

6a. $x + y \leq 40$

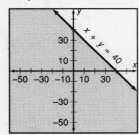

$x + y = 40$

b. possible combinations: 20 members and 20 non-members; 10 members and 15 nonmembers; 15 members and 10 non-members

7. $m = \frac{1}{3}$, $b = 3$ and the shading is above the solid line.

So, $y \geq \frac{1}{3}x + 3$.

PRACTICE

1. $5 > 5(3) - 10 \rightarrow 5 > 15 - 10 \rightarrow 5 > 5$

Not a solution

2. $2(1) \leq 0 \rightarrow 2 \leq 0$

Not a solution

3. $2(-1) + 4(3) \geq 7 \rightarrow -2 + 12 \geq 7 \rightarrow 10 \geq 7$

Yes, the ordered pair is a solution.

4. $3(0) - 0 \leq -7 \rightarrow 0 \leq -7$

Not a solution

5.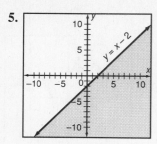

6.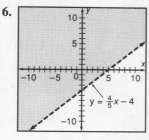

7.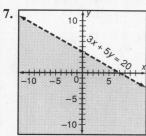

8.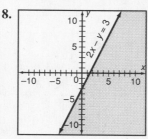

9.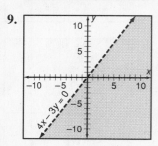

10.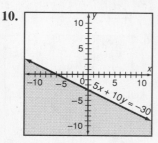

11. $m = 1$, $b = -4$, line is dashed and shaded below, so $y < x - 4$

12. $m = 0$, $b = -6$, line is dashed and shaded above, so $y > -6$

13. $m = -2$, $b = 2$, line is solid and shaded above, so $y \geq -2x + 2$

14. Vertical solid line at 2, shaded left, so $x \leq 2$

15. Possible answer: Same—you must determine whether a line or point is included in the graph and which half-plane or part of the number line contains the solution set; different—a linear inequality in two variables is graphed on a coordinate plane and the solutions are ordered pairs of real numbers; an inequality in one variable is graphed on a number line and the solutions are real numbers.

16a. $3x + 7y > 20$

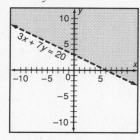

b. possible combinations: 6 field goals and 1 touchdown; 1 field goal and 4 touchdowns; 3 field goals and 3 touchdowns

17. not equivalent since lines are perpendicular

18. equivalent since second inequality is 3 times the first

19. equivalent, same line shaded on same side

20. not equivalent, lines have different slopes

21. Possible answer: Write both inequalities in the slope-intercept form to see if the coefficients, constants, and inequality symbols all match each other.

22a. $30x + 15y \leq 420$

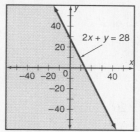

b. Possible combinations: 4 new and 19 returning, 14 new and 0 returning, 5 new and 10 returning, etc.

23a. $20x + 5y \geq 1000$

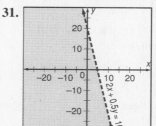

b. Possible combinations: 50 regular and 0 sale, 145 regular and 120 sale, 8 regular and 168 sale, etc.

24. dashed line, half plane below line

25. dashed line, half plane above line

26. solid line, half plane below line

27. solid line, half plane above line

28. y-coordinates are always positive and include zero on the x-axis, so $y \geq 0$.

29. x-coordinates are always positive, so $x > 0$.

30. The number line is the x-axis of the coordinate plane, and the graph on the number line is the shaded part of the graph of the coordinate palne.

31.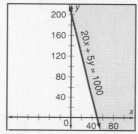

Possible problem: Ted and his friends bought hamburgers for $2 each and beverages for $0.50 each. They spent less than $10. What possible combinations of hamburgers and beverages could they have bought? Possible combinations: 2 hamburgers and 5 beverages; 1 hamburger and 10 beverages; 3 hamburgers and 6 beverages.

32. B; $19.625 = \pi r^2 \rightarrow \dfrac{19.625}{\pi} = r^2 \rightarrow \sqrt{\dfrac{19.625}{\pi}} = r \approx 2.5$

33.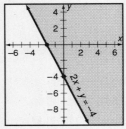

34.

35. $3x + 6y = 3 \rightarrow 15x + 30y = 15$

$2x - 5y = -16 \rightarrow \underline{12x - 30y = -96}$

$27x \qquad = -81$

$x = -3$

$3(-3) + 6y = 3$ Substitute -3 for x.

$y = 2$ Solve for y.

So $(-3, 2)$ solves the system.

36. $8x - 4y = 2 \rightarrow 16x - 8y = 4$

$4x + 8y = 26 \rightarrow \underline{4x + 8y = 26}$

$20x \qquad = 30$

$x = \dfrac{3}{2}$

$3\left(\dfrac{3}{2}\right) - 4y = 2$ Substitute $\dfrac{2}{3}$ for x.

$y = \dfrac{5}{2}$ Solve for y.

So $\left(\dfrac{3}{2}, \dfrac{5}{2}\right)$ solves the system.

37.

38.

Lesson 8.2, pages 379–385

EXPLORE

1. $15x + 25y \geq 15$

2. Possible solutions: $(1, 0)$, $(0, 1)$, $(2, 4)$, $(1.5, 3)$, $(4, 4)$

3. Cross out fractional values. Only whole numbered pairs are allowed, one of which is not zero.

4. $x + y \leq 6$

5. Only ordered pairs whose sum is 6 or less remain.

6. It must satisfy both inequalities.

TRY THESE

1.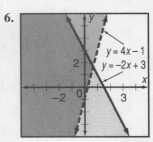

2.

3.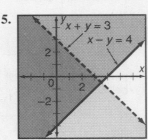

4.

5.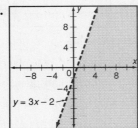

6.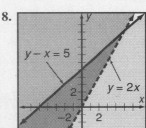

7.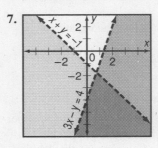

8.

9.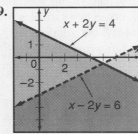

10. $m = -1$, $b = -2$; line is dashed and shaded above line

$m = \dfrac{1}{2}$, $b = 5$; line is solid and shaded above line

$\begin{cases} y > -x - 2 \\ y \geq \dfrac{x}{2} + 5 \end{cases}$

11a. No; not in the green-shaded area

  b. No; not in the green-shaded area

  c. Yes; point is in the green-shaded area

  d. Yes; point is in the green-shaded area

12. $\begin{cases} x + y \leq 8 \\ 60x + 40y \geq 400 \end{cases}$

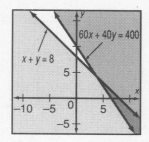

13. Possible combinations: 7 expressway hours and 0 highway hours, $6\frac{1}{2}$ expressway hours and 1 highway hour, $6\frac{1}{3}$ expressway hours and $1\frac{1}{4}$ highway hours

PRACTICE

1. no; not in green-shaded area

2. yes; point is in green-shaded area

3. no; not in green-shaded area

4. yes; point is in green-shaded area

5.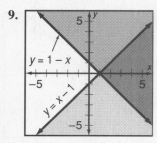

6.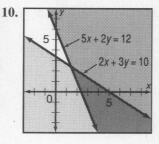

7.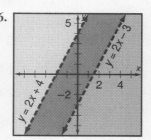

8.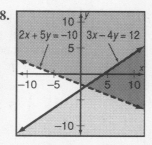

9.

10.

11.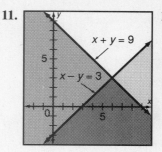

12.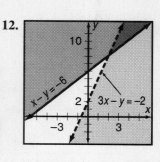

13.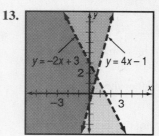

14.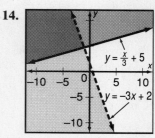

15.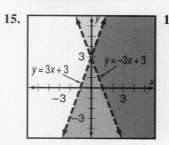

16.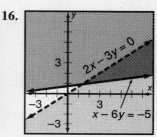

17. $m = 2$, $b = -1$, line is dashed and shaded above
$m = -2$, $b = 4$, line is dashed and shaded below
$\begin{cases} y > 2x - 1 \\ y < -2x + 4 \end{cases}$

18. solid vertical line at -4 and shaded right
dashed horizontal line at -3 and shaded above
$\begin{cases} x \geq -4 \\ y > -3 \end{cases}$

19. $m = 1$, $b = 2$, solid line and shaded above
$m = \frac{3}{5}$, $b = 3$, solid line and shaded below
$\begin{cases} y \geq x + 2 \\ y \leq \frac{3}{5}x + 3 \end{cases}$

20. Possible answer should include identifying the y-intercept, determining the slope, determining whether shading is above or below each line to indicate > or <, and determining whether each line is dotted or solid to indicate whether or not "or equal to" applies.

21a. $\begin{cases} \frac{3}{4}x + y \le 50 & \text{time constraint} \\ 50x + 80y > 3000 & \text{money constraint} \end{cases}$

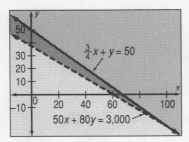

b. Possible combinations: 0 chairs and 50 tables, 10 chairs and 35 tables, 50 chairs and 10 tables

22. point above both lines so neither inequality

23. point below solid line, so satisfies $2x + y \le 3$

24. point below both lines, so satisfies both inequalities

25. point on solid line and below dashed line, so satisfies both inequalities

26. point above solid line and on dashed line, so satisfies neither inequalities

27. x and y are both positive in QI, so $\begin{cases} x > 0 \\ y > 0 \end{cases}$

28. x and y are both negative in QIII and zero on the axes, so $\begin{cases} x \le 0 \\ y \le 0 \end{cases}$

29. x is positive in QIV, y is negative in QIV, and zero on the x-axis, so $\begin{cases} x > 0 \\ y \le 0 \end{cases}$

30.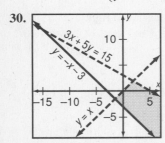

31.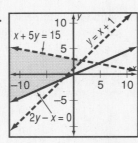

32.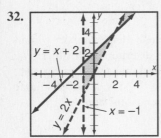

33.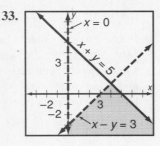

34a. $\begin{cases} \dfrac{x+y}{2} < 60 \\ x > 40 \\ x > y \end{cases}$

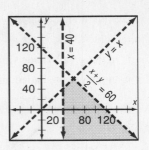

b. Possible combinations:
50°F and 60°F, 45°F and 55°F, 55°F and 62°F

35a. $\begin{cases} x + y \le 10,000 \\ 0.08x + 0.05y \ge 500 \\ x > y \end{cases}$

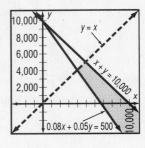

b. Possible combinations:

$5000 in mutual funds, $4000 in CD's
$6000 in mutual funds, $3000 in CD's
$9000 in mutual funds, $1000 in CD's

36. Lines need to be parallel with one $\le$ inequality and one $\ge$ inequality.

Possible answer: $\begin{cases} y \le x + 4 \\ y \ge x - 6 \end{cases}$

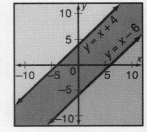

37. Lines must be parallel with one $\le$ inequality and one $\ge$ inequality but no intersecting shaded area.

Possible answer: $\begin{cases} y \ge x + 4 \\ y \le x - 6 \end{cases}$

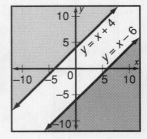

38. Use same line with one $\le$ and one $\ge$.

Possible answer: $\begin{cases} y \le x + 4 \\ y \ge x + 4 \end{cases}$

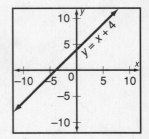

39. Not possible, because the coordinate plane extends to infinity in all directions.

PROJECT CONNECTION

3. Answers will vary. Amount earned or charged = interest – charge for checks – service charges

ALGEBRA WORKS

1. $y \le 3x$

2. $y \le 20 + 2x$; $y \ge 20$

3. Answers may vary. Possible ad: Do you watch more than two new-release videos a month? This club's for you.

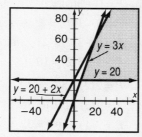

4. If the customer were to rent new releases only, the club would be the better plan. If the customer expected to rent both new and older releases, the set rate plan would probably be better.

5. Answers will vary. Possible answer: Yes; most people rent more than two videos per month, and they usually rent new releases, so the club plan would appeal to new customers.

Lesson 8.3, pages 386–387

THINK BACK

1.

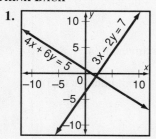

2. The point of intersection of the two graphs is $\left(2, -\frac{1}{2}\right)$.

3a. 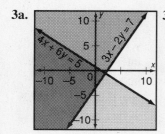 **3b.**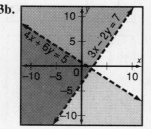

c.

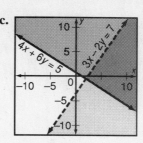

4. They all intersect at $\left(2, -\frac{1}{2}\right)$. It is a solution for a since both inequalities include the boundary lines, but not a solution for b or c.

5. Yes; the double-sided areas indicating the intersection of half-planes and any parts of the boundary lines that fall within the double-shaded areas.

6a. 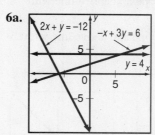 **b.**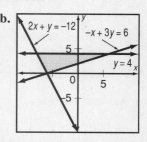

7. By inspection, they intersect at: $(-8, 4)$, $(6, 4)$, $(-6, 0)$.

8. No; any solution would be represented by a point at which all three lines intersect, but there is no such point in the graph of this system. Each intersection point is a solution to a different pairing of the three equations.

9. $(-8, 4)$, $(6, 4)$, $(-6, 0)$

10. Solutions are in the triple shaded region, including the boundary lines.

11. Graphs of corresponding systems consist of the same lines and points of intersection; graphs of systems of inequalities also consist of half-planes and intersections of half-planes. Boundary lines of systems of inequalities may be dashed; lines representing equations are never dashed.

12. Both types of representations are intersections; solutions of systems of equations are represented by the intersection of lines, and solutions of systems of inequalities are represented by the intersection of half-planes and, in some cases, boundary lines.

13. Provided that no equations or inequalities in the system are equivalent, both types of systems could have zero or one solution. Systems of inequalities could have an infinite number of solutions.

14. $m = -1$, $b = 2$, dashed line and shaded below line; vertical solid line at -6 and shaded to right; $m = \frac{1}{2}$, $b = -4$, solid line and shaded above line

$$\begin{cases} y < -x + 2 \\ x \geq -6 \\ y \geq \frac{1}{2}x - 4 \end{cases}$$

15. No; the intersection point of the graph of the system of equations is the intersection point of the boundary lines for the systems of inequalities. However, none of the points along the boundary line of $y < -\frac{1}{2}x$ are solutions of that inequality.

16. below line with positive slope and above line with negative slope; c

17. below line with negative slope and above line with positive slope; b

18. below both lines; a

19. Answers will vary.

Possible answer: $\begin{cases} y \leq x \\ y \geq 1 \\ x \leq 7 \end{cases}$

area $= \frac{1}{2}bh = \frac{1}{2}(6)(6) = 18$

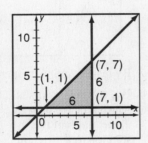

Lesson 8.4, pages 388–393

EXPLORE

1.
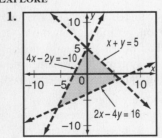

2. Answers will vary. The greatest sum can get close to 5 and least sum approaches -13.

3. The greatest sums are near the line $x + y = 5$. The lesser sums are near the intersection of the boundary lines for $4x - 2y > -10$ and $2x - 4y < 16$.

4. $(0,5)$, $(6,-1)$, $(-6,-7)$; Earlier sums may have been very close to vertex sums. No earlier sum was greater than the greatest vertex sum or less than the least vertex sum.

5. greatest sum is close to 6; least sum is close to -20; the sums of the coordinates of the vertices are 6, 2, and -20

6. Answers will vary. However, the answers for question 5 should be confirmed by this investigation.

TRY THESE

1. The point is not in the region since it makes the inequality $y \leq x - 3$ untrue.

2. The point lies in the region since it satisfies all inequalities.

3. $P = 0 + 3(0) = 0$ minimum at $(0,0)$
$P = 3 + 3(0) = 3$
$P = 0 + 3(7) = 21$ maximum at $(0,7)$
$P = 4 + 3(2) = 10$

4. $P = 2(2) + 5(3) = 19$ minimum at $(2,3)$
$P = 2(3) + 5(6) = 36$ maximum at $(3,6)$
$P = 2(1) + 5(6) = 32$
$P = 2(0) + 5(5) = 25$

5. $P = 4(0) + 2(1) = 2$ minimum at $(0,1)$
$P = 4(9) + 2(1) = 38$ maximum at $(9,1)$
$P = 4(2) + 2(7) = 22$
$P = 4(6) + 2(3) = 30$

6. $C = 5(10) + (10) = 70$ minimum at $(10,10)$
$C = 5(10) + (20) = 80$
$C = 5(20) + (20) = 120$
$C = 5(30) + (10) = 160$ maximum at $(30,10)$

7. $C = 3(0) + (0) = 0$ minimum at $(0,0)$
$C = 3(0) + (8) = 8$
$\left.\begin{array}{l} C = 3(4) + (3) = 15 \\ C = 3(5) + (0) = 15 \end{array}\right\}$ maximum at $(4,3)$ & $(5,0)$

8.
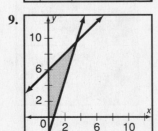
minimum at $(2,3)$
$C = 2 + 2(3) = 8$

9.
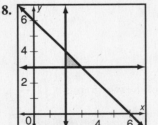
$y = x + 6$ and $y = 3x - 1$ intersect at $(3.5, 9.5)$.
maximum $P = 2(3.5) + 9.5 = 16.5$ occurs at that point.

10. The values represented by x and y are usually quantities that cannot be negative.

PRACTICE

1. The point lies in the region since it satisfies all inequalities.

2. The point is not in the region since it fails to satisfy $y \le -2x + 9$.

3. The point lies in the region since it satisfies all inequalities.

4. $P = (0) + 4(-1) = -4$ minimum at $(0,-1)$
$P = (4) + 4(0) = 4$
$P = (3) + 4(2) = 11$ maximum at $(3,2)$
$P = (0) + 4(1) = 4$

5. $P = 2(3) - 3(1) = 6$ maximum at $(3,1)$
$P = 2(2) - 3(4) = -8$
$P = 2(0) - 3(5) = -15$ minimum at $(0,5)$
$P = 2(0) - 3(0) = 0$

6. $P = -(-2) + 6(0) = 4$
$P = -(3) + 6(0) = -3$ minimum at $(3,0)$
$P = -(4) + 6(3) = 14$ maximum at $(4,3)$
$P = -(0) + 6(1) = 6$

7. $\left. \begin{array}{l} R = 3(0) - 2(0) = 0 \\ R = 3(-2) - 2(-3) = 0 \end{array} \right\}$ maximum at $(0,0)$ & $(-2,-3)$
$R = 3(-3) - 2(3) = -15$ minimum at $(-3,3)$
$R = 3(0) - 2(1) = -2$

8. $C = 4(0) + 5(2) = 10$ minimum at $(0,6)$
$C = 4(2) + 5(3) = 23$
$C = 4(4) + 5(0) = 16$
$C = 4(0) + 5(6) = 30$ maximum at $(0,6)$

9.
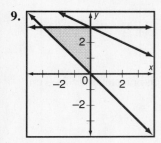

$P = -3 + 4(3) = 9$
$P = 0 + 4(0) = 0$
$P = 0 + 4(3) = 12$ maximum at $(0,3)$

10.
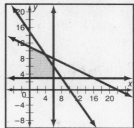

$C = 4(0) + 3(11) = 33$
$C = 4(4) + 3(9) = 43$
$C = 4(6) + 3(6) = 42$
$C = 4(6) + 3(3) = 33$
$C = 4(0) + 3(3) = 9$ minimum at $(0,3)$

11.
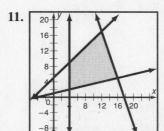

$P = 4(4) + (9) = 25$
$P = 4(12) + (17) = 65$
$P = 4(16) + (5) = 69$ maximum at $(16,5)$
$P = 4(4) + (2) = 18$

12.
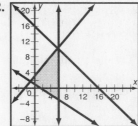

$C = (0) + 5(3) = 15$
$C = \left(5\frac{4}{7} \right) + 5\left(10\frac{3}{7} \right) = 57\frac{5}{7}$
$C = (6) + 5(10) = 56$
$C = (6) + 5(-3) = -9$ minimum at $(6,-3)$
$C = (0) + 5(1) = 5$

13.
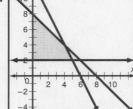
   $(0,8), (4,4), (5,2), (0,2)$

14. $C = 4(0) + 4(8) = 32$ $C = 3(0) + 2(8) = 16$

$C = 4(4) + 4(4) = 32$ $C = 3(4) + 2(4) = 20$

$C = 4(5) + 4(2) = 28$ $C = 3(5) + 2(2) = 19$

$C = 4(0) + 4(2) = 8$ $C = 3(0) + 2(2) = 4$ minimum;

so use $C = 3x + 2y$.

15.

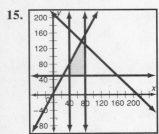

(40,50), (40,80), (70,140), (80,130), (80,50)

16. $P = 400(40) + 350(50) = 33,500$

$P = 400(40) + 350(80) = 44,000$

$P = 400(70) + 350(140) = 77,000$

$P = 400(80) + 350(130) = 77,500$ maximum

$P = 400(80) + 350(50) = 49,500$

Maximum is at 80 acres of x and 130 acres of y.

17.

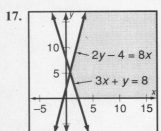

There is a minimum, but no maximum, because the graph of the constraints is unbounded.

18.

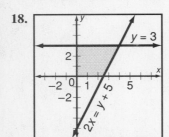

There are both minimum and maximum values because the graph is bounded.

19.

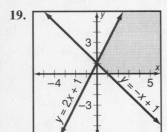

There is a minimum, but no maximum, because the graph of the constraints is unbounded.

20.

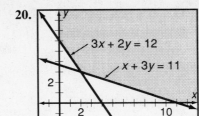

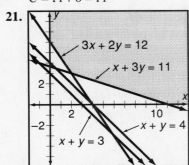

$C = 0 + 6 = 6$

$C = 2 + 3 = 5$ minimum at $(2,3)$

$C = 11 + 0 = 11$

21.

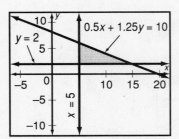

For any value less than M, the objective function misses the feasible region.

22. $\begin{cases} 0.50x + 1.25y \leq 10.00 \\ x \geq 5 \\ y \geq 2 \end{cases}$

23. The profit P is \$1.00 on each bracelet and \$2.00 on each necklace.

24. $P = 1(5) + 2(2) = 9$

$P = 1(5) + 2(6) = 17$

$P = 1(15) + 2(2) = 19$

maximum at 15 bracelets and 2 necklaces

25. $A: (-3)(-5) - (1)(-9) = 15 + 9 = 24$

$B: (-3)(-5) - (1)(-9) = 15 + 9 = 24$

$C: (-3)(5) - (-1)(9) = -15 + 9 = -6$

$D: (-3)(5) + (-1)(9) = 15 - 9 = 6$

26. $y = (-2)^2 - 4(-2) + 7 = 4 + 8 + 7 = 19$

27. $y = (-2)^4 - 3(-2)^3 + (-2) - 4 =$
$16 - 3(-8) - 2 - 4 = 16 + 24 - 2 - 4 = 34$

28. $y = (-2)^2 - 7(-2) + 12 =$
$3(4) + 14 + 12 = 12 + 14 + 12 = 38$

29. $y = (-2)^5 - 4(-2)^3 + 12(-2) - 6 =$
$2(-32) - 4(-8) - 24 - 6 =$
$-64 + 32 - 24 - 6 = -62$

Lesson 8.5, pages 394–397

EXPLORE

1. $y \geq x$
$x + y \geq 30$
$x + y \leq 50$
$y \leq 25$
$x \geq 0$ and $y \geq 0$

2. Cost $= 4x$ for short-sleeved shirts

3. Cost $= 5y$ for long-sleeved shirts

4. $C = 4x + 5y$

5. $(15, 15), (5, 25), (25, 25)$

6.

Vertex	$C = 4x + 5y$	Cost
(15,15)	$4(15) + 5(15)$	\$135
(5,25)	$4(5) + 5(25)$	\$145
(25,25)	$4(25) + 5(25)$	\$225

7. The cost is the minimum of \$135 at (15,15).

8. 15 short-sleeved and 15 long-sleeved shirts at a cost of \$135

9. Short - sleeved profit $= 6 - 4 = \$2$
Long - sleeved profit $= 8 - 5 = \$3$

10. $2x$ and $3y$

11. Profit $= 2x + 3y$

12.

Vertex	$C = 2x + 3y$	Profit
(15,15)	$2(15) + 3(15)$	\$75
(5,25)	$2(5) + 3(25)$	\$85
(25,25)	$2(25) + 3(25)$	\$125

13. Profit is maximum of \$125 at (25,25).

14. The minimum cost \$135 produces a profit of \$75, but to earn the maximum profit of \$125, the club must spend \$225 to buy more shirts.

15a. Let $x =$ number of cross country skis
       $y =$ number of slalom skis
$6x + 4y \leq 96$ manufacturing constraint
   $x + y \leq 20$ finishing constraint
     $x \geq 0$ and $y \geq 0$

b.

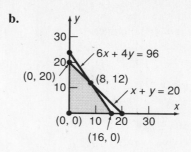

$(0, 0), (0, 20), (16, 0), (8, 12)$

c.

Vertex	$P = 45x + 30y$	Profit
(0, 0)	$P = 45(0) + 30(0)$	\$ 0
(0,20)	$P = 45(0) + 30(20)$	\$600
(16, 0)	$P = 45(16) + 30(0)$	\$720
(8,12)	$P = 45(8) + 30(12)$	\$720

d. Maximum profit is \$720 by producing 16 cross country and no slalom skis or 8 cross country and 12 slalom skis.

16a. Let $x =$ number of buses
       $y =$ number of vans
$60x + 10y \geq 300$
   $4x + 2y \leq 36$
     $x \geq 0$ and $y \geq 0$

b.

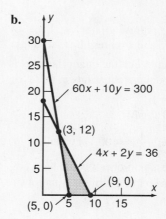

$(5, 0), (9, 0), (3, 12)$

c.

Vertex	$C = 1000x + 100y$	Cost
(5, 0)	$C = 1000(5) + 100(0)$	\$5000
(9, 0)	$C = 1000(9) + 100(0)$	\$9000
(3,12)	$C = 1000(3) + 100(12)$	\$4200

d. A minimum cost of \$4200 requires 3 buses and 12 vans.

17. Let $x =$ number of bicycles
$y =$ number of rowing machines
$$x + y \leq 110$$
$$75x + 125y \leq 10,000$$
$$x \geq 0 \text{ and } y \geq 0$$
Profit $= P = (125 - 75)x + (200 - 125)y = 50x + 75y$

Vertex	$P = 50x + 75y$	Profit
$(\ 0, 80)$	$P = 50(\ \ 0) + 75(80)$	$6000
$(110,\ 0)$	$P = 50(110) + 75(\ 0)$	$5500
$(\ 75, 35)$	$P = 50(\ 75) + 75(35)$	$6375

Maximum profit is at 75 bicycles and 35 rowers.

18. Answers will vary. Students should realize that, in general, for a business to produce two products or offer two services, higher production costs for one of the products or services must be offset by higher profits from that product or service.

REVIEW PROBLEM SOLVING STRATEGIES

1. Students may draw diagrams or make a physical model with paper slips; shortest solution: Lane to 102, Kim to 103, Raintree to 106, Lane to 105, Gonzalez to 102, Ziff to 101, Lane to 104, Raintree to 105, Kim to 106, Gonzalez to 103, Raintree to 102, Lane to 105, Ziff to 104, Raintree to 101, Gonzalez to 102, Kim to 103, Lane to 106 (17 moves)

2a. 3; Possible answer: let l represent large dogs, m represent medium dogs, and s represent small dogs

b. There are three facts that relate the quantities; this is enough to solve because three equations in three unknowns can be written.

c. 5 large, 25 medium, 70 small;
$$s + m + l = 100, \quad 3l + 2m + \tfrac{1}{2}s = 100, \quad m = 51$$

d. Answers will vary; guess-and-check could be used, but many students may feel that writing equations is more efficient.

3a. $\dfrac{tx}{4}$; $x - \dfrac{tx}{4}$

b. $\dfrac{tx}{5}$; $x - \dfrac{tx}{5}$

c. four times the length of candle A;
$$x - \frac{tx}{5} = 4\left(x - \frac{tx}{4}\right)$$

d. $3\tfrac{3}{4}h$; candle A was $\dfrac{1}{16}$ its original length, candle B was $\dfrac{1}{4}$ its original length.

Lesson 8.6, pages 398–403

EXPLORE
1. Answers will vary.

2. Possible answers: what the probability is of rolling a particular number on the die and on the spinner; what the probability is of the number rolled on the die being greater than the number spun on the spinner

3. Answers will vary. Some students may want to play 50 to 100 times or more before making a guess.

4. Answers will vary; probable answer: the spinner.

TRY THESE
1. $2 \cdot 52 = 104$

2. 26: head and any of 13 hearts plus head and any of 13 diamonds

(H, ♥A) (H, ♥2) (H, ♥3) (H, ♥4)(H, ♥5) (H, ♥6)
(H, ♥7) (H, ♥8) (H, ♥9) (H, ♥10)(H, ♥J) (H, ♥Q)
(H, ♥K) (H, ♦A) (H, ♦2) (H, ♦3)(H, ♦4) (H, ♦5)
(H, ♦6) (H, ♦7) (H, ♦8) (H, ♦9)(H, ♦10) (H, ♦J)
(H, ♦Q) (H, ♦K)

3. $\dfrac{26}{104}$ or $\dfrac{1}{4}$ or 25%

4. $\dfrac{1}{104}$

5. $2 \cdot 6 \cdot 3 \cdot 10 = 360$ (sleeve length • color • stripe • size)

6a. $6 \cdot 3 \cdot 10 = 180$ (color • stripe • size)

b. $2 \cdot 3 \cdot 10 = 60$ (sleeve length • stripe • size)

c. $3 \cdot 10 = 30$ (stripe • size)

d. 10 (size)

7. $\dfrac{20}{360}$ (20 different sleeve length and sizes, i.e. $2 \cdot 10$)
$= \dfrac{1}{18}$

PRACTICE
1. $26 \cdot 26 \cdot 26 = 17,576$

2. $9 \cdot 10 = 90$

3. $9 \cdot 10 \cdot 26 \cdot 26 \cdot 26 = 1,581,840$

4. There are $26 \cdot 26 = 676$ possibilities, so $\dfrac{1}{676}$.

5. $9 \cdot 10 \cdot 26 \cdot 1 \cdot 1 = 2340$

6. $28 \cdot 12 = 336$

7. $9 \cdot 2 \cdot 10 = 180$

8. There are $10 \cdot 10 \cdot 10 = 1000$ possible in sample space, so probability is $\dfrac{180}{1000}$ or $\dfrac{9}{50}$.

9. $6 \cdot 8 \cdot 5 = 240$

10. $6 \cdot (8 - x) \cdot 5 \le 150$ \quad Divide by 30.

$\qquad x = 3$ \qquad Solve for x.

It can remove 3 choices.

11. $\left(\dfrac{1}{2}\right)^5 = \dfrac{1}{32}$

12. Fair game; Of the 36 outcomes, 18 sums are odd and 18 are even, so each has a probability of $\dfrac{18}{36}$ or $\dfrac{1}{2}$ of winning.

13. Unfair game; Of the 36 outcomes, 27 products are even and 9 are odd, so:

$$P(A \text{ wins}) = \dfrac{27}{36} = \dfrac{3}{4} \text{ and } P(B \text{ wins}) = \dfrac{9}{36} = \dfrac{1}{4}$$

14. Fair game; Of the 8 outcomes, only 2 show all heads or all tails and 6 have mixed outcomes and tails.

$$P(A \text{ wins}) = \dfrac{2}{8} = \dfrac{1}{4} \text{ and } P(B \text{ wins}) = \dfrac{6}{8} = \dfrac{3}{4}$$

Since $P(B) = 3 \cdot P(A)$, A should get 3 times as many points to make the game fair.

Use the following display for Exercises 15-19.

$$bbbb \begin{vmatrix} & & & & b & b & g & g \\ b & b & b & g & b & g & b & g & g & g & g & b \\ b & b & g & b & g & b & b & g & g & g & b & b \\ b & g & b & b & b & g & g & b & g & b & g & g \\ g & b & b & b & g & b & g & b & b & g & g & g \\ & & & & g & g & b & b \end{vmatrix} gggg$$

15. $\left(\dfrac{1}{2}\right)^4 = \dfrac{1}{16}$

16. $\dfrac{1}{16} + \dfrac{1}{16} = \dfrac{2}{16} = \dfrac{1}{8}$

17. $\dfrac{4}{16} = \dfrac{1}{4}$; $\dfrac{4}{16} = \dfrac{1}{4}$

18. $\dfrac{6}{16} = \dfrac{3}{8}$

19. Exercises 16-18 include the sample space for all possible outcomes for a family with four children. The sum of all possible outcomes in a probability experiment is exactly 1.

PROJECT CONNECTION

1. $240 + 12(90) = 240 + 1080 = 1320$; $1320 - 1200 = 120$

2. $x \cdot 1200 = 120$ so $x = 0.10$ or 10%;

APR $= (24 \cdot 120) \div [1200 \cdot (12 + 1)] = 2880 \div 15{,}600$ ≈ 0.1846 or 18.46%

ALGEBRAWORKS

1. $3(9 \cdot 6 \cdot 2) = 324$

2. There are 3 large, blue, hooded shirts, so $\dfrac{3}{324} = \dfrac{1}{108}$.

3. There are $3 \cdot 6$ or 18 red crewneck shirts, so $\dfrac{18}{324} = \dfrac{1}{18}$.

4. Answers will vary but should reflect proportions of styles, sizes, and colors similar to those in the sales report. Yellow, pink, and purple were not included in the sales report, possibly because these colors were not in inventory or because none sold. Possible recommendations for these colors may be cautious or may take into consideration the popularity of purple.

Chapter Review, pages 404–405

1. d

2. c

3. b

4. a

5. $1 \le 3(1) - 6$

$\quad 1 \le -3$ \qquad false, not a solution

6. $3(0) > 0$

$\quad 0 > 0$ \qquad false, not a solution

7. $2(4) + 6 > 4$

$\quad 14 > 4$ \qquad True, $(4,6)$ is a solution.

8.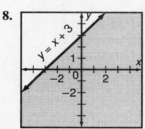

9.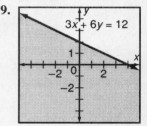

10.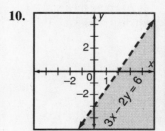

11.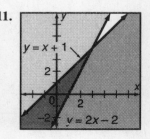

12.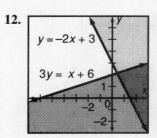

13.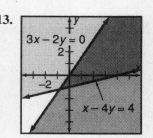

14. 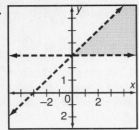   **15.**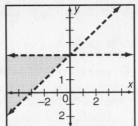

16.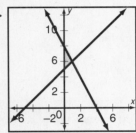

17.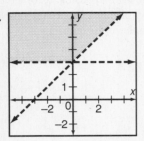

$P = 2(0) + 3(5) = 15$

$P = 2(0) + 3(8) = 24$ maximum at $(0,8)$

$P = 2(1) + 3(6) = 20$

18a. $P = 3x + 2y$ where $x =$ number of apple pies and $y =$ number of blueberry pies

b. $x + y \geq 20,\ x + y \leq 40,\ y = x \quad y \leq 20$

c. 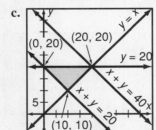   Maximum profit of $100 at 20 apple pies and 20 blueberry pies.

19. $6 \cdot 6 \cdot 6 = 216$

Chapter Assessment, pages 406–407

1. $1 > 4(4) - 3$

$\quad 1 > 13 \qquad$ false, not a solution

2. $3(2) - 2 < 4$

$\quad\quad 4 < 4 \qquad$ false, not a solution

3. 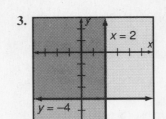   **4.**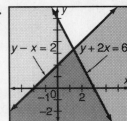

5. 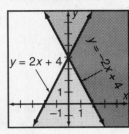   **6.**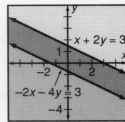

7. 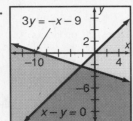   **8.**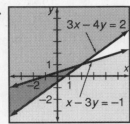

9a. $\begin{cases} x + y \geq 16 \\ 12x + 16y \leq 288 \\ x \geq 0,\ \ y \geq 0 \end{cases}$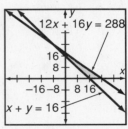

b. 10 CD's and 8 books, 16 CD's and 2 books, 8 CD's and 8 books

10. 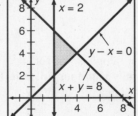   **11.**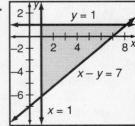

12. Answers will vary, but should include identifying variables for the number of CD's and the number of books to buy. Use these variables with the costs and limitations of the constraints to write inequalities.

13. $7 \cdot 9 \cdot 7 = 441$

14. By testing points, regions are 1 and 4.

15. By testing points, regions are 5 and 7.

16. By testing points, regions are 2 and 3.

17. By testing points, regions is 6.

18. By testing points, regions is 4.

19. By testing points, regions is 3.

20.
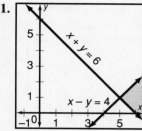

$P = 5(0) + 3(6) = 18$

$P = 5(5) + 3(1) = 28$ maximum at (5,1)

$P = 5(4) + 3(0) = 20$

$P = 5(0) + 3(0) = 0$

21.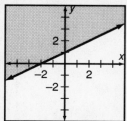

$C = 2(6) + (0) = 12$

$C = 2(5) + (1) = 11$ minimum at (5,1)

22a. $C = 40x + 60y$ where x is the number of five-ton trucks and y is the number of ten-ton trucks

$x \le 8,\ y \le 7,\ x + y \le 12,\ 40x + 60y \ge 420,\ x \ge 0,\ y \ge 0$

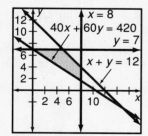

b. Since all 12 drivers are to work, only whole number values of x and y adding up to 12 can be considered. There are vertices at (5, 7) and (8, 4). So, 8 five-ton trucks and 4 ten-ton trucks must be used.

Cumulative Review, page 408

1.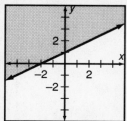

2.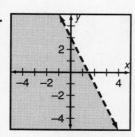

---

3. $\frac{2}{10} = \frac{1}{5}$ or 20%   **4.** $\frac{3}{10} = 0.3$ **5.** $\frac{0}{10} = 0$

6. No; 0.4 is Maria's experimental probability of choosing a red marble on her next selection. This probability will change after each successive selection.

7. $4x + 3y = 4.65 \rightarrow 20x + 15y = 23.25$

$5x + 2y = 4.85 \rightarrow \underline{20x + 8y = 19.40}$

$7y = 3.85$

$y = 0.55\ =\ \text{cost of fries}$

$4x + 3(0.55) = 4.65$

$x = 0.75\ =\ \text{cost of hamburger}$

So, the cost of the hamburger and fries is $0.55 + 0.75$ or \$1.30.

8. $y = kx$ so $30 = k \cdot 8$ and $k = 3.75$

Therefore $135 = 3.75x$

$36 = x$

He needs 36 customers.

9. Saturday

10. $\frac{40 + 30 + 55 + 45 + 25 + 50 + 65}{7} = \frac{310}{7} \approx 44$

11. 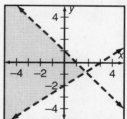   **12.**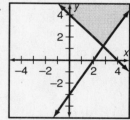

13. Answers will vary. Many graphing utilities will not automatically shade the half-plane.

14. $n + 18 = 15$

$n = -3$ Subtract 18.

15. $\frac{h}{-12} = -6$

$\phantom{\frac{h}{-12}}h = 72$ Multiply by -12.

16. $4(2d + 5) = 12$

$8d + 20 = 12$ Distribute.

$8d = -8$ Subtract 20.

$d = -1$ Divide by 8.

17. $\frac{1}{4}x + 3 = 16 - \frac{2}{5}x$

$\frac{13}{20}x = 13$ Add $\frac{2}{5}x$ and subtract 3.

$\phantom{\frac{13}{20}}x = 20$ Multiply by $\frac{20}{13}$.

18. $0.26x \ge 2000.00$

$x \ge 7693$ units   Divide by 0.26.

19. Since $x > 3$ is a single-variable inequality, the graph would be on a number line. The other is a two-variable inequality and would be graphed on a coordinate plane.

20. $D = \begin{vmatrix} 3 & -7 \\ 5 & 2 \end{vmatrix} = 3(2) - (-7)(5) = 41$

$D_x = \begin{vmatrix} 13 & -7 \\ 8 & 2 \end{vmatrix} = 13(2) - (-7)(8) = 82$

$x = \dfrac{D_x}{D} = \dfrac{82}{41} = 2$

21. E, since both have a slope of $\dfrac{2}{5}$, are parallel, so do not intersect.

Standardized Test, page 409

1. B

Col. 1: $m = \dfrac{4 - (-2)}{-2 - 4} = \dfrac{6}{-6} = -1$

Col. 2: $m = 0$

So, Col. 1 < Col. 2.

2. C

Col. 1:   Point on x-axis is $(x, 0)$.
Product is $x \cdot 0 = 0$.

Col. 2:   Point on y-axis is $(0, y)$.
Product is $0 \cdot y = 0$.

So, Col. 1 = Col. 2.

3. A

$\begin{cases} 3b - a = 5 & \rightarrow \quad 6b - 2a = 10 \\ 4a + 2b = 0 & \rightarrow \quad 4b + 2a = \;\; 0 \end{cases}$

$\quad \overline{10b = 10}$

$\; b = 1$

$3(1) - a = 5$

$a = -2$

Col. 1: $-2 + 1 = -1$

Col. 2: $(-2)(1) = -2$

Col. 1 > Col. 2

4. B

Col 1: $-3^2 = -9$

Col 2: $(-3)^2 = 9$

Col. 1 < Col. 2.

5. A

Col 1: $\dfrac{6}{36} = \dfrac{1}{6}$

Col 2: $\dfrac{3}{36} = \dfrac{1}{12}$

Col. 1 > Col. 2

6. B, since b is steeper in absolute value than a, so Col. 1 < Col. 2.

7. D

Col. 1: Weekly income ranges from $230 to $310.
Col. 2: Weekly income ranges from $220 to $340.
Ranges overlap so a comparison is impossible.

8. A; Two parallel lines have the same slope, so the product will be positive. The product of perpendicular slopes is always -1.

9. B; since $y = x - 3$, $x > y$

10. B; since the line inside box is further to the right

11. D

12. C

Col. 1: Weakest correlation = 0
Col. 2: Probability of an impossible event = 0
Col. 1 = Col. 2

13. D

Col. 1: $3m + 2n$
Col. 2: $-2m - 3n$
Without values for m and n, a relationship cannot be determined.

14. B

Col. 1: all real numbers, so least number is less than -3
Col. 2: all real numbers ≥ -3
Col. 1 < Col. 2

15. B

Col. 1: $2^2 = 4$
Col. 2: $25 - 9 = 16$
Col. 1 < Col. 2

Chapter 9

Absolute Value and the Real Number System

Data Activity, page 410

1. The sum of daily vehicle miles is 320,140 thousand. Dividing by 10 gives a mean of 32,014 thousand or 32,140,000 miles. The mean is much higher than the data values for eight of the ten cities due to the extreme values of NY and LA. However, it is also useful to see how these large cities affect the national average.

2. Answers will vary. The ratio of hours of delay to miles driven is higher for Phoenix than New York.

3. Answers will vary. Students may round to the nearest billion or half billion or use clustering. Accept a range of $18-20 billion.

4. & 5. Answers will vary.

Lesson 9.1, pages 413–415

THINK BACK/WORKING TOGETHER

1a. $|4| = 4$ **b.** $|-3.5| = 3.5$ **c.** $|0| = 0$

d. $\left|-\frac{1}{2}\right| = \frac{1}{2}$ **e.** $|-\pi| = \pi$

2. $|x| = x$ if $x \geq 0$

$|x| = -x$ if $x < 0$

3a. $x = 5$ if $x \geq 0$ or $-x = 5$ if $x < 0$

                $x = -5$

So, $x = 5$ or -5.

b. $x = 0$ if $x \geq 0$ or $-(-x) = 0$ if $x < 0$

So, $x = 0$.

c. No solution; absolute value is non-negative so it can never be -2

d. $-x = \frac{1}{2}$ if $-x \geq 0$ or $-(-x) = \frac{1}{2}$ if $-x < 0$

   $x = -\frac{1}{2}$ $x = \frac{1}{2}$

So, $x = \frac{1}{2}$ or $-\frac{1}{2}$.

EXPLORE

4. the letter V

5. The y-axis is a line of symmetry.

6. Absolute value is always non-negative and the y-coordinates are positive in Quadrants I and II.

7.    vertex is at $(0, 0)$

8. 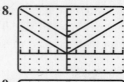 It has the same V-shape; it is shifted up 3 units; its vertex is at $(0, 3)$; its line of symmetry is the y-axis.

9. 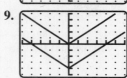 It has the same V-shape; it is shifted down 4 units; its vertex is $(0, -4)$; its line of symmetry is the y-axis.

10. The graph is shifted up if c is positive and down if c is negative.

11. Answers will vary.

12. 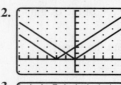 It has the same V-shape; it is shifted left 2 units; its vertex is $(-2, 0)$; its line of symmetry is $x = -2$.

13. 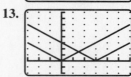 It has the same V-shape; it is shifted right 3 units; its vertex is $(3, 0)$; its line of symmetry is $x = 3$.

14. The graph is shifted right if b is negative and left if b is positive.

15. Answers will vary.

16. Answers will vary, but should include both vertical and horizontal shifts depending on the sign of b and c.

17. 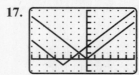 It has the same V-shape; it is shifted 3 units left and 1 unit down; its vertex is $(-3, -1)$; its line of symmetry is $x = -3$.

18. Vertex of $y = |x + b| \neq c$ will be at $(-b, c)$.

MAKE CONNECTIONS

19. b and c are both negative, so shift is right and down.

20. b is positive, so shift is left.

21. b and c are both positive, so shift is left and up.

22. Shift is 3 right and 2 up, so $y = |x - 3| + 2$.

23. Shift is 1 left and 5 down, so $y = |x + 1| - 5$.

24. Shift is 1.5 right, so $y = |x - 1.5|$.

25. b is 4, c is -3, so $y = |x + 4| - 3$.

26. b is -7.5, c is 1.5, so $y = |x - 7.5| - 1.5$.

27. b is -4.5, c is -2.5, so $y = |x - 4.5| - 2.5$.

28. b is 4.8, c is 1.8, so $y = |x + 4.8| + 1.8$.

SUMMARIZE

29. Answers will vary, but should include a shift horizontally of b units, right if $b < 0$, left if $b > 0$ and a vertical shift of c units, up if $c > 0$ and down if $c < 0$.

30. Answers will vary. The rays that form the V are steeper or farther from the x-axis than those of $y = |x|$ if $a > 1$. The rays that form the V are closer to the x-axis or less steep than those of $y = |x|$ if $0 < a < 1$.

31. It is upside down; a reflection over the x-axis.

32. Graph would be different if the order were changed, so the order of shifting and reflecting makes a difference.

Lesson 9.2, pages 416–423

EXPLORE

1. Answers will vary.

2. The graph should show a reflection of the original triangle over the y-axis.

3. A and A′ are equal distances from the y-axis. B and B′ are equidistant from the y-axis. C and C′ are equidistant from the y-axis.

TRY THESE

1. $(2.5, 0)$; $x = 2.5$ **2.** $(-4, 0)$; $x = -4$

3. $(0, -1.5)$; $x = 0$ **4.** $(0, 2.5)$; $x = 0$

5. $(-4, -9)$; $x = -4$ **6.** $(3, -3)$; $x = 3$

7. $(-3, -1)$; $x = -3$ **8.** $(1, 2)$; $x = 1$

9. The rays that form the V in the graph of $y = 4|x|$ are farther from the x-axis than those of $y = |x|$.

10. The rays that form the V in the graph of $y = \frac{1}{4}|x|$ are closer to the x-axis than those of $y = |x|$.

11. The rays that form the V in the graph of $y = -2|x|$ are farther from the x-axis than those of $y = |x|$. The graph opens downward, unlike the graph of $y = |x|$.

12. The rays that form the V in the graph of $y = -\frac{1}{2}|x|$ are closer to the x-axis than those of $y = |x|$. The graph opens downward, unlike the graph of $y = |x|$.

13. d; shift left 2 units since $b = -2$

14. c; shift up 2 units since $c = 2$

15. a; shift right 2 units since $b = -2$

16. b; shift down 2 units since $c = -2$

17. 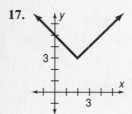   **18.**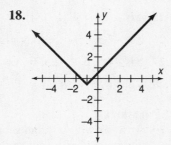

19. 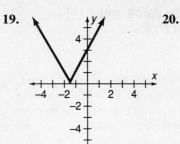   **20.**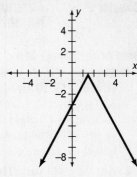

21. $b = -2$, $c = -8$ for right and down shift, so $y = |x - 2| - 8$.

22. Graph will be shifted, stretched and reflected according to the values and signs of a, b, and c.

PRACTICE

1. Vertex at $(0, -4)$; axis of symmetry: $x = 0$

2. Vertex at $(0, 5)$; axis of symmetry: $x = 0$

3. Vertex at $(1.5, 0)$; axis of symmetry: $x = 1.5$

4. Vertex at $(-0.5, 0)$; axis of symmetry: $x = -0.5$

5. Vertex at $(4, -3)$; axis of symmetry: $x = 4$

6. Vertex at $(-3, -3)$; axis of symmetry: $x = -3$

7. Vertex at $(-1, -2)$; axis of symmetry: $x = -1$

8. Vertex at $(3, 2)$; axis of symmetry: $x = 3$

9. The rays that form the V in the graph of $y = 3|x|$ are farther from the x-axis than those of $y = |x|$.

10. The rays that form the V in the graph of $y = \frac{1}{3}|x|$ are closer to the x-axis than those of $y = |x|$.

11. The rays that form the V in the graph of $y = -\frac{1}{5}|x|$ are closer to the x-axis than those of $y = |x|$. The graph opens downward, unlike the graph of $y = |x|$.

12. The rays that form the V in the graph of $y = -5|x|$ are farther from the x-axis than those of $y = |x|$. The graph opens downward, unlike the graph of $y = |x|$.

13. b; shift left 1 unit since $b = -1$

14. d; shift up 1 unit since $c = 1$

15. c; shift right 1 unit since $b = 1$

16. a; shift down 1 unit since $c = -1$

17. d; shift left 2 units since $b = -2$, and down 3 units since $c = -3$

18. b; reflect through x-axis, shift right 3 units since $b = 3$, and up 2 units since $c = 2$

19. a; reflect through x-axis, shift left 2 units since $b = -2$ and up 3 units since $c = 3$

20. c; shift left 3 units since $b = -3$, and down 2 units since $c = -2$

21.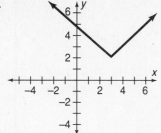

22.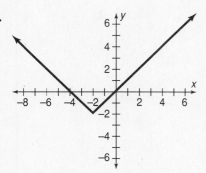

23. 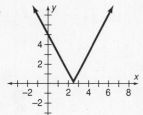 **24.**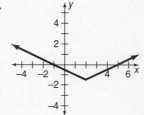

25. 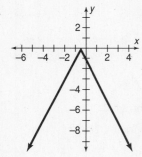 **26.**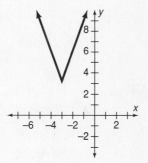

27. 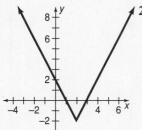 **28.**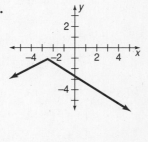

29. Reflect so $a = -1$, right 5 units so $b = 5$, down 2 units so $c = -2$, $y = -|x - 5| - 2$.

30. Angle will be obtuse when $a < 1$ because the rays of an obtuse angle will be closer to the x-axis. Angle will be acute when $a > 1$ because the rays that form an acute angle will be farther from the x-axis.

31. Slopes are 1 and -1, graph opens downward so $a = -1$, shift right and up 3 units so $b = c = 3$, $y = -|x - 3| + 3$

32. Slopes are 1 and -1 so $a = 1$, shift left 1 unit so $b = -1$ and down 1 unit so $c = -1$, $y = |x + 1| - 1$

33. All are shifted down 3 units so $c = -3$, slope of Tie A is ± 1, so $y = |x| - 3$; slope of Tie B = ± 2, so $y = 2|x| - 3$; slope of Tie C = ± 3, so $y = 3|x| - 3$.

EXTEND

34. $(-0.5, 1)$ and $(0.5, 1)$

35. $(-1, 1)$ and $(1, 1)$

36. Since star has 5 points, the only symmetry is about the y-axis.

37. Since star has 6 points (even number), symmetry is about both x- and y-axes.

38. Since vertical blue stripe is not centered, the only symmetry is about the x-axis.

39. Square flag, everything centered, symmetry is about x-axis, y-axis, $y = x$ and $y = -x$.

40. Because of the letter, there is no symmetry.

41. Since the white stripe is centered, symmetry is about both x- and y-axes.

42.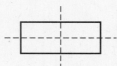

43. Any line passing through the circle's center is a line of symmetry.

44. 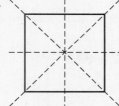 **45.**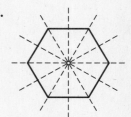

46.

Think Critically

47. The part in Quadrant IV is reflected through the x-axis into Quadrant I.

48. The part in Quadrant II is reflected through the x-axis into Quadrant III.

49. No; $y = |-2x| = |2x| = 2|x|$

50. Yes; possible answers: $y = -|3x|$ and $y = |3x|$; $y = -|3x|$ overlaps with $y = -3x$ when $x \geq 0$ and $y = |3x|$ overlaps with $y = -3x$ when $x < 0$.

51. Answers will vary; possible answer: $y = -|x| + 1$.

52. Answers will vary; possible answer: $y = -|x| + 1$.

Lesson 9.3, pages 424-429

Explore

1. & 2. Student activity, so no solution

3. 1 and -3

Try These

1. $2x - 3 = 12$ or $2x - 3 = -12$

2. $3x - 2 = 10$ or $3x - 2 = -10$

3. $2x + 5 = 17$ or $2x + 5 = -17$

4.
$$
\begin{aligned}
x - 2 = 6 \quad &\text{or} \quad x - 2 = -6 \quad \text{Write a disjunction.} \\
x = 8 \quad &\text{or} \quad x = -4 \quad \text{Solve.} \\
|8 - 2| \overset{?}{=} 6 \quad &\quad |-4 - 2| \overset{?}{=} 6 \quad \text{Check.} \\
6 = 6 ✔ \quad &\quad 6 = 6 ✔
\end{aligned}
$$

5.
$$
\begin{aligned}
x - 3 = 9 \quad &\text{or} \quad x - 3 = -9 \quad \text{Write a disjunction.} \\
x = 12 \quad &\text{or} \quad x = -6 \quad \text{Solve.} \\
|12 - 3| \overset{?}{=} 9 \quad &\quad |-6 - 3| \overset{?}{=} 9 \quad \text{Check.} \\
9 = 9 ✔ \quad &\quad 9 = 9 ✔
\end{aligned}
$$

6.
$$
\begin{aligned}
6 - x = 4 \quad &\text{or} \quad 6 - x = -4 \\
2 = x \quad &\text{or} \quad 10 = x \\
|6 - 2| \overset{?}{=} 4 \quad &\quad |6 - 10| \overset{?}{=} 4 \\
4 = 4 ✔ \quad &\quad 4 = 4 ✔
\end{aligned}
$$

7.
$$
\begin{aligned}
8 - x = 2 \quad &\text{or} \quad 8 - x = -2 \\
6 = x \quad &\quad 10 = x \\
|8 - 6| \overset{?}{=} 2 \quad &\quad |8 - 10| \overset{?}{=} 2 \\
2 = 2 ✔ \quad &\quad 2 = 2 ✔
\end{aligned}
$$

8.
$$
\begin{aligned}
2x - 7 = 5 \quad &\text{or} \quad 2x - 7 = -5 \\
2x = 12 \quad &\text{or} \quad 2x = 2 \\
x = 6 \quad &\text{or} \quad x = 1 \\
|2(6) - 7| \overset{?}{=} 5 \quad &\quad |2(1) - 7| \overset{?}{=} 5 \\
5 = 5 ✔ \quad &\quad 5 = 5 ✔
\end{aligned}
$$

9.
$$
\begin{aligned}
2x - 9 = 11 \quad &\text{or} \quad 2x - 9 = -11 \\
2x = 20 \quad &\text{or} \quad 2x = -2 \\
x = 10 \quad &\text{or} \quad x = -1 \\
|2(10) - 9| \overset{?}{=} 11 \quad &\quad |2(-1) - 9| \overset{?}{=} 11 \\
11 = 11 ✔ \quad &\quad 11 = 11 ✔
\end{aligned}
$$

10.
$$
\begin{aligned}
3|x - 5| &= 6 \\
|x - 5| &= 2 \\
x - 5 = 2 \quad &\text{or} \quad x - 5 = -2 \\
x = 7 \quad &\text{or} \quad x = 3 \\
3|7 - 5| \overset{?}{=} 6 \quad &\quad 3|3 - 5| \overset{?}{=} 6 \\
6 = 6 ✔ \quad &\quad 6 = 6 ✔
\end{aligned}
$$

11.
$$
\begin{aligned}
4|x - 3| &= 16 \\
|x - 3| &= 4 \\
x - 3 = 4 \quad &\text{or} \quad x - 3 = -4 \\
x = 7 \quad &\text{or} \quad x = -1 \\
4|7 - 5| \overset{?}{=} 16 \quad &\quad 4|-1 - 3| \overset{?}{=} 16 \\
16 = 16 ✔ \quad &\quad 16 = 16 ✔
\end{aligned}
$$

12.
$$
\begin{aligned}
2|3z + 1| - 3 &= 9 \\
2|3z + 1| &= 12 \quad \text{Add 3.} \\
|3z + 1| &= 6 \quad \text{Divide by 2.} \\
3z + 1 = 6 \quad &\text{or} \quad 3z + 1 = -6 \\
3z = 5 \quad &\text{or} \quad 3z = -7 \\
z = \frac{5}{3} \quad &\text{or} \quad z = -\frac{7}{3} \\
2\left|3\left(\frac{5}{3}\right) + 1\right| - 3 \overset{?}{=} 9 \quad &\quad 2\left|3\left(-\frac{7}{3}\right) + 1\right| - 3 \overset{?}{=} 9 \\
9 = 9 ✔ \quad &\quad 9 = 9 ✔
\end{aligned}
$$

13.
$$
\begin{aligned}
2|4w - 1| + 2 &= 12 \\
2|4w - 1| &= 10 \quad \text{Subtract 2.} \\
|4w - 1| &= 5 \quad \text{Divide by 2.} \\
4w - 1 = 5 \quad &\text{or} \quad 4w - 1 = -5 \\
4w = 6 \quad &\text{or} \quad 4w = -4 \\
w = \frac{3}{2} \quad &\text{or} \quad w = -1 \\
2\left|4\left(\frac{3}{2}\right) - 1\right| + 2 \overset{?}{=} 12 \quad &\quad 2|4(-1) - 1| + 2 \overset{?}{=} 12 \\
12 = 12 ✔ \quad &\quad 12 \overset{?}{=} 12
\end{aligned}
$$

14.
$$
\begin{aligned}
-|x + 4| - 3 &= 12 \\
-|x + 4| &= 15 \quad \text{Add 3.} \\
|x + 4| &= -15 \quad \text{Multiply by } -1.
\end{aligned}
$$

No solution since absolute value is always non-negative

164 Solutions Manual **Algebra 1: An Integrated Approach**

15. $-|x+5|-5=5$

$\qquad -|x+5|=10 \qquad$ Add 5.

$\qquad |x+5|=-10 \qquad$ Multiply by -1.

No solution since absolute value is never negative

16. 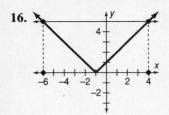 **17.**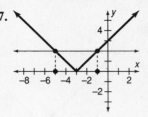

18.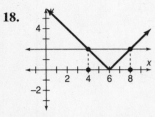

19. 5% of 30 ml = 0.05(30) = 1.5 ml so $|x-30|=1.5$

20. $|x-15|=3$

$\qquad x-15=3 \quad$ or $\quad x-15=-3$

$\qquad\qquad x=18 \quad$ or $\qquad x=12$

minimum is 12%
and maximum is 18%

21. Answers will vary.

PRACTICE, PAGE 427

1. $4x=12$ or $4x=-12$

2. $2x=10$ or $2x=-10$

3. $3x+4=11$ or $3x+4=-11$

4. $2x+1=8$ or $2x+1=-8$

5. $5x-1.5=8.5$ or $5x-1.5=-8.5$

6. $6x-4.1=9.1$ or $6x-4.1=-9.1$

7. $\quad 2w=16 \qquad$ or $\qquad 2w=-16 \qquad$ Write a disjunction.

$\qquad w=8 \qquad$ or $\qquad w=-8 \qquad$ Solve.

$\quad |2(8)| \overset{?}{=} 16 \qquad |2(-8)| \overset{?}{=} 16 \qquad$ Check.

$\qquad 16=16 \ \checkmark \qquad\quad 16=16 \ \checkmark$

8. $\qquad -5t=15 \qquad$ or $\qquad -5t=-15 \qquad$ Write a disjunction.

$\qquad\quad t=-3 \qquad$ or $\qquad t=3 \qquad$ Solve.

$\quad |-5(-3)| \overset{?}{=} 15 \qquad |-5(3)| \overset{?}{=} 15 \qquad$ Check.

$\qquad 15=15 \ \checkmark \qquad\quad 15=15 \ \checkmark$

9. $\left|-\dfrac{1}{2}q\right|=-16$, No solution since absolute value is non-negative

10. $\quad x-2=4 \qquad$ or $\qquad x-2=-4 \qquad$ Write a disjunction.

$\qquad x=6 \qquad\qquad\qquad x=-2 \qquad$ Solve.

$\quad |6-2| \overset{?}{=} 4 \qquad\quad |-2-2| \overset{?}{=} 4 \qquad$ Check.

$\qquad 4=4 \ \checkmark \qquad\qquad 4=4 \ \checkmark$

11. $\quad 8-x=5 \qquad$ or $\qquad 8-x=-5$

$\qquad x=3 \qquad$ or $\qquad x=13$

$\quad |8-3| \overset{?}{=} 5 \qquad\qquad |8-13| \overset{?}{=} 5$

$\qquad 5=5 \ \checkmark \qquad\qquad\quad 5=5 \ \checkmark$

12. $\quad 7-x=3 \qquad$ or $\qquad 7-x=-3$

$\qquad x=4 \qquad$ or $\qquad x=10$

$\quad |7-4| \overset{?}{=} 3 \qquad\qquad |7-10| \overset{?}{=} 3$

$\qquad 3=3 \ \checkmark \qquad\qquad\quad 3=3 \ \checkmark$

13. $\quad 2x-5=7 \qquad$ or $\qquad 2x-5=-7$

$\qquad x=6 \qquad\qquad\qquad x=-1$

$\quad |2(6)-5| \overset{?}{=} 7 \qquad |2(-1)-5| \overset{?}{=} 7$

$\qquad 7=7 \ \checkmark \qquad\qquad\quad 7=7 \ \checkmark$

14. $3|x-4|=10$

$\qquad |x-4|=\dfrac{10}{3}$

$\qquad x-4=\dfrac{10}{3} \quad$ or $\quad x-4=-\dfrac{10}{3}$

$\qquad x=\dfrac{22}{3} \quad$ or $\qquad x=\dfrac{2}{3}$

$3\left|\left(\dfrac{22}{3}\right)-4\right| \overset{?}{=} 10 \qquad 3\left|\left(\dfrac{2}{3}\right)-4\right| \overset{?}{=} 10$

$\qquad 10=10 \ \checkmark \qquad\qquad 10=10 \ \checkmark$

15. $8|3z+1|-6=18$

$\qquad 8|3z+1|=24$

$\qquad |3z+1|=3$

$\qquad 3z+1=3 \qquad$ or $\qquad\qquad 3z+1=-3$

$\qquad\quad 3z=2 \qquad$ or $\qquad\qquad\quad 3z=-4$

$\qquad\quad z=\dfrac{2}{3} \qquad$ or $\qquad\qquad\quad z=-\dfrac{4}{3}$

$8\left|3\left(\dfrac{2}{3}\right)+1\right|-6 \overset{?}{=} 18 \qquad 8\left|3\left(\dfrac{-4}{3}\right)+1\right|-6 \overset{?}{=} 18$

$\qquad 18=18 \ \checkmark \qquad\qquad\qquad 18=18 \ \checkmark$

16. $3|2w+1|+3=12$

$\qquad 3|2w+1|=9$

$\qquad |2w+1|=3$

$\qquad 2w+1=3 \qquad$ or $\qquad\qquad 2w+1=-3$

$\qquad\quad 2w=2 \qquad$ or $\qquad\qquad\quad 2w=-4$

$\qquad\quad w=1 \qquad$ or $\qquad\qquad\quad w=-2$

$3|2(1)+1|+3 \overset{?}{=} 12 \qquad 3|2(-2)+1|+3 \overset{?}{=} 12$

$\qquad 12=12 \ \checkmark \qquad\qquad 12=12 \ \checkmark$

17. $-|x+4|+12=3$

$\qquad -|x+4|=-9$

$\qquad |x+4|=9$

$\qquad x+4=9 \quad$ or $\qquad x+4=-9$

$\qquad x=5 \quad$ or $\qquad x=-13$

$\qquad -|5+4|+12\overset{?}{=}3 \qquad -|-13+4|+12\overset{?}{=}3$

$\qquad 3=3 \ ✔ \qquad\qquad 3=3 \ ✔$

18. $2|2x+3|+3=3$

$\qquad 2|2x+3|=0$

only true if $2x+3=0$, so $x=-\dfrac{3}{2}$

19. $2|x+3|+3=9$

$\qquad 2|x+3|=6$

$\qquad |x+3|=3$

$\qquad x+3=3 \quad$ or $\qquad x+3=-3$

$\qquad x=0 \quad$ or $\qquad x=-6$

$\qquad 2|0+3|+3\overset{?}{=}9 \qquad 2|-6+3|+3\overset{?}{=}9$

$\qquad 9=9 \ ✔ \qquad\qquad 9=9 \ ✔$

20. $4|x+2|-5=3$

$\qquad 4|x+2|=8$

$\qquad |x+2|=2$

$\qquad x+2=2 \quad$ or $\qquad x+2=-2$

$\qquad x=0 \quad$ or $\qquad x=-4$

$\qquad 4|0+2|-5\overset{?}{=}3 \qquad 4|-4+2|-5\overset{?}{=}3$

$\qquad 3=3 \ ✔ \qquad\qquad 3=3 \ ✔$

21. $\dfrac{1}{2}|4x-2|+6=12$

$\qquad \dfrac{1}{2}|4x-2|=6$

$\qquad |4x-2|=12$

$\qquad 4x-2=12 \quad$ or $\qquad 4x-2=-12$

$\qquad x=\dfrac{7}{2} \qquad\qquad\qquad x=-\dfrac{5}{2}$

$\dfrac{1}{2}\left|4\left(\dfrac{7}{2}\right)-2\right|+6\overset{?}{=}12 \qquad \dfrac{1}{2}\left|4\left(\dfrac{-5}{2}\right)-2\right|+6\overset{?}{=}12$

$\qquad 12=12 ✔ \qquad\qquad 12=12 ✔$

22.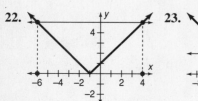

23.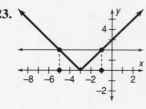

24.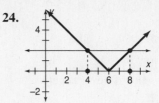

25.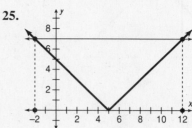

26.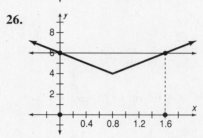

27.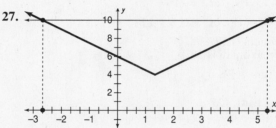

28. $|x-2000|=50$

$\qquad x-2000=50 \quad$ or $\quad x-2000=-50$

$\qquad\qquad x=2050 \quad$ or $\qquad\qquad x=1950$

Bonus is paid for minimum of 1950 hours to maximum of 2050 hours.

29. $|x-9|=3$

$\qquad x-9=3 \quad$ or $\quad x-9=-3$

$\qquad x=12 \quad$ or $\qquad x=6$

The kitten is between a minimum of 6 months old and a maximum of 12 months old.

30.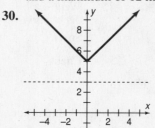

The graphs of y_1 and y_2 would not intersect. Equations will vary; possible equation: $|x+5|=3$

31.
$$x - 5 = 5x \quad \text{or} \quad x - 5 = -5x$$
$$-5 = 4x \quad \text{or} \quad -5 = -6x$$
$$-\frac{5}{4} = x \quad \text{or} \quad \frac{5}{6} = x$$

$$\left|-\frac{5}{4} - 5\right| \overset{?}{=} 5\left(-\frac{5}{4}\right) \qquad \left|\left(\frac{5}{6}\right) - 5\right| \overset{?}{=} 5\left(\frac{5}{6}\right)$$

$$\frac{25}{4} \neq -\frac{25}{4} \qquad\qquad \frac{25}{6} = \frac{25}{6} \; ✔$$

Solutions were $-\frac{5}{4}$ and $\frac{5}{6}$. $-\frac{5}{4}$ was rejected because it did not check. So, $\frac{5}{6}$ is the only solution.

32.
$$x - 4 = 2x \quad \text{or} \quad x - 4 = -2x$$
$$-4 = x \quad \text{or} \quad -4 = -3x$$
$$\frac{4}{3} = x$$

$$\left|-4 - 4\right| \overset{?}{=} 2(-4) \qquad \left|\left(\frac{4}{3}\right) - 4\right| \overset{?}{=} 2\left(\frac{4}{3}\right)$$

$$8 \neq -8 \qquad\qquad \frac{8}{3} = \frac{8}{3} \; ✔$$

Solutions were -4 and $\frac{4}{3}$. -4 was rejected because it did not check. So, $\frac{4}{3}$ is the only solution.

33.
$$x - 3 = 4x \quad \text{or} \quad x - 3 = -4x$$
$$-3 = 3x \quad \text{or} \quad -3 = -5x$$
$$-1 = x \quad \text{or} \quad \frac{3}{5} = x$$

$$\left|-1 - 3\right| \overset{?}{=} 4(-1) \qquad \left|\frac{3}{5} - 3\right| \overset{?}{=} 4\left(\frac{3}{5}\right)$$

$$4 \neq -4 \qquad\qquad \frac{12}{5} = \frac{12}{5} \; ✔$$

Solutions were -1 and $\frac{3}{5}$. -1 was rejected because it did not check. So, $\frac{3}{5}$ is the only solution.

34.
$$\frac{1}{4}\left|5 - 2x\right| = 2 - x$$
$$\left|5 - 2x\right| = 8 - 4x$$
$$5 - 2x = 8 - 4x \quad \text{or} \quad 5 - 2x = -8 + 4x$$
$$2x = 3 \quad \text{or} \quad 13 = 6x$$
$$x = \frac{3}{2} \quad \text{or} \quad \frac{13}{6} = x$$

$$\frac{1}{4}\left|5 - 2\left(\frac{3}{2}\right)\right| \overset{?}{=} 2 - \frac{3}{2} \qquad \frac{1}{4}\left|5 - 2\left(\frac{13}{6}\right)\right| \overset{?}{=} 2 - \frac{13}{6}$$

$$\frac{1}{2} = \frac{1}{2} \; ✔ \qquad\qquad \frac{1}{6} \neq -\frac{1}{6}$$

Solutions were $\frac{3}{2}$ and $\frac{13}{6}$. $\frac{13}{6}$ was rejected because it did not check. So, $\frac{3}{2}$ is the only solution.

35. Estimates will vary. Correct answers are: -2 and 6.2

36. Estimates will vary. Correct answers are: -7 and 12.8

37. Estimates will vary. Correct answers are: $-3.\overline{4}$ and $14.\overline{3}$

38.
$$\left|x - 700\right| = 100$$
$$x - 700 = 100 \quad \text{or} \quad x - 700 = -100$$
$$x = 800 \quad \text{or} \quad x = 600$$

Scores accepted are between a minimum of 600 and a maximum of 800.

39.
$$\left|x - 3.5\right| = 0.5$$
$$x - 3.5 = 0.5 \quad \text{or} \quad x - 3.5 = -0.5$$
$$x = 4.0 \quad \text{or} \quad x = 3.0$$

GPA's accepted are between a minimum of 3.0 and a maximum of 4.0.

40. Level 1: min is 18,000; max is 32,000;
$$\left|x - 25,000\right| = 7000$$
Level 2: min is 23,000; max is 39,000;
$$\left|x - 31,000\right| = 8000$$
Level 3: min is 36,000; max is 54,000;
$$\left|x - 45,000\right| = 9000$$
Level 4: min is 50,000; max is 72,000;
$$\left|x - 61,000\right| = 11,000$$

THINK CRITICALLY

41. Answers will vary; possible answer: $\left|5x - 4\right| = 3x + 4$

42. Answers will vary; possible answer: $\left|5x - 4\right| = 6x + 3$

43. y_1 would be V-shaped and y_2 would be a line that intersects y_1 in exactly one point.

44.
$$ax = b \quad \text{or} \quad ax = -b$$
$$x = \frac{b}{a} \quad \text{or} \quad x = -\frac{b}{a}$$

45.
$$cx - d = e \quad \text{or} \quad cx - d = -e$$
$$cx = d + e \quad \text{or} \quad cx = d - e$$
$$x = \frac{d + e}{c} \qquad\qquad x = \frac{d - e}{c}$$

46.
$$-\left|x + c\right| = d$$
$$\left|x + c\right| = -d$$

There is no solution since the absolute value is always non-negative and $-d$ is negative.

47.
$$x - 4 = 2x + c \quad \text{or} \quad x - 4 = -2x - c$$
$$-4 - c = x \quad \text{or} \quad 3x = 4 - c$$
$$x = \frac{4 - c}{3}$$

PROJECT CONNECTION

1. Equations should be of the form $\left|x - M\right| = M$, where M is the midpoint of the actual distance.

2. Equations should be of the form $\left|x - M'\right| = M'$, where M' is the midpoint of the map distance.

3. The scale of the map is $\dfrac{M'}{M}$, the ratio of map distance to actual distance.

4. Answers will vary.

Lesson 9.4, pages 430–436

EXPLORE

1. Set $A = [-2, -1, 0, 1, 2, 3, 4, 5, 6, 7]$
Set $B = [-4, -3, -2, -1, 0, 1, 2, 3, 4, 5]$
A but not $B = \{6, 7\}$
B but not $A = \{-4, -3\}$
Intersection of A and $B = \{-2, -1, 0, 1, 2, 3, 4, 5\}$

2. 8 dots, so 8 solutions

3. Because -2 and 5 are in the solution set

4. -2 and 5 would not be included in solution set

5. Yes, there would be an infinite number of solutions because all real numbers ≥ -2 and ≤ 5 would be included as solutions.

TRY THESE

1. $-6 < x + 3$ and $x + 3 < 6$ **2.** $x - 4 < -7$ or $x - 4 > 7$

3. $2x - 4 \leq -6$ or $2x - 4 \geq 6$ **4.** $-9 \leq 3x + 5$ and $3x + 5 \leq 9$

5.

6.

7.

8.

9. c; inequality is $\leq$, midpoint is zero

10. a; inequality is $>$, midpoint is 2

11. d; inequality is $\geq$, midpoint is 2

12. b; inequality is $<$, midpoint is 3

13. $2q - 7 \leq -1$ or $2q - 7 \geq 1$
  $2q \leq 6$ or $2q \geq 8$
  $q \leq 3$ $q \geq 4$

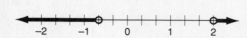

14. $9x - 6 < -12$ or $9x - 6 > 12$
  $9x < -6$ or $9x > 18$
  $x < -\dfrac{2}{3}$ $x > 2$

15. $-7 < 3t + 5$ and $3t + 5 < 7$
  $-12 < 3t$ and $3t < 2$
  $-4 < t$ and $t < \dfrac{2}{3}$

16. $-6 \leq 4x - 2$ and $4x - 2 \leq 6$
  $-4 < 4x$ and $4x \leq 8$
  $-1 \leq x$ and $x \leq 2$

17. all real numbers; $|2x| \geq 0$ is always true since absolute value is always non-negative

18. no solution since absolute value is always non-negative

19. Midpoint is 85, distance to end points is 3; $|x - 85| \leq 3$

20. Answers will vary. They are similar because they both have open circles at $-c + b$ and $c + b$. They are different because the graph of $|x - b| < c$ has a line connecting $-c + b$ and $c + b$, whereas the graph of $|x - b| > c$ has a ray extending left from $-c + b$ and another ray extending right from $c + b$.

21. The equation has no solution. Absolute value cannot be negative.

PRACTICE

1. $-5 < x + 1$ and $x + 1 < 5$ **2.** $-3 < x + 2$ and $x + 2 < 3$

3. $x - 4 < -6$ or $x - 4 > 6$ **4.** $x - 3 < -5$ or $x - 3 > 5$

5. $-6 \leq 2x - 2$ or $2x - 2 \leq 6$ **6.** $-15 \leq 3x - 3$ or $3x - 3 \leq 15$

7. $2x + 5 \leq -3$ or $2x + 5 \geq 3$ **8.** $3x + 5 \leq -13$ or $3x + 5 \geq 13$

9. $-3 < z + 1$ and $z + 1 < 3$
  $-4 < z$ and $z < 2$

10. no solution since absolute value is always non-negative

11. $x - 5 < -2$ or $x - 5 > 2$
  $x < 3$ or $x > 7$

12. $x - 4 < -2$ or $x - 4 > 2$
  $x < 2$ or $x > 6$

13. $-3.5 \le x - 0.5$ or $x - 0.5 \le 3.5$

$\quad -3 \le x \quad$ or $\quad x \le 4$

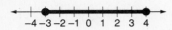

14. $-4.5 \le t - 1.5$ and $t - 1.5 \le 4.5$

$\quad -3 \le t \quad$ and $\quad t \le 6$

15. $w + 2.5 \le -3.5$ or $w + 2.5 \ge 3.5$

$\quad w \le -6 \quad$ or $\quad w \ge 1$

16. The solution is all real numbers, since the absolute value is always non-negative.

17. b; inequality is $\le$, midpoint is 2

18. d; inequality is $>$, midpoint is 3

19. c; inequality is $\ge$, midpoint is 3

20. a; inequality is $<$, midpoint is zero

21. $2q - 4 \le -6$ or $2q - 4 \ge 6$

$\quad 2q \le -2 \quad$ or $\quad 2q \ge 10$

$\quad q \le -1 \quad$ or $\quad q \ge 5$

22. $3x - 9 < -6$ or $3x - 9 > 6$

$\quad 3x < 3 \quad$ or $\quad 3x > 15$

$\quad x < 1 \quad$ or $\quad x > 5$

23. $-10 < 3t + 5$ and $3t + 5 < 10$

$\quad -15 < 3t \quad$ and $\quad 3t > 5$

$\quad -5 < t \quad$ and $\quad t > \dfrac{5}{3}$

24. $-10 \le 4x - 2$ and $4x - 2 \le 10$

$\quad -8 \le 4x \quad$ and $\quad 4x \le 12$

$\quad -2 \le x \quad$ and $\quad x \le 3$

25. $3z - 6 < -12$ or $3z - 6 > 12$

$\quad 3z \le -6 \quad$ or $\quad 3z > 18$

$\quad z < -2 \quad$ or $\quad z > 6$

26. The solution is all real numbers, since the absolute value is always non-negative.

27. $-9 \le 2q + 5$ and $2q + 5 \le 9$

$\quad -14 \le 2q \quad$ and $\quad 2q \le 4$

$\quad -7 \le q \quad$ and $\quad q \le 2$

28. $-5 < 5z - 10$ and $5z - 10 < 5$

$\quad 5 < 5z \quad$ and $\quad 5z < 15$

$\quad 1 < z \quad$ and $\quad z < 3$

29. Since absolute value is always non-negative, the solution is all real numbers except 0.

30. Since absolute value is always non-negative, the solution is all real numbers.

31. Since absolute value is always non-negative, there is no solution.

32. Since absolute value is always non-negative, there is no solution.

33. Since absolute value is always non-negative, the solution is all real numbers except where $\frac{1}{2}x - 1 = 0$ or $x = 2$.

34. Since absolute value is always non-negative, the solution is all real numbers except where $\frac{1}{4}x - 2 = 0$ or $x = 8$.

35. Since absolute value is always non-negative, the solution is all real numbers.

36. $4 - \frac{1}{2}x \le -2$ or $4 - \frac{1}{2}x \ge 2$

$\quad -\frac{1}{2}x \le -6 \quad$ or $\quad -\frac{1}{2}x \ge -2$

$\quad x \ge 12 \quad$ or $\quad x \le 4$

37. $-12 < 6 - 2x$ and $6 - 2x < 12$

$\quad -18 < -2x \quad$ and $\quad -2x < 6$

$\quad 9 > x \quad$ and $\quad x > -3$

38. $-7 \le 5 - 2x$ and $5 - 2x \le 7$

$\quad -12 \le -2x \quad$ and $\quad -2x \le 2$

$\quad 6 \ge x \quad$ and $\quad x \ge -1$

39. Since absolute value is always non-negative, the only solution is when $2x + 4 = 0$ or $x = -2$.

40. Since absolute value is always non-negative, the only solution is where $3x + 9 = 0$ or $x = -3$.

41. $2|1+2z|+5 \le 9$

$2|1+2z| \le 4$ Subtract 5.

$|1+2z| \le 2$ Divide by 2.

$-2 \le 1+2z \le 2$ Write as a conjunction.

$-3 \le 2z \le 1$ Solve for z.

$-\dfrac{3}{2} \le z \le \dfrac{1}{2}$

42. $3|1+3x|-6 < 12$

$3|1+3x| < 18$ Add 6.

$|1+3x| < 6$ Divide by 3.

$-6 < 1+3x < 6$ Write as a conjunction.

$-7 < 3x < 5$ Solve for x.

$-\dfrac{7}{3} < x < \dfrac{5}{3}$

43. $4-|5-t| \ge 1$

$-|5-t| \ge -3$ Subtract 4.

$|5-t| \le 3$ Multiply by -1.

$-3 \le 5-t \le 3$ Write as a conjunction.

$-8 \le -t \le -2$ Solve for t.

$8 \ge t \ge 2$

44. $3-|4-t| > 1$

$-|4-t| > -2$ Subtract 3.

$|4-t| < 2$ Multiply by -1.

$-2 < 4-t < 2$ Write as a conjunction.

$-6 < -t < -2$ Solve for t.

$6 > t > 2$

45. $-2 \le \dfrac{8-4x}{3} < 2$ Write as a conjunction.

$-6 < 8-4x < 6$ Multiply by 3.

$-14 < -4x < -2$ Subtract 8.

$\dfrac{7}{2} > x > \dfrac{1}{2}$ Solve for x.

46. $\dfrac{5-3x}{2} < -4$ or $\dfrac{5-3x}{2} > 4$ Write as a disjunction.

$5-3x < -8$ or $5-3x > 8$ Multiply by 2.

$-3x < -13$ or $-3x > 3$ Subtract 5.

$x > \dfrac{13}{3}$ or $x < -1$ Divide by -3.

47. Midpoint is 21 and values are between 11 and 31, so $|x-21| \le 10$

48. Midpoint is 30 and values are between 25 and 35, so $|x-30| \le 5$

49. Possible answer: need 8 ± 2 liters of fluid per day; $|x-8| \le 2$

50. $-x-2 \le x+2$ and $x+2 \le x+2$

$-4 \le 2x$ always true

$-2 \le x$

So, the solution is $x \ge -2$.

51. $-5x < x-2$ and $x-2 < 5x$

$-6x \le -2$ and $-2 < 4x$

$x > \dfrac{1}{3}$ and $-\dfrac{1}{2} < x$

So, the solution is $x > \dfrac{1}{3}$.

52. $3-x < -3+x$ or $3-x > 3-x$

$-2x < -6$ This is never true.

$x > 3$

So, the solution is $x > 3$.

53. $5-x \le -3x$ or $5-x \ge 3x$

$2x \le -5$ or $5 \ge 4x$

$x \le -\dfrac{5}{2}$ or $\dfrac{5}{4} \ge x$

So, the solution is $x \le \dfrac{5}{4}$.

54. Estimates will vary. Correct answer is: $-4 \le x \le 10.2$

55. Estimates will vary. Correct answer is: $-7 \le x \le 10.8$

56. Estimates will vary. Correct answer is: $x < 3.\overline{4}$ or $x > 12.\overline{1}$

57. Estimates will vary. Correct answer is: $x \le 1.\overline{72}$ or $x \ge 7.\overline{36}$

58. Midpoint of $70-116$ is 93 so: $|x-93| \le 23$

59. Midpoint of $0.7-1.4$ is 1.05 so: $|x-1.05| \le 0.35$

60. Midpoint of $0-45$ is 22.5 so: $|x-22.5| \le 22.5$

THINK CRITICALLY

61. $|x| > 0$

62. Answers will vary; possible answer: $|x| = x-3$

63. Since $<$ includes all the numbers that $\ge$ does not, any real number that does not satisfy $|x| < c$ must satisfy $|x| \ge c$.

64a. if a and b are both positive, if a and b are both negative, or if $a = 0$ or $b = 0$

b. for all values of a and b

c. if a and b are both positive and $a > b$, if a and b are both negative and $b > a$, if $b = 0$, or if a and b are both 0

d. for all values of a and b except $b = 0$

65. C; $2(100 - 52)2 - 3(8 + 2)3 = 2(75)2 - 3(10)3 =$
$2(5625) - 3(1000) = 11250 - 3000 = 8,250$

66. $3x - 7 = 4y + 3$
$3x - 10 = 4y$
$\frac{3}{4}x - \frac{5}{2} = y$
$y = \frac{3}{4}x - \frac{5}{2}$

67. $5y - 5 = 2x + 11$
$5y = 2x + 16$
$y = \frac{2}{5}x + \frac{16}{5}$

68. $\begin{cases} x + 3y = 15 \\ \quad 4x = 24 \end{cases} \rightarrow x = 6.$
$6 + 3y = 15$
$3y = 9$
$y = 3$

So, the solution is $(6,3)$.

69. $\begin{cases} x + 7y = 30 \\ \quad -2y = -8 \end{cases} \rightarrow y = 4.$
$x + 7(4) = 30$
$x = 2$

So, the solution is $(2,4)$.

70. $2x + 6 < -15 \quad \text{or} \quad 2x + 6 > 15$
$2x < -21 \quad \text{or} \quad 2x > 9$
$x < -\frac{21}{2} \qquad \qquad x > \frac{9}{2}$

71. $-21 < 3x - 7 < 21$
$-14 < 3x < 28$
$-\frac{14}{3} < x < \frac{28}{3}$

ALGEBRAWORKS

1. Turbo with 205

2. Regular sedan with 60

3. Turbo with aspect ratio of 50

4. Diameter is 1" more so radius is $\frac{1}{2}"$ more

5. Midpoint of $1100 - 1400$ is 1250, so: $|x - 1250| \le 150$

6. For 4 tires multiply values by 4, so: $|x - 5000| \le 600$

7. $60,000 \text{ mi} = (60,000 \text{ mi})\left(\frac{5280 \text{ ft}}{\text{mi}}\right)\left(\frac{12 \text{ in.}}{\text{ft}}\right) =$

$3,801,600,000$ in. The radius of the wheel is
$14 \div 2 = 7$ in. The radius of the wheel plus the
height of the tire sidewall is $7 + 5.7 = 12.7$ in.
In one rotation, a point on the tire travels
$(2)(\pi)(12.7 \text{ in.}) = 25.4\pi$ in.
So, $3,801,600,000 \div 25.4\pi \approx 47,641,215$ rotations

Lesson 9.5, pages 437-444

EXPLORE

1. Student activity, so no solution

2. $0, \frac{1}{3}, \frac{2}{3}, \frac{3}{3}$ or 1, $\frac{4}{3}, \frac{5}{3}, \frac{6}{3}$ or 2

3. $0, \frac{1}{4}, \frac{2}{4}$ or $\frac{1}{2}, \frac{3}{4}, \frac{4}{4}$ or 1, $\frac{5}{4}, \frac{6}{4}$ or $\frac{3}{2}, \frac{7}{4}, \frac{8}{4}$ or 2

4. Numbers of the form $0, \frac{1}{q}, \frac{2}{q}, \frac{3}{q}, \frac{4}{q}, \cdots, \frac{pq}{q}$

5. Every point cannot be located this way because not
every number is rational, or in the form $\frac{p}{q}$ where p
and q are integers.

TRY THESE,

1a–e.

2. rationals and reals

3. irrationals and reals

4. integers, rationals, and reals

5. wholes, integers, rationals, and reals

6. rationals and reals

7. $-\frac{3}{1}$

8. $4 + \frac{11}{100} = \frac{411}{100}$

9. $5 + \frac{3}{7} = \frac{38}{7}$

10. $6 + \frac{6}{100} = \frac{606}{100}$

11. $\frac{1}{100}$

12. $n = 0.77777\ldots$
$\underline{10n = 7.77777\ldots}$
$9n = 7.00000\ldots$ Subtract 1st equation from 2nd.
$n = \frac{7}{9}$

13. $n = 0.353535\ldots$
$\underline{100n = 35.353535\ldots}$
$99n = 35.000000\ldots$ Subtract 1st equation from 2nd.
$n = \frac{35}{99}$

14. $n = 0.22222\ldots$
$\underline{10n = 2.22222\ldots}$
$9n = 2.00000\ldots$ Subtract 1st equation from 2nd.
$n = \frac{2}{9}$

15. $n = 0.123123123\ldots$
$\underline{1000n = 123.123123123\ldots}$
$999n = 123.000000000\ldots$ Subtract equations.
$n = \frac{123}{999} = \frac{41}{333}$

16.

$$n = 0.753753753\ldots$$
$$1000n = 753.753753753\ldots$$
$$\overline{999n = 753.000000000\ldots} \quad \text{Subtract equations.}$$
$$n = \frac{753}{999} = \frac{251}{333}$$

17. Divide 9 by 20 to get 0.45

18. Divide 451 by 8 to get 56.375

19. Divide −7 by 5 to get −1.4

20. Divide −336 by 21 to get −16

21. Divide 27 by 33 to get $0.\overline{81}$

22. $\sqrt{\dfrac{225}{256}} = \sqrt{\dfrac{15 \cdot 15}{16 \cdot 16}} = \dfrac{15}{16}$

23. not a real number; No real number multiplied by itself is equal to −36.

24. $-\sqrt{0.81} = -\sqrt{(0.9)(0.9)} = -0.9$

25. $\sqrt{2704} = -\sqrt{52 \cdot 52} = 52$

26. $\sqrt{3.61} = -\sqrt{(1.9)(1.9)} = 1.9$

27. $\sqrt{82} \approx 9.06$ **28.** $\sqrt{179} \approx 13.38$

29. $-\sqrt{47} \approx -6.86$ **30.** $-\sqrt{111} \approx -10.54$

31. $\sqrt{0.88} \approx 0.94$ **32.** $\sqrt{0.69} \approx 0.83$

33. $0.263 = \dfrac{263}{1000}$; $\dfrac{263}{1000} \cdot 617 = \dfrac{162{,}271}{1000} \approx 162$

34. If $0 < x < 1$, then $\sqrt{x} > x$.

PRACTICE

1a–e.

2. terminating decimal; so, rationals and reals

3. number = 0; so, wholes, integers, rationals, and reals

4. positive integer; so naturals, wholes, integers, rationals, and reals

5. 1.69 is perfect square decimal; so, rationals and reals

6. fraction; so, rationals and reals

7. repeating decimal; so, rationals and reals

8. 25 is perfect square; so, naturals, wholes, integers, rationals, and reals

9. 26 is not perfect square, so irrationals and reals

10. 6.0 = 6; so, naturals, wholes, integers, rationals, and reals

11. fraction; so, rationals and reals

12. $-4 = \dfrac{-4}{1}$ **13.** $5 + \dfrac{22}{100} = \dfrac{522}{100}$

14. $10 + \dfrac{1}{6} = \dfrac{61}{6}$ **15.** $0.078 = \dfrac{78}{1000} = \dfrac{39}{500}$

16.

$$n = 0.5555\ldots$$
$$10n = 5.5555\ldots$$
$$\overline{9n = 5.0000\ldots} \quad \text{Subtract 1st equation from 2nd.}$$
$$n = \frac{5}{9}$$

17.

$$n = 0.12121212\ldots$$
$$100n = 12.12121212\ldots$$
$$\overline{99n = 12.00000000\ldots} \quad \text{Subtract equations.}$$
$$n = \frac{12}{99} = \frac{4}{33}$$

18.

$$n = -0.969696\ldots$$
$$100n = -96.969696\ldots$$
$$\overline{99n = -96.000000\ldots} \quad \text{Subtract equations.}$$
$$n = -\frac{96}{99} = -\frac{32}{33}$$

19.

$$n = 0.167167167\ldots$$
$$1000n = 167.167167167\ldots$$
$$\overline{999n = 167.000000000\ldots} \quad \text{Subtract equations.}$$
$$n = \frac{167}{999}$$

20.

$$n = 0.779779779\ldots$$
$$1000n = 779.779779779\ldots$$
$$\overline{999n = 779.000000000\ldots} \quad \text{Subtract equations.}$$
$$n = \frac{779}{999}$$

21.

$$n = 0.157157157\ldots$$
$$1000n = 157.157157157\ldots$$
$$\overline{999n = 157.157157157\ldots} \quad \text{Subtract equations.}$$
$$n = \frac{157}{999}$$

22. Divide 21 by 50 to get 0.42

23. Divide 756 by 400 to get 1.89

24. Divide −17 by 8 to get −2.125

25. Divide 1664 by 32 to get 52

26. Divide 568 by 1111 to get $0.\overline{5112}$

27. $\sqrt{\dfrac{900}{529}} = \sqrt{\dfrac{30 \cdot 30}{23 \cdot 23}} = \dfrac{30}{23}$

28. Not a real number, because no real number multiplied by itself equals −144.

29. $-\sqrt{0.64} = -\sqrt{(0.8)(0.8)} = -0.8$

30. $\sqrt{1764} = \sqrt{42 \cdot 42} = 42$

31. $\sqrt{2.89} = \sqrt{(1.7)(1.7)} = 1.7$

32. $\sqrt{77} \approx 8.77$ **33.** $\sqrt{151} \approx 12.29$

34. $-\sqrt{59} \approx -7.68$ **35.** $-\sqrt{166} \approx -12.88$

36. $\sqrt{0.67} \approx 0.82$ **37.** $\sqrt{0.56} \approx 0.75$

38. $\sqrt{0.05} \approx 0.22$ **39.** $\sqrt{0.09} = 0.3$

40. $-\sqrt{1122} \approx 33.50$ **41.** $-\sqrt{2233} \approx -47.25$

42. $100 = \frac{1}{2}(9.8)t^2$

$100 = 4.9t^2$

$\frac{100}{4.9} = t^2$

$\sqrt{\frac{100}{4.9}} = t \approx 4.5 \text{ s}$

43. $100 \text{ cm} = 1 \text{ m}$ So, $1 = \frac{1}{2}(9.8)t^2$

$\sqrt{\frac{1}{4.9}} = t^2 \approx 0.45 \text{ s}$

44. The square root of a real number is rational if the real number is a perfect square.

EXTEND

45. $\sqrt[3]{27} = 3.00$ **46.** $\sqrt[3]{-27} = -3.00$

47. $-\sqrt[3]{0.27} = -0.30$ **48.** $\sqrt[3]{-990} \approx -9.97$

49. $\sqrt[3]{75} \approx 4.22$

50. $A = l \cdot w = 2w \cdot w = 2w^2$

$16 = 2w^2$

$\sqrt{8} = w \approx 2.83$ So, $l = 2w$ or about 5.66

51. $A = \pi r^2 = 122$

$r = \sqrt{\frac{122}{\pi}} \approx 6.23$

52. $V = \frac{4}{3}\pi r^3 = 905$

$r^3 = \frac{3 \cdot 905}{4\pi}$

$r = \sqrt[3]{\frac{3 \cdot 905}{4\pi}} \approx 6.00$

53. $9^2 = 81$, $10^2 = 100$, $11^2 = 121$

THINK CRITICALLY

54. $\sqrt{21}$, since the greatest of the radicands is 21.

55. $\sqrt{\frac{1}{7}}$, since the greatest of the radicands is $\frac{1}{7}$.

56. $\sqrt{64}$, since $64 > 27$ and $\sqrt{x} > \sqrt[3]{x}$ if $x > 1$.

57. $\sqrt[3]{0.064}$, since $0.064 > 0.027$ and $\sqrt[3]{x} > \sqrt[2]{x}$ if $0 < x < 1$.

58. false; $\sqrt{9+16} \neq \sqrt{9} + \sqrt{16}$

59. false; $\sqrt{25-16} \neq \sqrt{25} + \sqrt{16}$

60. The Square of every real number ≥ 0, therefore negative numbers have no square roots in the real number system.

61. A rational number k is a perfect square if there is a number x so that $x^2 = k$

62. When $3x - 6 \geq 0$

$3x \geq 6$

$x \geq 2$

63. When $4x + 12 \geq 0$

$4x \geq -12$

$x \geq -3$

ALGEBRAWORKS

1. Using pattern from table, reaction distance $= x$

2. $10 = \sqrt{20(5)}$; $20 = \sqrt{20(20)}$

$30 = \sqrt{20(45)}$; $40 = \sqrt{20(80)}$

$50 = \sqrt{20(125)}$; $60 = \sqrt{20(180)}$

3. stopping distance = braking distance + thinking distance, or $\frac{s^2}{20} + 5$

4. Since $s = \sqrt{20b}$, from solutions for 2 and 3,

$s^2 = 20b$ and $\frac{s^2}{20} = b$.

If s doubles to $2s$, then $\frac{(2s)^2}{20} = \frac{4s^2}{20} = \frac{s^2}{5}$. This

has increased from $\frac{s^2}{20}$ by a factor of 4.

5. Breaking distance is $\frac{45^2}{20} = \frac{2025}{20} = 101.25 \text{ ft}$

Stopping distance is $\frac{45^2}{20} + 45 = 146.25 \text{ ft}$

6. $65 = \sqrt{30 \cdot 0.5 \cdot d}$

$65^2 = 15d$

$\frac{65^2}{15} = d \approx 281.67 \text{ ft}$

7. Since distance is related to speed by a square, multiply by $\left(\frac{1}{2}\right)^2$ or $\frac{1}{4}$.

8. Answers will vary; for each type of road surface condition, the length of the skid mark shows that the speed was greater than 35 mi/h.

Lesson 9.6, pages 445-451

EXPLORE,

1.
$c \quad a \quad b$

2. Point b will be to the right of point c.

3. c, a, b

4. $b > a, a > c, c < a < b$

5.
$d \quad f \quad e$

6. Point d will be to the left of point e.

7. d, f, e

8. $d < f, f < e, d < f < e$

1.

$-3, -1, 0, 2, \sqrt{5}, 4$

2.

$-3.5, -2, -\sqrt{2}, 1, 2.5, 3$

3.

$\dfrac{3}{4}, 1.2, \dfrac{5}{4}, \dfrac{3}{2}, 1.8$

4.

$-2, -\dfrac{7}{4}, -\sqrt{3}, -1.5, -1.25, -\dfrac{1}{2}$

5. Answers will vary; examples include $1.11, 1.111, 1.111$

6. Answers will vary; examples include $0.173, 0.175, 0.177$

7. symmetric property of equality

8. transitive property of equality

9. reflexive property of equality

10. substitution property of equality

11. $a = b$ Given

$ac = ac$ reflexive property of equality

$ac = bc$ substitution property of equality

12. Drive alone was the only means that experienced an increase of 8.8%.

13. Carpool experienced $19.7 - 13.4 = 6.3$ percent decrease.

14. Answers will vary but may include graphing the numbers on a number line and ordering them from left to right.

PRACTICE

1.

$-4, -\sqrt{6}, -2, 0, 3, 5$

2.
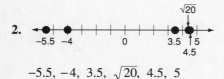

$-5.5, -4, 3.5, \sqrt{20}, 4.5, 5$

3.

$\dfrac{6}{5}, 1.3, \dfrac{8}{5}, 1.8, \dfrac{19}{10}$

4.

$-1.75, -1.5, -\dfrac{5}{4}, -1, -\dfrac{3}{4}$

5. $\dfrac{4}{7}, \dfrac{5}{8}, \dfrac{6}{9}$ **6.** $\dfrac{13}{16}, \dfrac{14}{17}, \dfrac{16}{19}$

7. $1\dfrac{6}{7}, 2.1, \dfrac{15}{7}$ **8.** $2.5, 2\dfrac{5}{9}, 2\dfrac{2}{3}$

9. $-0.1511, -0.151, -0.1501$ **10.** $-3.434, -3.43, -3.343$

11. Answers will vary; examples include $3.105, 3.107, 3.109$

12. Answers will vary; examples include $5.71, 5.725, 5.74$

13. Answers will vary; examples include $4.1181, 4.11811,$ 4.118118

14. Answers will vary; examples include $3.3453, 3.34534,$ 3.345345

15. reflexive property of equality

16. substitution property of equality

17. symmetric property of equality

18. transitive property of equality

19. $x = y$ Given

$x + z = x + z$ reflexive property of equality

$x + z = y + z$ substitution property of equality

20. $x + y = z$ Given

$x + y - y = z - y$ subtraction property of equality

$x + 0 = z - y$ additive inverse property

$x = z - y$ zero property of addition

21. Flora, with a mean distance of 204.4 million miles.

22. Pallas, with a mean distance of 257.4 million miles.

23.

Flora	Vesta	Iris	Metis	Hebe
3.27	3.63	3.68	3.69	3.78

Astraea	Juno	Ceres	Pallas	Hygeia
4.14	4.36	4.60	4.61	5.59

24. Answers will vary.

EXTEND

25. Answers will vary. Example: if $5 < 10$ and $10 < 20$, then $5 < 20$.

26. Answers will vary. Example: if $20 > 10$ and $10 > 5$, then $20 > 5$.

27. No, a number is not greater than itself; $5 \ngtr 5$

28. No, if $5 < 10$, then $10 \not< 5$.

29. No, if $5 < 10$ and $5 + 12 < 20$, then $10 + 12 < 20$ is false.

30. true

31. false

32. true

33. true

THINK CRITICALLY

34. No, the trichotomy property states that only one statement can be true at a time.

35. For all real numbers a and b, one and only one of the following is true: either $a = b$ or $a \neq b$

36. No; $5 \neq 5$ is false.

37. Yes; if $5 \neq 7$, then $7 \neq 5$.

38. No; if $5 \neq 7$ and $7 \neq 5$, then $5 \neq 5$ is false.

39. No; if $5 \neq 7$ and $5 + 6 \neq 13$, then $7 + 6 \neq 13$ is a false statement.

40. No; if $x = -5$, then $-(-5)$ is not less than -5.

MIXED REVIEW

41. $\dfrac{89 + 92 + 88 + 77 + 92 + 83 + 69 + 98 + 72 + 95}{10} =$

$\dfrac{855}{10} = 85.5$

42. 92 occurs twice, so the mode is 92.

43. 69, 72, 77, 83, 88, 89, 92, 92, 95, 98

$\dfrac{88 + 89}{2} = 88.5$

44. $98 - 69 = 29$

45. $f(8) = \dfrac{8^2 - 8}{4} = \dfrac{64 - 8}{4} = \dfrac{56}{4} = 14$; B

46. $\dfrac{17}{21}, \dfrac{13}{16}, \dfrac{17}{19}$

47. $-0.1611, -0.161, 0.1601$

ALGEBRAWORKS

1. Phone plan A, Phone plan B, TelCo Plan B, ABC, Inc. Plan B, TelCo Plan A, ABC, Inc. Plan A

2. Phone plan A, Phone plan B and ABC, Inc. Plan B (tied), Telco Plan A, TelCo Plan B, ABC, Inc. Plan A

3. Answers will vary.

4. Answers will vary; possible answer: discounts on total amount of bill; discounts on calls made to certain locations frequently called by the customer

5. Answers will vary.

Lesson 9.7, pages 452-455

EXPLORE THE PROBLEM

1. (70, 164) and (90, 604) for 1970; graph is linear

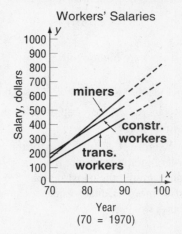

Workers' Salaries

2. Compute slope: $m = \dfrac{604 - 164}{90 - 70} = \dfrac{440}{20} = \$22$ per year

3. First year income exceeds \$300 is 1997. This agrees with the graph.

Year	Average Weekly Income	Year	Average Weekly Income
1970	164	1976	296
1971	186	1977	318
1972	208	1978	340
1973	230	1979	362
1974	252	1980	384
1975	274		

4. $164 = 22(70) + b$

   $164 = 1540 + b$

   $-1376 = b$

   So, $y = 22x - 1376$.

5. Continue the table to the year 2000, and extend the graph to show income for the year 2000. Graph: Substitute 100 for 2000 in the equation and solve. The weekly income of miners in the year 2000 will be \$824.

Year	Wkly Avg Income	Year	Wkly Avg Income	Year	Wkly Avg Income
1981	406	1988	560	1995	714
1982	428	1989	582	1996	736
1983	450	1990	604	1997	758
1984	472	1991	626	1998	780
1985	494	1992	648	1999	802
1986	516	1993	670	2000	824
1987	538	1994	692		

6. Answers will vary. Students may feel that the equation gives a single answer most directly; a graph presents information and trends visually; a table might be helpful when information needs to be available for ready reference.

INVESTIGATE FURTHER

7a. Construction workers were higher with $195 weekly income.

b. Miners were higher in 1990 with $604 weekly income.

c. The rate for miners increased more rapidly than the rate for construction workers.

8. See the graph for Exercise 1. The y-value for the construction workers' graph is greater than the y-value for the miner's graph; the construction workers' graph rises more gradually because the rate of change per year is smaller.

9a. Rate is slope: $m = \dfrac{526 - 195}{90 - 20} = \dfrac{331}{20} = \$16.55$ per year

Year	Average Weekly Income	Year	Average Weekly Income
1970	195	1976	294.30
1971	211.55	1977	310.85
1972	228.10	1978	327.40
1973	244.65	1979	343.95
1974	261.20	1980	360.50
1975	277.75		

b. Find the year in both tables where the values are closest to being equal (1975). The two graphs should intersect close to this point.

10a.
$$195 = 16.55(70) + b$$
$$195 = 1158.50 + b$$
$$-963.50 = b$$
So, $y = 16.55x - 963.5$

b. Answers will vary: possible range values
x min = 65 y min = 0
x max = 105 y max = 1000
x scl = 10 y scl = 100

11. Use substitution to obtain $22x - 1376 = 16.55x - 963.50$; Calculations could be tedious without a calculator.

12. $y = 16.55(100) - 962.50 = 1655 - 963.50 = \$691.50$

APPLY THE STRATEGY

13a. Compute slope : $m = \dfrac{442 - 132}{90 - 70} = \dfrac{309}{20} = 15.45$

So, it would rise more slowly because the average increase is less.

b. Slope is 15.45; $442 = 15.45(90) + b$ so $b = -948.50$
So, $y = 15.45x - 948.50$

c. $y = 15.45(100) - 948.50 = \$596.50$

14a.

R	B	B	R
B	G	G	B
B	G	G	B
R	G	G	R

4 Red
8 Blue
4 Green

b.

Square	Red	Blue	Green
2 x 2	4	0	0
3 x 3	4	4	1
4 x 4	4	8	4
5 x 5	4	12	9
6 x 6	4	16	16
7 x 7	4	20	25
8 x 8	4	24	36
9 x 9	4	28	49
10 x 10	4	32	64

c.

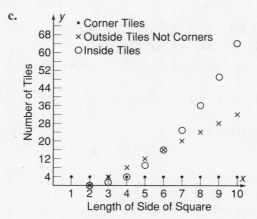

Each point represents a discrete whole number. A connecting line would have no meaning.

d. Green increases most quickly. Red is always constant 4, blue is a linear increase of 4 for each larger square, green is a quadratic increase for each square going up by perfect squares.

e. $R = 4$ or $4(n^0)$
$B = 4(n - 2)$
$G = (n - 2)^2$

f. 4 red, 392 blue, and 9604 green; The formulas in 14e were most useful in this problem.

g. 10,000 tiles; check by $100 \cdot 100$ which is area or $4 + 392 + 9604$ which is the sum of the tiles from 14f.

15. Answers will vary.

REVIEW PROBLEM SOLVING STRATEGIES,

1a. No; the pearl at the $100 end must be worth more than the one at the $150 end in order for the values to be equal by the center.

b. Answers will vary. Use two variables, x and y, to represent the value of each end pearl.

c. answers will vary; $x + 16(150) = y + 16(100)$

d. $17x + 136(150) + 16y + 120(100) = 65,000$

Students must be careful to include the center pearl only once in the equation. The value of each pearl from one end, including the center pearl, is x, $x + 150$, $x + 2(150) \ldots, x + 16(150)$, and so on. The sum of these terms is $17x + (136)150$. From the other end, the value of the pearls is the sum of y, $y + 100$, $y + 2(100), \ldots, y + 15(100)$.

e. $3,000; the pearl at the \$100 end is worth \$1400; the pearl at the \$150 end is worth \$600

f. $43,550

2. 228 numbers; if the list had started with $3(3, 6, 9, \ldots, 684)$, the value of any term would be $3n$; solve the equation $3n = 684$, since the list starts with 2 and lists every third number, the value would be found by $3n - 1 = 683$

3.
```
X  X        X  X    Answers will vary; possible
X  X  X  X  X  X    arrangement is shown.
X        X  X  X
X     X     X  X
```

Chapter Review, page 456

1. e   **2.** c   **3.** a   **4.** d   **5.** b

6.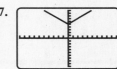

7.

8.

9.

10. Vertex at $(-4, -1)$;
Axis of symmetry: $x = -4$

11. Vertex at $(4, 1)$;
Axis of symmetry: $x = 4$

12. Vertex at $(2, 3)$;
Axis of symmetry: $x = 2$

13. Vertex at $(2, 3)$;
Axis of symmetry: $x = 2$

14. $6x = 36$ or $6x = -36$ Write a disjunction.

$x = 6$ or $x = -6$ Solve for x.

$|6(6)| \overset{?}{=} 36$ $|6(-6)| \overset{?}{=} 36$ Check.

$36 = 36$ ✔ $36 = 36$ ✔

15. $x - 5 = 8$ or $x - 5 = -8$ Write a disjunction.

$x = 13$ or $x = -3$ Solve for x.

$|13 - 5| \overset{?}{=} 8$ $|-3 - 5| \overset{?}{=} 8$ Check.

$8 = 8$ ✔ $8 = 8$ ✔

16. $3 + x = 9$ or $3 + x = -9$ Write a disjunction.

$x = 6$ or $x = -12$ Solve for x.

$|3 + 6| \overset{?}{=} 9$ $|3 + (-12)| \overset{?}{=} 9$ Check.

$9 = 9$ ✔ $9 = 9$ ✔

17. $-3|4g + 2| - 3 = 7$

$-3|4g + 2| = 10$

$|4g + 2| = -\dfrac{10}{3}$

There is no solution since the absolute value is always non-negative.

18. $-7 < x - 2$ and $x - 2 < 7$ Write a conjunction.

$-5 < x$ and $x < 9$ Solve for x.

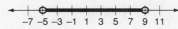

19. $z + 1 < -4$ or $z + 1 > 4$ Write a disjunction.

$z < -5$ or $z > 3$ Solve for x.

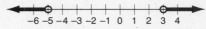

20. $-12 \le 3w + 6$ and $3w + 6 \le 12$ Write a conjunction.

$-18 \le 3w$ and $3w \le 6$ Solve for x.

$-6 \le w$ and $w \le 2$

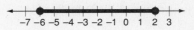

21. Since the absolute value is always non-negative, the answer is all real numbers.

22. The midpoint from 94 to 100 is 97, so $|x - 97| \le 3$.

23. Divide 2 by 11 to get $0.\overline{18}$.

24. Divide 5 by 9 to get $0.\overline{5}$.

25. Divide 11 by 20 to get 0.55.

26. Divide 323 by 19 to get 17.

27. Divide 5 by 18 to get $0.2\overline{7}$.

28. $-\sqrt{0.49} = -\sqrt{(0.7)(0.7)} = -0.7$

29. It is not a real number, since no real number times itself can be -36.

30. $\sqrt{\dfrac{625}{961}} = \sqrt{\dfrac{25 \cdot 25}{31 \cdot 31}} = \dfrac{25}{31}$ **31.** $\sqrt{4096} = \sqrt{64 \cdot 64} = 64$

32.

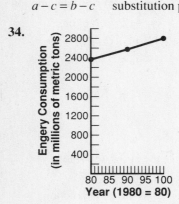

$-\sqrt{5}, \ -2, \ -\sqrt{2}, \ -\dfrac{2}{3}, \ -\sqrt{\dfrac{1}{9}}$

33. $a = b$ Given

$a - c = a - c$ reflexive property of equality

$a - c = b - c$ substitution property of equality

34.

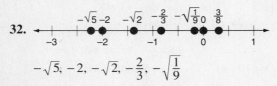

Estimate 2800 million metric tons from graph.

Slope of line: $m = \dfrac{2572.3 - 2364.4}{1990 - 1980} =$

$\dfrac{207.9}{10} = 20.79$ million metric tons per year

If trend increases to year 2000 (10 years past 1990) then $2572.3 + 207.9 = 2780.2$ should be the energy consumption.

Chapter Assessment, page 458

1. The rays that form the graph of $y = 3|x-1| + 4$ are farther from the x-axis than those of $y = |x|$. The graph is shifted 1 unit to the right and 4 units upward.

2. $b = 0$ and $c = 0$ so vertex is at $(0, 0)$ and axis of symmetry is $x = 0$

3. $b = -3$ and $c = -7$ so vertex is at $(-3, -7)$ and axis of symmetry is $x = -3$

4. $b = -1$ and $c = 5$ so vertex is at $(-1, 5)$ and axis of symmetry is $x = -1$

5. $b = 2$ and $c = 3$ so vertex is at $(2, 3)$ and axis of symmetry is $x = 2$

6. C; three units left $\rightarrow b = -3$, 5 units down $\rightarrow c = -5$

7. $x + 3 = 2$ or $x + 3 = -2$ Write a disjunction.

$x = -1$ or $x = -5$ Solve for x.

8. $x - 8 = 3$ or $x - 8 = -3$ Write a disjunction.

$x = 11$ or $x = 5$ Solve for x.

9. $3|4x - 1| - 7 = 5$ Add 7.

$3|4x - 1| = 12$ Divide by 3.

$4x - 1 = 4$ or $4x - 1 = -4$ Write a disjunction.

$4x = 5$ or $4x = -3$ Solve for x.

$x = \dfrac{5}{4}$ or $x = -\dfrac{3}{4}$

10. $4|3x + 2| + 6 = 2$ Subtract 6.

$4|3x + 2| = -4$ Divide by 4.

$|3x + 2| = -1$

No solution; absolute value is always non-negative.

11. $t + 7 < -2$ or $t + 7 > 2$ Write a disjunction.

$t < -9$ or $t > -5$ Solve for t.

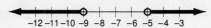

12. $-1 < w - 5$ and $w - 5 < 1$ Write a conjunction.

$4 < w$ and $w < 6$ Solve for w.

13. $2x + 5 \le -1$ or $2x + 5 \ge 1$ Write a disjunction.

$2x \le -6$ or $2x \ge -4$ Solve for x.

$x \le -3$ or $x \ge -2$

14. Solution is all real numbers since absolute value is always non-negative

15. D

16. Slope is $m = \dfrac{9.7 - 17.2}{1792 - 1975} = \dfrac{-7.5}{17}$

So, $y - 9.7 = -\dfrac{7.5}{17}(x - 1992)$ or $y = -\dfrac{7.5}{17}x + 888.5$.

In the year 2000: $y = -\dfrac{7.5}{17}(2000) + 888.5 \approx 6.15$

17. A   **18.** C

19. $\sqrt{1.44} = \sqrt{(1.2)(1.2)} = 1.2$

20. $-\sqrt{121} = -\sqrt{(11)(11)} = -11$

21. $\sqrt{484} = \sqrt{(22)(22)} = 22$

22. $\sqrt{-1681}$ is not a real number since no real number times itself can be negative

23. $-4\dfrac{17}{100} = -\dfrac{417}{100}$

24. $\dfrac{91}{1000}$

25. $n = 0.88888\ldots$

$10n = 8.88888\ldots$

$\overline{}$

$9n = 8.00000\ldots$ Subtract 1st equation from 2nd.

$n = \dfrac{8}{9}$

26. $n = 0.16161616\ldots$

$100n = 16.16161616\ldots$

$\overline{}$

$99n = 16.00000\ldots$ Subtract 1st equation from 2nd.

$n = \dfrac{16}{99}$

27. Divide 17 by 6 to get $2.8\overline{3}$; C

28. The midpoint of interval is 75,000,
so $|x - 75,000| \le 5000$

Cumulative Review, page 460

1.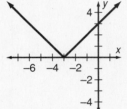

2.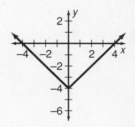

3.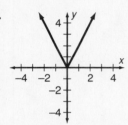

4. $4x + 2 \le -10$ or $4x + 2 \ge 10$ Write a disjunction.

$4x \le -12$ or $4x \ge 8$ Solve for x.

$x \le -3$ or $x \ge 2$

5. $-7 < 2x - 5$ and $2x - 5 < 7$ Write a conjunction.

$-2 < 2x$ and $2x < 12$ Solve for x.

$-1 < x$ and $x < 6$

6. $12 - 3(6n + 5) + 4n < \dfrac{1}{2}(10n - 6)$

$12 - 18n - 15 + 4n < 5n - 3$

$0 < 19n$

$0 < n$

7. $\begin{cases} 9x - 5y = 22 \\ y = 3x - 8 \end{cases}$

$9x - 5(3x - 8) = 22$ Substitute $3x - 8$ for y in Equation 1.

$9x - 15x + 40 = 22$

$-6x = -18$

$x = 3$

$y = 3(3) - 8$ Substitute 3 for x in Equation 2.

$y = 1$

So, $(3, 1)$ solves the system.

8. $7x - 4y = 11 \quad \rightarrow \quad 14x - 8y = 22$

$-14x + 8y = -11 \quad \rightarrow \quad \dfrac{-14x + 8y = -11}{0 = 11}$ Add equations.

The statement is false. So, the system is inconsistent. There is no solution.

9. Scatter plot B has the stronger linear relationship.

10. I, III and IV are all true, so D

11. No. Students did not really have to study in order to do well, and those who studied longer did not do too well.

12. $x =$ number of hot dogs sold
$y =$ number of hamburgers sold
Profit: $P = 1.2x + 1.4y$ (Profit = Sale Price − Cost)
Constraints: $x + y \le 500$

$0.3x + 0.6y \le 240$

$x \ge 0$ and $y \ge 0$

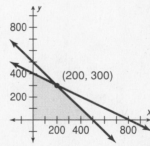

Vertices at $(0, 0)$; $(0, 400)$; $(200, 300)$; $(500, 0)$

$P = 1.2(0) + 1.4(0) = 0$

$P = 1.2(0) + 1.4(400) = 560$

$P = 1.2(200) + 1.4(300) = 660$

$P = 1.2(500) + 1.4(0) = 600$

Maximum profit at 200 hot dogs and 300 hamburgers

13. $A = P + Prt$

$A - P = Prt$ Subtract P.

$\dfrac{A - P}{Pt} = r$ Divide by Pt.

14. $SA = 2\pi r^2 + 2\pi rh$

$SA - 2\pi r^2 = 2\pi rh$ Subtract $2\pi r^2$.

$\dfrac{SA - 2\pi r^2}{2\pi r} = h$ Divide by $2\pi r$.

15. transitive property of equality

16. associative property of addition

17. multiplicative inverse property

18. reflexive property of equality

19. They are similar in that they both involve a change of order. They are different in that the commutative properties apply to an expression with an operation, and the symmetric property applies to both sides of the equation.

20.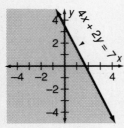

(2, −3) is a solution

21.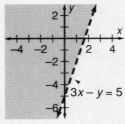

(2, −3) is not a solution

22. $(5-3)^2$ ☐ $5^2 - 3^2$

2^2 ☐ $25 - 9$

4 ☐ 16

So, the correct symbol is $<$.

23. $-(-3)^3$ ☐ $-(-2)^6$

$-(-27)$ ☐ $-(64)$

27 ☐ -64

So, the correct symbol is $>$.

24. $|-5-7|$ ☐ $|-5| + |-7|$

$|-12|$ ☐ $5 + 7$

12 ☐ 12

So, the correct symbol is $=$.

1. C; $3x - 2 \le -5$ or $3x - 2 \ge 5$ Write a disjunction.

$3x \le -3$ or $3x \ge 7$ Solve for x.

$x \le -1$ or $x \ge \dfrac{7}{3}$

2. A: $6 / 4 * 8 = 12$

B: $8 / 4 - 6 = 12$

C: $6 * 8 / 4 = 12$

D: $6 * 4 - 8 = 16$

3. B is rational since it repeats.

4. E; $\dfrac{86 + 86 + 86 + x}{4} = 90$

So, $258 + x = 360$ and $x = 102$.

5. C; Cannot have more than one solution to a system of linear equations

6. D

7. D

8. A; domain of $A = \{x : x \ge 0\}$

9. B; $3(3 - 2x) > 4x - 1$

$9 - 6x > 4x - 1$

$-10x > -10$

$x < 1$

Data Activity, page 462–463

1.

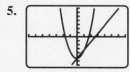

(Graph: Total Number of Businesses (in 1000's) vs Year)

Y-axis: 9,000; 8,000; 7,000; 6,000; 5,000; 4,000; 3,000; 2,000; 1,000; 0

X-axis (Year): '81 '82 '83 '84 '85 '86 '87 '88 '89 '90 '91 '92

2. 1983-1984, 1988-1989

3. Yes; Answers will vary. There were more failures, but there were also more businesses in '92 than in '86. A type of rate is needed to make a comparison.

4. 1981: 61; 1982: 88; 1983: 110; 1984: 107; 1985: 115; 1986: 120; 1987: 102; 1988: 98; 1989: 65; 1990: 76; 1991: 107; 1992: 110

5. Answers will vary.

Lesson 10.1, pages 465–467

THINK BACK

1.

x	y	Change in y
−2	−9	
−1	−7	2
0	−5	2
1	−3	2
2	−1	2
3	1	2
4	3	2
10	15	12
25	45	30

2a. y changes by 2 for each unit increase in x

b. constant change

c. x changes by 15 units from 10 to 25, so y changes by 2 • 15 or 30 units

3. Read slope from equation $y = 2x - 5$, $m = 2$

4. $m = \dfrac{45 - 15}{25 - 10} = \dfrac{30}{15} = 2$

EXPLORE

5.

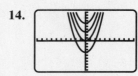

One equation is first degree and linear, second equation is second degree and graph is a parabola.

6. the letter U

7.

x	y	Difference Column 1 (Change in y)	Difference Column 2 (Change in the change in y)
−4	27		
−3	13	−14	
−2	3	−10	4
−1	−3	−6	4
0	−5	−2	4
1	−3	2	4
2	3	6	4
3	13	10	4
4	27	14	4

8. It is not constant. It can be both positive or negative.

9. Answers will vary, but should include the fact that there is an x^2 in this equation.

10. It is a constant change of 4.

MAKE CONNECTIONS

11a. $y_1 = x^2 - x - 6$

x	y	Difference Column 1 (Change in y)	Difference Column 2 (Change in the change in y)
−4	14		
−3	6	−8	
−2	0	−6	2
−1	−4	−4	2
0	−6	−2	2
1	−6	0	2
2	−4	2	2
3	0	4	2
4	6	6	2
5	14	8	2

b. $y_2 = x^2 - 2x - 3$
Table will resemble the one in 11a, except the change in the change in y will be 2.

c. $y_3 = 3x^2 + x + 2$
Table will resemble the one in 11a, except the change in the change in y will be 60.

d. $y_4 = -2x^2 + 4$
Table will resemble the one in 11a, except the change in the change in y will be −4.

12. For a, b, and c, y decreases first, then increases. For d, y increases first, then decreases.

13. Column 2 must remain constant, but column 1 may increase or decrease.

14.

Graphs are identical in shape with a vertical translation.

15. Answers will vary, but students should recognize c as a vertical translation of the graph of $y = x^2$.

16. Set I: Set II:

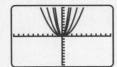 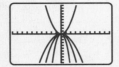

When $a > 0$, the graph opens upward; when $a < 0$, it opens downward. The graph also changes shape. When $|a| > 1$, the graph narrows; when $|a| < 1$, graph widens.

17. Activity, so no solution

SUMMARIZE

18. Answers will vary, but should include the differences between lines and curves and the similarities of vertical translations.

19. Graph is an upward opening parabola with vertex at $(3, -5)$. This is the minimum.

20. No; second equation is translated up 2 units.

21.

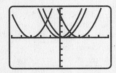

Adding a constant prior to squaring shifts the graph horizontally.

Lesson 10.2, pages 468–475

EXPLORE

1. Graph is upward opening parabola with vertex at $(3, -1)$.

2. No; answers will vary.

3. Yes; answers will vary.

4. Vertex of $y = 2x^2 - 12x + 16$ is $(3, -2)$.
   Vertex of $y = 3x^2 - 18x + 24$ is $(3, -3)$.

5. Vertex of $y = 4x^2 - 24x + 32$ is $(3, -4)$.
   Vertex of $y = -x^2 + 6x - 8$ is $(3, 1)$.

TRY THESE

1. $a < 0$, so parabola opens downward.

2. $a > 0$, so parabola opens upward.

3. When multiplied out, $a > 0$, so opens upward.

4. $x = \dfrac{-b}{2a} = \dfrac{-0}{2(5)} = 0$, $y = 5(0)^2 = 0$, vertex at $(0, 0)$; $x = 0$

5. $x = \dfrac{-b}{2a} = \dfrac{-0}{2(-7)} = 0$, $y = -7(0)^2 = 0$, vertex at $(0, 0)$;
   $x = 0$

6. $x = \dfrac{-b}{2a} = \dfrac{-0}{2(3)} = 0$, $y = 3(0)^2 + 4 = 4$, vertex at $(0, 4)$;
   $x = 0$

7. $x = \dfrac{-b}{2a} = \dfrac{-0}{2(-3)} = 0$, $y = -3(0)^2 + 15 = 15$, vertex at $(0, 15)$; $x = 0$

8. $x = \dfrac{-b}{2a} = \dfrac{-(-2)}{2(4)} = \dfrac{1}{4}$, $y = 4\left(\dfrac{1}{4}\right)^2 - 2\left(\dfrac{1}{4}\right) + 15 = \dfrac{59}{4}$,
   vertex at $\left(\dfrac{1}{4}, \dfrac{59}{4}\right)$; $x = \dfrac{1}{4}$.

9. $x = \dfrac{-b}{2a} = \dfrac{-(-5)}{2(-3)} = -\dfrac{5}{6}$, $y = 3\left(-\dfrac{5}{6}\right)^2 - 5\left(-\dfrac{5}{6}\right) - 9 = -\dfrac{83}{12}$,
   vertex at $\left(-\dfrac{5}{6}, -\dfrac{83}{12}\right)$; $x = -\dfrac{5}{6}$.

10. 7 is subtracted after x is squared, so shift is 7 units down.

11. 6 is added after x is squared, so shift is 6 units up.

12. x^2 is multiplied by 4, so graph is narrower.

13. x^2 is multiplied by $\dfrac{1}{4}$, so graph is wider.

14. 5 is subtracted before x is squared, so shift is 5 units right.

15. 8 is added before x is squared, so shift is 8 units left.

16. No shift, narrower graph; c

17. Shift up 3; a

18. Shift right 3; b

19.

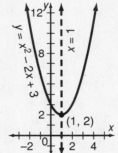

20.

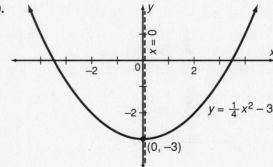

21.

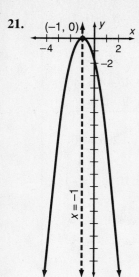

22. $x = \dfrac{-b}{2a} = \dfrac{-8}{2(-1)} = 4,\ y = -(4)^2 + 8(4) - 4 = 12$

so, price is $4.00, profit is $12,000.00.

23. Since the equation of the axis of symmetry is $x = \dfrac{-b}{2a}$,

which is a vertical line, it must be parallel to the y-axis.

If $b = 0$, it is the y-axis.

PRACTICE

1. $a > 0$, so parabola opens upward.

2. $a < 0$. so parabola opens downward.

3. $a > 0$, so parabola opens upward.

4. $x = \dfrac{-b}{2a} = \dfrac{-0}{2(-66)} = 0,\ y = -66(0)^2 = 0,$

vertex is $(0, 0); x = 0$

5. $x = \dfrac{-b}{2a} = \dfrac{-0}{82} = 0,\ y = 82(0)^2 = 0,$ vertex is $(0, 0); x = 0$

6. $x = \dfrac{-b}{2a} = \dfrac{-0}{2(7)} = 0,\ y = 7(0)^2 + 3 = 3,$ vertex is $(0, 3);$

$x = 0$

7. $x = \dfrac{-b}{2a} = \dfrac{-0}{2(-2)} = 0,\ y = -2(0)^2 + 14 = 14,$

vertex is $(0, 14); x = 0$

8. $x = \dfrac{-b}{2a} = \dfrac{-(-4)}{2(3)} = \dfrac{2}{3},\ y = 3\left(\dfrac{2}{3}\right)^2 - 4\left(\dfrac{2}{3}\right) - 11 = -12\dfrac{1}{3},$

vertex is $\left(\dfrac{2}{3}, -12\dfrac{1}{3}\right); x = \dfrac{2}{3}$

9. $x = \dfrac{-b}{2a} = \dfrac{-(-10)}{2(-5)} = -1,\ y = -5(-1)^2 - 10(-1) - 15 = -10,$

vertex is $(-1, -10); x = -1$

10. $x = \dfrac{-b}{2a} = \dfrac{-(-6)}{2\left(\frac{1}{2}\right)} = 6,\ y = \dfrac{1}{2}(6)^2 - 6(6) - 2 = -20,$

vertex is $(6, -20); x = 6$

11. $x = \dfrac{-b}{2a} = \dfrac{-(-4)}{2\left(-\frac{1}{2}\right)} = -4,\ y = -\dfrac{1}{2}(-4)^2 - 4(-4) + 14 = 22,$

vertex is $(-4, 22); x = -4$

12. $x = \dfrac{-b}{2a} = \dfrac{-6}{2\left(\frac{1}{4}\right)} = -12,\ y = \dfrac{1}{4}(-12)^2 - 5 + 14 = -41,$

vertex is $(-12, -41); x = -12$

13. 10 is subtracted after squaring x, so shift is down 10 units.

14. 16 is added after squaring x, so shift is up 16 units.

15. x^2 is multiplied by 3, so graph is narrower.

16. x^2 is multiplied by $\dfrac{1}{3}$, so graph is wider.

17. 25 is added before squaring, so shift is left 25 units.

18. 18 is subtracted before squaring, so shift is right 18 units.

19. No shift, multiplied by 4, so graph is narrower; c

20. Shift is down by 4; b

21. Left shift of 4; a

22. $x = \dfrac{-b}{2a} = \dfrac{-4}{2(2)} = -1;$ b

23. $x = \dfrac{-b}{2a} = \dfrac{-(-2)}{2\left(-\frac{1}{2}\right)} = -2;$ a

24. $x = \dfrac{-b}{2a} = \dfrac{-2}{2\left(-\frac{1}{4}\right)} = 4;$ c

25.

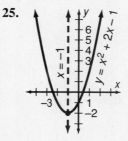

26.

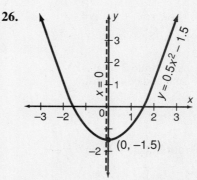

27.

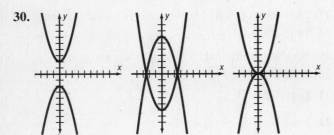

28. $x = \dfrac{-b}{2a} = \dfrac{-200}{2(-50)} = 2,$

2 decreases of $0.05
is a $0.10 decrease
from $1.10; so charge
$1.00 for the fare.

29. $x = \dfrac{-b}{2a} = \dfrac{-0.025}{2(-0.00004)} = 312.5,$

$y = -0.00004(312.5)^2 + 0.025(312.5) + 100 \approx 103.91$

So, the summit is 312.5 feet from point A and about 3.91 feet higher in elevation than point A.

30.

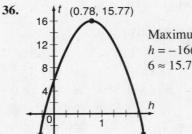

Graphs can intersect in 0, 1, or 2 points.

EXTEND

31. I. $a(0)^2 + b(0) + c = 4$, so $c = 4$

II. $a(1)^2 + b(1) + c = 5$, so $a + b + c = 5$

III. $a(-1)^2 + b(-1) + c = 9$, so $a - b + c = 9$

$$
\begin{array}{l}
a + b + 4 = 5 \\
\underline{a - b + 4 = 9} \\
2a \quad + 8 = 14 \\
\quad\quad 2a = 6 \\
\quad\quad a = 3
\end{array}
$$

Substitute 4 for c and add equations.

$3 + b + 4 = 5$ from Equation II, so $b = -2$;
$y = 3x^2 - 2x + 4$

32. I. $a(0)^2 + b(0) + c = 2$, so $c = 2$

II. $a(1)^2 + b(1) + c = -4$, so $a + b + c = -4$

III. $a(-1)^2 + b(-1) + c = 12$, so $a - b + c = 12$

$$
\begin{array}{l}
a + b + 2 = -4 \\
\underline{a - b + 2 = 12} \\
2a \quad + 4 = 8 \\
\quad\quad 2a = 4 \\
\quad\quad a = 2
\end{array}
$$

Substitute 2 for c and add equations.

$2 + b + 2 = -4$ from Equation II, so $b = -8$;
$y = 2x^2 - 8x + 2$

33. I. $a(-1)^2 + b(-1) + c = 1$, so $a - b + c = 1$

II. $a(4)^2 + b(4) + c = 26$, so $16a + 4b + c = 26$

III. $a(0)^2 + b(0) + c = -2$, so $c = -2$

$$
\begin{array}{l}
a - b - 2 = 1 \quad \rightarrow \quad 4a + 4b - 8 = 4 \\
16a + 4b - 2 = 26 \rightarrow \underline{16a + 4b - 2 = 26} \\
\quad\quad\quad\quad\quad\quad\quad 20a \quad -10 = 30 \\
\quad\quad\quad\quad\quad\quad\quad 20a = 40 \\
\quad\quad\quad\quad\quad\quad\quad a = 2
\end{array}
$$

$2 - b - 2 = 1$ from Equation II, so $b = -1$;
$y = 2x^2 - x - 2$

34. I. $a(4)^2 + b(4) + c = 12$, so $16a + 4b + c = 12$

II. $a(-2)^2 + b(-2) + c = 9$, so $4a - 2b + c = 9$

III. $a(0)^2 + b(0) + c = 6$, so $c = 6$

$$
\begin{array}{l}
16a + 4b + 6 = 12 \rightarrow 16a + 4b + 6 = 12 \\
4a - 2b + 6 = 9 \rightarrow \underline{8a - 4b + 12 = 18} \\
\quad\quad\quad\quad\quad\quad\quad 24a \quad\quad +18 = 30 \\
\quad\quad\quad\quad\quad\quad\quad 24a = 12 \\
\quad\quad\quad\quad\quad\quad\quad a = \dfrac{1}{2}
\end{array}
$$

$16\left(\dfrac{1}{2}\right) + 4b + 6 = 12$ from Equation I, so $b = -\dfrac{1}{2}$;

$y = \dfrac{1}{2}x^2 - \dfrac{1}{2}x + 6$

35. $h = \dfrac{1}{2}(-32)t^2 + 25t + 6 = -16t^2 + 25t + 6$

36.

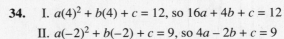

Maximum occurs at vertex;
$h = -16(0.78)^2 + 25(0.78) + 6 \approx 15.77$

37. $C = 104.143(94)^2 - 19{,}026.286(94) + 869{,}356.657$
$C = 1093.321$ or about 1.093 billion offers

38. Intersect at $(-1.66, -1.5)$, and $(1.66, -1.5)$

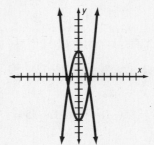

39.

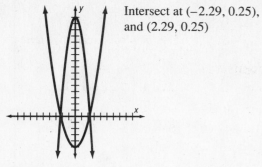

Intersect at $(-2.29, 0.25)$, and $(2.29, 0.25)$

THINK CRITICALLY

40. Answers will vary. Possible answers: $y = x^2$, $y = 2x^2$, $y = -x^2$, $y = -4x^2$

41. $x = \dfrac{-b}{2a} = \dfrac{0}{2a} = 0$, $y = a(0)^2 + c = c$, vertex at $(0, c)$

42. Both shift the graph left if they are positive and right if they are negative.

43. Graph shifts up if c is positive and down if c is negative.

44. $u > v$

45. $x = \dfrac{-b}{2a} = \dfrac{0}{2a} = 0$, and $y = a(0)^2 = 0$;

So all have vertex at $(0, 0)$ and axis of symmetry $x = 0$.

MIXED REVIEW

46. $3x - 7 = 5x + 20$

$\qquad -27 = 2x$ Subtract 20 and $3x$.

$\qquad -13.5 = x$ Divide by 2.

47. $\dfrac{1}{2}(6x - 4) = 4(2x + 5)$

$\qquad 3x - 2 = 8x + 20$ Distribute.

$\qquad -22 = 5x$ Subtract $3x$ and 20.

$\qquad \dfrac{-22}{5} = x$ Divide by 5.

48. $3y - 4 = 6x$

$\qquad 3y = 6x + 4$

$\qquad y = 2x + \dfrac{4}{3}$; $m = 2$, $b = \dfrac{4}{3}$;   C

49. Graph is shifted right 1 unit, up 3 units and is narrower.

50. Graph is shifted left 3 units, down 1 unit, and is wider.

51. $x = \dfrac{-b}{2a} = \dfrac{-(-8)}{2(2)} = 2$;

$y = 2(2)^2 - 8(2) - 17 = -25$;

So, vertex is $(2, -25)$ and line of symmetry is $x = 2$.

52. $x = \dfrac{-b}{2a} = \dfrac{-12}{2(-4)} = \dfrac{3}{2}$;

$y = -4\left(\dfrac{3}{2}\right)^2 + 12\left(\dfrac{3}{2}\right) - 22 = -13$;

So, vertex is $\left(\dfrac{3}{2}, -13\right)$ and line of symmetry is $x = \dfrac{3}{2}$.

ALGEBRAWORKS

1. $A = \pi r^2 = \pi(9)^2 = 254$ in.^{2}

2. $A = \pi r^2$; the graph is a parabola.

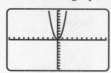

3. $\pi(10)^2 - \pi(8)^2 = 100\pi - 64\pi = 36\pi \approx 113$ in.^{2}

4. $\pi(5)^2 = 25\pi \approx 79$ in.^{2}

5. No; Best Pizza gives away 79 in.^{2}, and Rossi's gives away 113 in.^{2} of pizza.

6. Adding 2 in. to the radius of a 14 in. pizza adds $\pi(9)^2 - \pi(7)^2 \approx 101$ in.^{2}
Adding 2 in. to the radius of an 18 in. pizza adds $\pi(11)^2 - \pi(9)^2 \approx 126$ in.^{2}

Lesson 10.3, pages 476–482

EXPLORE

1a. 0 and 4

  b. 5 and -5

  c. 1

  d. No solution since the least value for x^2 is 0 so $x^2 > 1 > 0$.

2. The graph crosses x-axis twice, once at $(0, 0)$ and again at $(4, 0)$. The x-coordinates are the solutions to the equation $x^2 - 4x = 0$

3. Twice; at $(5, 0)$ and at $(-5, 0)$; the x-coordinates match the solutions to the equation $x^2 - 25 = 0$.

4. The graph touches the x-axis once at $(1, 0)$ and the x-coordinate matches the solution to the equation $x^2 - 2x + 1 = 0$.

5. The graph does not cross the x-axis and there are no solutions to the equation $x^2 + 1 = 0$.

TRY THESE

1. $7 = 3x^2 - 4x + 15$

$\quad 0 = 3x^2 - 4x + 8$ Subtract 7.

2. $4 = 8 - 2x^2 + 5x$

$\quad 0 = -2x^2 - 5x + 4$ Subtract 4 and rearrange terms.

3. $2 = 2 - 3x + \dfrac{1}{2}x^2$

$\quad 0 = \dfrac{1}{2}x^2 - 3x$ Subtract 2 and rearrange terms.

4. Not equivalent since all terms except -2 are opposite signs.

5. $8 = 4x^2 - 6x - 2$

$\quad 0 = 4x^2 - 6x - 10$ Subtract 8.   Equivalent.

6. $8 = 4x^2 - 6x - 2$

   $10 = 4x^2 - 6x$ Add 2.   Equivalent.

7. Graph equation and use trace feature; x-intercepts are -5 and 3.

8. Graph equation and use the trace feature; x-intercepts are -1 and 4.

9. Graph equation and use trace feature; x-intercept is 5.

10. $6 = x^2 + 4x - 6$

   $0 = x^2 + 4x - 12$ Subtract 6.

   Graph equation and use trace feature; x-intercepts are -6 and 2.

11. $8 = -x^2 + 6x + 1$

   $0 = -x^2 + 6x - 7$ Subtract 8.

   Graph equation and use trace feature; x-intercepts are approximately 1.59 and 4.41.

12. $-25 = -x^2 + x - 5$

   $0 = -x^2 + x + 20$ Add 25.

   Graph equation and use trace feature; x-intercepts are -4 and 5.

13. Graph $y = 12$ and $y = x^2 - 2x - 3$ on same axes. Use trace feature to find intersection points at $x = -3$ and 5.

14. Graph $y = -5$ and $y = -x^2 + 2x + 1$ on the same axes. Use the trace feature to find intersection points at $x \approx -1.65$ and 3.65.

15. Graph $y = 1$ and $y = 2x^2 - 2x - 8$ on the same axes. Use the trace feature to find intersection points at $x \approx -1.68$ and 2.68.

16. Graph of $y = 3x^2 - 6x + 2$ crosses x-axis twice, so it has 2 roots.

17. Graph of $y = 4x^2 + 5x + 2$ never crosses x-axis, so it has no roots.

18. Vertex of $y = x^2 - 8x + 16$ is at $(4, 0)$. Graph only touches x-axis once at the vertex, so it has only one root.

19. $h = \frac{1}{2}(-5.32)t^2 + 120t = 150$

   $-2.66t^2 + 120t - 150 = 0$

   Graph and use trace feature to find x-intercept of 1.29 and 43.83

20. Answers will vary, but should include graphing the equation and finding the x-coordinates where the graph crosses the x-axis.

PRACTICE

1. $9 = 2x^2 - 5x + 12$

   $0 = 2x^2 - 5x + 3$ Subtract 9.

2. $3 = 6 - x^2 + 3x$

   $0 = -x^2 + 3x + 3$ Subtract 3 and rearrange terms.

3. $5 = 5 - 5x + \frac{1}{4}x^2$

   $0 = \frac{1}{4}x^2 - 5x$ Subtract 5 and rearrange terms.

4. $5 = 3x^2 - 4x - 4$

   $9 = 3x^2 - 4x$ Add 4.   Equivalent.

5. $5 = 3x^2 - 4x - 4$

   $0 = 3x^2 - 4x - 9$ Subtract 5.   Equivalent.

6. $5 = 3x^2 - 4x - 4$

   $-5 = 3x^2 - 4x - 14$ Subtract 10.   Equivalent.

7. Graph equation and use the trace feature; x-intercepts are -4 and 6.

8. Graph equation and use the trace feature; x-intercepts are -5 and 6.

9. Graph equation and use the trace feature; x-intercepts are -5 and 2.

10. Graph equation and use the trace feature; x-intercepts are -2 and 6.

11. Graph equation and use the trace feature; x-intercept is 6.

12. Graph equation and use the trace feature; x-intercept is -3.

13. Graph equation and use the trace feature; x-intercepts are 0.31 and 3.19.

14. Graph equation and use the trace feature; x-intercepts are -0.23 and 2.90.

15. Graph equation and use the trace feature; x-intercepts are -0.78 and 1.28.

16. $4 = x^2 + 2x - 4$

   $0 = x^2 - 2x - 8$ Subtract 4.

   Graph equation and use the trace feature; x-intercepts are 4 and -2.

17. $48 = x^2 - 8x - 17$

   $0 = x^2 - 8x - 65$ Subtract 48.

   Graph equation and use the trace feature; x-intercepts are -5 and 13.

18. $9 = x^2 - 18x - 31$

   $0 = x^2 - 18x - 40$ Subtract 9.

   Graph equation and use the trace feature; x-intercepts are -2 and 20.

19. $-10 = -x^2 + 2x - 6$

   $0 = -x^2 + 2x + 4$; -1.24 and 3.24 Graph the equation $(-1.24, 3.24)$.

20. $2 = 3x^2 + 4x - 1$

   $0 = 3x^2 + 4x - 3$; -1.87 and 0.54 Graph equation $(-1.87, 0.54)$.

21. $5 = 4x^2 - 2x + 1$

   $0 = 4x^2 - 2x - 4$; -0.78 and 1.28 Graph equation $(-1.87, 0.54)$.

22. Graph $y = 5$ and $y = x^2 - 4x$ and then use the trace feature to find intersections at -1 and 5.

23. Graph $y = -6$ and $y = -x^2 + 3x + 6$ and then use the trace feature to find intersections at -2.27 and 5.27.

24. Graph $y = 1$ and $y = 3x^2 - x - 4$ on same axes, then use trace to find points of intersection at -1.14 and 1.47.

25. Graph touches x-axis once so there is 1 solution.

26. Graph never crosses x-axis, so no solutions

27. Graph crosses x-axis twice, so 2 solutions

28. Graph never crosses x-axis, so no solutions

29. Graph crosses x-axis twice, so 2 solutions

30. Graph touches x-axis once so there is 1 solution.

31. $2 = 0.052x^2 - 0.551x + 17$

   $0 = 0.0052x^2 - 0.551x - 0.3$ Subtract 2.
   Graph $y = 0.0052x^2 - 0.551x - 0.3$ and use the trace feature to find x-intercepts near -0.52 and 11.52. -0.52 makes no sense in the problem, so month is first month past 11.12, or April '94 (month 12).

32. $h = \frac{1}{2}(-12.2)t^2 + 20t + 5 = -6.1t^2 + 20t + 5 = 10$

   $-6.1t^2 + 20t - 5 = 0$
   Graph $y = 6.1t^2 + 20t - 5$ and use the trace feature; x-intercepts are 0.27 and 3.01.

33. The graph of the parabola will have no x-intercepts.

EXTEND

34. $371 = -x^2 + 60x$ or $0 = -x^2 + 60x - 371$;
   Graph $y = -x^2 + 60x - 37$ and use the trace feature; x-intercepts are 7 and 53.

35. $704 = -x^2 + 60x$ or $0 = -x^2 + 60x - 704$;
   Graph and use the trace feature; x-intercepts are 16 and 44.

36. Maximum profit is at vertex: $x = \frac{-b}{2a} = \frac{-60}{2(-1)} = 30$ and
   $y = -(30)^2 + 60(30) = -900 + 1800 = \$900$ profit with 30 people.

37. $0 = -x^2 + 60x$; Graph and use the trace feature; x-intercept are 0 and 60.
   If more than 60 people go, the agency loses money.

THINK CRITICALLY

38. To have one solution, vertex must be on x-axis, so $y = 0$,
   $x = \frac{-b}{2a} = \frac{4}{2} = 2$, $0 = 2^2 - 4(2) + c$ or $0 = -4 + c$ or $c = 4$

39. Answers will vary, any value > 4 works.

40. Answers will vary, may use equation from 38 with $c < 4$.

41. Answers will vary, depending on answer to 40.

42. The solutions are equidistant from the axis of symmetry.

MIXED REVIEW

43. C; $9x + 3y = 12$
   $\qquad 3y = -9x + 12$
   $\qquad\ y = -3x + 4$

44. $\quad 2x + 3y = 6 \quad \rightarrow 4x + 6y = 12$
   $\underline{-12 + 4x = -6y \rightarrow 4x + 6y = 12}$
   $\qquad\qquad\qquad\qquad\ \ 0 = 0$ Subtract.
   Dependent systems, so infinite solutions

45. $5x + 4y = 13$ Slope is $\frac{-5}{4}$.

   $4x + 6y = 16$ Slope is $\frac{-4}{6}$.

   Slopes are different, so only one solution

46. Answers will vary; Possible answer: $-8 = \frac{-8}{1}$

47. Answers will vary; Possible answer: $5.33 = \frac{533}{100}$

48. Answers will vary; Possible answer: $-9\frac{1}{2} = \frac{-19}{2}$

49. Answers will vary; Possible answer: $0.095 = \frac{95}{1000}$

50. $-6 = x^2 + 5x$
   $\quad 0 = x^2 + 5x + 6$ Add 6.
   Graph $y = x^2 + 5x + 6$ and use the trace feature; x-intercepts are -2 and -3.

51. $-2 = -x^2 + 3x + 1$
   $\quad 0 = -x^2 + 3x + 3$ Add 2.
   Graph $y = x^2 + 3x + 3$ and use the trace feature; x-intercepts are -0.79 and 3.79.

Lesson 10.4, pages 483–489

EXPLORE/WORKING TOGETHER

7. When $c - b$ is < 0 the equation has no real solutions.

TRY THESE

1. $x^2 = 81$
   $x = \pm\sqrt{81} = \pm 9$

2. $x^2 = 100$
   $x = \pm\sqrt{100} = \pm 10$

3. $x^2 = 48$
   $x = \pm\sqrt{48} = \pm\sqrt{16 \cdot 3} = \pm 4\sqrt{3}$

4. $x^2 - 3 = 72$
   $\ x^2 = 75$
   $\quad x = \pm\sqrt{75} = \pm\sqrt{25 \cdot 3} = \pm 5\sqrt{3}$

5. $\frac{1}{2}x^2 + 3 = 23$
   $\ \ \frac{1}{2}x^2 = 20$
   $\quad\ x^2 = 40$
   $\qquad x = \pm\sqrt{40} = \pm\sqrt{4 \cdot 10} = \pm 2\sqrt{10}$

6. $\frac{1}{2}x^2 - 8 = 44$

$\qquad \frac{1}{2}x^2 = 52$

$\qquad x^2 = 104$

$\qquad x = \pm\sqrt{104} = \pm\sqrt{4 \cdot 26} = \pm 2\sqrt{26}$

7. $(x-1)^2 = 25$

$\qquad x - 1 = \pm\sqrt{25} = -4, 6$

$\qquad x = 1 \pm 5 = 6 \text{ or } -4$

8. $(x-1)^2 = 64$

$\qquad x - 1 = \pm\sqrt{64} = \pm 8$

$\qquad x = 1 \pm 8 = -7 \text{ or } 9$

9. $2(x+1)^2 = 162$

$\qquad (x+1)^2 = 81$

$\qquad x + 1 = \pm\sqrt{81} = \pm 9$

$\qquad x = -1 \pm 9 = -10 \text{ or } 8$

10. $3(x-4)^2 = 108$

$\qquad (x-4)^2 = 36$

$\qquad x - 4 = \pm\sqrt{36} = \pm 6$

$\qquad x = 4 \pm 6 = -2 \text{ or } 10$

11. $3\left(x+\frac{2}{3}\right)^2 = \frac{1}{3}$

$\qquad \left(x+\frac{2}{3}\right)^2 = \frac{1}{9}$

$\qquad x + \frac{2}{3} = \pm\sqrt{\frac{1}{9}} = \pm\frac{1}{3}$

$\qquad x = -\frac{2}{3} \pm \frac{1}{3} = -1 \text{ or } -\frac{1}{3}$

12. $2\left(x+\frac{3}{4}\right)^2 = \frac{1}{8}$

$\qquad \left(x+\frac{3}{4}\right)^2 = \frac{1}{16}$

$\qquad x + \frac{3}{4} = \pm\sqrt{\frac{1}{16}} = \pm\frac{1}{4}$

$\qquad x = -\frac{3}{4} \pm \frac{1}{4} = -1 \text{ or } -\frac{1}{2}$

13. $x^2 - 5 = -6$

$\qquad x^2 = -1 \quad$ No real solution

14. $x^2 - 10 = -10$

$\qquad x^2 = 0$

$\qquad x = \pm\sqrt{0} = \pm 0 = 0$

15. $4x^2 - 3 = 18$

$\qquad 4x^2 = 21$

$\qquad x^2 = \frac{21}{4}$

$\qquad x = \pm\sqrt{\frac{21}{4}} \approx \pm 2.29$

16. $2x^2 - 7 = 13$

$\qquad 2x^2 = 20$

$\qquad x^2 = 10$

$\qquad x = \pm\sqrt{10} \approx \pm 3.16$

17. $3x^2 - 4 = 16$

$\qquad 3x^2 = 20$

$\qquad x^2 = \frac{20}{3}$

$\qquad x = \pm\sqrt{\frac{20}{3}} \approx \pm 2.58$

18. $3(x-3)^2 = 17$

$\qquad (x-3)^2 = \frac{17}{3}$

$\qquad x - 3 = \sqrt{\frac{17}{3}}$

$\qquad x = 3 \pm\sqrt{\frac{17}{3}} \approx 3 \pm 2.58 = 0.62 \text{ or } 5.38$

19. $5(x-2)^2 = 12$

$\qquad (x-2)^2 = \frac{12}{5}$

$\qquad x - 2 = \pm\sqrt{\frac{12}{5}}$

$\qquad x = 2 \pm\sqrt{\frac{12}{5}} \approx 2 \pm 1.55 = 0.45 \text{ or } 3.55$

20. $6\left(x-\frac{1}{2}\right)^2 = 3$

$\qquad \left(x-\frac{1}{2}\right)^2 = \frac{1}{2}$

$\qquad x - \frac{1}{2} = \pm\sqrt{\frac{1}{2}}$

$\qquad x = \frac{1}{2} \pm\sqrt{\frac{1}{2}} \approx \frac{1}{2} \pm 0.71 = -0.21 \text{ or } 1.21$

21. $5(x+0.5)^2 = 7.5$

$\qquad (x+0.5)^2 = 1.5$

$\qquad x + 0.5 = \pm\sqrt{1.5}$

$\qquad x = -0.5 \pm\sqrt{1.5} \approx -0.5 \pm 1.22 = -1.72 \text{ or } 0.72$

22.
$$441 = 400\left(1 + \frac{r}{100}\right)^2$$
$$\frac{441}{400} = \left(1 + \frac{r}{100}\right)^2$$
$$\pm\sqrt{\frac{441}{400}} = 1 + \frac{r}{100}$$
$$\pm\sqrt{\frac{441}{400}} - 1 = \frac{r}{100}$$
$$100\left(\pm\sqrt{\frac{441}{100}} - 1\right) = r$$
$$5 \text{ or } -205 = r$$

-205 makes no sense in the context of the problem. Interest rate is 5%.

23. Two real solutions when $\dfrac{c-b}{a} > 0$

One real solution when $\dfrac{c-b}{a} = 0$

No real solutions when $\dfrac{c-b}{a} < 0$, or when $a = 0$

PRACTICE

1. $x^2 = 49$
$$x = \pm\sqrt{49} = \pm 7$$

2. $x^2 = 121$
$$x = \pm\sqrt{121} = \pm 11$$

3. $x^2 = \dfrac{25}{81}$
$$x = \pm\sqrt{\frac{25}{81}} = \pm\frac{5}{9}$$

4. $x^2 - \dfrac{16}{64} = 0$
$$x^2 = \frac{16}{64}$$
$$x = \pm\sqrt{\frac{16}{64}} = \pm\frac{4}{8} = \pm\frac{1}{2}$$

5. $x^2 = \dfrac{361}{484}$
$$x = \pm\sqrt{\frac{361}{484}} = \pm\frac{19}{22}$$

6. $x^2 + 6 = 0$
$$x^2 = -6$$
No real solution since x^2 is never negative.

7. $x^2 + 3 = 0$
$$x^2 = -3$$
No real solution since x^2 is never negative.

8. $x^2 + 4 = 112$
$$x^2 = 108$$
$$x = \pm\sqrt{108} = \pm\sqrt{36 \cdot 3} = \pm 6\sqrt{3}$$

9. $x^2 + 6 = 86$
$$x^2 = 80$$
$$x = \pm\sqrt{80} = \pm\sqrt{16 \cdot 5} = \pm 4\sqrt{5}$$

10. $4x^2 - 5 = 283$
$$4x^2 = 288$$
$$x^2 = 72$$
$$x = \pm\sqrt{72} = \pm\sqrt{36 \cdot 2} = \pm 6\sqrt{2}$$

11. $3x^2 - 4 = 290$
$$3x^2 = 294$$
$$x^2 = 98$$
$$x = \pm\sqrt{98} = \pm\sqrt{49 \cdot 2} = \pm 7\sqrt{2}$$

12. $(x - 1)^2 = 64$
$$x - 1 = \pm\sqrt{64} = \pm 8$$
$$x = 1 \pm 8 = -7 \text{ or } 9$$

13. $(x - 4)^2 = 225$
$$x - 4 = \pm\sqrt{225} = \pm 15$$
$$x = 4 \pm 15 = -11 \text{ or } 19$$

14. $\dfrac{1}{2}x^2 + 4 = 158$
$$\frac{1}{2}x^2 = 154$$
$$x^2 = 308$$
$$x = \pm\sqrt{308} = \pm\sqrt{4 \cdot 77} = \pm 2\sqrt{77}$$

15. $\dfrac{1}{2}x^2 + 5 = 250$
$$\frac{1}{2}x^2 = 245$$
$$x^2 = 490$$
$$x = \pm\sqrt{490} = \pm\sqrt{49 \cdot 10} = \pm 7\sqrt{10}$$

16. $3(x + 5)^2 = 75$
$$(x + 5)^2 = 25$$
$$x + 5 = \pm\sqrt{25} = \pm 5$$
$$x = -5 \pm 5 = -10 \text{ or } 0$$

17. $5(x + 6)^2 = 80$
$$(x + 6)^2 = 16$$
$$x + 6 = \pm\sqrt{16} = \pm 4$$
$$x = -6 \pm 4 = -10 \text{ or } -2$$

18. $3(2x + 5)^2 = 300$
$$(2x + 5)^2 = 100$$
$$2x + 5 = \pm\sqrt{100} = \pm 10$$
$$2x = -5 \pm 10$$
$$x = \frac{-5 \pm 10}{2} = -7.5 \text{ or } 2.5$$

19. $2(3x+4)^2 = 180.5$

$\qquad (3x+4)^2 = 90.25$

$\qquad 3x+4 = \pm\sqrt{90.25} = \pm 9.5$

$\qquad 3x = -4 \pm 9.5$

$\qquad x = \dfrac{-4 \pm 9.5}{3} = -4\dfrac{1}{2}$ or $1\dfrac{5}{6}$

20. $2\left(x-\dfrac{5}{6}\right)^2 = \dfrac{1}{18}$

$\qquad \left(x-\dfrac{5}{6}\right)^2 = \dfrac{1}{36}$

$\qquad x-\dfrac{5}{6} = \pm\sqrt{\dfrac{1}{36}} = \pm\dfrac{1}{6}$

$\qquad x = \dfrac{5}{6} \pm \dfrac{1}{6} = \dfrac{2}{3}$ or 1

21. $5(x+0.3)^2 = 4.05$

$\qquad (x+0.3)^2 = 0.81$

$\qquad x+0.3 = \pm\sqrt{0.81} = \pm 0.9$

$\qquad x = -0.3 \pm 0.9 = -1.2$ or 0.6

22. $\qquad 1690 = 1440\left(1+\dfrac{r}{100}\right)^2$

$\qquad\quad \dfrac{1690}{1440} = \left(1+\dfrac{r}{100}\right)^2$

$\qquad \pm\sqrt{\dfrac{1690}{1440}} = \pm\dfrac{13}{12} = 1+\dfrac{r}{100}$

$\qquad\quad -1 \pm \dfrac{13}{12} = \dfrac{r}{100}$

$\qquad 100\left(-1 \pm \dfrac{13}{12}\right) = r = 8\dfrac{1}{3}$ or $-208\dfrac{1}{3}$

$-208\dfrac{1}{3}$ makes no sense, so interest rate is $8\dfrac{1}{3}\%$.

23. $3x^2 - 5 = 16$

$\qquad 3x^2 = 21$

$\qquad x^2 = 7$

$\qquad x = \pm\sqrt{7} \approx \pm 2.65$

24. $4x^2 - 10 = 25$

$\qquad 4x^2 = 35$

$\qquad x^2 = \dfrac{35}{4}$

$\qquad x = \pm\sqrt{\dfrac{35}{4}} \approx \pm 2.96$

25. $5x^2 + 9 = 38$

$\qquad 5x^2 = 29$

$\qquad x^2 = \dfrac{29}{5}$

$\qquad x = \pm\sqrt{\dfrac{29}{5}} \approx \pm 2.41$

26. $3x^2 + 7 = 70$

$\qquad 3x^2 = 63$

$\qquad x^2 = 21$

$\qquad x = \pm\sqrt{21} \approx \pm 4.58$

27. $x^2 + 8 = 81$

$\qquad x^2 = 73$

$\qquad x = \pm\sqrt{73} \approx \pm 8.54$

28. $2(x-3)^2 = 55$

$\qquad (x-3)^2 = \dfrac{55}{2}$

$\qquad x-3 = \pm\sqrt{\dfrac{55}{2}}$

$\qquad x = 3 \pm\sqrt{\dfrac{55}{2}} \approx -2.24$ or 8.24

29. $5(x-6)^2 = 82$

$\qquad (x-6)^2 = \dfrac{82}{5}$

$\qquad x-6 = \pm\sqrt{\dfrac{82}{5}}$

$\qquad x = 6 \pm\sqrt{\dfrac{82}{5}} \approx 1.95$ or 10.05

30. $7(x-0.6)^2 = 3$

$\qquad (x-0.6)^2 = \dfrac{3}{7}$

$\qquad x-0.6 = \pm\sqrt{\dfrac{3}{7}}$

$\qquad x = 0.6 \pm\sqrt{\dfrac{3}{7}} \approx -0.05$ or 1.25

31. $5(x-0.8)^2 = 12$

$\qquad (x-0.8)^2 = \dfrac{12}{5}$

$\qquad x-0.8 = \pm\sqrt{\dfrac{12}{5}}$

$\qquad x = 0.8 \pm\sqrt{\dfrac{12}{5}} \approx -0.75$ or 2.35

32. $d = \dfrac{1}{2}at^2 = \dfrac{1}{2}(-9.8)t^2 = -4.9t^2$

$\qquad -222 = -4.9t^2$

$\qquad \dfrac{222}{4.9} = t^2$

$\qquad \pm\sqrt{\dfrac{222}{4.9}} = t \approx 6.73$ sec $\quad$ (negative value makes no sense in problem)

33. Answers will vary, but as in Exercise 32, time may not be negative so disregard one solution.

34. $18\pi = \dfrac{1}{3}\pi r^2(6)$

$\qquad 18\pi = 2\pi r^2 \qquad$ Simplify.

$\qquad 9 = r^2 \qquad$ Divide by 2π.

$\qquad 3 = r \qquad$ Take square root.

$\qquad\qquad\qquad$ -3 since radius cannot be negative.

35. $x^2 > 144$; Solve $x^2 = 144$; $x = \pm 12$;

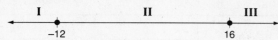

I: Try -13, $(-13)^2 > 144$ which is true.
II: Try 0, $(0)^2 > 144$ which is false.
III: Try 13, $13^2 > 144$ which is true.
$\quad x < -12$ or $x > 12$.

36. $x^2 < 225$; Solve $x^2 = 225$; $x = \pm 15$;

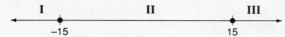

I: Try -16, $(-16)^2 < 225$ which is false.
II: Try 0, $(0)^2 < 225$ which is true.
III: Try 16, $16^2 < 225$ which is false.
$\quad -15 < x < 15$.

37. $2x^2 - 6 > 66$; Solve $2x^2 - 6 = 66$

$\quad 2x^2 = 72$

$\quad x^2 = 36$

$\quad x = \pm 6$

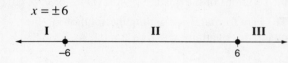

I: Try -7, $2(-7)^2 - 6 > 66$ which is true.
II: Try 0, $(0)^2 - 6 > 66$ which is false.
III: Try 7, $2(7)^2 - 6 > 66$ which is true.
$\quad x < -6$ or $x > 6$.

38. $4x^2 - 8 < 392$; Solve $4x^2 - 8 = 392$;

$\quad 4x^2 = 400$

$\quad x^2 = 100$

$\quad x = \pm 10$

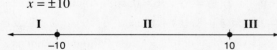

I: Try -11, $4(-11)^2 - 8 < 392$ which is false.
II: Try 0, $4(0)^2 - 8 < 392$ which is true.
III: Try 11, $4(11)^2 - 8 < 392$ which is false.
$\quad -10 < x < 10$.

39. $4(x-2)^2 < 16$; Solve $4(x-2)^2 = 16$

$\quad (x-2)^2 = 4$

$\quad x - 2 = \pm\sqrt{4} = \pm 2$

$\quad x = 2 \pm 2 = 0$ or 4

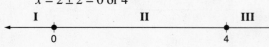

I: Try -1, $4(-1-2)^2 < 16$ which is false.
II: Try 1, $4(1-2)^2 < 16$ which is true.
III: Try 5, $4(5-2)^2 < 16$ which is false.
$\quad 0 < x < 4$.

40. $3(x+1)^2 > 147$; Solve $3(x+1)^2 = 147$

$\quad (x+1)^2 = 49$

$\quad x + 1 = \pm\sqrt{49} = \pm 7$

$\quad x = -1 \pm 7 = -8$ or 6

I: Try -9, $3(-9+1)^2 > 147$ which is true.
II: Try 0, $3(0)^2 > 147$ which is false.
III: Try 7, $3(7+1)^2 > 147$ which is true.
$\quad x < -8$ or $x > 6$.

41. $\quad 100 = 4(3.14)r^2$

$\quad 100 = 12.56r^2$

$\quad \dfrac{100}{12.56} = r^2$

$\quad \sqrt{\dfrac{100}{12.56}} = r \approx 2.82$ m

42. $S = \dfrac{1}{2}\left(4\pi r^2\right) + \pi r^2 = 3\pi r^2$

$\quad 100 = 3(3.14)r^2 = 9.42r^2$

$\quad \dfrac{100}{9.42} = r^2$

$\quad \sqrt{\dfrac{100}{9.42}} = r \approx 3.26$ m

43. $\quad 350 = 6x^2$

$\quad \dfrac{350}{6} = x^2$

$\quad \sqrt{\dfrac{350}{6}} = x \approx 7.64$ m

THINK CRITICALLY

44. Surface area is multiplied by 4 since the radius is squared. $S = 4\pi r^2$; If r is doubled, then $4\pi(2r)^2 = 4\pi \cdot 4r^2 = 16\pi r^2 = 4(4\pi r^2) = 4S$.

45. Surface area is multiplied by 4 since the side is squared.

46. Since height is given as a formula with time squared, height increases as a square root. If height is doubled, time is multiplied by $\sqrt{2} \approx 1.41$.

47. $A = P\left(1 + \dfrac{r}{100}\right)^4$

48. $A = P\left(1 + \dfrac{r}{100}\right)^4$

49.

Simple Interest

$A = P + P\left(\dfrac{r}{100}\right)t$

$A = 800 + 800\left(\dfrac{8.1}{100}\right)2$

$A = \$929.60$

Compound Interest

$A = P\left(1 + \dfrac{r}{100}\right)^2$

$A = 800\left(1 + \dfrac{8}{100}\right)^2$

$A = \$933.12$

Less money in simple interest account

50. No; it can be any number such that $c - b \geq 0$.

51. Six possible numbers, only 2 divisible by 3, so probability is $\frac{2}{6} = \frac{1}{3}$; D

52. multiplicative inverse

53. multiplicative property of zero

54. $b = 5$, $c = 1$; Vertex is (5, 1), Axis of symmetry: $x = 5$.

55. $b = -4$, $c = -2$; Vertex is (−4, −2), Axis of symmetry: $x = -4$.

56. $3(x + 3)^2 = 18$

$(x + 3)^2 = 6$

$x + 3 = \pm\sqrt{6}$

$x = -3 \pm \sqrt{6} \approx -5.45$ or -0.55

57. $4(x - 5)^2 = 29$

$(x - 5)^2 = \dfrac{29}{4}$

$x - 5 = \pm\sqrt{\dfrac{29}{4}}$

$x = 5 \pm \sqrt{\dfrac{29}{4}} \approx 2.31$ or 7.69

ALGEBRAWORKS

1. $-36 = -16t^2$

$\dfrac{36}{16} = t^2$

$\sqrt{\dfrac{36}{16}} = t = \dfrac{3}{2}$ or 1.5 sec

2. Speed $= 32t = 32(1.5) = 48 \dfrac{ft}{sec}$.

3. $-72 = -16t^2$

$\dfrac{72}{16} = t^2$

$\sqrt{\dfrac{72}{16}} = t \approx 2.12$ sec

4. Speed $= 32t = 32(2.12) = 67.84 \dfrac{ft}{sec}$.

5. $\dfrac{1\ mi}{1\ min} \cdot \dfrac{5280\ ft}{mi} \cdot \dfrac{1\ min}{60\ sec} = \dfrac{88\ ft}{sec}$

6. Speed is $32 \dfrac{ft}{sec}$ after 1 sec.

$\dfrac{x}{60} = \dfrac{32}{88}$

$x = \dfrac{1920}{88} \approx 21.82 \dfrac{mi}{hr}$

7. Speed is $32(2.5) = 80 \dfrac{ft}{sec}$ after 2.5 sec.

$\dfrac{x}{60} = \dfrac{80}{88}$

$x = \dfrac{4800}{88} \approx 54.55 \dfrac{mi}{hr}$

8. Answers will vary; between 1 and 2 minutes

Lesson 10.5, pages 490–497

EXPLORE

1.

a	b	c	$\dfrac{-b + \sqrt{b^2 - 4ac}}{2a}$	$\dfrac{-b - \sqrt{b^2 - 4ac}}{2a}$
1	−7	12	4	3
1	5	6	−2	−3
1	6	5	−1	−5
1	−1	2	1	−2

2. Graph $y = x^2 - 7x + 12$ and use the trace feature; x-intercepts are 3 and 4.

3. Graph $y = x^2 + 5x + 6$ and use the trace feature; x-intercepts are −3 and −2.

4. Graph $y = x^2 + 6x + 5$ and use the trace feature; x-intercepts are −5 and −1.

5. Graph $y = x^2 + x - 2$ and use the trace feature; x-intercepts are −2 and 1.

6. For each row, the values in the last two columns are the solutions of the quadratic with values of a, b, c.

7. They are formulas for finding the solutions to the quadratic equation $ax^2 + bx + c = 0$.

TRY THESE

1. $b^2 - 4ac = (4)^2 - 4(2)(9) = 16 - 72 = -56$.

2. $b^2 - 4ac = (-6)^2 - 4(-1)(12) = 36 + 48 = 84$.

3. $b^2 - 4ac = (16)^2 - 4(16)(4) = 256 - 256 = 0$.

4. $(-4)^2 - 4(1)(-5) = 16 + 20 = 36 > 0$; so, 2 solutions

5. $(3)^2 - 4(1)(52) = 9 - 208 = -199 < 0$; so, no solutions

6. $(4)^2 - 4(1)(4) = 16 - 16 = 0$; so, 1 solution

7. $(6)^2 - 4(9)(1) = 36 - 36 = 0$; so, 1 solution

8. $(-2)^2 - 4(6)(-8) = 4 + 192 = 196 > 0$; so, 2 solutions

9. $(2)^2 - 4(2)(4) = 4 - 32 = -28 < 0$; So, no solutions

10. $x = \dfrac{9 \pm \sqrt{(-9)^2 - 4(1)(8)}}{2(1)} = \dfrac{9 \pm \sqrt{49}}{2} = \dfrac{9 \pm 7}{2} = 1$ or 8

11. $x = \dfrac{7 \pm \sqrt{(-7)^2 - 4(1)(-8)}}{2(1)} = \dfrac{7 \pm \sqrt{81}}{2} = \dfrac{7 \pm 9}{2} = -1$ or 8

12. $x = \dfrac{6 \pm \sqrt{(-6)^2 - 4(1)(8)}}{2(1)} = \dfrac{6 \pm \sqrt{4}}{2} = \dfrac{6 \pm 2}{2} = 2$ or 4

13. $x = \dfrac{-3 \pm \sqrt{(3)^2 - 4(1)(-10)}}{2(1)} = \dfrac{-3 \pm \sqrt{49}}{2} =$

$\dfrac{-3 \pm 7}{2} = -5$ or 2

14. $x = \dfrac{-3 \pm \sqrt{(3)^2 - 4(4)(-5)}}{2(4)} = \dfrac{-3 \pm \sqrt{89}}{8} \approx -1.55$ or 0.80

15. $x = \dfrac{-4 \pm \sqrt{(4)^2 - 4(2)(-7)}}{2(2)} = \dfrac{-4 \pm \sqrt{72}}{4} \approx -3.12 \text{ or } 1.12$

16. $x = \dfrac{2 \pm \sqrt{(-2)^2 - 4(3)(-2)}}{2(3)} = \dfrac{2 \pm \sqrt{28}}{6} \approx -0.55 \text{ or } 1.22$

17. $x = \dfrac{1 \pm \sqrt{(-1)^2 - 4(5)(-3)}}{2(5)} = \dfrac{1 \pm \sqrt{61}}{10} \approx -0.68 \text{ or } 0.88$

18. $x = \dfrac{6 \pm \sqrt{(-6)^2 - 4(2)(3)}}{2(2)} = \dfrac{6 \pm \sqrt{12}}{4} \approx 0.63 \text{ or } 2.37$

19. $b^2 - 4ac = 0$, so,

$$(12)^2 - 4(1)(k) = 0$$
$$144 - 4k = 0$$
$$144 = 4k$$
$$36 = k$$

20. $b^2 - 4ac = 0$, so,

$$(12)^2 - 4(k)(9) = 0$$
$$144 - 36k = 0$$
$$144 = 36k$$
$$4 = k$$

$k = 0$ also works, making the equation linear.

21. $b^2 - 4ac = 0$, so,

$$k^2 - 4(16)(1) = 0$$
$$k^2 - 64 = 0$$
$$k^2 = 64$$
$$k = \pm 8$$

22. $16t^2 + 300t + 60 = 20$

$$-16t^2 + 300t + 40 = 0$$

$$t = \dfrac{-300 \pm \sqrt{300^2 - 4(-16)(40)}}{2(-16)} =$$

$$\dfrac{-300 \pm \sqrt{92,560}}{-32} \approx 19 \text{ sec}$$

23. Answers will vary, but should include: Positive discriminant indicates 2 solutions, zero discriminant indicates 1 solution, and negative discriminant indicates no solutions.

PRACTICE

1. $b^2 - 4ac = (3)^2 - 4(3)(8) = 9 - 96 = -87$

2. $b^2 - 4ac = (-8)^2 - 4(-1)(6) = 64 + 24 = 88$

3. $b^2 - 4ac = (20)^2 - 4(25)(4) = 400 - 400 = 0$

4. $(1)^2 - 4(1)(1) = -3 < 0$, so no real solutions

5. $(4)^2 - 4(1)(-6) = 16 + 24 = 40 > 0$, so 2 real solutions

6. $(10)^2 - 4(1)(25) = 100 - 100 = 0$, so 1 real solution

7. $(12)^2 - 4(36)(1) = 144 - 144 = 0$, so 1 real solution

8. $(-3)^2 - 4(6)(-5) = 9 + 120 = 129 > 0$, so 2 real solutions

9. $(3)^2 - 4(2)(6) = 9 - 48 = -39 < 0$, so no real solutions

10. $x = \dfrac{10 \pm \sqrt{(-10)^2 - 4(1)(9)}}{2(1)} = \dfrac{10 \pm \sqrt{64}}{2} = \dfrac{10 \pm 8}{2} = 1 \text{ or } 9$

11. $x = \dfrac{6 \pm \sqrt{(-6)^2 - 4(1)(-7)}}{2(1)} = \dfrac{6 \pm \sqrt{64}}{2} = \dfrac{6 \pm 8}{2} = -1 \text{ or } 7$

12. $x = \dfrac{9 \pm \sqrt{(-9)^2 - 4(1)(18)}}{2(1)} = \dfrac{9 \pm \sqrt{9}}{2} = \dfrac{9 \pm 3}{2} = 3 \text{ or } 6$

13. $x = \dfrac{11 \pm \sqrt{(-11)^2 - 4(1)(30)}}{2(1)} = \dfrac{11 \pm \sqrt{1}}{2} = \dfrac{11 \pm 1}{2} = 5 \text{ or } 6$

14. $x = \dfrac{-2 \pm \sqrt{2^2 - 4(1)(-35)}}{2(1)} = \dfrac{-2 \pm \sqrt{144}}{2} =$

$$\dfrac{-2 \pm 12}{2} = -7 \text{ or } 5$$

15. $x = \dfrac{7 \pm \sqrt{(-7)^2 - 4(1)(-30)}}{2(1)} = \dfrac{7 \pm \sqrt{169}}{2} =$

$$\dfrac{7 \pm 13}{2} = -3 \text{ or } 10$$

16. $x = \dfrac{4 \pm \sqrt{(-4)^2 - 4(1)(-45)}}{2(1)} = \dfrac{4 \pm \sqrt{196}}{2} =$

$$\dfrac{4 \pm 14}{2} = -5 \text{ or } 9$$

17. $x = \dfrac{-4 \pm \sqrt{4^2 - 4(1)(-21)}}{2(1)} = \dfrac{-4 \pm \sqrt{100}}{2} =$

$$\dfrac{-4 \pm 10}{2} = -7 \text{ or } 3$$

18. $x = \dfrac{-1 \pm \sqrt{1^2 - 4(6)(-35)}}{2(6)} = \dfrac{-1 \pm \sqrt{841}}{12} =$

$$\dfrac{-1 \pm 29}{12} = -\dfrac{5}{2} \text{ or } \dfrac{7}{3}$$

19. $10,000 = -16p^2 + 320p + 20,000$

$$0 = -16p^2 + 320p + 10,000$$

$$p = \dfrac{-320 \pm \sqrt{320^2 - 4(-16)(10,000)}}{2(-16)} =$$

$$\dfrac{-320 \pm \sqrt{742,400}}{-32} \approx \$37$$

20. $x = \dfrac{3 \pm \sqrt{(-3)^2 - 4(10)(-1)}}{2(10)} = \dfrac{3 \pm \sqrt{49}}{20} =$

$$\dfrac{3 \pm 7}{20} = -\dfrac{1}{5} \text{ or } \dfrac{1}{2}$$

21. $x = \dfrac{3 \pm \sqrt{(-3)^2 - 4(14)(-5)}}{2(14)} = \dfrac{3 \pm \sqrt{289}}{28} =$

$$\dfrac{3 \pm 17}{28} = -\dfrac{1}{2} \text{ or } \dfrac{5}{7}$$

22. $x = \dfrac{4 \pm \sqrt{(-4)^2 - 4(20)(-24)}}{2(20)} = \dfrac{4 \pm \sqrt{1936}}{40} =$

$\dfrac{4 \pm 44}{40} = -1$ or $\dfrac{6}{5}$

23. $x = \dfrac{-5 \pm \sqrt{5^2 - 4(1)(-2)}}{2(1)} = \dfrac{-5 \pm \sqrt{33}}{2} \approx -5.37$ or 0.37

24. $x = \dfrac{-4 \pm \sqrt{4^2 - 4(1)(-1)}}{2(1)} = \dfrac{-4 \pm \sqrt{20}}{2} \approx -4.24$ or 0.24

25. $x = \dfrac{4 \pm \sqrt{(-4)^2 - 4(2)(-12)}}{2(2)} = \dfrac{4 \pm \sqrt{112}}{4} \approx -1.65$ or 3.65

26. $x = \dfrac{6 \pm \sqrt{(-6)^2 - 4(6)(-7)}}{2(6)} = \dfrac{6 \pm \sqrt{204}}{12} \approx -0.69$ or 1.69

27. $x = \dfrac{-10 \pm \sqrt{10^2 - 4(3)(4)}}{2(3)} = \dfrac{-10 \pm \sqrt{52}}{6} \approx$

-2.87 or -0.46

28. $x = \dfrac{-12 \pm \sqrt{12^2 - 4(5)(-2)}}{2(5)} = \dfrac{-12 \pm \sqrt{184}}{10} \approx$

-2.56 or -0.16

29. $b^2 - 4ac = 14^2 - 4(1)(k) = 0$

$196 - 4k = 0$

$196 = 4k$

$49 = k$

30. $b^2 - 4ac = (-30)^2 - 4(k)(25) = 0$

$900 - 100k = 0$

$900 = 100k$

$9 = k$ or $k = 0$

31. $b^2 - 4ac = (k)^2 - 4(25)(16) = 0$ For 1 solution.

$k^2 - 1600 = 0$

$k^2 = 1600$

$k = \pm 40$

32. $x^2 = 2x + 80$

$x^2 - 2x - 80 = 0$

$x = \dfrac{2 \pm \sqrt{(-2)^2 - 4(1)(-80)}}{2(1)} =$

$\dfrac{2 \pm \sqrt{324}}{2} = \dfrac{2 \pm 18}{2} = -8$ or 10

33. It is a parabola with no x-intercepts.

34. $x^2 - [5 + (-6)]x + (5)(-6) = 0$

$x^2 + x - 30 = 0$

35. $x^2 - [7 + (-3)]x + (7)(-3) = 0$

$x^2 - 4x - 21 = 0$

36. $x^2 - \left[\dfrac{1}{2} + \left(-\dfrac{1}{4}\right)\right]x + \left(\dfrac{1}{2}\right)\left(-\dfrac{1}{4}\right) = 0$

$x^2 - \dfrac{1}{4}x - \dfrac{1}{8} = 0$ or $8x^2 - 2x - 1 = 0$

37. $x^2 - \left[7 + \left(-\dfrac{1}{2}\right)\right]x + (7)\left(-\dfrac{1}{2}\right) = 0$

$x^2 - \dfrac{13}{2}x - \dfrac{7}{2} = 0$ or $2x^2 - 13x - 7 = 0$

38. $x^2 - [10.5 + (-10.5)]x + (10.5)(-10.5) = 0$

$x^2 - 110.25 = 0$

39. $x^2 - [c + (-c)]x + (c)(-c) = 0$

$x^2 - c^2 = 0$

40. $x^2 - 30x + 216 = 0$

$x = \dfrac{30 \pm \sqrt{(-30)^2 - 4(1)(216)}}{2(1)}$

$= \dfrac{30 \pm \sqrt{36}}{2} = \dfrac{30 \pm 6}{2} = 12$ or 18

41. $x^2 + 37x - 650 = 0$

$x = \dfrac{-37 \pm \sqrt{37^2 - 4(1)(-650)}}{2(1)} =$

$= \dfrac{-37 \pm \sqrt{3969}}{2} = \dfrac{-37 \pm 63}{2} = -50$ or 13

42. $x^2 + 26x + 153 = 0$

$x = \dfrac{-26 \pm \sqrt{26^2 - 4(1)(153)}}{2(1)}$

$= \dfrac{-26 \pm \sqrt{64}}{2} = \dfrac{-26 \pm 8}{2} = -17$ or -9

43. $b^2 - 4ac = 6^2 - 4(1)(k) > 0$

$36 - 4k > 0$

$-4k > -36$

$k < 9$

44. $b^2 - 4ac = (-5)^2 - 4(-1)(k) < 0$; For no solutions.

$25 + 4k < 0$

$4k < -25$

$k < -6.25$

45. Equation is quadratic for variable $(x - a)$, so:

$x - a = \dfrac{1 \pm \sqrt{(-1)^2 - 4(1)(-5)}}{2(1)}$

$x - a = \dfrac{1 \pm \sqrt{21}}{2}$

$x = a + \dfrac{1 \pm \sqrt{21}}{2} = \dfrac{2a + 1 \pm \sqrt{21}}{2}$

46. Equation is quadratic for variable $(x - a)$, so:

$$x - a = \frac{1 \pm \sqrt{(-1)^2 - 4(1)(-6)}}{2(1)}$$

$$x - a = \frac{1 \pm \sqrt{25}}{2}$$

$$x - a = -2 \text{ or } 3$$

$$x = a - 2 \text{ or } a + 3$$

47. Since $b^2 - 4ac$ is a perfect square, solutions will be rational.

48. Set $2x^2 - 7x - 4 = 0$, where it crosses the x-axis.

$$x = \frac{7 \pm \sqrt{(-7)^2 - 4(2)(-4)}}{2(2)} = \frac{7 \pm \sqrt{81}}{4} = \frac{7 \pm 9}{4} = -\frac{1}{2} \text{ or } 4$$

Graph crosses x-axis at $-\frac{1}{2}$ and 4.

49. $\bar{x} = \frac{18 + 44 + 76 + 6 + 16}{5} = \frac{160}{5} = 32$; C

50. Function, since each x has a unique y value

51. Not a function, since 2 is paired with both 5 and 1

52. $4 - 3x = 21$ or $4 - 3x = -21$

53. $2x - \frac{1}{2} = \frac{1}{4}$ or $2x - \frac{1}{2} = -\frac{1}{4}$

54. $x = \frac{37 \pm \sqrt{(-37)^2 - 4(2)(105)}}{2(2)} = \frac{37 \pm \sqrt{529}}{4} =$

$$\frac{37 \pm 23}{4} = \frac{7}{2} \text{ or } 15$$

55. $x = \frac{-2 \pm \sqrt{(2)^2 - 4(12)(-2)}}{2(12)} = \frac{-2 \pm \sqrt{100}}{24} =$

$$\frac{-2 \pm 10}{24} = -\frac{1}{2} \text{ or } \frac{1}{3}$$

ALGEBRA WORKS

1.

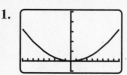

2. $C = 2\pi r$ and $r = 9$, so $C = 2\pi(9) = 18\pi$ ft

3. $A = \pi r^2 = \pi(9)^2 = 81\pi$ ft^2

4. $f(x) = 0.04\,(9)^2 = 3.24$ ft

5. Greater depth with $g(x) = 0.05x^2$

Lesson 10.6, pages 498–501

EXPLORE

1. Perimeter is 200 ft $= 2l + 2w = 2l + 20$
  $180 = 2l$, so $l = 90$ ft
  Area $= l \cdot w = 90 \cdot 10 = 900$ ft^2

2. $200 = 2l + 2w = 2l + 40$
  $160 = 2l$, so $l = 80$
  Area $= lw = 80 \cdot 20 = 1600$ ft^2
  Yes, dimensions have effect on area.

3. $200 = 2l + 2w$ or $100 = l + w$

4.

l	w	Area	l	w	Area
5	95	475	55	45	2475
10	90	900	60	40	2400
15	85	1275	65	35	2275
10	80	1600	70	30	2100
25	75	1875	75	25	1875
30	70	2100	80	20	1600
35	65	2275	85	15	1275
40	60	2400	90	10	900
45	55	2475	95	5	475
50	50	2500	99	1	99

5. Areas increase as width goes from 5 to 50 ft.

6. For each rectangle with a width greater than 50, there is an equal area rectangle with width less than 50.

7. a square with area of 2500 ft^2

8.

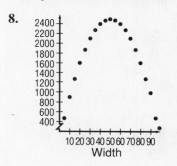

9. Maximum area is with a square, so $\frac{350}{4} = 87.5$ ft sides with area of $(87.5)^2 = 7656.25$ ft^2

10. It is a quadratic function.

11.

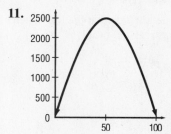

The graph of $-w^2 + 100w$ is the graph of w^2 reflected over the x-axis, and translated right 50 units and up 2500 units.

12. $w = \frac{-b}{2a} = \frac{-100}{2(-1)} = 50$, $A(50) = -50^2 + 100(50) = 2500$; vertex is (50, 2500); agrees with table

13a. $120 = l + 2w$, so $120 - 2w = l$

  b. $A = lw = (120 - 2w)w = 120w - 2w^2$

  c. Vertex at $w = \frac{-b}{2a} = \frac{-120}{2(-2)} = 30$
  So, $l = 120 - 2(30) = 60$ and $A = 30(60) = 1800$ ft^2

  d. Divide 120 ft of fence by 3, each side is 40 ft, area $= 40(40) = 1600$ ft^2

14a. $120 = l + 3w$, so $120 - 3w = l$

$A = lw = (120 - 3w)w = 120w - 3w^2$

b. Vertex at $w = \dfrac{-b}{2a} = \dfrac{-120}{2(-3)} = 20$,

so $l = 120 - 3(20) = 60$

$A = 20 \cdot 60 = 1200 \ \text{ft}^2$

For each section to have equal area, make two 20 x 30 rectangles.

c. Divide 120 ft of fence by 5, so each square is 24 x 24, two squares with an area of $24 \cdot 24 = 576 \ \text{ft}^2$ each. Dimensions of garden is 24 x 48.

d. Divide 120 ft of fence by 4, so each section is 30 x 15, for a total dimension of 30 by 30. Each section's area is $30(15) = 450 \ \text{ft}^2$

15a. In previous problems, perimeter was constant. Now area is constant.

b. Yes; the amount of fencing will change.
for a 4 x 100 rectangle, $P = 208$;
for a 5 x 80 rectangle, $P = 170$;
for an 8 x 50 rectangle, $P = 116$;
for a 10 x 40 rectangle, $P = 100$;
for a 20 x 20 rectangle, $P = 80$.

c. Tables will vary with students choice of dimension, but the most efficient shape is a 20 ft square.

d. $A = 400 = lw$, so, $\dfrac{400}{w} = l$

e. $P = 2l + 2w = 2\left(\dfrac{400}{w}\right) + 2w = \dfrac{800}{w} + 2w$

Descriptions of graphs will vary.

16. Answers will vary, but may include that diagrams help in understanding and analyzing the problem.

REVIEW PROBLEM SOLVING STRATEGIES

1a. 70 in.

b. $70 - x$

c. $2(70 - x)$

d. no; $70 + 2$ to represent the final depth of the well

e. 17.5 in.; 52.5 in.; 105 in. or 8 ft 8 in.

2. The juice from the small glass was $\dfrac{1}{6}$ of the total contents of the pitcher, and the juice from the larger glass was $\dfrac{2}{9}$ of the total. Add: $\dfrac{7}{18}$ of the total was juice, so $\dfrac{11}{18}$ was water.

3. 4 marks are needed; one possible set includes marks at 2, 5, 8, and 11.

Lesson 10.07, pages 502–507

TRY THESE

1. quadratic

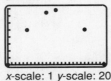

x-scale: 1 *y*-scale: 20

2. linear

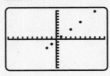

3. neither

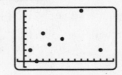

4. quadratic

5. quadratic

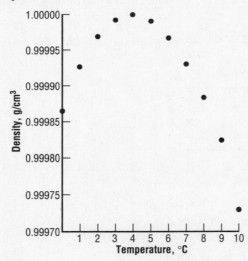

6. Density increases from 0°C to 4°C, to a maximum of $\dfrac{1 \text{g}}{\text{cm}^3}$.

7. $y = -0.000008(4)^2 + 0.000065(4) + 0.999867$

$= -0.000128 + 0.00026 + 0.999867$

$= 0.999999$ for error of $\dfrac{0.000001}{1}$ or 0.0001%

8. $y = -0.000008(10)^2 + 0.000065(10) + 0.999867$

$\quad = -0.0008 + 0.00065 + 0.999867$

$\quad = 0.999717 \dfrac{\text{g}}{\text{cm}^3}$

PRACTICE

1. Answers will vary; possible answers:

A: quadratic   B: linear   C: linear

D: quadratic   E: quadratic

2. $y = 0.006x^2 + 0.456x + 74.121$

3. $y = 0.006(0)^2 + 0.456(0) + 74.121 = 74.121$ for $0°$

$y = 0.006(100)^2 + 0.456(100) + 74.121 = 179.21$ for $100°$

4. For $0°$: $\left|\dfrac{73 - 74.121}{73}\right| = \left|\dfrac{-1.121}{73}\right| \approx 0.015$ or 1.5%

For $100°$: $\left|\dfrac{180 - 179.72}{180}\right| = \dfrac{0.279}{180} \approx 0.0016$ or 0.16%

Model seems acceptable, but answers will vary.

5. $y = 0.783x + 128.857$ (used Linreg on calculator)

6. For $60°$ $y = 0.783(60) + 128.857 = 175.837$ is predicted value, actual is 176.

Percent error is:

$\dfrac{176 - 175.837}{176} = \dfrac{0.163}{176} = 0.00093$ or 0.093% error

Linear model looks like a good fit.

7. Answers will vary, but may include computing several predicted values for both models and determining which has the lowest percent error.

8. $179 = a(0)^2 + b(0) + c \quad \rightarrow \quad 179 = c$

$238 = a(40)^2 + b(40) + c \quad \rightarrow \quad 238 = 1600a + 40b + c$

$362 = a(80)^2 + b(80) + c \quad \rightarrow \quad 362 = 6400a + 80b + c$

9. $238 = 1600a + 40b + 179 \quad \rightarrow \quad 118 = 3200a + 80b$

$362 = 6400 + 80b + 179 \quad \rightarrow \quad \underline{183 = 6400a + 80b}$

$\qquad\qquad\qquad\qquad\qquad\qquad -65 = -3200a$

$\qquad a = \dfrac{65}{3200} = 0.203125$

$\qquad 238 = 1600(0.0203125) + 40b + 179$

$\qquad 26.5 = 40b$

$\qquad 0.6625 = b$

10. $y = 0.0203125x^2 + 0.6625x + 179$

11. $y = 0.0203125(60)^2 + 0.6625(60) + 179 = 291.875$ for $60°$

$y = 0.0203125(100)^2 + 0.6625(100) + 179 = 448.375$ for $100°$

Not a good fit for high temperatures

12. $y = 0.030x^2 - 0.017x + 185$

$y = 0.030(60)^2 - 0.017(60) + 185 = 291.98$ for $60°$

$y = 0.030(100)^2 - 0.017(100) + 185 = 483.3$ for $100°$

13. False, it could be exponential or no relationship

14. True, but one is probably better than the other

15. associative property; D

16. $-2m + 3 > -5$

$\quad -2m > -8 \qquad$ Subtract 3.

$\quad\quad m < 4 \qquad$ Divide by -2 and reverse the inequality symbol.

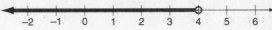

17. $m = \dfrac{-9 - (-12)}{3 - 4} = \dfrac{3}{-1} = -3$

18. $m = \dfrac{6 - (-4)}{9 - 4} = \dfrac{10}{5} = 2$

19. $25x - 10y = 51 \quad \rightarrow \quad 75x - 30y = 153$

$5x + 15y = 60 \quad \rightarrow \quad \underline{10x + 30y = 120}$

$\qquad\qquad\qquad\qquad\qquad 85x = 273$ Add equations.

$\qquad\qquad\qquad\qquad\qquadx \approx 3.2$

$\qquad 5(3.2) + 15y = 60$

$\qquad\qquad 15y = 44$

$\qquad\qquad\qquad y \approx 2.9$ So, the solution is $(3.2, 2.9)$.

20. $\quad m - n = 16$

$\quad \underline{-m - n = 0}$

$\qquad -2n = 16$

$\qquadn = -8$ So, $m = 8$ and solution is $(8, -8)$.

21. Vertex at $x = \dfrac{-b}{2a} = \dfrac{-6}{2(1)} = -3$;

$y = (-3)^2 + 6(-3) + 8 = -1; (-3, -1)$

22. Axis of symmetry: $x = -3$.

Chapter Review, pages 508–509

1. b   **2.** d   **3.** c   **4.** a

5. Vertical shift of 4; c

6. Since $a > 0$, graph is narrow; a

7. Horizontal shift of 4; b

8. Vertex at $x = \dfrac{-b}{2a} = \dfrac{-0}{2(1)} = 0$, $y = 0^2 + 5$, so vertex is $(0, 5)$; axis of symmetry: $x = 0$

9. Vertex at $x = \dfrac{-b}{2a} = \dfrac{4}{2(2)} = 1$, $y = 2(1)^2 - 4(1) + 11 = 9$, so vertex is $(1, 9)$; axis of symmetry: $x = 1$

10. Vertex at $x = \dfrac{-b}{2a} = \dfrac{-6}{2(1)} = -3$, $y = (-3)^2 + 6(-3) - 1 = -10$, so vertex is $(-3, -10)$; axis of symmetry: $x = -3$

11. Graph equation and use the trace feature; intercepts are $x = 1$ and $x = 2$.

12. Graph equation and use the trace feature; intercepts are $x = -4$ and $x = 5$.

13. Graph equation and use the trace feature; intercepts are $x = 1.32$ and $x = -5.32$.

14. $x^2 - 49 = 0$

$\qquad x^2 = 49$

$\qquad x = \pm\sqrt{49} = \pm 7$

15. $2x^2 = 64$

$\qquad x^2 = 32$

$\qquad x = \pm\sqrt{32} = \pm\sqrt{16 \cdot 2} = \pm 4\sqrt{2}$

16. $3x^2 - 1 = 47$

$\qquad 3x^2 = 48$

$\qquad x^2 = 16$

$\qquad x = \pm 4$

17. $x = \dfrac{-7 \pm \sqrt{7^2 - 4(1)(10)}}{2(1)} = \dfrac{-7 \pm \sqrt{9}}{2} = \dfrac{-7 \pm 3}{2} = -5 \text{ or } -2$

18. $x = \dfrac{12 \pm \sqrt{(-12)^2 - 4(4)(9)}}{2(4)} = \dfrac{12 \pm \sqrt{0}}{8} = 1.5$

19. $x = \dfrac{1 \pm \sqrt{(-1)^2 - 4(1)(-5)}}{2(1)} = \dfrac{1 \pm \sqrt{21}}{2} \approx -1.79 \text{ and } 2.79$

20. $b^2 - 4ac = (-4)^2 - 4(1)(-1) = 20 > 0$, so 2 solutions

21. $b^2 - 4ac = (3)^2 - 4(1)(6) = -15 < 0$, so no solutions

22. $b^2 - 4ac = (7)^2 - 4(2)(-9) = 121 > 0$, so 2 solutions

23. $P = 12 = 2l + 2w$ or $6 = l + w$ or $6 - w = l$

$\quad$ Area $= A(w) = (6 - w)w = 6w - w^2$

$\quad w = \dfrac{-b}{2a} = \dfrac{-6}{2(-1)} = 3$ and $l = 6 - 3 = 3$,

$\quad$ Dimensions are $3y \times 3y$.

24. $y = 0.286(2)^2 - 2.857(2) + 0.071 = -4.499$

$\quad$ Percent error $= \dfrac{-5 - (-4.499)}{-5} = 0.100$ or 10%

Chapter Test, pages 510–511

1. $x = \dfrac{-b}{2a} = \dfrac{4}{2(1)} = 2,\quad y = 2^2 - 4(2) + 1 = -3$,

$\quad$ so vertex is $(2, -3)$; axis of symmetry: $x = 2$

2. $x = \dfrac{-b}{2a} = \dfrac{-8}{2(2)} = -2,\quad y = 2(-2)^2 + 8(-2) - 5 = -13$,

$\quad$ so vertex is $(-2, -13)$; axis of symmetry: $x = -2$

3. $x = \dfrac{-b}{2a} = \dfrac{8}{2(1)} = 4,\quad y = 4^2 - 8(4) + 15 = -1$,

$\quad$ so vertex is $(4, -1)$; axis of symmetry: $x = 4$

4. $x = \dfrac{-b}{2a} = \dfrac{-6}{2(1)} = -3,\quad y = (-3)^2 + 6(-3) + 2 = -7$,

$\quad$ so vertex is $(-3, -7)$; axis of symmetry: $x = -3$

5. $x = \dfrac{3 \pm \sqrt{(-3)^2 - 4(1)(2)}}{2(1)} = \dfrac{3 \pm \sqrt{1}}{2} = 1 \text{ or } 2$

6. $x = \dfrac{8 \pm \sqrt{(-8)^2 - 4(1)(16)}}{2(1)} = \dfrac{8 \pm 0}{2} = 4$

7. $x^2 - 70 = 11$

$\qquad x^2 = 81$

$\qquad x = \pm 9$

8. $x = \dfrac{12 \pm \sqrt{(-12)^2 - 4(3)(4)}}{2(3)} = \dfrac{12 \pm \sqrt{96}}{6} = \dfrac{12 \pm 4\sqrt{6}}{6} = \dfrac{6 \pm 2\sqrt{6}}{3}$

9. $x = \dfrac{2 \pm \sqrt{(-2)^2 - 4(1)(-48)}}{2(1)} = \dfrac{2 \pm \sqrt{196}}{2} = -6 \text{ and } 8$

10. $x = \dfrac{-6 \pm \sqrt{6^2 - 4(1)(-9)}}{2(1)} = \dfrac{-6 \pm \sqrt{72}}{2} = -3 \pm 3\sqrt{2}$

11. $x = \dfrac{-6 \pm \sqrt{6^2 - 4(1)(-6)}}{2(1)} = \dfrac{-6 \pm \sqrt{60}}{2} = -3 \pm \sqrt{15}$

12. $x = \dfrac{14 \pm \sqrt{(-14)^2 - 4(1)(49)}}{2(1)} = \dfrac{14 \pm \sqrt{0}}{2} = 7$

13. $x^2 + 6 = 48$

$\qquad x^2 = 42$

$\qquad x^2 = \pm\sqrt{42}$

14. $x = \dfrac{2 \pm \sqrt{(-2)^2 - 4(1)(-2)}}{2(1)} = \dfrac{2 \pm \sqrt{12}}{2} = 1 \pm \sqrt{3}$

15. Graph is shifted down 4 units; B

16. $x = \dfrac{7 \pm \sqrt{(-7)^2 - 4(1)(12)}}{2(1)} = \dfrac{7 \pm \sqrt{1}}{2} = 3 \text{ or } 4$

17. $x = \dfrac{-4 \pm \sqrt{4^2 - 4(1)(-14)}}{2(1)} = \dfrac{-4 \pm \sqrt{72}}{2} = -2 \pm 3\sqrt{2} \approx$

$\quad -6.24 \text{ or } 2.24$

18. $x = \dfrac{-1 \pm \sqrt{1^2 - 4(2)(-10)}}{2(2)} = \dfrac{-1 \pm \sqrt{81}}{4} = -2.5 \text{ or } 2$

19. $x = \dfrac{2 \pm \sqrt{(-2)^2 - 4(3)(-16)}}{2(3)} = \dfrac{2 \pm \sqrt{196}}{6} =$

$\quad -2 \text{ or } 2.67$

20. $x^2 = 11$

$\qquad x = \pm\sqrt{11} \approx \pm 3.32$

21. $2x^2 - 108 = 20$

$\qquad 2x^2 = 128$

$\qquad x^2 = 64$

$\qquad x = \pm 8$

22. A: $x^2 - 3x = 4 \rightarrow x^2 - 3x - 4 = 0$

$\quad$ B: $x^2 - 3x + 1 = -3 \rightarrow x^2 - 3x + 4 = 0$

$\quad$ C: $x^2 - 4 = 3x \rightarrow x^2 - 3x - 4 = 0$

$\quad$ D: $x^2 - 3x - 2 = 2 \rightarrow x^2 - 3x - 4 = 0$

23. Answers will vary, one possibility is to write equation as $ax^2 + bx + (c - d) = 0$, and use the quadratic formula.

24. $b^2 - 4ac = (-10)^2 - 4(1)(25) = 0$, so 1 solution

25. $b^2 - 4ac = 5^2 - 4(1)(-12) = 73 > 0$, so 2 solutions

26. $b^2 - 4ac = (-1)^2 - 4(2)(14) = -111 < 0$, so no solutions

27. $b^2 - 4ac = 3^2 - 4(1)(-12) = 57 > 0$, so 2 solutions

28. $x = \dfrac{-b}{2a} = \dfrac{-250}{2(-5)} = 25 = \$25$

29. $A = lw$, $l + 2w = 50$

So, $l = 50 - 2w$ and $A = (50 - 2w)w = 50w - 2w^2$

Maximum is at vertex: $w = \dfrac{-b}{2a} = \dfrac{-50}{2(-2)} = 12.5$

So, $l = 50 - 2(12.5) = 25$

The dimensions are 25 ft $\times$ 12.5 ft.

30. $x = \dfrac{90 \pm \sqrt{90^2 - 4(-16)(0)}}{2(-16)} = \dfrac{-90 \pm \sqrt{8100}}{-32} = 5.625$ sec

31a. $y = 0.198(5)^2 - 0.396(5) + 1.102 = 4.072$

b. $\left| \dfrac{4 - 4.072}{4} \right| = \left| \dfrac{0.072}{4} \right| = 0.018$ or 1.8%

Cumulative Review, page 512

1.

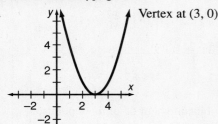

Vertex at (3, 0)

2.

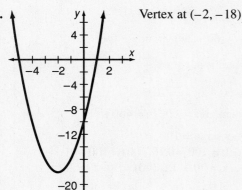

Vertex at (−2, −18)

3.

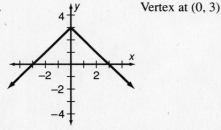

Vertex at (0, 3)

4.

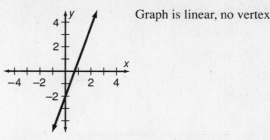

Graph is linear, no vertex

5. Graph $y = x^2 - 3x + 1$ and $y = 7$ on same axes. Use the trace feature to find intersection at $x = -1.37$ and 4.37.

6. Graph $y = 5x^2 + 2x - 3$ and $y = 4x - 1$ on same axes. Use the trace feature to find intersection at $x = -0.46$ and 0.86.

7. Answers will vary, may include 5.064, 5.065, 5.066

8. Answers will vary, may include $\dfrac{2}{3}, \dfrac{3}{5}, \dfrac{4}{7}$

9. $c + a = 20 \rightarrow c = 20 - a;$

$6.50c + 2.25a = 4.50(20)$

$6.50(20 - a) + 2.25a = 90$

$130 - 6.50a + 2.25a = 90$

$-4.25a = -40$

$a \approx 9.4$

So, $c = 20 - 9.4 = 10.6$

10. $-\sqrt{17}, -3.943, -3.94, -\dfrac{11}{3}$

11. $x(5) = 6$

$x = \dfrac{6}{5} = 1.20$ or 120%

12. $0.28(800) = 224$

13. $m = \dfrac{1 - (-4)}{3 - (-2)} = \dfrac{5}{5} = 1$ and $y - 1 = 1(x - 3)$

or $y = x - 2$

14. Same as y-coordinate, so horizontal line; $y = 2$

15. Parallel to x-axis is vertical line, $x = 5$

16. sleep   **17.** 2 hours

18. Circle graph; the data show 100% of the day. The number of hours can be converted to a percent of 24 h.

19. $3(x + 1)^2 = 24$

$(x + 1)^2 = 8$

$x + 1 = \pm\sqrt{8} = \pm 2\sqrt{2}$

$x = -1 \pm 2\sqrt{2}$

20. $9x^2 + 5 = 45$

$9x^2 = 40$

$x^2 = \dfrac{40}{9}$

$x = \pm\sqrt{\dfrac{40}{9}} = \pm\dfrac{2\sqrt{10}}{3}$

21. $26 - 7 \le n \le 26 + 7$

$\quad\quad 19 \le n \le 33 \text{ or } |n - 26| \le 7$

22. $5 \cdot 3 \cdot 6 = 90$

23. $3x - 5 > -2$

$\quad\quad 3x > 3$

$\quad\quad\quad x > 1$

24. $-3x \le 9 \quad$ and $\quad x + 2 < 5$

$\quad\quad x \ge -3 \quad$ and $\quad\quad x < 3$

$\quad\quad -3 \le x < 3$

Standardized Test, page 513

1. $-12 = -4x \quad \rightarrow \quad x = 3$

$\quad -3y = 15 \quad\quad \rightarrow \quad y = -5$

$\quad x + y = z \quad\quad \rightarrow \quad -2 = z$

$\quad w = 2z \quad\quad\quad \rightarrow \quad w = -4$

2. $5m + 8n = 13 \quad \rightarrow \quad 25m + 40n = \;\; 65$

$\quad 4m - 5n = 56 \quad \rightarrow \quad \dfrac{32m - 40n = 448}{57m \quad\quad\;\; = 513}$

$\quad\quad\quad\quad\quad\quad\quad\quad\quad\quad\quad m = 9$

So, $4(9) - 5n = 56 \rightarrow n = -4$

Product is $(9)(-4) = -36$

3. $-2 < x - 2 \quad$ and $\quad x - 2 < 2$

$\quad\quad 0 < x \quad\quad$ and $\quad\quad x < 4$

Integer solutions are 1, 2, 3; Product is $(1)(2)(3) = 6$.

4. $b^2 - 4ac = 8^2 - 4(1)(-5) = 64 + 20 = 84$

5. $f(-2) + g(4) = 4(-2) + 5 + 2(4)^2 - 7(4) + 3 =$
$-8 + 5 + 32 - 28 + 3 = 4$

6. $\dfrac{0 \cdot 6 + 1 \cdot 15 + 2 \cdot 25 + 3 \cdot 42 + 4 \cdot 12}{6 + 15 + 25 + 42 + 12} = \dfrac{239}{100} = 2.39$

7. $x = \dfrac{-10 \pm \sqrt{10^2 - 4(-3)(-5)}}{2(-3)} = \dfrac{-10 \pm \sqrt{40}}{-6} \approx$

$\quad 2.72$ or -0.89, greater solution is 2.72

8. $3.50 + 0.32(3.50) = 3.50 + 1.12 = 4.62$

9. $2x - 9y = 14$

$\quad\quad -9y = -2x + 14$

$\quad\quad\quad y = \dfrac{2}{9}x - \dfrac{14}{9},$

Slope is $\dfrac{2}{9}$; Perpendicular slope is $-\dfrac{9}{2}$.

10. $3 + (-4) \cdot 3 - (-4) = 3 + (-12) - (-4) = -5$

11. Total of 8 balls. $P(\text{first is blue}) = \dfrac{3}{8}$,

$\quad P(\text{second is also blue}) = \dfrac{2}{7}$,

$\quad P(\text{third is also blue}) = \dfrac{1}{6}$,

$\quad$ So, $P(\text{all 3 balls are blue}) = \dfrac{3}{8} \cdot \dfrac{2}{7} \cdot \dfrac{1}{6} = \dfrac{1}{56}$

12. $4n + \dfrac{1}{2}(5 - 2n) = 5(n + 2)$

$\quad\quad 4n + \dfrac{5}{2} - n = 5n + 10$

$\quad\quad\quad -\dfrac{15}{2} = 2n$

$\quad\quad\quad -\dfrac{15}{4} = n,$ So $|n| = \dfrac{15}{4}$

13. $\dfrac{31 + 27 + 35 + 24 + 14 + x + x}{7} \ge 28$

$\quad\quad\quad\quad\quad\quad 131 + 2x \ge 196$

$\quad\quad\quad\quad\quad\quad\quad\quad 2x \ge 65$

Need a combined total of 65 or more points.

14. $P = 60 = l + 2w$

So, $60 - 2w = l$.

$A = lw = (60 - 2w)w = 60w - 2w^2$

Maximum is at vertex: $\dfrac{-b}{2a} = \dfrac{-60}{2(-2)} = 15 = w$

So, $l = 60 - 2(15) = 30$

Maximum area $= 30(15) = 450 \text{ ft}^2$

15. Two weeks ago $= 100$
Last week $= 100 - 0.10(100) = 100 - 10 = 90$
This week $= 90 + 0.10(90) = 90 + 9 = 99$

Chapter 11   Polynomials and Exponents

Data Activity, page 515

1. $28.95 - 22.55 = 6.4$ m   **2.** $\dfrac{6.67}{2.83} \approx 2.4$

3. $\dfrac{0.5}{2} = \dfrac{19.66}{x}$ **4.** $22.55\,\text{m} \cdot \dfrac{3.28\,\text{ft}}{\text{m}} \approx 74.0\,\text{ft}$

    $0.5x = 39.32$

       $x = 78.6$ cm

5. $7623\,\text{kg} \cdot \dfrac{2.2\,\text{lb}}{\text{kg}} \approx 16{,}770$ lbs, so closer to 16,800 lbs

6. Answers will vary.

Lesson 11.1, pages 517–520

THINK BACK

1. $-2(3)$

2. -6

3. Models will vary; Possible answer: $-2(-3) = 6$

EXPLORE

4. $2(x)$

5. $2x$

6. $-3(x) = -3x$ **7.** $2x(-3) = -6x$

8. $-2x(-2) = 4x$ **9.** $-2y(x) = -2xy$

10. 3^2 **11.** 2^3

12. Nine units were needed for 3^2, but only eight for 2^3, so $3^2 > 2^3$

13. two x^2 blocks in Quadrant I, so product is $2x^2$

14. six y^2 blocks in Quadrant IV, so product is $-6y^2$

15. four xy blocks in Quadrant III, so product is $4xy$

16. one xy block in Quadrant IV, so product is $-xy$

17. Place three x-blocks on the negative part of the horizontal axis and two x-blocks on the negative part of the vertical axis. Form a rectangular area using six x^2-blocks in the quadrant bounded by the blocks. Read the answer, $6x^2$, from the mat.

18. $(2x)(5x) = 10x^2$ **19.** $(-3x)(3x) = -9x^2$

20. $(-4y)(2y) = -8y^2$ **21.** $(3x)(-2y) = -6xy$

22. $(-x)(2y) = -2xy$ **23.** $(-y)(-2y) = 2y^2$

24. Correctly place the blocks that represent the length of the rectangle $-2y$.

25. Place two x-blocks on the positive side of the quadrant bordering the divisor. Complete the vertical axis for the rectangle. Read the answer, x, from the vertical axis.

26.

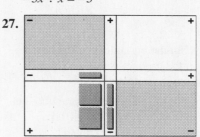

$-3x \div x = -3$

27.

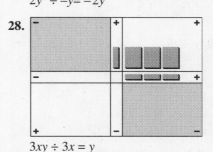

$2y^2 \div -y = -2y$

28.

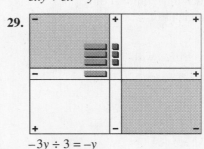

$3xy \div 3x = y$

29.

$-3y \div 3 = -y$

30. $\dfrac{-6y}{3y} = -2$ **31.** $\dfrac{2xy}{x} = 2y$ **32.** $\dfrac{-4x^2}{-2x} = 2x$

MAKE CONNECTIONS

33. square; 2 dimensions of length and width; product is x^2

34. cube; 3 dimensions of length; width and height; product is x^3

35. no, cannot model 4th dimension effectively

36. Write each factor of $2x \cdot 3x^3$. You get $2 \cdot x \cdot 3 \cdot x \cdot x \cdot x$. When you use the commutative and associative properties, you see that the product is $6x^4$. That is, multiply the numerals multiply the variables, then multiply these together.

37. Write each factor of the numerator and denominator of the fraction $\frac{5x^6}{10x}$. You get $\frac{5 \cdot x \cdot x \cdot x \cdot x \cdot x \cdot x}{10 \cdot x}$. Divide the numerator and denominator by the common factor $5x$; $\frac{x^5}{2}$. That is, divide the numerals and the variables to get the final quotient.

SUMMARIZE
38. $x(-3x) = -3x^2$

39. $2x^2 \div (-x) = -2x$ or $2x^2 \div (-2x) = -x$

40. Answers will vary, but the variable is in the product and the exponent in the product is the sum of the exponents of the factors.

41. Answers will vary, but the variable is in the quotient and the exponent in the quotient is the difference of the dividend exponent and the divisor exponent.

42. $V = 2x \cdot 2 \cdot 3y = 12xy$

Lesson 11.2, pages 521–526

EXPLORE
1. One fold produces 2 regions.

2.

Folds	1	2	3	4	5
Regions	2	4	8	16	32

3. Regions double each time; number of regions equals 2^f, where f is the number of folds.

4. $2^{10} = 1024$

5. Number of regions triples with each fold.

TRY THESE
1. 10   **2.** 2

3. 1   **4.** $1 + 3 + 2 = 6$

5. $3a(-2a^3) = -6a^{1+3} = -6a^4$

6. $-2b^4(b^8) = -2b^{4+8} = -2b^{12}$

7. $-6c^3d(-3c^8d^2) = 18c^{3+8}d^{1+2} = 18c^{11}d^3$

8. $10ef^2(-ef^5) = -10e^{1+1}f^{2+5} = -10e^2f^7$

9. $(3^2)^3 = 3^{2 \cdot 3} = 3^6 = 729$ **10.** $(x^8)^4 = x^{8 \cdot 4} = x^{32}$

11. $(z^{14})^2 = z^{14 \cdot 2} = z^{28}$ **12.** $(4a^3)^2 = 4^2a^{3 \cdot 2} = 16a^6$

13. $(-3b)^3 = (-3)^3b^3 = -27b^3$

14. $(-2c^2d)^4 = (-2)^4c^{2 \cdot 4}d^4 = 16c^8d^4$

15. $-(2m)^6 = -2^6m^6 = -64m^6$

16. $-3(x^3y)^4 = -3x^{3 \cdot 4}y^4 = -3x^{12}y^4$

17. (1) The product of powers: add exponents when multiplying variables with the same base; example: $y^4 \cdot y^5 = y^9$. (2) The power of a power: multiply exponents when raising a variable with an exponent to a power; example: $(a^2)^4 = a^8$. (3) The power of a product: each factor of a monomial raised to a power must be raised to that power; example: $(3x^3y^2)^3 = 27x^9y^6$.

PRACTICE
1. 2   **2.** 3

3. 8   **4.** 5

5. 1   **6.** $1 + 1 = 2$

7. $2 + 4 = 6$ **8.** $4 + 5 = 9$

9. $a^3(a^4) = a^{3+4} = a^7$ **10.** $b^6(b^2) = b^{6+2} = b^8$

11. $2x^3(-5x^2) = 2(-5)x^{3+2} = -10x^5$

12. $8y^4(3y^4) = 8(3)y^{4+4} = 24y^8$

13. $5g(-4g^9) = 5(-4)g^{1+9} = -20g^{10}$

14. $-6h^3(-7h) = -6(-7)h^{3+1} = 42h^4$

15. $2a^2b(9ab^5) = 2(9)a^{2+1}b^{1+5} = 18a^3b^6$

16. $(-3p^5q^2)(-pq^7) = -3(-1)p^{5+1}q^{2+7} = 3p^6q^9$

17. $(-2m^3n^3)(7m^4n) = -2(7)m^{3+4}n^{3+1} = -14m^7n^4$

18. $(-5x^6y^9)(3x^9y^6) = -5(3)x^{6+9}y^{9+6} = -15x^{15}y^{15}$

19. $3xyz(-4x^2z^3) = 3(-4)x^{1+2}yz^{1+3} = -12x^3yz^4$

20. $(-8a^4b^2)(-9a^3b^2) = -8(-9)a^{4+3}b^{2+2} = 72a^7b^4$

21. $(a^3)^4 = a^{3 \cdot 4} = a^{12}$ **22.** $(b^2)^4 = b^{2 \cdot 4} = b^8$

23. $(3c^2)^2 = 3^3c^{2 \cdot 3} = 27c^6$ **24.** $(5d^7)^2 = 5^2d^{7 \cdot 2} = 25d^{14}$

25. $(6a^2b^3)^3 = 6^3a^{2 \cdot 3}b^{3 \cdot 3} = 216a^6b^9$

26. $(-4x^9y^3)^2 = (-4)^2x^{9 \cdot 2}y^{3 \cdot 2} = 16x^{18}y^6$

27. $(-2gh^7)^3 = (-3)^3g^3h^{7 \cdot 3} = -27g^3h^{21}$

28. $(-2j^4k)^5 = (-2)^5j^{4 \cdot 5}k^5 = -32j^{20}k^5$

29. $(-2x^4yz^2)^3 = (-2)^3x^{4 \cdot 3}y^3z^{2 \cdot 3} = -8x^{12}y^3z^6$

30. $-2(-3x^6y^{12}z^5)^4 = -2(-3)^4x^{6 \cdot 4}y^{12 \cdot 4}z^{5 \cdot 4}$
$= -162x^{24}y^{48}z^{20}$

31. $-1(-4a^5b^3)^3 = -1(-4)^3a^{5 \cdot 3}b^{3 \cdot 3} = 64a^{15}b^9$

32. $3(-2m^6n^5)^4 = 3(-2)^4m^{6 \cdot 4}n^{5 \cdot 4} = 48m^{24}n^{20}$

33. $d = |-16(5)^2| = |-16(25)| = |-400| = 400$ ft

34. $V = 2x \cdot x \cdot 3x = 2 \cdot 3 \cdot x \cdot x \cdot x \cdot = 6x^3$

EXTEND
35. $(3x^4y)^2(-4xy^3) = (3^2x^{4 \cdot 2}y^2)(-4xy^3) = 9(-4)x^{8+1}y^{2+3} = -36x^9y^5$

36. $(-ab^5)^5(7a^4b^2) = (-1)^5a^5b^{5 \cdot 5}(7a^4b^2) = -1(7)a^{5+4}b^{25+2} = -7a^9b^{27}$

37. $(p^2q^7)(-2p^2q^4)^3 = (p^3q^7)(-2)^3p^{2 \cdot 3}q^{4 \cdot 3} = -8p^{3+6}q^{7+12} = -8p^9q^{19}$

38. doubling n times $= 2n$, so 2^nx

39. $W = EI = (IR)I = I^{1+1}R = I^2R$

40. $n = 1$, probability is $\frac{1}{2}$

$n = 2$, probability is $\frac{1}{2} \cdot \frac{1}{2} = \frac{1}{2^2}$

$n = 3$, probability is $\frac{1}{2} \cdot \frac{1}{2} \cdot \frac{1}{2} = \frac{1}{2^3}$

$n = n$, probability is $\frac{1}{2} \cdot \frac{1}{2} \cdots \frac{1}{2} = \frac{1}{2^n}$, or n factors of $\frac{1}{2}$

THINK CRITICALLY

41. $n + 5 = 8$, so $n = 3$ **42.** $n \cdot 2 = 8$, so $n = 4$

43. $5 \cdot n = 25$, so $n = 5$

44. yes, $(a^m)^n = a^{m \cdot n} = a^{n \cdot m} = (a^n)^m$

45. $p + q = 2p + 1$ or $q = p + 1$

46. The volume will be 8 times greater; The volume of a cube of side x is x^3. The volume of a cube of side $2x$ is $(2x)^3$ or $8x^3$, which is 8 times the volume of the other.

PROJECT CONNECTION

1. harder to run; more surface area for air resistance

2. $A = \frac{(b+1)^2}{s} = \frac{b^2 + 2b + 1}{s} = \frac{b^2}{s} + \frac{2b+1}{s}$,

so aspect ratio increases by $\frac{2b+1}{s}$

3. $A = \frac{b^2}{s} + \frac{b^2}{bc} = \frac{b}{c}$ **4.** increase b or decrease c

ALGEBRAWORKS

1. $P = \frac{\left(\frac{1}{16}\right)^2 (2)(8)}{3.78} \approx 0.017$ hp

2. $P = \frac{t^2 dN}{3.78}$ Multiply by 3.78.

$3.78P = t^2 dN$ Divide by $t^2 d$.

$\frac{3.78P}{t^2 d} = N$

$N = \frac{3.78(0.1)}{\left(\frac{3}{16}\right)^2 \left(\frac{3}{2}\right)} \rightarrow 7$ holes

3. 7 holes   **4.** Decrease t, or d, or both.

Lesson 11.3, pages 527-531

EXPLORE

1.
Expression	2^6	2^5	2^4	2^3	2^2	2^1
Value	64	32	16	8	4	2

2. Exponent decreases by one each time.

3. Value is divided by 2 or halved each time.

4. 2^0; 1

5.
Expression	2^{-1}	2^{-2}	2^{-3}
Value	$\frac{1}{2}$	$\frac{1}{4}$	$\frac{1}{8}$

6. $\frac{1}{2^1}, \frac{1}{2^2}, \frac{1}{2^3}; 2^{-n} = \frac{1}{2^n}$

TRY THESE

1. $-\frac{-15x^7}{5x^5} = 3x^{7-5} = 3x^2$ **2.** $\frac{-24a^4}{-3a^3} = 8a^{4-3} = 8a$

3. $\frac{6a^{12}b^4}{2a^4b^2} = 3a^{12-4}b^{4-2} = 3a^8b^2$

4. $\frac{20p^9q}{-4p^3q} = -5p^{9-3}q^{1-1} = -5p^6q^0 = -5p^6$

5. $100^0 = 1$ **6.** $W^0 = 1$

7. $4^{-2} = \frac{1}{4^2} = \frac{1}{16}$ **8.** $3^{-3} = \frac{1}{3^3} = \frac{1}{27}$

9. $\frac{-3g^3h^2}{-9g^5h^4} = \frac{1}{3}g^{3-5}h^{2-4} = \frac{1}{3}g^{-2}h^{-2} = \frac{1}{3g^2h^2}$

10. $\frac{15km^4}{5k^2m^3} = 3k^{1-2}m^{4-3} = 3k^{-1}m^1 = \frac{3m}{k}$

11. $\frac{6p^3q^4}{3p^3q^3} = 2p^{3-3}q^{4-3} = 2p^0q^1 = 2q$

12. $\frac{-r^2s^5}{rs^6} = -r^{2-1}s^{5-6} = -r^1s^{-1} = -\frac{r}{s}$

13. $\left(\frac{4}{5}\right)^3 = \frac{4^3}{5^3} = \frac{64}{125}$ **14.** $\left(\frac{-3}{4}\right)^{-2} = \frac{(-3)^{-2}}{4^{-2}} = \frac{16}{9}$

15. $\left(\frac{2x^4}{y^3}\right)^5 = \frac{2^5x^{4 \cdot 5}}{y^{3 \cdot 5}} = \frac{32x^{20}}{y^{15}}$

16. $\left(\frac{-3a}{2b^4}\right)^3 = \frac{(-3)^3a^{1 \cdot 3}}{2^3b^{4 \cdot 3}} = \frac{-27a^3}{8b^{12}}$

17. $4x^2 \div -2x = -2x$

18. *Quotient Rule*: When dividing variables with the same base, subtract the exponents. Example: $\frac{a^4}{a^2} = a^2$.

Zero Property: Any variable or number raised to the zero power is 1. Example: $3^0 = 1$.

Property of Negative Exponents: A variable or number with a negative exponent may be written as a reciprocal of the base raised to the positive power.

Example: $r^{-4} = \frac{1}{r^4}$.

Power of a Quotient Rule: To raise a quotient to a power, raise both the numerator and the denominator to

that power. Example: $\left(\frac{a}{3}\right)^3 = \frac{a^3}{4^3} = \frac{a^3}{64}$.

PRACTICE

1. $\frac{-16a^7}{-2a^6} = 8a^{7-6} = 8a$ **2.** $\frac{9b^4}{-3b^2} = -3b^{4-2} = -3b^2$

3. $\frac{12a^6b^4}{-4a^5b} = -3a^{6-5}b^{4-1} = -3ab^3$

4. $\dfrac{-5c^4d^6}{7c^2d^5} = -\dfrac{5}{7}c^{4-2}d^{6-5} = -\dfrac{5}{7}c^2d$ or $\dfrac{-5c^2d}{7}$

5. $\dfrac{6x^0y^3}{y^2} = 6x^0y^{3-2} = 6y$ **6.** $\dfrac{-7cd^0}{14c} = -\dfrac{1}{2}c^{1-1}d^0 = -\dfrac{1}{2}$

7. $\dfrac{62yz^4}{14y^5z^4} = \dfrac{31}{7}y^{1-5}z^{4-4} = \dfrac{31}{7}y^{-4}z^0 = \dfrac{31}{7y^4}$

8. $\dfrac{-2p^2q^7}{8p^4q} = -\dfrac{1}{4}p^{2-4}q^{7-1} = -\dfrac{1}{4}p^{-2}q^6 = \dfrac{-q^6}{4p^2}$

9. $2^4 \cdot 5 - 7^0 = 16 \cdot 5 - 1 = 80 - 1 = 79$

10. $8^2 + 2 \cdot 5^0 = 64 + 2 \cdot 1 = 64 + 2 = 66$

11. $9^0 \cdot 2^{-3} \cdot 6 = 1 \cdot \dfrac{1}{8} \cdot 6 = \dfrac{6}{8} = \dfrac{3}{4}$

12. $5^{-2} \cdot 3^{-2} \cdot 4^0 = \dfrac{1}{25} \cdot \dfrac{1}{9} \cdot 1 = \dfrac{1}{225}$

13. $\dfrac{-52x^7y}{-4x^3y^3} = 13x^{7-3}y^{1-3} = 13x^4y^{-2} = \dfrac{13x^4}{y^2}$

14. $\dfrac{35rs^4}{-7rs^{10}} = -5r^{1-1}s^{4-10} = -5r^0s^{-6} = \dfrac{-5}{s^6}$

15. $\dfrac{12s^2t^3}{-4st^4} = -3s^{2-1}t^{3-4} = -3s^1t^{-1} = \dfrac{-3s}{t}$

16. $\dfrac{-8t^2v}{-10tv^4} = \dfrac{4}{5}t^{2-1}v^{1-4} = \dfrac{4}{5}t^1v^{-3} = \dfrac{4t}{5v^3}$

17. $\left(\dfrac{2}{3}\right)^4 = \dfrac{2^4}{3^4} = \dfrac{16}{81}$ **18.** $\left(\dfrac{1}{2}\right)^{-3} = \dfrac{1^{-3}}{2^{-3}} = \dfrac{8}{1} = 8$

19. $\left(\dfrac{2c^4}{5d}\right)^2 = \dfrac{2^2c^{4\cdot2}}{5^2d^2} = \dfrac{4c^8}{25d^2}$

20. $\left(\dfrac{-3g^5}{2h^3}\right)^3 = \dfrac{(-3)^3g^{5\cdot3}}{2^3h^{3\cdot3}} = \dfrac{-27g^{15}}{8h^9}$

21. $V = b \cdot w \cdot h$

     $4x^2y = 4x \cdot y \cdot h$

     $\dfrac{4x^2y}{4y} = h$

     $x = h$

22. $V = \dfrac{4\pi r^3}{3}$ since $d = 2r$, then $\dfrac{d}{2} = r$

     $V = \dfrac{4\pi\left(\dfrac{d}{2}\right)^3}{3} = 3^{-1} \cdot 4\pi\left(\dfrac{d^3}{2^3}\right)$

     $= 3^{-1} \cdot 2^2 \cdot 2^{-3}\pi d^3 = \dfrac{\pi d^3}{6}$

23. $\dfrac{(2a^3b)^4(4ab^3)}{(3b)^2} = \dfrac{16a^{12}b^4 \cdot 4ab^3}{9b^2} = \dfrac{64a^{13}b^7}{9b^2} = \dfrac{64a^{13}b^5}{9}$

24. $\dfrac{(4c^2)(-cd^4)^5}{8c^4} = \dfrac{4c^2(-c^5d^{20})}{8c^4} = \dfrac{-4c^7d^{20}}{8c^4} = \dfrac{-c^3d^{20}}{2}$

25. $\dfrac{(g^2h^4)^2(-2gh)^3}{(4g^3h^5)^2} = \dfrac{g^4h^8(-8g^3h^3)}{16g^6h^{10}} = \dfrac{-8g^7h^{11}}{16g^6h^{10}} = \dfrac{-gh}{2}$

26. $\dfrac{(3s^4t^5)^2(-st)^4}{(5s^4t^5)^2} = \dfrac{9s^8t^{10}(s^4t^4)}{25s^8t^{10}} = \dfrac{9s^{12}t^{14}}{25s^8t^{10}} = \dfrac{9}{25}s^4t^4$

27. $\left(\dfrac{2wx}{5w^4x^3}\right)^{-2} = \left(\dfrac{2}{5w^3x^2}\right)^{-2} = \left(\dfrac{5w^3x^2}{2}\right)^2 = \dfrac{25w^6x^4}{4}$

28. $\left(\dfrac{-y^2z^5}{3yz^3}\right)^{-3} = \left(\dfrac{-yz^2}{3}\right)^{-3} = \left(\dfrac{3}{yz^2}\right)^3 = -\dfrac{27}{y^3z^6}$

29. Possible response: Substitute $\dfrac{2}{3}$ for K, use the reciprocal of $\dfrac{2}{3}$ to get a positive exponent, $6\left(\dfrac{3}{2}\right)^2$. Now evaluate $6\left(\dfrac{3^2}{2^2}\right) = 6 \cdot \dfrac{9}{4} = \dfrac{54}{4} = 13\dfrac{1}{2}$.

30. If a positive exponent represents the future, a negative exponent represents the past, so $p(1.022)^{-t}$

31. $7,800,000(1.022)^{-2} \approx 7,467,802$

32. The force is multiplied by 1.5. Replacing m by $2m$, M by $3M$, and r by $2r$, you get $F = \dfrac{G(2m)(3M)}{(2r)^2}$. When you simplify it, you get $F = \dfrac{3GmM}{2r^2}$. So, F is $\dfrac{3}{2}$ or 1.5 times more than the original.

33. $h^{-3+n} = 1 = h^0$, so $-3 + n = 0$ and $n = 3$

34. $(ab^3)^n = 1 = (ab^3)^0$, so $n = 0$

35. $\left(\dfrac{cd^6}{c^5d^n}\right)^2 = \dfrac{d^8}{c^8}$

     $\dfrac{c^2d^{12}}{c^{10}d^{2n}} = \dfrac{d^8}{c^8}$

     $\dfrac{d^{12-2n}}{c^8} = \dfrac{d^8}{c^8}$ so, $12 - 2n = 8$ and $n = 2$.

36. If $\dfrac{p^a}{p^b} = p^2$, then $p^{a-b} = p^2$ and $a - b = 2$ or $a = b + 2$

37. $p < q$; p and q are positive and $n(p - q)$ is negative, so p must be less than q

38. $6x + 5 = 3x - 13$
    $3x = -18$
    $x = -6$

39. $7y + 2(y - 1) = 4y + 18$
    $7y + 2y - 2 = 4y + 18$
        $5y = 20$
        $y = 4$

40. $2z = 9z + 28$
    $-7z = 28$
     $z = -4$

41.

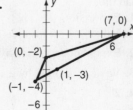

42. Yes; The points $(-1, -4)$, $(1, -3)$, and $(7, 0)$ appear to result in one side of a triangle. To justify that the points are in a line, use the point slope formula to find the equation of the lines connecting them. Then substitute the values $(1, -3)$ in that equation to see if they are a solution of the equation.

43. $2x - 1 > -7$
$2x > -6$
$x > -3$

44. $5 - 3x \geq 26$
$-3x \geq 21$
$x \leq -7$

45. $4x + 1 < -1$
$4x < -2$
$x < -\dfrac{1}{2}$

46. $4 > 5x - 1$
$5 > 5x$
$1 > x$ or $x < 1$

47. C; by substituting $(9, 2)$ in equations

48. $\dfrac{-14b^9}{-7b^7} = 2b^{9-7} = 2b^2$

49. $\dfrac{-5x^6y^4}{x^8y^5} = -5x^{6-8}y^{4-5} = -5x^{-2}y^{-1} = \dfrac{-5}{x^2y}$

50. $\left(\dfrac{4a^2}{3b^3}\right)^3 = \dfrac{4^3a^{2 \cdot 3}}{3^3b^{3 \cdot 3}} = \dfrac{64a^6}{27b^9}$ **51.** $\left(\dfrac{1}{3}\right)^{-4} = 3^4 = 81$

Lesson 11.4, pages 532–537

EXPLORE

1.
Exponential Expression	Value
10^5	100,000
10^4	10,000
10^3	1,000
10^2	100
10^1	10
10^0	1
10^{-1}	0.1
10^{-2}	0.01
10^{-3}	0.001
10^{-4}	0.0001
10^{-5}	0.00001

2. Possible pattern: Bases are all 10 and each exponent names the number of places the decimal point moves to the right of 1 (if exponent > 0) or to the left of 1 (if exponent < 0).

TRY THESE

1. $3\,250,000,000 = 3.25 \times 10^9$
9 places

2. $1\,09,000,000 = 1.09 \times 10^8$
8 places

3. $7\,2,000,000 = 7.2 \times 10^7$
7 places

4. $9\,23,000,000 = 9.23 \times 10^8$
8 places

5. $0.003\,15 = 3.15 \times 10^{-3}$
3 places

6. $0.00007\,2 = 7.2 \times 10^{-5}$
5 places

7. $0.000005\,4 = 5.4 \times 10^{-6}$
6 places

8. $0.0004\,32 = 4.32 \times 10^{-4}$
4 places

9. $(3.5)(2.1) \times 10^{4+5} = 7.35 \times 10^9$

10. $(1.4)(4.5) \times 10^{7+8} = 6.3 \times 10^{15}$

11. $\dfrac{9.6}{1.2} \times 10^{9-2} = 8 \times 10^7$ **12.** $\dfrac{8.4}{2.1} \times 10^{6-3} = 4 \times 10^3$

13. $7.6^2 \times 10^{4 \cdot 2} = 5\,7.76 \times 10^8 = 5.776 \times 10^9$
1 place

14. $(6.8)(9) \times 10^{6+9} = 6\,1.2 \times 10^{15} = 6.12 \times 10^{16}$
1 place

15. $6\,7,200,000 = 6.72 \times 10^7$
7 places

16. Move decimal point 4 places to the right. Write the number as 6.2×10^{-4}.

PRACTICE

1. $5.12 \times 10^8 = 5\,12000000 = 512,000,000$
8 places

2. $9.7 \times 10^{10} = 9\,7000000000 = 97,000,000,000$
10 places

3. $1.2 \times 10^{-5} = 0\,00001\,2 = 0.000012$
5 places

4. $6.5 \times 10^{-4} = 0\,0006\,5 = 0.00065$
4 places

5. $3\,14,000 = 3.14 \times 10^5$
5 places

6. $4,300,000 = 4.3 \times 10^6$
6 places

7. $2\,3,000,000 = 2.3 \times 10^7$
7 places

8. $6\,10,000,000 = 6.1 \times 10^8$
8 places

9. $0.0004\,15 = 4.15 \times 10^{-4}$
4 places

10. $0.001\,03 = 1.03 \times 10^{-3}$
3 places

11. $0.000008 = 8 \times 10^{-6}$
6 places

12. $0.00003\,2 = 3.2 \times 10^{-5}$
5 places

13. $(2.4)(1.3) \times 10^{5+9} = 3.12 \times 10^{14}$

14. $(4)(2.1) \times 10^{6+8} = 8.4 \times 10^{14}$

15. $\frac{9}{3} \times 10^{7-4} = 3 \times 10^3$ **16.** $\frac{7.4}{2} = 10^{10-6} = 3.7 \times 10^4$

17. $2^2 \times 10^{5 \cdot 2} = 4 \times 10^{10}$

18. $(3 \times 10^6)^2 = 3^2 \times 10^{6 \cdot 2} = 9 \times 10^{12}$

19. $(4.6)(9) \times 10^{4+15} = 4\underset{\text{1 place}}{\underset{\uparrow}{1}}.4 \times 10^{19} = 4.14 \times 10^{20}$

20. $(8.1)(7.5) \times 10^{5+11} = 6\underset{\text{1 place}}{\underset{\uparrow}{0}}.75 \times 10^{16} = 6.075 \times 10^{17}$

21. $5\underset{\text{7 places}}{\underline{7,000,000}} = 5.79 \times 10^7$

22. myriad myriad $= 10,000 (10,000) = (1 \times 10^4)(1 \times 10^4) =$
$(1)(1) \times 10^{4+4} = 1 \times 10^8$

23. $0\underset{\text{6 places}}{\underline{.000006}}5 = 6.5 \times 10^{-6}$

24. $(1.5 \times 10^5)(5 \times 10^9) = (1.5)(5) \times 10^{5+9} = 7.5 \times 10^{14}$

25. Area of head of pin $= \pi r^2 \approx \pi \left(\frac{2.25}{2}\right)^2 \approx 3.976 \text{ mm}^2$

$3.976 \div (1.5 \times 10^{-25}) = (3.976 \div 1.5) \times 10^{25} \approx$

2.65×10^{25} molecules

In **26-31**, estimates will vary but should be close to the following:

26. $(9.23 \times 10^7)^2 \approx 81 \times 10^{7 \cdot 2}$ or 8.1×10^{15}

27. $(1.09 \times 10^9)^4 \approx 1 \times 10^{9 \cdot 4}$ or 1×10^{36}

28. $\frac{2.03 \times 10^6}{3.98 \times 10^3} \approx 0.5 \times 10^{6-3}$ or 5×10^2

29. $\frac{3.13 \times 10^{10}}{8.6 \times 10^4} \approx 0.3 \times 10^{10-4}$ or 3×10^5

30. $(7.35 \times 10^4)(8.19 \times 10^9) \approx 60 \times 10^{4+9}$ or 6×10^{14}

31. $(4.7 \times 10^7)(8.3 \times 10^9) \approx 40 \times 10^{7+9}$ or 4×10^{17}

32. $C = 2\pi r = 2\pi(4100) = 25,748$ miles

$\frac{25,748}{600} = 42.913 = 4.2913 \times 10^1$ hours

33. $\frac{700 - 330}{1000 - 0} = 0.37 = 3.7 \times 10^{-1} \frac{\frac{\text{m}}{\text{s}}}{°\text{C}}$

34. Answers will vary.

35. Decimal point moved right two spaces so $p = 6 - 2 = 4$

36. Decimal point moved left two spaces so $p = -7 + 2 = -5$

37. Decimal point moved right one space so $p = 5.3(10) = 53$

38. Decimal point moves right four so $p = 6.7(10,000) = 67,000$

39. $(a \times 10^m)(b \times 10^n) = ab \times 10^{m+n} = c \times 10^{m+n+1}$
so $ab > 10$ because decimal for $c \times 10^{m+n+1}$ had to be moved 1 space left

ALGEBRA WORKS

1. $150 \text{ mm} \cdot \frac{1 \text{ m}}{1000 \text{ mm}} = 0.150 \text{ m}$

2. $0.96 \text{ km} \cdot \frac{1000 \text{ m}}{1 \text{ km}} = 960 \text{ m}$

3. $0.150 = 1.5 \times 10^{-1}$; $960 = 9.6 \times 10^2$

4. $\frac{1.5 \times 10^{-1}}{9.6 \times 10^2} = 0.15625 \times 10^{-1-2} = 1.5625 \times 10^{-4}$

5. $225 \text{ m} = 0.225 \text{ mm}$; $\frac{0.225}{AB} = 1.5625 \times 10^{-4} \rightarrow$

$\frac{0.225}{1.5625 \times 10^{-4}} = AB = 1440 \text{ m or } 1.44 \text{ km}$

6. $\frac{10^{-6}}{AB} = 1.5625 \times 10^{-4}$ or $\frac{10^{-6}}{1.5625 \times 10^{-4}} =$

$AB = 0.0064 \text{ m}$

7. Answers will vary.

Lesson 11.5, pages 538–540

THINK BACK

1.

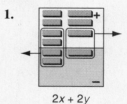

$2x + 2y$

2. Two kinds of unlike terms were present, one with x and one with y; simplified algebraic expression had 2 terms.

3.

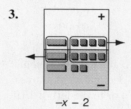

$-x - 2$

4. two kinds of unlike terms; two terms

5. They both had the same number of unlike terms and same number of terms after combining them.

6.

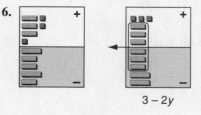

$3 - 2y$

7. three kinds of unlike terms; simplified expression had 2 terms since x terms summed to zero and were eliminated.

EXPLORE

8.
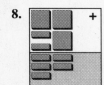

9. two kinds of blocks; x^2 blocks and x blocks

10. There are three x^2 blocks left on positive mat and three x blocks on negative mat so $3x^2 - 3x$ or $3x^2 + (-3x)$

11.

$3x^2 + 3x = 1$

12.

$x^3 + 2x + 6$

13.
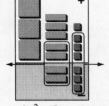

$xy + 2x - 3y - 1$

14. Answers will vary: Possible answer: $(2x + y) + (x - 5)$

15.
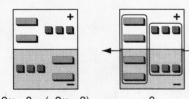
$2x - 3 + (-2x + 3)$ 0

16. $-x - 2$

17. $-2xy + y$

18. $-x^2 - y^2$

19. $x - 3$

20. $2x + 1$

21. opposite of $2x^2 + x - 3$ is $-2x^2 - x + 3$; result of combining the two is zero.

22. $x^2 + 3x + 5$

23. $(3x - 2) - (x + 4) = (3x - 2) + (-x - 4) = 2x - 6$

24. $(2xy + y) - (3xy - x) = (2x + y) + (-3xy + x) = -xy + y + x$

25. $(3x^2 + 4x - 1) - (2x^2 - 7) = (3x^2 + 4x - 1) + (2x^2 + 7) = x^2 + 4x + 6$

26. $(2y^2 - 3y) - (y^2 + 3y) = (2y^2 - 3y) + (-y^2 - 3y) = y^2 - 6y$

MAKE CONNECTIONS

27. $3x + 2x = (3 + 2)x = 5x$ **28.** Yes

29. No   **30.** Yes

31. No

32. Like terms have same variables and exponents.

33. $4 \times 10^2 + 2 \times 10^2 = (4 + 2) \times 10^2 = 6 \times 10^2$; They are the same if you let $x = 10$.

34. North America: $2\,90,000,000 = 2.9 \times 10^8$

Latin America: $4\,70,000,000 = 4.7 \times 10^8$

Europe: $7\,28,000,000 = 7.28 \times 10^8$

Asia: $3\,392,000,000 = 3.392 \times 10^9$

Africa: $7\,00,000,000 = 7 \times 10^8$

Oceania: $2\,8,000,000 = 2.8 \times 10^7$

35. change Asia to 33.92×10^8 and Oceana to 0.28×10^8. Now add first factor of each number.

36. $(2.9 + 4.7 + 7.28 + 33.92 + 7 + 0.28) \times 10^8 = 56.08 \times 10^8 = 5.608 \times 10^9$

SUMMARIZE

37a. Two mats are different because of the x^2 term.

b. $3x + 2x = 5x$ since they are like terms; right side cannot be simplified because terms are unlike terms.

38. Answers will vary, but should include adding like terms by combining coefficients of the variables.

39. The opposite of an algebraic expression is the opposite of each term. If you add an algebraic expression and its opposite, you will get 0.

40. Answers will vary, but should include adding the opposite of $x - y$ and combining like terms to get $6x^2 - x$.

41. like terms are: $3x^4$ and $5x^4$, $2x^3$ and $4x^3$, $2x$ and $3x$; $8x^4 + 6x^3 + 5x + 12$

Lesson 11.6, pages 541–547

EXPLORE

1. $9 + 10 + 11 + 16 + 17 + 18 + 23 + 24 + 25 = 153$

2. $9 \cdot 14 = 153 =$ sum of the nine numbers

3. arrangement will work for any 9 numbers in the same month, so just multiply middle number by 9.

4. $x = 17$, $x - 8 = 9$, $x - 7 = 10$, $x - 6 = 11$, $x - 1 = 16$, $x + 1 = 18$, $x + 6 = 23$, $x + 7 = 24$, $x + 8 = 25$

5. The sum of column 1 is $3x - 3$, the sum of column 2 is $3x$, and the sum of column 3 is $3x + 3$.
So, $3x - 3 + 3x + 3x + 3 = 9x$.

TRY THESE

1. Highest power is 6, so degree is 6.

2. Highest power is 2, so degree is 2.

3. Highest sum of powers is 3, so degree is 3.

4. $2x^3 + 3x^2 + 4x + 8$

5. $-2x^4y + 3x^2y^2 + 6xy^3 + 9$

6.
$$\begin{array}{r} 5x - 11 \\ + \ 6x - 1 \\ \hline 11x - 12 \end{array}$$

7.
$$\begin{array}{r} -2^2 + 6x - 9 \\ + \ 4x^2 - 3x + 4 \\ \hline 2x^2 + 3x - 5 \end{array}$$

8.
$$\begin{array}{r} 8x + 2y \\ + \quad 7y - 8 \\ \hline 8x + 9y - 8 \end{array}$$

9.
$$\begin{array}{r} 7x^2 - 3xy \\ + 2x^2y + 4xy + 8y \\ \hline 9x^2y + \ xy + 8y \end{array}$$

10. $-(-3x + 9) = 3x - 9$

11. $-(8a^2 + 2a - 7) = -8a^2 - 2a + 7$

12. $-(9b^4 - 1) = -9b^4 + 1$

13. $-(a + 2b + c) = -a - 2b - c$

14.
$$\begin{array}{rcl} 7a - 2 & \to & 7a - 2 \\ -(4a + 1) & \to & -4a - 1 \\ \hline & & 3a - 3 \end{array}$$

15.
$$\begin{array}{rcl} 3b^2 + 2b - 1 & \to & 3b^2 + 2b - 1 \\ (-2b^2 - 7b + 4) & \to & 2b^2 + 7b - 4 \\ \hline & & 5b^2 + 9b - 5 \end{array}$$

16.
$$\begin{array}{rcl} 6h + 2m & \to & 6h + 2m \\ -(4h + 5m - 1) & \to & -4h - 5m + 1 \\ \hline & & 2h - 3m + 1 \end{array}$$

17.
$$\begin{array}{rcl} 9p^2 + 5p + 3q & \to & 9p^2 + 5p + 3q \\ -(\quad\quad + 7q + 6q^2) & \to & \quad\quad\quad -7q - 6q^2 \\ \hline & & 9p^2 + 5p - 4q - 6q^2 \end{array}$$

18. On a basic mat, start with 2 x^2 blocks, 2 x blocks, and 1 unit block in the positive section, and 3 x^2 blocks and 4 unit blocks in the negative section. Remove 2 zero pairs of x's and 1 zero pair of 1's. The sum is $2x^2 - x - 3$.

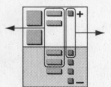

19. Answers will vary, but should include that like terms have the same variables and exponents. Add and subtract polynomials by combining like terms.

20. $(3x - 3y) + (3x + y) + x + 2y + 6x =$
$3x + 3x + x + 6x - 3y + y + 2y = 13x$

PRACTICE

1. Highest power is 2, so degree is 2.

2. Highest power is 5, so degree is 5.

3. Highest sum of powers is 4, so degree is 4.

4. Highest sum of powers is 7, so degree is 7.

5. Highest sum of powers is 7, so degree is 7.

6. Highest power is 4, so degree is 4.

7.
$$\begin{array}{r} 4a - 3b + 5c \\ + 8a + 3b - 7c \\ \hline 12a \quad\quad - 2c \end{array}$$

8.
$$\begin{array}{r} 5r + 3s + t \\ + \ r + 9s + 7t \\ \hline 6r + 12s + 8t \end{array}$$

9.
$$\begin{array}{r} -4p + q - 8 \\ + \ p + 5q + 17 \\ \hline -3p + 6q + 9 \end{array}$$

10.
$$\begin{array}{r} 4x^2 - 8x \\ + \ 9x^2 - 2x \\ \hline 13x^2 - 10x \end{array}$$

11.
$$\begin{array}{rcl} 2a + 4b \ - c & \to & 2a + 4b \ - c \\ -(6a + 3b + 5c) & \to & -6a - 3b - 5c \\ \hline & & -4a + b - 6c \end{array}$$

12.
$$\begin{array}{rcl} x - y + z & \to & x - y + z \\ -(x + y - z) & \to & -x - y + z \\ \hline & & -2y + 2z \end{array}$$

13.
$$\begin{array}{rcl} 8y^4 + 7y^2 - 3 & \to & 8y^4 + 7y^2 - 3 \\ -(2y^4 - y^2 + 7) & \to & -2y^4 + y^2 - 7 \\ \hline & & 6y^4 + 8y^2 - 10 \end{array}$$

14.
$$\begin{array}{rcl} z^3 + 4z^2 - z & \to & z^3 + 4z^2 - z \\ -(3z^3 - 2z^2 + 5z) & \to & -3z^3 + 2z^2 - 5z \\ \hline & & -2z^3 + 6z^2 - 6z \end{array}$$

15.
$$\begin{array}{r} 5x + 7y - 10 \\ + \quad 3y - 14 \\ \hline 5x + 10y - 24 \end{array}$$

16.
$$\begin{array}{r} a^2 - 2ab + b^2 \\ + \ 3a^2 \quad\quad - 6b^2 \\ \hline 4a^2 - 2ab - 5b^2 \end{array}$$

17.
$$\begin{array}{r} a + 2b \\ b - c \\ + \ 3a \quad\quad + 5c \\ \hline 4a + 3b + 4c \end{array}$$

18.
$$\begin{array}{r} 4r^2 + 3r \\ 5r^3 \quad\quad - r \\ + \ r^3 \quad\quad + 4r \\ \hline 6r^3 + 4r^2 + 6r \end{array}$$

19.
$$\begin{array}{rcl} 7d - 8d^2 + d^3 & \to & 7d - 8d^2 + d^3 \\ -(3d + d^2 - 4d^3) & \to & -3d - d^2 + 4d^3 \\ \hline & & 4d - 9d^2 + 5d^3 \end{array}$$

20.
$$8k - 3k^2 \quad \rightarrow \quad -3k^2 + 8k$$
$$\underline{-(9k \quad\quad + 2) \rightarrow \quad \underline{\quad -9k - 2}}$$
$$\qquad\qquad\qquad\qquad -3k^2 - k - 2$$

21.
$$8x^2 + 5xy - y^2 \quad \rightarrow \quad 8x^2 + 5xy - y^2$$
$$\underline{-(6x^2 - 3xy + y^2) \rightarrow \underline{-6x^2 + 3xy - y^2}}$$
$$\qquad\qquad\qquad\qquad 2x^2 + 8xy - 2y^2$$

22.
$$4b^2 - 3b \quad \rightarrow \quad 4b^2 - 3b$$
$$\underline{-(\quad\quad 7b + 6) \rightarrow \quad \underline{\quad -7b - 6}}$$
$$\qquad\qquad\qquad\quad 4b^2 - 10b - 6$$

23. $A = 7x(5x) - 3x(x) = 35x^2 - 3x^2 = 32x^2$

24. Surface area $= 2(3x)(4y) + 2(3x)(2x) + 2(2x)(4y) = 24xy + 12x^2 + 16xy = 12x^2 + 40xy$

EXTEND

25. $(x + 3y) + (8x - 2y) - (4x + y)$
$x + 3y + 8x - 2y - 4x - y$
$(x + 8x - 4x) + (3y - 2y - y)$
$5x$

26. $(2x + 3) + (4x^2 + x - 7) - (x^2 + x + 2)$
$2x + 3 + 4x^2 + x - 7 - x^2 - x - 2$
$(4x^2 - x^2) + (2x + x - x) + (3 - 7 - 2)$
$3x^2 + 2x - 6$

27. $2(0.6)^2 + 7(0.6) - 3.2 = 1.72$

28. $(0.6)^3 - 5(0.6)^2 + 4(0.6) - 7 = -6.184$

29. $(-2g)^2 + 2(-2g) - 3 = 4g^2 - 4g - 3$

30. $2(-2g)^3 - 5(-2g)^2 + (-2g) - 3$
$2(-8g^3) - 5(4g^2) - 2g - 3$
$-16g^3 - 20g^2 - 2g - 3$

31. structural weight $= W_{st}$ and engine weight $= W_e$

32. Payload weight, (W_p) must be decreased.

33. $40 + \dfrac{40^2}{20} = 40 + \dfrac{1600}{20} = 40 + 80 = 120$ ft

34. no; a car at 40 mph needs 120 ft; rule of thumb says 80 ft

THINK CRITICALLY

35. $2x + \square = -5x$, so $\square = -7xy$

36. $5xy - \square = 9xy$, so $\square = -4xy$

37. True: $(a - b) + (b - a) = (a - a) + (-b + b) = 0$, so they are opposites.

MIXED REVIEW

38. $z = \sqrt{121} = \sqrt{11 \cdot 11} = 11$ **39.** $z = \sqrt{289} = \sqrt{17 \cdot 17} = 17$

40. $z = \sqrt{47} \approx 6.86$ **41.** $z = \sqrt{103} \approx 10.15$

42. $0.04x = 450$
$$x = \dfrac{4.50}{0.04} = \$11,250$$

43. $20,000 - 0.025(20,000) = 20,000 - 500 = 19,500$

44. $3.4 \times 10^6 = 3{,}400{,}000$ — 6 places

45. $1.02 \times 10^8 = 102{,}000{,}000$ — 8 places

46. $4.8 \times 10^{-7} = 0.00000048$ — 7 places

47. $3.03 \times 10^{-5} = 0.0000303$ — 5 places

PROJECT CONNECTION

1. Activity, so no solution   **2.** Answers will vary.

3. The area of a circle is $A = \pi r^2$, so $r = \sqrt{\dfrac{A}{\pi}}$.

The radius of the circle whose area equals that of an 8×8 in. square is $r = \sqrt{\dfrac{64}{\pi}}$ or about 4.5 in.

4-6. Answers will vary.

7. Circular. The square chute allow air to escape out the sides making it less stable. The circular chute holds air more uniformly making it more stable.

8. The hole will make the chute more stable and it will fall more slowly.

Lesson 11.7, pages 548–549

THINK BACK

1. $(3x)(-2y)$ **2.** $-6xy$

3. Answers will vary; possible answer: $2x(-x)$.

EXPLORE

4. the two regions to the right of the vertical axis

5. $2x^2$ and $-4x$ **6.** $2x^2 + (-4x) = 2x^2 - 4x$

7. $x(x + 2) = x^2 + 2x$ **8.** $y(x + 3) = xy + 3y$

9.

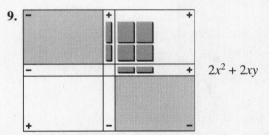

$2x^2 + 2xy$

10.

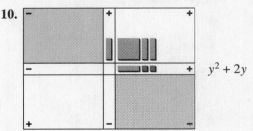

$y^2 + 2y$

11. 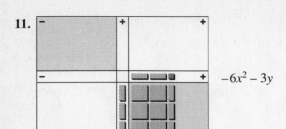 $-6x^2 - 3y$

12. 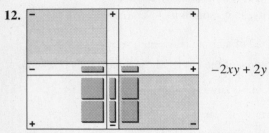 $-2xy + 2y$

13. $2x^2 - 4x$. The Algeblocks model shows $2x(x)$ in Quadrant I and $(2x)(-2)$ in Quadrant IV.

14. Activity, so no solution

15. $2x^2, -6x, x, -3$

16. $2x^2 + (-6x) + x + (-3)$ simplifies to $2x^2 - 5x - 3$ by combining like terms

17. 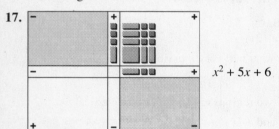 $x^2 + 5x + 6$

18. 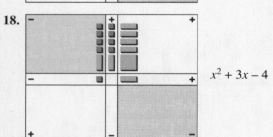 $x^2 + 3x - 4$

19. 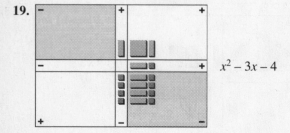 $x^2 - 3x - 4$

20. 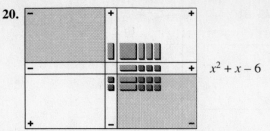 $x^2 + x - 6$

21. 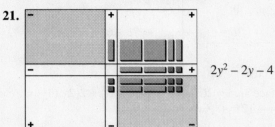 $2y^2 - 2y - 4$

22. 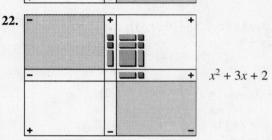 $x^2 + 3x + 2$

23a. $(x)(x); (x)(-2); (1)(x); (1)(-2)$

 b. $x^2; -2x; x; -2$

 c. $x^2 + (-2x) + x + (-2) = x^2 - x - 2$

24. Each term of the second factor is multiplied by each term of the first factor and the result is the sum of these products.

25a. $2x^2 + (-6x) + x + (-3)$

 b. $2x^2 - 5x - 3$

26. $x(x - 4) - 3(x - 4) = x^2 - 4x - 3x + 12 = x^2 - 7x + 12$

27. $x(x - 5) + 7(x - 5) = x^2 - 5x + 7x - 35 = x^2 + 2x - 35$

28. $2y(y + 2x) - x(y + 2x) = 2y^2 + 4xy - xy - 2x^2 = 2y^2 + 3xy - 2x^2$

29. $(5 - 3)(5 - 4) = (2)(1) = 2$

30. $(5)^2 - 7(5) + 12 = 25 - 35 + 12 = 2$; the expressions are equivalent

31. $-y(y + 2) = -y^2 - 2y$

32. Answers will vary, but might include using the distributive property.

33. the same number of terms as the polynomial

34. $(x + 1)(x + 2) = x^2 + 3x + 2$

35. Answers will vary but should include multiplying each term in the second factor by each term in the first factor and then simplifying.

36. $(x + 3)(x^2 - 2x + 5) = x(x^2 - 2x + 5) + 3(x^2 - 2x + 5) =$
$x(x^2) + x(-2x) + x(5) + 3(x^2) + 3(-2x) + 3(5) =$
$x^3 - 2x^2 + 5x + 3x^2 - 6x + 15 = x^3 + x^2 - x + 15$

Lesson 11.8, pages 551–557

EXPLORE

1.

	1	**2**	**3**	**4**	**5**
1 × 1 squares	1	4	9	16	25
2 × 2 squares	0	1	4	9	16
3 × 3 squares	0	0	1	4	9
4 × 4 squares	0	0	0	1	4
5 × 5 squares	0	0	0	0	1
Total squares	1	5	14	30	55

2. by trial and error, a, b, and d

3. They are equivalent to one another.

4. $\dfrac{8(8+1)(2 \cdot 8+1)}{6} = \dfrac{8(9)(17)}{6} = \dfrac{1224}{6} = 204$ squares

TRY THESE

1. $x(3x) + x(5) = 3x^2 + 5x$

2. $c(c^2) + c(9c) = c^3 + 9c^2$

3. $-4(y) - 4(2) = -4y - 8$

4. $-3(2z) - 3(-7) = -6z + 21$

5. $x(x + 8) + 7(x + 8) = x(x) + x(8) + 7(x) + 7(8) =$
$x^2 + 8x + 7x + 56 = x^2 + 15x + 56$

6. $g(g - 6) + 4(g - 6) = g(g) + g(-6) + 4(g) + 4(-6) =$
$g^2 - 6g + 4g - 24 = g^2 - 2g - 24$

7. $2x(x - 4) + 1(x - 4) = 2x(x) + 2x(-4) + 1(x) + 1(-4) =$
$2x^2 - 8x + x - 4 = 2x^2 - 7x - 4$

8. $z(2z - 3) - 3(2z - 3) = z(2z) + z(-3) - 3(2z) - 3(-3) =$
$2z^2 - 3z - 6z + 9 = 2z^2 - 9z + 9$

9. $a(a^2 - 3a + 2) - 2(a^2 - 3a + 2) =$
$a(a^2) + a(-3a) + a(2) - 2(a^2) - 2(-3a) - 2(2) =$
$a^3 - 3a^2 + 2a - 2a^2 + 6a - 4 = a^3 - 5a^2 + 8a - 4$

10. $k(k^2 - 4k + 1) + 4(k^2 - 4k + 1) =$
$k(k^2) + k(-4k) + k(1) + 4(k^2) + 4(-4k) + 4(1) =$
$k^3 - 4k^2 + k + 4k^2 - 16k + 4 = k^3 - 15k + 4$

11. $(a)^2 + 2(a)(2) + (2)^2 = a^2 + 4a + 4$

12. $(b)^2 + 2(b)(-6) + (-6)^2 = b^2 - 12b + 36$

13. $(x)^2 - (5)^2 = x^2 - 25$

14. $(2y)^2 - (1)^2 = 4y^2 - 1$

15. $(x + y)(2x + 1) = 2x^2 + 5x + 2$

16. Answers will vary; possible answer: The square of a binomial should be a trinomial.

PRACTICE

1. $x(x) + x(-5) = x^2 - 5x$

2. $2y(y) + 2y(-3) = 2y^2 - 6y$

3. $-5z(2z) - 5z(3) = -10z^2 - 15z$

4. $-2k(k) - 2k(9) = -2k^2 - 18k$

5. $-3m(2m) - 3m(-4) = -6m^2 + 12m$

6. $-7g(3g) - 7g(-1) = -21g^2 + 7g$

7. $2xy(x^2) + 2xy(-y^2) = 2x^3y - xy^3$

8. $3ab(a^2) + 3ab(-2b^2) = 3a^3b - 6ab^3$

9. $x^2 + 5x + 2x + 10 = x^2 + 7x + 10$

10. $y^2 + 3y + 7y + 21 = y^2 + 10y + 21$

11. $a^2 - 4a - 6a + 24 = a^2 - 10a + 24$

12. $b^2 - 4b - 2b + 8 = b^3 - 6b + 8$

13. $c^2 + 3c - 5c - 15 = c^2 - 2c - 15$

14. $d^2 + 4dh + 9dh + 36h^2 = d^2 + 13dh + 36h^2$

15. $g^4 - 2g^2 + 7g^2 - 14 = g^4 + 5g^2 - 14$

16. $h^6 - h^3 + 8h^3 - 8 = h^6 + 7h^3 - 8$

17. $2m^2 + 6m - m - 3 = 2m^2 + 5m - 3$

18. $3n^2 + 15n + 4n + 20 = 3n^2 + 19n + 20$

19. $6p^2 - 4pq + 3pq - 2q^2 = 6p^2 - pq - 2q^2$

20. $20q^2 + 4qr + 5qr + r^2 = 20q^2 + 9qr + r^2$

21. $(x - 2)(x + 2) = (x)^2 - (2)^2 = x^2 - 4$

22. $(s)^2 - (9)^2 = s^2 - 81$

23. $(3r)^2 - (2)^2 = 9r^2 - 4$

24. $(4t)^2 - (5)^2 = 16t^2 - 25$

25. $(v)^2 + 2(v)(-6) + (-6)^2 = v^2 - 12v + 36$

26. $(g)^2 + 2(g)(8) + (8)^2 = g^2 + 16g + 64$

27. $(2w)^2 + 2(2w)(8) + (8)^2 = 4w^2 + 32w + 64$

28. $(3x)^2 + 2(3x)(-2) + (-2)^2 = 9x^2 - 12x + 4$

29. $x(x^2 + 4x - 3) + 3(x^2 + 4x - 3) =$
$x^3 + 4x^2 - 3x + 3x^2 + 12x - 9 = x^3 + 7x^2 + 9x - 9$

30. $y(y^2 - 3y - 8) + 5(y^2 - 3y - 8) =$
$y^3 - 3y^2 - 8y + 5y^2 - 15y - 40 = y^3 + 2y^2 - 23y - 40$

31. $a(a^2 - 2ab - 3b^2) - b(a^2 - 2ab - 3b^2) =$
$a^3 - 2a^2b - 3ab^2 - a^2b + 2ab^2 + 3b^3 =$
$a^3 - 3a^2b - ab^2 + 3b^3$

32. $p(p^2 - 5p - 1) + 2(p^2 - 5p - 1) =$
$p^3 - 5p^2 - p + 2p^2 - 10p - 2 = p^3 - 3p^2 - 11p - 2$

33. $(x + 5)(2x) + x(3x - 2x) = 2x^2 + 10x + x^2 = 3x^2 + 10x$

34a. $A = 1000(1 + 0.04)^3 = 1000(1.04)^3 = 1000(1.12486) =$
$\$1124.86$

b. $A = 2000(1 + r)^2 = 2000(1 + 2r + r^2) =$
$2000 + 4000r + 2000r^2$

35. $(q-2)(q-2)(q-2) = (q^2 - 4q + 4)(q-2) =$
$q(q^2 - 4q + 4) - 2(q^2 - 4q + 4) =$
$q^3 - 4q^2 + 4q - 2q^2 + 8q - 8 = q^3 - 6q^2 + 12q - 8$

36. $(a+5)(a+5)(a+5) = (a^2 + 10a + 25)(a+5) =$
$a(a^2 + 10a + 25) + 5(a^2 + 10a + 25) =$
$a^3 + 10a^2 + 25a + 5a^2 + 50a + 125 =$
$a^3 + 15a^2 + 75a + 125$

37. $(g+3)(g^2 - 2g + 1) = g(g^2 - 2g + 1) + 3(g^2 - 2g + 1) =$
$g^3 - 2g^2 + g + 3g^2 - 6g + 3 = g^3 + g^2 - 5g + 3$

38. $(k-1)(k^2 + 4k + 4) = k(k^2 + 4k + 4) - 1(k^2 + 4k + 4) =$
$k^3 + 4k^2 + 4k - k^2 - 4k - 4 = k^3 + 3k^2 - 4$

39. $54 = \dfrac{1}{2}(3 + 2x + 3)(2x + 2x) - (2x)(2x)$

$54 = \dfrac{1}{2}(6 + 2x)(4x) - 4x^2$

$54 = (3 + x)(4x) - 4x^2$

$54 = 12x + 4x^2 - 4x^2$

$54 = 12x$

$4.5 = x$

40. $255 = \dfrac{1}{2}(x + 15)(x + x + 2) - (x)(x)$

$255 = \dfrac{1}{2}(x + 15)(2x + 2) - x^2$

$255 = (x + 15)(x + 1) - x^2$

$255 = x^2 + x + 15x + 15 - x^2$

$255 = 16x + 15$

$240 = 16x$

$15 = x$

41. $66 = x(x + 5); \ x = 6$

42. Shaded area $= (a + b)(a - b) = a(a - b) + b(a - b)$
$= a^2 - ab + ab - b^2 = a^2 - b^2$

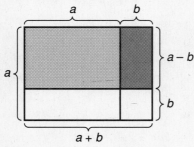

43. $19,800,000\left(1 = \dfrac{3}{100}\right)^2 = 19,800,000(1.03)^2 =$
$19,800,000(1.0609) = 21,005,820$

44. $n(-8) = 32$, so $n = -4$

45. $-6(n) = -42$, so $n = 7$

46. If either $a = 0$, $b = 0$ or both $= 0$, the statement is true.

47. Since $(a + b)^2 = a^2 + 2ab + b^2$, it is $2ab$ bigger than $a^2 + b^2$.

1. The wing rises.

2. $P = \dfrac{1}{2}(1.29)(59.2^2 - 55.7^2)$

$= \dfrac{1}{2}(1.29)(3504.64 - 3102.49)$

$\approx 259.4 \ \dfrac{\text{N}}{\text{m}^2}$

3. $P = \dfrac{1}{2}d(v_1 - v_2)(v_1 + v_2) = \dfrac{1}{2}d(v_1^2 - v_2^2)$

Lesson 11.9, pages 558–561

1. $t + 1$ hours

2.

	Rate	Time	Distance
Dan	55	$t + 1$	$55(t + 1)$
Fran	45	t	$45t$

3. Dan's distance plus Fran's distance will equal 200 miles;
$55(t + 1) + 45t = 200$

4. $55t + 55 + 45t = 200$

$100t = 145$

$t = 1.45$ hr, or $1\dfrac{9}{20}$ hr, or 1 hr 27 min,
so trucks pass each other at 9:27 A.M.

5. Dan's distance: $55\left(1\dfrac{9}{20} + 1\right) =$

$55\left(2\dfrac{9}{20}\right) = \ 134\dfrac{3}{4}$ miles

Fran's distance: $45\left(1\dfrac{9}{20}\right) = \ 65\dfrac{1}{4}$ miles

Total distance $= \ $ 200 miles

6. 42 minutes need to be changed to hours;

$42 \text{ min} \cdot \dfrac{1 \text{ hr}}{60 \text{ min}} = 0.7$ hr

7. $t - 0.7$ hr

8.

	Rate	Time	Distance
Going	500	t	$500t$
Returning	600	$t - 0.7$	$600(t - 0.7)$

9. $500t = 600(t - 0.7)$

10. $500t = 600t - 420$

$420 = 100t$

$4.2 = t$, so distance $= 500(4.2) = 2100$ miles

11. $600(4.2 - 0.7) = 600(3.5) = 2100$ miles

12. Answers will vary

13.
```
camp ──► Darrell at 4 km/h  ⎫
     ──► Ivan at 6 km/h      ⎬ going same direction
         leaving ½ hour later ⎭
```

14. If t is Ivan's time, Darrell's time is longer by $\dfrac{1}{2}$ hour, so $t + 0.5$

15. Their distances are equal.

$$4(t + 0.5) = 6t$$
$$4t + 2 = 6t$$
$$2 = 2t$$
$$1 = t$$

Ivan hikes 1 hr and covers 6 km, Darrell hikes $1\frac{1}{2}$ hours and also cover 6 km.

16. x = Mrs. Gonzalez's speed, $x - 8$ = Mr. Gonzalez's speed, time = 3 hours, distance = 300 miles:

$$300 = 3(x) + 3(x - 8)$$
$$300 = 3x + 3x - 24$$
$$324 = 6x$$
$$54 = x$$

Mr. Gonzalez's distance is $3(54 - 8) = 138$ miles

17. passenger train travels t hours at 62 mph for distance of $62t$

freight train travels $(t - 2)$ hours at 48 mph for distance of $48(t - 2)$

Total distance:
$$355 = 62t + 48(t - 2)$$
$$355 = 62t + 48t - 96$$
$$451 = 110t$$
$$4.1 = t$$

So, trains meet 4.1 hours or 4 hours 6 minutes after noon at 4:06 P.M.

18. Distance to park and return are equal.

t = time returning, so, $t - \frac{1}{3}$ hr = time going to park

$$54\left(t - \frac{1}{3}\right) = 46t$$
$$54t - 18 = 46t$$
$$8t = 18 \;\rightarrow\; t = 2.25 \text{ hours}$$

So, distance = $46(2.25) = 103.5$ miles

19. Faster runner is Jackson at 8 mph

$$t = \frac{d}{r} = \frac{26}{8} = 3.25 \text{ h}$$

In 3.25 hours, Jason has gone 5 (3.25) or 16.25 miles and still has $26 - 16.25$ or 9.75 miles to go.

20. Let t = time when they are 5.1 miles apart

$$5.1 = 8t - 5t$$
$$5.1 = 3t$$
$$1.7 = t, \text{ or } 1 \text{ h } 42 \text{ min}$$

If they start at 9:30, then they are 5.1 miles apart at 11:12 A.M.

REVIEW PROBLEM SOLVING STRATEGIES

1. The envelopes all weigh the same, except the empty one is lighter. Divide them into 3 groups of 9. Weigh one group of 9 against another group of 9. If they balance, the empty envelope is in the leftover group of 9. If they do not, the empty one is in the group in the higher pan. Take the group with the empty envelope. Divide it into 3 groups of 3. Weigh one group of 3 against another group of 3. If they balance, the empty envelope is in the leftover group of 3. If they do not, the empty one is in the group in the higher pan. Take the group with the empty envelope. Weigh one envelope against another. If they balance, the empty one is in the leftover envelope; if they do not, the empty is in the higher pan. Exactly three weighings are necessary.

2. Joel is in Lane 3, the third lane from the left. Methods will vary, but the problem can be solved by making a table as follows. 12 is 3 less than 15 and Lane 6 is 3 places to the right of Lane 3.

Lane	To right	To left	Product
1	8	0	0
2	7	1	7
3	6	2	12
4	5	3	15
5	4	4	16
6	3	5	15
7	2	6	12
8	1	7	7
9	0	8	0

3. Under $5, you save money at Bill's and not at Good Buys. You also save at Bill's for purchases of $7-9 and $14. For all other amounts over $5, you save more at Good Buys.

Chapter Review, pages 562–563

1. b **2.** c **3.** a

4. 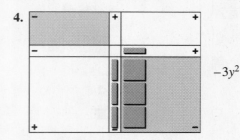 $-3y^2$

5. 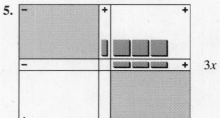 $3x$

6. 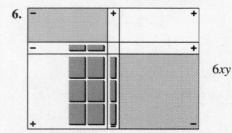 $6xy$

7. 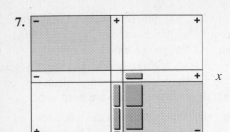 x

8. $z^{4+2} = z^6$

9. $-6a^{4+3} = -6a^7$

10. $-12x^{2+1}y^{1+2} = -12x^3y^7$

11. $6a^{3+1}b^{1+3}c^{1+2} = 6a^4b^4c^3$

12. $3x^{5-3} = 3$

13. $-6a^{8-5} = -6a^3$

14. $\frac{1}{4}p^{6-1}q^{5-7} = \frac{1}{4}p^5q^{-2} = \frac{p^5}{4q^2}$

15. $-5x^{3-2}y^0 = -5x$

16. $\frac{3^3g^{2\cdot3}}{2^3h^{4\cdot3}} = \frac{27g^6}{8h^{12}}$

17. $(1.8 \times 10^6)(2.5 \times 10^4) = 4.5 \times 10^{6+4} = 4.5 \times 10^{10}$

18. $\frac{6.3 \times 10^8}{2 \times 10^5} = 3.15 \times 10^{8-5} = 3.15 \times 10^3$

19. $1{,}200{,}000{,}000 \cdot 2 = 2\,444{,}000{,}000 = 2.4 \times 10^9$

 9 places

20. $0.000000023 \cdot 3 = 0{,}00000006\,9 = 6.9 \times 10^{-8}$

 8 places

21.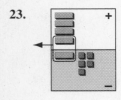

 $2x + y$

22.

 $-2x^2 + 3y$

23.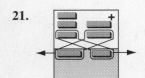

 $3x - 5$

24.

 $4x - y$

25.
$$\begin{array}{r} 4r + 2s - t \\ + \ 7r - 3s + 2t \\ \hline 11r - \ s + \ t \end{array}$$

26.
$$\begin{array}{rl} 7x^2 - 2x - 8 & \to \quad 7x^2 + 2x - \ 8 \\ -(4x^2 - 5x + 2) & \to \ \underline{-4x^2 + 5x - \ 2} \\ & \qquad \ \ 3x^2 + 7x - 10 \end{array}$$

27.
$$\begin{array}{r} 3a + 2b \\ - \ a - 4b + 6c \\ \hline 2a - 2b + 6c \end{array}$$

28.
$$\begin{array}{rl} 4x^2 - xy - 6 & \to \quad 4x^2 + \ xy - 6 \\ -(x^2 - 2xy - 9) & \to \ \underline{- \ x^2 + 2xy + 9} \\ & \qquad \ \ 3x^2 + 3xy + 3 \end{array}$$

29. 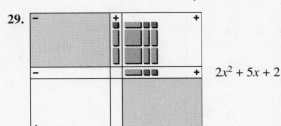 $2x^2 + 5x + 2$

30. 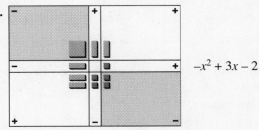 $-x^2 + 3x - 2$

31. 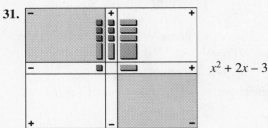 $x^2 + 2x - 3$

32. $k(k) + k(4) + k^2 + 4k$

33. $-5x(3x) - 5x(-y) = -15x^2 + 5xy$

34. $2ab(3a^2) + 2ab(-4b^2) = 6a^3b - 8ab^3$

35. $a^2 + a + 3a + 3 = a^2 + 4a + 3$

36. $x^2 + 4x - 2x - 8 = x^2 + 2x - 8$

37. $2r^2 - 2rs + rs - s^2 = 2r^2 - rs - s^2$

38. $(y)^2 - (6)^2 = y^2 - 36$

39. $(b)^2 + 2(b)(2) + (2)^2 = b^2 + 4b + 4$

40. $y(y^2 + 3y - 4) - 2(y^2 + 3y - 4) =$
$y^3 + 3y^2 - 4y - 2y^2 - 6y + 8 = y^3 + y^2 - 10y + 8$

41.
Boston	→ Lizzie left at 10:30 driving 54 mi/h
	→ José left at 10:50 driving 60 mi/h

$54t = 60\left(t - \frac{1}{3}\right)$

$54t = 60t - 20$

$20 = 6t$

$3\frac{1}{3} = t$, or 3 h 20 min after Lizzie left.

José overtakes Lizzie at 1 : 50 P.M.

1. I: Highest sum of powers = 4, so degree is 4
    II: Highest sum of powers = 7, so degree is 7
    III: Highest powers is 4, so degree is 4
    I and III, so D

2. $g^{2+3}h^{1+4} = g^5h^5$

3. $(2r)^2 + 2(2r)(s) + (s)^2 = 4r^2 + 4rs + s^2$

4. $(3x^2 - 2y) - (2x^2 - 3y) = 3x^2 - 2y - 2x^2 + 3y = x^2 + y$

5. Answers will vary, but should include at least two of the following methods: distributive, FOIL, area model, Algeblocks, or vertical.

6. $(-2x^3y)^2(-xy^3) = (4x^6y^2)(-xy^3) = -4x^7y^5$
    A: $-4x^7y^5$; equivalent
    B: $(4x^6y^2)(-xy^3) = -4x^7y^5$; equivalent
    C: $(-2)^2(x^3)^2(y^2)(-x)^3(y^3) = (4)(x^6)(y^2)(-x^3)(y^3) = -4x^9y^5$; not equivalent
    D: $(-2)^2(x^3)^2(y^2)(-xy^5) = (4)(x^6)(-xy^5) = -4x^7y^5$; equivalent

7. $SA = 2(x+3)(2x-4) + 2(2x-4)(x) + 2(x+3)(x)$
    $SA = (2x+6)(2x-4) + (4x-8)(x) + (2x+6)(x)$
    $SA = 4x^2 - 8x + 12x - 24 + 4x^2 - 8x + 2x^2 + 6x$
    $SA = 10x^2 + 2x - 24$

8. $0.0000002\underset{\text{7 places}}{\underbrace{\,15}} = 2.15 \times 10^{-7}$; B

9. $4\underset{\text{10 places}}{\underbrace{3,000,000,000}} = 4.3 \times 10^{10}$

10a. $3.78 \times 10^{6+8} = 3.78 \times 10^{14}$

b. $33.58 \times 10^{5+4} = 33.58 \times 10^9 = 3.358 \times 10^{10}$

c. $3^2 \times 10^{7 \cdot 2} = 9 \times 10^{14}$

d. $2 \times 10^{9-3} = 2 \times 10^6$

11. $\left(\dfrac{2x^3y}{-xy^4}\right)^2 = \left(-\dfrac{2x^2}{y^3}\right)^2 = \dfrac{4x^4}{y^6}$

    I: $4\left(\dfrac{x^3y}{-xy^4}\right)^2 = 4\left(-\dfrac{x^2}{y^3}\right)^2 = \dfrac{4x^4}{y^6}$; equivalent

    II: $-4\left(\dfrac{x^3y}{-xy^4}\right)^2 = -4\left(-\dfrac{x^2}{y^3}\right)^2 = -\dfrac{4x^4}{y^6}$; not equivalent

    III: $\dfrac{4x^6y^2}{-x^2y^8} = -\dfrac{4x^4}{y^6}$; not equivalent

    IV: $\dfrac{4x^4}{y^6}$; equivalent

    V: $4x^4y^{-6} = \dfrac{4x^4}{y^6}$; equivalent

    I, IV, and V are equivalent, so B

12. $A = 4(x+1)(x+1) = (4x+4)(x+1) = 4x^2 + 4x + 4x + 4 = 4x^2 + 8x + 4$

13. w = width, $3w$ = length
    $A = 3w(w) = 3w^2$
    $48 = 3w^2 \to 16 = w^2 \to w = 4$
    The width is 4 cm and the length is $3 \cdot 4$ or 12 cm.

14. $(4xy)(x)(?) = 8x^2y^2$
    $(4x^2y)(?) = 8x^2y^2$
    $(?) = \dfrac{8x^2y^2}{4x^2y} = 2y$

15. Joelle drives for t hours at 58 mph for a distance of $58t$. Jason drives for $\left(t - 1\dfrac{1}{2}\right)$ hours at 50 mph for distance of $50\left(t - 1\dfrac{1}{2}\right)$.
    Combined distance =
    $465 \text{ miles} = 58t + 50\left(t - 1\dfrac{1}{2}\right)$
    $465 = 58t + 50t - 75$
    $540 = 108t \to t = 5$
    They meet at $6:30 + 5:00$ or $11:30$ A.M.

Cumulative Review, pages 566–567

1. $y^{3+5+2} = y^{10}$

2. $k^{16-8} = k^8$

3. $t^{5 \cdot 8} = t^{40}$

4. $3^3 m^{4 \cdot 3} n^3 = 27m^{12}n^3$

5. $x^{7-10} = x^{-3} = \dfrac{1}{x^3}$

6. $\dfrac{(-2)^3 a^{5 \cdot 3}}{b^{2 \cdot 3}} = \dfrac{-8a^{15}}{b^6}$

7. $2x^2 - 5x - 11 = 0$
    $x = \dfrac{-(-5) \pm \sqrt{(-5)^2 - 4(2)(-11)}}{2(2)} = \dfrac{5 \pm \sqrt{113}}{4} \approx -1.41 \text{ or } 3.91$

8. $\dfrac{1}{2}x^2 + 6 = 13x$
    $\dfrac{1}{2}x^2 - 13x + 6 = 0$
    $x^2 - 26x + 12 = 0$
    $x = \dfrac{26 \pm \sqrt{(-26)^2 - 4(1)(12)}}{2(1)}$
    $= \dfrac{26 \pm \sqrt{628}}{2} \approx 0.47 \text{ or } 25.53$

9. $4n - 3 < 2n + 9$ and $1 - 2n < 5$
    $2n < 12$ and $-2n < 4$
    $n < 6$ and $n > -2$

    ![number line from -3 to 7 with open circles at -2 and 6, shaded between]

10. $6(2w - 1) \le 3(3w - 5)$ or $w - 25 \ge -23$
    $12w - 6 \le 9w - 15$ or $w \ge 2$
    $3w \le -9$ or $w \ge 2$
    $w \le -3$ or $w \ge 2$

    ![number line from -4 to 4 with closed circles at -3 and 2, shaded left of -3 and right of 2]

11. $\frac{1}{4}z - 3 \le 1 - \frac{3}{4}z$ and $2z - 2.7 > 5.3$

$\qquad z \le 4$ and $2z > 8$

$\qquad z \le 4$ and $z > 4$

impossible, so no solution

12. $x^3 + 5x^2 + 3x - 36$

13. $(5v^3 - 7v^2 + v - 4) + (-8v^3 + 7v^2 + 3v - 10) = -3v^3 + 4v - 14$

14. $n = 0.363636...$

$\underline{100n = 36.363636...}$

$99n = 36$

$n = \frac{36}{99} = \frac{4}{11}$

15. $8.75 = 8\frac{3}{4} = \frac{35}{4}$

16. $n = 2.166666...$

$10n = 21.66666...$

$\underline{100n = 216.66666...}$

$90n = 195$

$n = \frac{195}{90} = \frac{13}{6}$

17. $4 = a + b + c$

$25 = (-2)^2 a + (-2)b + c$

$16 = 49a + 7b + c$

$\begin{array}{l} 4 = a + b + c \\ \underline{25 = 4a - 2b + c} \\ 21 = 3a - 3b \\ b = a - 7 \end{array}$ $\quad \begin{array}{l} 25 = 4a - 2b + c \\ \underline{16 = 49a + 7b + c} \\ 9 = -45a - 9b \\ b = -5a - 1 \end{array}$

$a - 7 = -5a - 1 \rightarrow a = 1$, so $b = -6$ and $c = 9$

$y = x^2 - 6x + 9$

18. $x + (x + 1) + (x + 2) < 70$

$\qquad 3x + 3 < 70$

$\qquad 3x < 67$

$\qquad x < 22\frac{1}{3}$

$x = 22, x + 1 = 23,$ and $x + 2 = 24$

19. $x^2 - 7x + 3x - 21 = x^2 - 4x - 21$

20. $(3t)^2 - (4)^2 = 9t^2 - 16$

21. $(2z)^2 + 2(2z)(5) + (5)^2 = 4z^2 + 20z + 25$

22. $-12m^{1+4}n^{2+1} - 8mn^{2+3} = -12m^5n^3 - 8mn^5$

23. $x(2x^2 + 7x - 4) - 3(2x^2 + 7x - 4) = 2x^3 + 7x^2 - 4x - 6x^2 - 21x + 12 = 2x^3 + x^2 - 25x + 12$

24. FOIL represents each of the four products to be found when multiplying binomials. Excluding monomials, any other combination of polynomials will yield more than four products, thus rendering FOIL useless as an aid.

25. $80 = l + 2w$, so $80 - 2w = l$

Area $= l \cdot w = (80 - 2w)w = 80w - 2w^2$

Maximum at vertex: $w = \frac{-b}{2a} = \frac{-80}{2(-2)} = 20$,

so $l = 80 - 2(20)$

Dimensions are 40 by 20 for 800 ft.

26. additive inverse property

27. symmetric property of equality

28. commutative property of addition

29. I: $(2x + 3)(x - 1) = 2x^2 - 2x + 3x - 3 = 2x^2 + x - 3;$ not equivalent

II: $(x^3 + 4x - 5) - (x^3 - 2x^2 + 5x - 2)$
$= (x^3 + 4x - 5) + (-x^3 + 2x^2 - 5x + 2)$
$= 2x^2 - x - 3;$ equivalent

III: $(x^2 + 3x + 1) + (x^2 - 2x - 4) = 2x^2 + x - 3;$ not equivalent
only II is equivalent, so B

30. If their original salary was $1000, then after the first month Jorge made $1100 and Consuela $900. After the second month, Jorge's salary was $990, as was Consuela's. So, yes, they still earn the same amount, but their salary is slightly less than it was originally.

31. $4\,0{,}500{,}000 = 4.05 \times 10^7$

7 places

32. $0{,}000000003\,2 = 3.2 \times 10^{-9}$

9 places

33. $23.4 \times 10^{6+6} = 2\,3.4 \times 10^{12} = 2.34 \times 10^{13}$

1 place

Standardized Test, page 567

1. home → Boy leaves at 10:00 at 4 mi/h; Sister leaves at 10:30 at 7 mi/h } distance is equal

$4t = 7\left(t - \frac{1}{2}\right)$

$4t = 7t - 3\frac{1}{2}$

$3\frac{1}{2} = 3t$

$t = 1\frac{1}{6}$ h or 1 h 10 min

So, sister catches brother at 11:10 A.M.; D

2. Slope $= m = \frac{y_2 - y_1}{x_2 - x_1} = \frac{d - b}{c - a};$ C

3. Check discriminant: $(-b)^2 - 4(1)(-c) = b^2 + 4c$ which is > 0, so 2 solutions; D

4. $b = 3$, $c = -4$, so vertex is at $(3, -4)$; C

5. Sorted data: 7, 10, 13, 14, 16, 21, 24, 27, 35, 38, 42
$Q_1 = 13$, $Q_2 = 21$, $Q_3 = 35$; $Q_3 - Q_1 = 35 - 13 = 22$; A

6. A: $(n + 3)(n - 2) = n^2 - 2n + 3n - 6 = n^2 + n - 6$; No
B: $(n - 5)(n - 1) = n^2 - n - 5n + 5 = n^2 - 6n + 5$; No
C: $(n - 6)(n + 1) = n^2 + n - 6n - 6 = n^2 - 5n - 6$; No
D: $(n - 3)(n + 2) = n^2 + 2n - 3n - 6 = n^2 - n - 6$; No
E: $(n - 1)(n + 6) = n^2 + 6n - n - 6 = n^2 + 5n - 6$; Yes

7. $\$42.50 - 0.20(42.50) = 42.50 - 8.50 = \$34.00$; B

8. I: True, since product of 2 negative numbers is positive
II: False, since y is always opposite of x
III: False, since 2 negative numbers can never total 4; A

9. $6x + 5y = 1.35 \rightarrow 36x + 30y = 8.10$
$\underline{5x + 6y = 1.40} \rightarrow \underline{25x + 30y = 7.00}$
$ \overline{11x = 1.10}$
$x = 0.10$ and $5(.10) + 6y = 1.40$, so $y = 0.15$
$0.15 - 0.10 = 0.05$; B

10. $21 \times 10^{8 + (-3)} = 2.1 \times 10^5 = 2.1 \times 10^6$; C

1 place

Data Activity, page 568

1. $1150 - 27.2 = 1122.8$ mi^2

2. $4(27.2) = 108.8$, Snake River

3. $0.40x = 1150$

$$x = \frac{1150}{0.40} = 2875 \text{ thousand mi}^2 \text{ or } 2.875 \times 10^6 \text{ mi}^2$$

4.

Drainage Area of U.S. Rivers

27.2   42.75 108   293   1150

5. Answers will vary.

Lesson 12.1, page 571

THINK BACK

1. $(x + 1)(x + 2)$

2. Blocks showing $x + 2$ are placed along the horizontal axis. Blocks showing $x + 1$ are placed along the vertical axis. Within the quadrant, blocks are used to form a rectangle that has those dimensions.

3. $x^2 + 3x + 2$

EXPLORE

4. The common side can be either 1, 2, or 4. The largest common side is 4.

5.

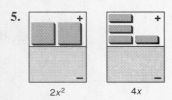

$2x^2$ $4x$

6. The common side can be either $1x$ or $2x$. The largest common side is $2x$.

7.

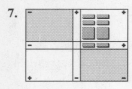

8. $2x$ and $x + 2$

9. $2x^2 + 4x = 2x(x + 2)$

10. Answers will vary, but the above shows $2x^2 + 4x$ as the product of the two factors $2x$ and $x + 2$.

11. $y^2 + 2y = y(y + 2)$

12. $-4x^2 + 2x = 2x(-2x + 1)$

13.

14. $(x + 2)$ and $(x + 3)$

15. $x^2 + 4x + 3 = (x + 3)(x + 1)$

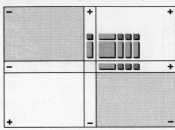

16. $y^2 + 6y + 8 = (y + 4)(y + 2)$

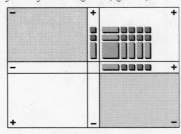

17. $y^2 + 9y + 8 = (y + 1)(y + 8)$

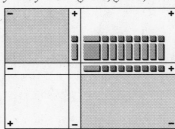

18. The coefficient of squared variable is 1. All terms are positive.

19. Arrangements will vary.   **20.** $(x - 3)$ and $(x - 1)$

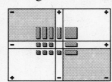

21. Students might place 3 negative x-blocks in Quadrant II and 1 negative x-block in Quadrant IV, or the reverse. The answer is not affected, but the order of the factors may be different.

22. $x^2 - 2x + 1 = (x - 1)(x - 1)$

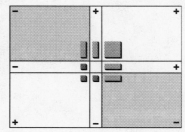

23. $x^2 - 7x + 6 = (x - 6)(x - 1)$

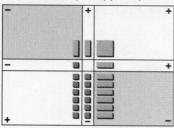

24. $y^2 - 3y + 2 = (y - 2)(y - 1)$

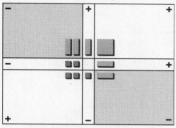

25. The coefficient of the squared variable is 1; the middle term is negative, and the constant term is positive.

26.

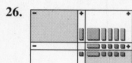

27. $(x - 1)$ and $(x + 4)$

28. $y^2 + y - 2 = (y + 2)(y - 1)$

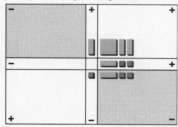

29. $x^2 + 2x - 3 = (x + 3)(x - 1)$

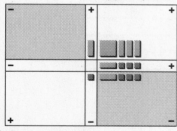

30. $y^2 + 2y - 8 = (y + 4)(y - 2)$

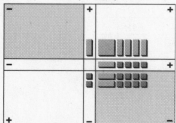

31. The coefficient of the squared variable is 1; the middle term is positive and the constant is negative.

32. Arrangements will vary. The 4 x-blocks and 4 unit blocks may be in Quadrant IV instead of in Quadrant II.

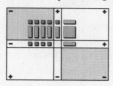

33. $(x + 1)$ and $(x - 4)$

34. $x^2 - x - 6 = (x + 2)(x - 3)$

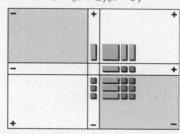

35. $y^2 - 4y - 5 = (y + 1)(y - 5)$

36. $x^2 - 3x - 10 = (x + 2)(x - 5)$

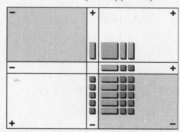

37. The coefficient of the squared variable is 1; the middle term and constant are both negative.

MAKE CONNECTIONS

38. The coefficient of the x^2 term is 1; the constant term is ±6. The factors all include ±3 and ±2.

39. The first and last terms are positive, and the three terms in each are the same. The sign of the middle term is opposite.

40. The first terms are positive, the last terms are negative, and the three terms in each are the same. The sign of the middle term is opposite.

41. c; negative constant so signs are different, negative middle term

42. a; all positive terms

43. d; negative constant so signs are different, positive middle term

44. b; positive constant so signs are the same, negative middle term

45. Answers will vary. Possible answer: If the constant term of the trimonial is positive, the constants in the binomial factors are both the same sign, If the constant term of the trinomial is negative, the constants in the binomial factors have different signs.

SUMMARIZE

46. Answers will vary. Possible answer: Model $3x^2$ and $-6x$. Make rectangles of each so they have the greatest possible common dimension. The factored form of $3x^2 - 6x$ is $3x(x - 2)$.

47. Find two factors that give the constant term when multiplied and the coefficient of the middle term when added algebraically. Pay attention to the signs of the factors.

48. $2x^2 + 7x + 3 = (2x + 1)(x + 3)$; since all terms in the trinomial are positive, all signs are positive, 2 and 3 are prime so use trial and error to factor.

Lesson 12.2, page 575

EXPLORE

1. 1×24 and 1×80
2×12 and 2×40
4×6 and 4×20
8×3 and 8×10
Common factors of 24 and 80 are 1, 2, 4, 8.

2. $2(1) + 2(24) + (1) + 2(80) = 211$ ft
$2(2) + 2(12) + (2) + 2(40) = 110$ ft
$2(4) + 2(6) + (4) + 2(20) = 64$ ft
$2(8) + 2(3) + (8) + 2(10) = 50$ ft

3. 50 ft for 8×3 and 8×10 pens

4. They are the common factors of the two areas.

TRY THESE

1. GCF of $9x^2 + 3x$ is $3x$, so factored form is $3x(3x + 1)$

2. GCF of $15y^4z + 10y2^{z2} - 20yz$ is $5yz$, so factored form is $5yz(3y^3 + 2yz - 4)$

3. GCF of $2y^3 - 16y^2$ is $2y^2$, so factored form is $2y^2(y - 8)$

4. Determine what factors of 3 have a sum of -4.

5. $x^2 + 5x + 6 = (x + ?)(x + ?)$

Factors of 6	Sum of factors
1, 6	7
−1, −6	−7
2, 3	5
−2, −3	−5

So $x^2 + 5x + 6 = (x + 2)(x + 3)$

6. $x^2 + 6x + 8 = (x + ?)(x + ?)$

Factors of 8	Sum of factors
1, 8	9
−1, −8	−9
2, 4	6
−2, −4	−6

So $x^2 + 6x + 8 = (x + 2)(x + 4)$

7. $x^2 - 4x + 3 = (x + ?)(x + ?)$

Factors of 3	Sum of factors
1, 3	4
−1, −3	−4

So $x^2 - 4x + 3 = (x - 1)(x - 3)$

8. $x^2 - 11x + 30 = (x + ?)(x + ?)$

Factors of 30	Sum of factors
1, 30	31
−1, −30	−31
2, 15	17
−2, −15	−17
5, 6	11
−5, −6	−11

So $x^2 - 11x + 30 = (x - 5)(x - 6)$

9. $x^2 - 2x - 8 = (x + ?)(x + ?)$

Factors of −8	Sum of factors
−1, 8	7
1, −8	−7
−2, 4	2
2, −4	−2

So $x^2 - 2x - 8 = (x + 2)(x - 4)$

10. $x^2 - x - 12 = (x + ?)(x + ?)$

Factors of −12	Sum of factors
−1, 12	11
1, −12	−11
−2, 6	4
2, −6	−4
−3, 4	1
3, −4	−1

So $x^2 - x - 12 = (x + 3)(x - 4)$

11. $x^2 + 3x - 28 = (x + ?)(x + ?)$

Factors of −28	Sum of factors
−1, 28	27
1, −28	−27
−2, 14	12
2, −14	−12
−4, 7	3
4, −7	−3

So $x^2 + 3x - 28 = (x - 4)(x + 7)$

12. $x^2 + 5x - 14 = (x + ?)(x + ?)$

Factors of –14	Sum of factors
–1, 14	13
1, –14	–13
–2, 7	5
2, –7	–5

So $x^2 + 5x - 14 = (x - 2)(x + 7)$

13. $2x^2 + 5x + 3 = (2x + ?)(x + ?)$

Trial factors	Product
$(2x + 1)(x + 3)$	$2x^2 + 7x + 3$
$(2x - 1)(x - 3)$	$2x^2 - 7x + 3$
$(2x + 3)(x + 1)$	$2x^2 + 5x + 3$
$(2x - 3)(x - 1)$	$2x^2 - 5x + 3$

So $2x^2 + 5x + 3 = (2x + 3)(x + 1)$

14. $x^2 + 7x + 6 = (x + ?)(x + ?)$

Factors of 6	Sum of factors
1, 6	7
–1, –6	–7
2, 3	5
–2, –3	–5

So dimensions are $(x + 1)$ and $(x + 6)$

15. Position of some of the blocks may vary.

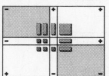

PRACTICE

1. GCF of $2a^4 + 8a$ is $2a$, so factored form is $2a(a^3 + 4)$

2. GCF of $7b^3 + 21b$ is $7b$, so factored form is $7b(b^2 + 3)$

3. GCF of $8ab^2 - 12a^2b^3$ is $4ab^2$, so factored form is $4ab^2(2 - 3ab)$

4. GCF of $10c^3d^2 - 15cd^2$ is $5cd^2$, so factored form is $5cd^2(2c^2 - 3d)$

5. GCF of $6c^2 - 9d^2$ is 3, so factored form is $3(2c^2 - 3d^2)$

6. GCF of $15f - 20g^2$ is 5, so factored form is $5(3f - 4g^2)$

7. GCF of $4x^3 - 2x^2 + 14x$ is $2x$, so factored form is $2x(2x^2 - x + 7)$

8. GCF of $3y^4 + 9y^2 - 15$ is 3, so factored form is $3(y^4 + 3y^2 - 5)$

9. GCF of $2z^3 + 3z^2 + 4z$ is z, so factored form is $z(2z^2 + 3z + 4)$

10. GCF of $9mn - 3m^2 + 4mn^2$ is m, so factored form is $(9n - 3m + 4n^2)$

11. GCF of $8abc^2 - 4b^2c + 12a^2bc$ is $4bc$, so factored form is $4bc(2ac - b + 3a^2)$

12. GCF of $6x^2yz + 2xy^2z - 4xyz$ is $2xyz$, so factored form is $2xyz(3x + y - 2)$

13. GCF of $12a^4b^3c^2 - 4a^3bc^2 + 8a^2c - 16ab$ is $4a$, so factored form is $4a(3a^3b^3c^2 - a^2bc^2 + 2ac - 4b)$

14. GCF of $9x^3yz^2 - 6x^2yz^2 + 12xyz^2 - 21yz^2$ is $3yz^2$, so factored form is $3yz^2(3x^3 - 2x^2 + 4x - 7)$

15. $x^2 + 5x + 4 = (x + ?)(x + ?)$

Factors of 4	Sum of factors
1, 4	5
–1, –4	–5
2, 2	4
–2, –2	–4

So $x^2 + 5x + 4 = (x + 1)(x + 4)$

16. $y^2 + 5y + 6 = (y + ?)(y + ?)$

Factors of 6	Sum of factors
1, 6	7
–1, –6	–7
2, 3	5
–2, –3	–5

So $y^2 + 5y + 6 = (y + 2)(x + 3)$

17. $z^2 + 8z + 15 = (z + ?)(z + ?)$

Factors of 15	Sum of factors
1, 15	16
–1, –15	–16
3, 5	8
–3, –5	–8

So $z^2 + 8z + 15 = (z + 3)(z + 5)$

18. $y^2 - 4y + 4 = (y + ?)(y + ?)$

Factors of 4	Sum of factors
1, 4	5
–1, –4	–5
2, 2	4
–2, –2	–4

So $y^2 - 4y + 4 = (y - 2)(y - 2)$

19. $x^2 - 6x + 9 = (x + ?)(x + ?)$

Factors of 9	Sum of factors
1, 9	10
–1, –9	–10
3, 3	6
–3, –3	–6

So $x^2 - 6x + 9 = (x - 3)(x - 3)$

20. $x^2 - 10x + 9 = (x + ?)(x + ?)$

Factors of 9	Sum of factors
1, 9	10
–1, –9	–10
3, 3	6
–3, –3	–6

So $x^2 - 10x + 9 = (x - 1)(x - 9)$

21. $z^2 - 10z + 9 = (z - 1)(z - 9)$ Identical to Exercise 20 with $z = x$

22. $z^2 - 11z + 28 = (z + ?)(z + ?)$

Factors of 28	Sum of factors
1, 28	29
−1, −28	−29
2, 14	16
−2, −14	−16
4, 7	11
−4, −7	−11

So $z^2 - 11z + 28 = (z - 4)(z - 7)$

23. $y^2 + 7y - 8 = (y + ?)(y + ?)$

Factors of −8	Sum of factors
1, −8	−7
−1, 8	7
2, −4	−2
−2, 4	2

So $y^2 + 7y - 8 = (y - 1)(y + 8)$

24. $x^2 + x - 6 = (x + ?)(x + ?)$

Factors of −6	Sum of factors
1, −6	−5
−1, 6	5
2, −3	−1
−2, 3	1

So $x^2 + x - 6 = (x - 2)(x + 3)$

25. $y^2 + 11y - 12 = (y + ?)(y + ?)$

Factors of −12	Sum of factors
1, −12	−11
−1, 12	11
2, −6	−4
−2, 6	4
3, −4	−1
−3, 4	1

So $y^2 + 11y - 12 = (y - 1)(y + 12)$

26. $a^2 + 6a - 7 = (a + ?)(a + ?)$

Factors of −7	Sum of factors
1, −7	−6
−1, 7	6

So $a^2 + 6a - 7 = (a - 1)(a + 7)$

27. $b^2 + 3b - 4 = (b + ?)(b + ?)$

Factors of −4	Sum of factors
1, −4	−3
−1, 4	3
2, −2	0

So $b^2 + 3b - 4 = (b - 1)(b + 4)$

28. $x^2 - x - 12 = (x + ?)(x + ?)$

Factors of −12	Sum of factors
1, −12	−11
−1, 12	11
2, −6	−4
−2, 6	4
3, −4	−1
−3, 4	1

So $x^2 - x - 12 = (x + 3)(x - 4)$

29. $x^2 - 2x - 35 = (x + ?)(x + ?)$

Factors of −35	Sum of factors
1, −35	−34
−1, 35	34
5, −7	−2
−5, 7	2

So $x^2 - 2x - 35 = (x + 5)(x - 7)$

30. $n^2 - 4n - 12 = (n + ?)(n + ?)$

Factors of −12	Sum of factors
1, −12	−11
−1, 12	11
2, −6	−4
−2, 6	4
3, −4	−1
−3, 4	1

So $n^2 - 4n - 12 = (n + 2)(n - 6)$

31. $c^2 - 3c - 18 = (c + ?)(c + ?)$

Factors of −18	Sum of factors
1, −18	−17
−1, 18	17
2, −9	−7
−2, 9	7
3, −6	−3
−3, 6	3

So $c^2 - 3c - 18 = (c + 3)(c - 6)$

32. $z^2 - 6z - 7 = (z + ?)(z + ?)$

Factors of −7	Sum of factors
1, −7	−6
−1, 7	6

So $z^2 - 6z - 7 = (z + 1)(z - 7)$

33. $5x^2 + 12x + 7 = (5x + ?)(x + ?)$
Factors of 7 are 1 and 7, −1 and −7

Trial factors	Product
$(5x + 1)(x + 7)$	$5x^2 + 36x + 7$
$(5x - 1)(x - 7)$	$5x^2 - 36x + 7$
$(5x + 7)(x + 1)$	$5x^2 + 12x + 7$
$(5x - 7)(x - 1)$	$5x^2 - 12x + 7$

So $5x^2 + 12x + 7 = (5x + 7)(x + 1)$

34. $2a^2 + 13a - 7 = (2a + ?)(a + ?)$

Factors of -7 are 1 and -7, -1 and 7

Trial factors	Product
$(2a + 1)(a - 7)$	$2a^2 - 13a - 7$
$(2a - 1)(a + 7)$	$2a^2 + 13a - 7$
$(2a + 7)(a - 1)$	$2a^2 + 5a - 7$
$(2a - 7)(a + 1)$	$2a^2 - 5a - 7$

So $2a^2 + 13a - 7 = (2a - 1)(a + 7)$

35. $2b^2 + 5b - 3 = (2b + ?)(b + ?)$

Factors of -3 are 1 and -3, -1 and 3

Trial factors	Product
$(2b + 1)(b - 3)$	$2b^2 - 5b - 3$
$(2b - 1)(b + 3)$	$2b^2 + 5b - 3$
$(2b + 3)(b - 1)$	$2b^2 + b - 3$
$(2b - 3)(b + 1)$	$2b^2 - b - 3$

So $2b^2 + 5b - 3 = (2b - 1)(b + 3)$

36. $(3r)(3r) - \pi r^2 = 9r^2 - \pi r^2 = r^2(9 - \pi)$

37. $3x^2 + 13x - 10 = (3x + ?)(x + ?)$

Factors of -10 are 1 and -10, -1 and 10, 2 and -5, -2 and 5

Trial factors	Product
$(3x + 1)(x - 10)$	$3x^2 - 29x - 10$
$(3x - 1)(x + 10)$	$3x^2 + 29x - 10$
$(3x + 10)(x - 1)$	$3x^2 + 7x - 10$
$(3x - 10)(x + 1)$	$3x^2 - 7x - 10$
$(3x + 2)(x - 5)$	$3x^2 - 13x - 10$
$(3x - 2)(x + 5)$	$3x^2 + 13x - 10$
$(3x + 5)(x - 2)$	$3x^2 - x - 10$
$(3x - 5)(x + 2)$	$3x^2 + x - 10$

So $3x^2 + 13x - 10 = (3x - 2)(x + 5)$

EXTEND

38. $a^{n+4} + a^n = a^n \cdot a^4 + a^n = a^n(a^4 + 1)$

39. $6b^{2n} + 15b^{2n+2} = 6b^{2n} + 15b^{2n}b^2 = 3b^{2n}(2 + 5b^2)$

40. $x^2 + 13x + 36 = (x + ?)(x + ?)$

Factors of 36	Sum of factors
1, 36	37
$-1, -36$	-37
2, 18	20
$-2, -18$	-20
3, 12	15
$-3, -12$	-15
4, 9	13
$-4, -9$	-13
6, 6	12
$-6, -6$	-12

So $x^2 + 13x + 36 = (x + 4)(x + 9)$ and $(x + 9) - (x + 4) = 5$ so the length is 5 more than width.

41. $Z\left(\dfrac{1}{2}\right) + N\left(-\dfrac{1}{2}\right)$ The common factor is $\dfrac{1}{2}$.

$\dfrac{1}{2}(Z - N)$

42a. Common factor is $v\pi$ so $q = \dfrac{v\pi(D^2 - d^2)}{4}$

b. $q = \dfrac{35\pi(3^2 - 0.75^2)}{4} = \dfrac{109.9(9 - 0.5625)}{4} \approx$

$\dfrac{109.9(8.4375)}{4} = \dfrac{927.28125}{4} \approx 232$

THINK CRITICALLY

43.

Factors of 12	Sum of factors
1, 12	13
$-1, -12$	-13
2, 6	8
$-2, -6$	-8
3, 4	7
$-3, -4$	-7

Possible values for b

44. $x^2 + bx + 24 = (x + 3)(x + s) = x^2 + sx + 3x + 3s = x^2 + (s + 3)x + 3s$

Since $3s = 24$, $s = 8$ and $b = s + 3 = 8 + 3 = 11$

ALGEBRAWORKS

1.

Linear Model				
x	1.13	1.08	1.03	0.97
$y = 57.6x - 12.3$	52.79	49.91	47.03	43.57

2.

Quadratic Model				
x	1.13	1.08	1.03	0.97
$y = 35.8x^2 - 17.4x + 26.9$	52.95	49.87	46.96	43.71

3. Answers will vary, but students may prefer the linear model because it is easier to compute.

4. The hydrologist may gather more information and use that data to refine the model.

Lesson 12.3, page 582

EXPLORE/WORKING TOGETHER

1. $(37)(25) = 925 = 961 - 36 = 31^2 - 6^2$

2. $(36)(28) = 1008 = 1024 - 16 = 32^2 - 4^2$

3. The product equals the difference of two squares. Find the mean of the two factors. The other number to be squared is the difference between the mean and either of the two original factors.

4. $\dfrac{(51 + 35)}{2} = 43$; $51 - 43 = 8$; $(43 + 8)(43 - 8) = 1785$

TRY THESE

1. $a^2 - 16$ is a difference of squares so it factors into $(a + 4)(a - 4)$

2. $b^2 - 81$ is a difference of squares so it factors into $(b + 9)(b - 9)$

3. $c^4 + 49$ is a sum of squares and does not factor

4. $d^4 - 36$ is a difference of squares so it factors into $(d^2 + 6)(d^2 - 6)$

5. $9e^4 - 16f^2$ is a difference of squares so it factors into $(3e^2 + 4f)(3e^2 - 4f)$

6. 21 is not a perfect square, so $25g^2 - 21h^2$ is not factorable

7. The first and last terms are perfect squares, and the square root of the first term is a, and the square root of the last term is 5. Twice the product of these is $10a$ so it factors into $(a - 5)^2$

8. The first and last terms are perfect squares and the square root of the first term is b, and the square root of the last term is 7. Twice the product of these is $14a$ so it factors into $(b + 7)^2$

9. The first and last terms are perfect squares and the square root of the first term is c^2, and the square root of the last term is 2. Twice the product of these is $4c^2$ so it factors into $(c^2 - 2)^2$

10. The first and last terms are perfect squares and the square root of the first term is $3y$, and the square root of the last term is 2. Twice the product of these is $12y$, which is not the middle term, so it is not a perfect square trinomial.

11. $x^2 - 6x + 9 = (x - 3)^2$

12. $a^2 - 25b^2$ is a difference of squares and factors into $(a + 5b)(a - 5b)$

13. The first and last terms are perfect squares and the middle term is twice the product of the square roots of the first and last terms.

PRACTICE

1. perfect square trinomial so it factors into $(a - 8)^2$

2. perfect square trinomial so it factors into $(b + 1)^2$

3. perfect square trinomial so it factors into $(c + 3)^2$

4. not a perfect square trinomial since $7d \neq 2(d)(7)$

5. perfect square trinomial so it factors into $(e - 10)^2$

6. perfect square trinomial so it factors into $(f^2 + 5)^2$

7. perfect square trinomial so it factors into $(9h + 2)^2$

8. not a perfect square trinomial since $15j \neq 2(5j)(3)$

9. perfect square trinomial so it factors into $(10p + 3q)^2$

10. difference of squares so it factors into $(t + 4)(t - 4)$

11. difference of squares so it factors into $(v + 11)(v - 11)$

12. sum of squares so it does not factor

13. difference of squares so it factors into $(y + 1)(y - 1)$

14. difference of squares so it factors into $(2z + 7)(2z - 7)$

15. GCF is 3, so it factors into $3(3a^2 - 4)$, cannot factor further

16. difference of squares so it factors into $(5b + 8)(5b - 8)$

17. difference of squares so it factors into $(6c^2 + d)(6c^2 - d)$

18. 21 is not a perfect square, so not factorable

19. Shaded area $= 4x(x) - 5(5) = 4x^2 - 25 = (2x + 5)(2x - 5)$ since $4x^2 - 25$ is the difference of squares.

EXTEND

20. perfect square trinomial so it factors into $(g^3 + 6)^2$

21. perfect square trinomial so it factors into $(4r^2 - 5)^2$

22. difference of squares so it factors into $(3g^5 + 10h^4)(3g^5 - 10h^4)$

23. does not factor since $60b \neq 2(20b)(3)$

24. sum of squares so it does not factor

25. difference of squares so it factors into $(11x + 17)(11x - 17)$

26. $y = x^2 - 6x + 9 = (x - 3)^2$ so it is shifted right 3 units.

27a. By trial and error 2 pairs are:
7 and 5: $49 - 25 = 24$
5 and 1: $25 - 1 = 24$

b. 12 and 2: $(7 + 5)(7 - 5) = 49 - 25 = 24$
6 and 4: $(5 + 1)(5 - 1) = 25 - 1 = 24$

THINK CRITICALLY

28. $(x + y)^2 - z^2$ difference of squares
$[(x + y) + z][(x + y) - z] =$
$(x + y + z)(x + y - z)$

29. $(a + b)^2 - (c + d)^2$ difference of squares
$[(a + b) + (c + d)][(a + b) - (c + d)] =$
$(a + b + c + d)(a + b - c - d)$

30. $(p^2 - 2pq + q^2) - s^2$ perfect square trinomial
$(p - q)^2 - s^2$ difference of squares
$[(p - q) + s][(p - q) - s] =$
$(p - q + s)(p - q - s)$

31. $4x^2 - 3Ny^4 = (2x + 9y^2)(2x - 9y^2) = 4x^2 - 81y^4$ so $3Ny^4 = 81y^4$ and $N = 27$

32. $16a^2 + 5Nb = (4a + 5b^3)(4a - 5b^3) = 16a^2 - 25b^6$ so $5Nb = -25b^6$ and $N = -5b^5$

MIXED REVIEW

33.

34.

35.

36. Substitute points in equations and check, B is only equation that works.

37. $\begin{cases} y = 2x \\ x + y = 24 \end{cases}$

$x + 2x = 24$ Substitute $2x$ for y in Equation 2.

    $x = 8$ and $y = 2(8) = 16$

So, $(8, 16)$ solves the system.

38. $\begin{cases} y = -x \\ 2x - 3y = 25 \end{cases}$

$2x - 3(-x) = 25$ Substitute $-x$ for y in Equation 2.

    $x = 5$ and $y = -5$

So, $(5, -5)$ solves the system.

39. $\begin{cases} y = 2x + 1 \\ 2x + 3y = 11 \end{cases}$

$2x + 3(2x + 1) = 11$ Substitute $2x + 1$ for y in Equation 2.

    $x = 1$ and $y = 2(1) + 1 = 3$

So, $(1, 3)$ solves the system.

Lesson 12.4, page 586

EXPLORE

1.

2. $\text{Area}_1 = xy$

$\text{Area}_2 = (x + 2)(y + 3) = xy + 3x + 2y + 6$

$\text{Area}_3 = (x + 3)(y + 2) = xy + 2x + 3y + 6$

3. no; $\text{Area}_2 = \text{Area}_3$ only if $x = y$

4. Compare the areas. Subtract equal parts from each area.

$$\begin{array}{ccc} xy + 2x + 3y + 6 & ? & xy + 3x + 2y + 6 \\ -xy - 2x - 2y - 6 & ? & -xy - 2x - 2y - 6 \\ \hline y & ? & x \end{array}$$

Since $y > x$, the rectangle with dimensions $x + 3$ and $y + 2$ is greater.

TRY THESE

1. $5(c + d) + 7(c + d)$

    $(5 + 7)(c + d)$ Factor out the binomial.

    $12(c + d)$

2. $13(f^2 + 8) - 9(f^2 + 8)$

    $(13 - 9)(f^2 + 8)$ Factor out the binomial.

    $4(f^2 + 8)$

3. $g(b + 3) - 4(b + 3)$

    $(g - 4)(b + 3)$ Factor out the binomial.

4. $(xz + 10x) + (yz + 10y)$

    $x(z + 10) + y(z + 10)$ Factor each group.

    $(x + y)(z + 10)$ Factor out the binomial.

5. $(2h - 2k) + (gh - gk)$

    $2(h - k) + g(h - k)$ Factor each group.

    $(2 + g)(h - k)$ Factor out the binomial.

6. $(x^2 - x) + (xy - y)$

    $x(x - 1) + y(x - 1)$ Factor each group.

    $(x + y)(x - 1)$ Factor out the binomial.

7. $(y^4 - 2y^3) + (3y - 6)$

    $y^3(y - 2) + 3(y - 2)$ Factor each group.

    $(y^3 + 3)(y - 2)$ Factor out the binomial.

8. $(3a - 3b) + (ab - a^2)$

    $3(a - b) - a(-b + a)$ Factor each group.

    $3(a - b) - a(a - b)$

    $(3 - a)(a - b)$ Factor out the binomial.

9. $(2wz - w) + (3 - 6z)$

    $w(2z - 1) - 3(-1 + 2z)$ Factor each group.

    $w(2z - 1) - 3(2z - 1)$

    $(w - 3)(2z - 1)$ Factor out the binomial.

10. $(ab - 3a) + (5b - 15)$

    $a(b - 3) + 5(b - 3)$ Factor each group.

    $(a + 5)(b - 3)$ Factor out the binomial.

So, the rectangle is $(a + 5)$ by $(b - 3)$.

11.

12. Yes; a polynomial can be factored by grouping if there are pairs of terms that have common factors, and factoring x from $(ax + bx)$ and $-y$ from $(-ay + by)$ leaves the common factor of $(a - b)$.

PRACTICE

1. $5(x + 1) + w(x + 1)$

    $(5 + w)(x + 1)$ Factor out the binomial.

2. $z(y - 3) + 2(y - 3)$

    $(z + 2)(y - 3)$ Factor out the binomial.

3. $(xy + 5x) + (2y + 10)$

    $x(y + 5) + 2(y + 5)$ Factor each group.

    $(x + 2)(y + 5)$ Factor out the binomial.

4. $(ab + 7a) + (4b + 28)$

    $a(b + 7) + 4(b + 7)$ Factor each group.

    $(a + 4)(b + 7)$ Factor out the binomial.

5. $(xy + 2x) - (7y - 14)$

    $x(y + 2) - 7(y + 2)$ Factor each group.

    $(x - 7)(y + 2)$ Factor out the binomial.

6. $(ab - 3a) + (9b - 27)$

    $a(b - 3) + 9(b - 3)$ Factor each group.

    $(a + 9)(b - 3)$ Factor out the binomial.

7. $(ps - 2pt) + (qs - 2qt)$

    $p(s - 2t) + q(s - 2t)$ Factor each group.

    $(p + q)(s - 2t)$ Factor out the binomial.

8. $(mw - mx) - (nw - nx)$

$m(w - x) - n(w - x)$ Factor each group.

$(m - n)(w - x)$ Factor out the binomial.

9. $(12ab + 15a) + (4b + 5)$

$3a(4b + 5) + 1(4b + 5)$ Factor each group.

$(3a + 1)(4b + 5)$ Factor out the binomial.

10. $(2xy - 8x) + (3y - 12)$

$2x(y - 4) + 3(y - 4)$ Factor each group.

$(2x + 3)(y - 4)$ Factor out the binomial.

11. $(3wz + 12w) - (z + 4)$

$3w(z + 4) - 1(z + 4)$ Factor each group.

$(3w - 1)(z + 4)$ Factor out the binomial.

12. $(cd - 8c) - (3d - 24)$

$c(d - 8) - 3(d - 8)$ Factor each group.

$(c - 3)(d - 8)$ Factor out the binomial.

13. $(mn - 4m) + (2n - 8)$

$m(n - 4) + 2(n - 4)$ Factor each group.

$(m + 2)(n - 4)$ Factor out the binomial.

So, the dimensions are $(m + 2)$ by $(n - 4)$.

14. $(2fg + 4f) - (7g + 14)$

$2f(g + 2) - 7(g + 2)$ Factor each group.

$(2f - 7)(g + 2)$ Factor out the binomial.

15. $(yx - 2y) + (8 - 4x)$

$y(x - 2) - 4(-2 + x)$ Factor each group.

$y(x - 2) - 4(x - 2)$

$(y - 4)(x - 2)$ Factor out the binomial.

16. $(gh - g) + (3 - 3h)$

$g(h - 1) - 3(-1 + h)$ Factor each group.

$g(h - 1) - 3(h - 1)$

$(g - 3)(h - 1)$ Factor out the binomial.

17. $(pq - 7p) + (35 - 5q)$

$p(q - 7) - 5(-7 + q)$ Factor each group.

$p(q - 7) - 5(q - 7)$

$(p - 5)(q - 7)$ Factor out the binomial.

18. $(st - 3s) + (18 - 6t)$

$s(t - 3) - 6(-3 + t)$ Factor each group.

$s(t - 3) - 6(t - 3)$

$(s - 6)(t - 3)$ Factor out the binomial.

19. $(xy + 3x) - (4y + 12)$

$x(y + 3) - 4(y + 3)$ Factor each group.

$(x - 4)(y + 3)$ Factor out the binomial.

20. $4x^2 + 4x + 1 - 36y^2$

$(2x + 1)^2 - 36y^2$ Factor the perfect square trinomial.

$(2x + 1 + 6y)(2x + 1 - 6y)$ Factor the difference of squares.

EXTEND

21. $(ax + bx + cx) + (2a + 2b + 2c)$

$x(a + b + c) + 2(a + b + c)$ Factor each group.

$(x + 2)(a + b + c)$ Factor out the trinomial.

22. $(xw + 2yw + 3zw) - (4x + 8y + 12z)$

$w(x + 2y + 3z) - 4(x + 2y + 3z)$ Factor each group.

$(w - 4)(x + 2y + 3z)$ Factor out the trinomial.

23. $(ap + aq - ar) - (bp + bq - br)$

$a(p + q - r) - b(p + q - r)$ Factor each group.

$(a - b)(p + q - r)$ Factor out the trinomial.

24. $(x^2 - cx) - (ax - ac) - (bx - bc)$

$x(x - c) - a(x - c) - b(x - c)$ Factor each group.

$(x - a - b)(x - c)$ Factor out the trinomial.

25. $(xy - 5x) + (4y - 20)$

$x(y - 5) + 4(y - 5)$ Factor each group.

$(x + 4)(y - 5)$ Factor out the trinomial.

Dimensions are $(x + 4)$ by $(y - 5)$.

Perimeter: $2(x + 4) + 2(y - 5) = 2x + 8 + 2y - 10 = 2x + 2y - 2$

26. $$R = \frac{P + Pnr}{12n}$$

$12Rn = P + Pnr$

$12Rn = P(1 + nr)$

$$\frac{12Rn}{1 + nr} = P$$

27a. $(xy + 4x) + (2y + 8)$

$x(y + 4) + 2(y + 4)$ Factor each group.

$(x + 2)(y + 4)$ Factor out the binomial.

b. $3xy + 9y = ya$ ka 3 bha ka 9

THINK CRITICALLY

28. Answers will vary. $-2jk$, $-7(j - 2)$; $7j$, $2(jk - 7)$; 14, $j(2k - 7)$

29. Answers will vary. $-16a$, $(a + 3)(b + 4)(b - 4)$; $-ab^2$, $(3)(b + 4)(b - 4)$

30. Answers will vary. $40q$, $(5q + 4 + p)(5q + 4 - p)$; $-40q$, $(5q - 4 + p)(5q - 4 - p)$; $-25q^2$, $(4 + p)(4 - p)$; and -16, $(5q - p)(5q + p)$

ALGEBRAWORKS

1. $\dfrac{1,407,300 - 1,249,800}{1,249,800} \approx 0.126$ or 12.6%

2. $\dfrac{2,217,000 - 1,249,800}{1,249,800} \approx 0.774$ or 77.4%

3. The capacity will increase at a faster rate.

4. No; the horizontal scale is not uniform; the change in capacity was caused by periodic upgradings of technology, which are not a function of time.

Lesson 12.5, page 591

EXPLORE

1. $756 = 2 \cdot 378 = 2 \cdot 2 \cdot 189 = 2 \cdot 2 \cdot 3 \cdot 63 =$
$2 \cdot 2 \cdot 3 \cdot 3 \cdot 21 = 2 \cdot 2 \cdot 3 \cdot 3 \cdot 3 \cdot 7 = 2^2 \cdot 3^3 \cdot 7$

2. Answers will vary.   **3.** Answers will vary.

TRY THESE

1. $4x^2 + 8x - 32$
$4(x^2 + 2x - 8)$ Factor out the GCF of 4.
$4(x + 4)(x - 2)$ Factor the trinomial.

2. $3y^2 - 12y + 9$
$3(y^2 - 4y + 3)$ Factor out the GCF of 3.
$3(y - 3)(y - 1)$ Factor the trinomial.

3. $5x^2 - 20$
$5(x^2 - 4)$ Factor out the GCF of 5.
$5(x + 2)(x - 2)$ Factor the difference of squares.

4. $7a^2 - 63$
$7(a^2 - 9)$ Factor out the GCF of 7.
$7(a + 3)(a - 3)$ Factor the difference of squares.

5. $(xy^2 - 16x) + (2y^2 - 32)$
$x(y^2 - 16) + 2(y^2 - 16)$ Factor each group.
$(x + 2)(y^2 - 16)$ Factor out the binomial.
$(x + 2)(y + 4)(y - 4)$ Factor the difference of squares.

6. $(ab^2 - a) + (9b^2 - 9)$
$a(b^2 - 1) + 9(b^2 - 1)$ Factor each group.
$(a + 9)(b^2 - 1)$ Factor out the binomial.
$(a + 9)(b + 1)(b - 1)$ Factor the difference of squares.

7. $ab^2 + 13ab + 40a.$
$a(b^2 + 13b + 40)$ Factor out the GCF of a.
$a(b + 5)(b + 8)$ Factor the trinomial.

8. Look for patterns, such as monomial factors, common factors, binomial factors, trinomials, perfect square trinomials, or a difference of squares. If you have been able to factor the polynomial as the product of two or more binomials, check to see whether you can factor still further.

9. Build a base of 1 x^2-block, 2 x-blocks, and one unit cube. Make three layers of this same base. The prism has dimensions of $x + 1$, $x + 1$, and 3.
$3x^2 + 6x + 3$ is its volume.

PRACTICE

1. $2x^2 + 24x + 70$
$2(x^2 + 12x + 35)$ Factor out the GCF of 2.
$2(x + 5)(x + 7)$ Factor the trinomial.

2. $3y^2 + 21y + 36$
$3(y^2 + 7y + 12)$ Factor out the GCF of 3.
$3(y + 3)(y + 4)$ Factor the trinomial.

3. $5z^2 - 15z - 90$
$5(z^2 - 3z - 18)$ Factor out the GCF of 5.
$5(z - 6)(z + 3)$ Factor the trinomial.

4. $2a^2 - 4a - 160$
$2(a^2 - 2a - 80)$ Factor out the GCF of 2.
$2(a - 10)(a + 8)$ Factor the trinomial.

5. $4bc^2 + 12bc - 40b$
$4b(c^2 + 3c - 10)$ Factor out the GCF of $4b$.
$4b(c - 2)(c + 5)$ Factor the trinomial.

6. $6gh^4 + 18gh^2 - 168g$
$6g(h^4 + 3h^2 - 28)$ Factor out the GCF of $6g$.
$6g(h^2 - 4)(h^2 + 7)$ Factor the trinomial.
$6g(h + 2)(h - 2)(h^2 + 7)$ Factor the difference of squares.

7. $3x^2 - 75$
$3(x^2 - 25)$ Factor out the GCF of 3.
$3(x + 5)(x - 5)$ Factor the difference of squares.

8. $4m^2 - 144$
$4(m^2 - 36)$ Factor out the GCF of 4.
$4(m + 6)(m - 6)$ Factor the difference of squares.

9. $4x^2 + 24x + 36$
$4(x^2 + 6x + 9)$ Factor out the GCF of 4.
$4(x + 3)^2$ Factor the trinomial.

10. $8y^2 - 160y + 800$
$8(y^2 - 20y + 100)$ Factor out the GCF of 8.
$8(y - 10)^2$ Factor the trinomial.

11. $5p^2q^2 - 50$
$5(p^2q^2 - 100)0$ Factor out the GCF of 5.
$5(pq + 10)(pq - 10)$ Factor the difference of squares.

12. $3r^2s^4 - 147$
$3(r^2s^4 - 49)$ Factor out the GCF of 3.
$3(rs^2 + 7)(rs^2 - 7)$ Factor the difference of squares.

13. $2xy^2 - 32x$
$2x(y^2 - 16)$ Factor out the GCF of $2x$.
$2x(y + 4)(y - 4)$ Factor the difference of squares.

14. $3a^3b^4 - 192a^3$
$3a^3(b^4 - 64)$ Factor out the GCF of $3a^3$.
$3a^3(b^2 + 8)(b^2 - 8)$ Factor the difference of squares.

15. $2xy^2 - 2x + 4y^2 - 4$
$2[(xy^2 - x) + (2y^2 - 2)]$ Factor out the GCF of 2.
$2[x(y^2 - 1) + 2(y^2 - 1)]$ Factor each group.
$2(x + 2)(y^2 - 1)$ Factor out the binomial.
$2(x + 2)(y + 1)(y - 1)$ Factor the difference of squares.

16. $5ab^2 - 20a + 30b^2 - 120$
$5[(ab^2 - 4a) + (6b^2 - 24)]$ Factor out the GCF of 5.
$5[a(b^2 - 4) + 6(b^2 - 4)]$ Factor each group.
$5(a + 6)(b^2 - 4)$ Factor out the binomial.
$5(a + 6)(b + 2)(b - 2)$ Factor the difference of squares.

17. $4cd^2 - 4c - 12d^2 + 12$
$4[(cd^2 - c) - (3d^2 - 3)]$ Factor out the GCF of 4.
$4[c(d^2 - 1) - 3(d^2 - 1)]$ Factor each group.
$4(c - 3)(d^2 - 1)$ Factor out the binomial.
$4(c - 3)(d + 1)(d - 1)$ Factor the difference of squares.

18. $3x^2 - 243$
$3(x^2 - 81)$ Factor out the GCF of 3.
$3(x + 9)(x - 9)$ Factor the difference of squares.
Dimensions are 3 by $(x + 9)$ by $(x - 9)$.

EXTEND

19. $(ab^2 + 8ab + 12a) + (3b^2 + 24b + 36)$
$a(b^2 + 8b + 12) + 3(b^2 + 8b + 12)$
$(a + 3)(b^2 + 8b + 12)$
$(a + 3)(b + 6)(b + 2)$

20. $(xy^2 - 12xy + 36x) + (4y^2 - 48y + 144)$
$x(y^2 - 12y + 36) + 4(y^2 - 12y + 36)$
$(x + 4)(y^2 - 12y + 36)$
$(x + 4)(y - 6)^2$

21. $6pq^2 - 54p - 12q^2 + 108$
$6[(pq^2 - 9p) - (2q^2 - 18)]$
$6[p(q^2 - 9) - 2(q^2 - 9)]$
$6(p - 2)(q^2 - 9)$
$6(p - 2)(q + 3)(q - 3)$

22. $(xy^2 + 7xy + 12x) + (2y^2 + 14y + 24)$
$x(y^2 + 7y + 12) + 2(y^2 + 7y + 12)$
$(x + 2)(y^2 + 7y + 12)$
$(x + 2)(y + 3)(y + 4)$

23. $(ab^2 + 7ab + 10a) + (b^2 + 7b + 10)$
$a(b^2 + 7b + 10) + 1(b^2 + 7b + 10)$
$(a + 1)(b^2 + 7b + 10)$
$(a + 1)(b + 2)(b + 5)$

24. $(ac^2 + 5ac + 6a) - (bc^2 + 5bc + 6b)$
$a(c^2 + 5c + 6) - b(c^2 + 5c + 6)$
$(a - b)(c^2 + 5c + 6)$
$(a - b)(c + 2)(c + 3)$

25. $(m^2n^2 + 9m^2n + 20m^2) - (4n^2 + 36n + 80)$
$m^2(n^2 + 9n + 20) - 4(n^2 + 9n + 20)$
$(m^2 - 4)(n^2 + 9n + 20)$
$(m + 2)(m - 2)(n + 4)(n + 5)$

26. $(a^4b^2 + 2a^4b + a^4) - (9b^2 + 18b + 9)$
$a^4(b^2 + 2b + 1) - 9(b^2 + 2b + 1)$
$(a^4 - 9)(b^2 + 2b + 1)$
$(a^2 + 3)(a^2 - 3)(b + 1)^2$

27a. $V = \pi R^2 h - \pi r^2 h$
$V = \pi h(R^2 - r)$
$V = \pi h(R + r)(R - r)$

b. $W \approx (3.14)(4)(3 + 2.5)(3 - 2.5)(168.5)$
$= (3.14)(4)(5.5)(0.5)(168.5) \approx 5820.0$

28. $x^3 + 2^3 = (x + 2)(x^2 - 2x + 4)$

29. $x^3 - 1^3 = (x - 1)(x^2 + x + 1)$

30. $x^3 + 64 = x^3 + 4^3 = (x + 4)(x^2 - 4x + 16)$

31. $y^6 - 2^6 = (y^3)^2 - (2^3)^2 =$
$(y^3 + 2^3)(y^3 - 2^3) =$
$(y + 2)(y^2 - 2y + 4)(y - 2)(y^2 + 2y + 4)$

THINK CRITICALLY

32. $(a + b)(a^2 - ab + b^2) = a^3 - a^2b + ab^2 + a^2b - ab^2 + b^3$
After like terms are combined, the result is $a^3 + b^3$.

33. $16x^3 + 54y^3$
$2[8x^3 + 27y^3]$
$2[(2x)^3 + (3y)^3]$
$2(2x + 3y)(4x^2 - 6xy + 9y^2)$

34. $(p + 2)(q + 1)(q - 3n) = (pq + p + 2q + 2)(q - 3n) =$
$pq^2 + pq + 2q^2 + 2q - 3npq - 3np - 6nq - 6n =$
$pq^2 + (1 - 3n)pq - 3np + 2q^2 + (2 - 6n)q - 6n =$
$pq^2 - 14pq - 3nq + 2q^2 - 28q - 30$
$(1 - 3n)pq = -14pq \rightarrow n = 5$

35. $2(2r^{2n + 3} + 5s)(2r^{2n + 3} - 5s) = 2(4r^{4n + 6} - 25s^2) =$
$8r^{4n + 6} - 50s^2 = 8r^{5n - 1} - 50s^2$
$4n + 6 = 5n - 1$ Equate exponents of like terms.
$n = 7$

MIXED REVIEW

36. B; $0.28x = 42$
$$x = \frac{42}{0.28} = 150$$

37. $(xy - 5x) + (2y - 10)$
$x(y - 5) + 2(y - 5)$ Factor each group.
$(x + 2)(y - 5)$ Factor out the binomial.

38. $(2ab - 2a) + (3b - 3)$
$2a(b - 1) + 3(b - 1)$ Factor each group.
$(2a + 3)(b - 1)$ Factor out the binomial.

39. $(4y^2z + 4y^2) - (9z + 9)$
$4y^2(z + 1) - 9(z + 1)$ Factor each group.
$(4y^2 - 9)(z + 1)$ Factor out the binomial.
$(2y + 3)(2y - 3)(z + 1)$ Factor the difference of squares.

Lesson 12.6, page 595

EXPLORE

1. $x = -6$ and $x = 1$

2. $x^2 + 5x - 6 = (x + 6)(x - 1)$

3. When $x^2 + 5x - 6 = 0$ then one or both factors are equal to zero.

4. $x^2 + 3x + 2 = 0$

$(x + 2)(x + 1) = 0$

Either $x + 2 = 0$ or $x + 1 = 0$

$x = -2$ $x = -1$

TRY THESE

1. $x(x + 3) = 0$

$x = 0$ or $x + 3 = 0$

$x = -3$

2. $y(y - 9) = 0$

$y = 0$ or $y - 9 = 0$

$y = 9$

3. $(x + 5)(x - 1) = 0$

$x + 5 = 0$ or $x - 1 = 0$

$x = -5$ or $x = 1$

4. $(x + 1)(x - 2) = 0$

$x + 1 = 0$ or $x - 2 = 0$

$x = -1$ or $x = 2$

5. $x^2 - 6x = 0$

$x(x - 6) = 0$

$x = 0$ or $x - 6 = 0$

$x = 6$

6. $y^2 + 2y = 0$

$y(y + 2) = 0$

$y = 0$ or $y + 2 = 0$

$y = -2$

7. $x^2 + 5x - 14 = 0$

$(x + 7)(x - 2) = 0$

$x + 7 = 0$ or $x - 2 = 0$

$x = -7$ or $x = 2$

8. $y^2 - 4y + 3 = 0$

$(y - 3)(y - 1) = 0$

$y - 3 = 0$ or $y - 1 = 0$

$y = 3$ or $y = 1$

9. $z^2 - z = 12$

$z^2 - z - 12 = 0$

$(z - 4)(z + 3) = 0$

$z - 4 = 0$ or $z + 3 = 0$

$z = 4$ or $z = -3$

10. $x^2 + 4x = 12$

$x^2 + 4x - 12 = 0$

$(x + 6)(x - 2) = 0$

$x + 6 = 0$ or $x - 2 = 0$

$x = -6$ or $x = 2$

11. $y^2 + 6y = -8$

$y^2 + 6y + 8 = 0$

$(y + 4)(y + 2) = 0$

$y + 4 = 0$ or $y + 2 = 0$

$y = -4$ or $y = -2$

12. $z^2 + 2z = 80$

$z^2 + 2z - 80 = 0$

$(z + 10)(z - 8) = 0$

$z + 10 = 0$ or $z - 8 = 0$

$z = -10$ or $z = 8$

13. $55 = (2w + 1)w$

$55 = 2w^2 + w$

$0 = 2w^2 + w - 55$

$0 = (2w + 11)(w - 5)$

$0 = 2w + 11$ or $0 = w - 5$

$-\dfrac{11}{2} = w$ or $5 = w$

w cannot be negative so dimensions are 5 by 2 • 5 + 1 or 11

14. $b^2 - 4ac = 13^2 - 4(1)(22) = 169 - 88 = 81$ is a perfect square.

$x^2 + 13x + 22 = (x + 2)(x + 11)$

15. $b^2 - 4ac = (-3)^2 - 4(1)(-24) = 9 + 96 = 105$ is not a perfect square, not factorable

16. $b^2 - 4ac = (-20)^2 - 4(1)(36) = 400 - 144 = 256$ is a perfect square.

$x^2 - 20x + 36 = (x - 18)(x - 2)$

17. The equation $ax^2 + bx + c = 0$ can be solved by factoring when $b^2 - 4ac$ is a perfect square. You can set each factor equal to zero and then solve for the unknown variable.

PRACTICE

1. $a(a + 1) = 0$

$a = 0$ or $a + 1 = 0$

$a = -1$

2. $b(b - 10) = 0$

$b = 0$ or $b - 10 = 0$

$b = 10$

3. $(c + 3)(c - 8) = 0$

$c + 3 = 0$ or $c - 8 = 0$

$c = -3$ or $c = 8$

4. $(d - 4)(d - 9) = 0$

$d - 4 = 0$ or $d - 9 = 0$

$d = 4$ or $d = 9$

5. $x^2 - 5x = 0$

$x(x - 5) = 0$

$x = 0$ or $x - 5 = 0$

$x = 5$

6. $y^2 + 7y = 0$

$y(y + 7) = 0$

$y = 0$ or $y + 7 = 0$

$y = -7$

7. $x^2 - 3x - 70 = 0$

$(x + 7)(x - 10) = 0$

$x + 7 = 0$ or $x - 10 = 0$

$x = -7$ or $x = 10$

8. $x^2 + 4x - 45 = 0$

$(x + 9)(x - 5) = 0$

$x + 9 = 0$ or $x - 5 = 0$

$x = -9$ or $x = 5$

9. $x^2 + 11x + 28 = 0$

$(x + 7)(x + 4) = 0$

$x + 7 = 0$ or $x + 4 = 0$

$x = -7$ or $x = -4$

10. $x^2 - 15x + 44 = 0$

$(x - 11)(x - 4) = 0$

$x - 11 = 0$ or $x - 4 = 0$

$x = 11$ or $x = 4$

11. $x^2 + 3x = 18$

$x^2 + 3x - 18 = 0$

$(x + 6)(x - 3) = 0$

$x + 6 = 0$ or $x - 3 = 0$

$x = -6$ or $x = 3$

12. $x^2 - 2x = 63$

$x^2 - 2x - 63 = 0$

$(x - 9)(x + 7) = 0$

$x - 9 = 0$ or $x + 7 = 0$

$x = 9$ or $x = -7$

13.
$$y^2 - 14 = 5y$$
$$y^2 - 5y - 14 = 0$$
$$(y - 7)(y + 2) = 0$$
$$y - 7 = 0 \quad \text{or} \quad y + 2 = 0$$
$$y = 7 \quad \text{or} \quad y = -2$$

14.
$$z^2 + 10 = 11z$$
$$z^2 - 11z + 10 = 0$$
$$(z - 10)(z - 1) = 0$$
$$z - 10 = 0 \quad \text{or} \quad z - 1 = 0$$
$$z = 10 \quad \text{or} \quad z = 1$$

15.
$$2x^2 + 7x = -3$$
$$2x^2 + 7x + 3 = 0$$
$$(2x + 1)(x + 3) = 0$$
$$2x + 1 = 0 \quad \text{or} \quad x + 3 = 0$$
$$x = -\frac{1}{2} \quad \text{or} \quad x = -3$$

16.
$$3x^2 + 14x = 5$$
$$3x^2 + 14x - 5 = 0$$
$$(3x - 1)(x + 5) = 0$$
$$3x - 1 = 0 \quad \text{or} \quad x + 5 = 0$$
$$x = \frac{1}{3} \quad \text{or} \quad x = -5$$

17.
$$x^2 - x = 3x + 12$$
$$x^2 - 4x - 12 = 0$$
$$(x + 2)(x - 6) = 0$$
$$x + 2 = 0 \quad \text{or} \quad x - 6 = 0$$
$$x = -2 \quad \text{or} \quad x = 6$$

18.
$$y^2 + 3y - 2 = y + 1$$
$$y^2 + 2y - 3 = 0$$
$$(y + 3)(y - 1) = 0$$
$$y + 3 = 0 \quad \text{or} \quad y - 1 = 0$$
$$y = -3 \quad \text{or} \quad y = 1$$

19. $18 = (w + 7)w$
$$18 = w^2 + 7w$$
$$0 = w^2 + 7w - 18$$
$$0 = (w + 9)(w - 2)$$
$$0 = w + 9 \quad \text{or} \quad 0 = w - 2$$
$$-9 = w \quad \text{or} \quad 2 = w$$
Width cannot be negative so dimensions are 2m by 9m.

20. $b^2 - 4ac = (-2)^2 - 4(1)(5) = -16$, not factorable

21. $b^2 - 4ac = (-12)^2 - 4(1)(32) = 16$, a perfect square
$x^2 - 12x + 32 = (x - 4)(x - 8)$

22. $b^2 - 4ac = (-13)^2 - 4(1)(-48) = 361$, a perfect square
$x^2 - 13x - 48 = (x + 3)(x - 16)$

23. $b^2 - 4ac = (-1)^2 - 4(1)(-90) = 361$, a perfect square
$x^2 - x - 90 = (x + 9)(x - 10)$

24. $b^2 - 4ac = (13)^2 - 4(2)(15) = 49$, a perfect square
$2x^2 + 13x + 15 = (2x + 3)(x + 5)$

25. $b^2 - 4ac = (1)^2 - 4(2)(3) = -23$, not factorable

26. $b^2 - 4ac = (-5)^2 - 4(1)(-84) = 361$, a perfect square
$x^2 - 5x - 84 = (x + 7)(x - 12)$

27. $b^2 - 4ac = (-1)^2 - 4(1)(4) = -15$, not factorable

28. $b^2 - 4ac = (-17)^2 - 4(1)(66) = 25$, a perfect square
$x^2 - 17x + 66 = (x - 6)(x - 11)$

29.
$$x(x + 1) = 132$$
$$x^2 + x - 132 = 0$$
$$(x + 12)(x - 11) = 0$$
$$x + 12 = 0 \quad \text{or} \quad x - 11 = 0$$
$$x = -12 \quad \text{or} \quad x = 11$$
So, the numbers are $-12, -11$ and $11, 12$.

30.
$$x(x + 2) = 143$$
$$x^2 + 2x - 143 = 0$$
$$(x + 13)(x - 11) = 0$$
$$x + 13 = 0 \quad \text{or} \quad x - 11 = 0$$
$$x = -13 \quad \text{or} \quad x = 11$$
So, the numbers are $-13, -11$ and $11, 13$.

EXTEND

31.
$$(x + 1)^2 + (x - 5)^2 = 20$$
$$x^2 + 2x + 1 + x^2 - 10x + 25 = 20$$
$$2x^2 - 8x + 6 = 0$$
$$2(x + 3)(x - 1) = 0$$
$$x - 1 = 0 \quad \text{or} \quad x - 3 = 0$$
$$x = 1 \quad \text{or} \quad x = 3$$

32.
$$(x + 3)^2 - 2(x + 1) = 3$$
$$x^2 + 6x + 9 - 2x - 2 = 3$$
$$x^2 + 4x + 4 = 0$$
$$(x + 2)^2 = 0$$
$$x + 2 = 0$$
$$x = -2$$

33. $40 = 2l + 2w \rightarrow l = 20 - w$
$$A = 96 = (20 - w)w = 20w - w^2$$
$$w^2 - 20w + 96 = 0$$
$$(w - 8)(w - 12) = 0$$
$$w - 8 = 0 \quad \text{or} \quad w - 12 = 0$$
$$w = 8 \quad \text{or} \quad w = 12$$
The dimensions are 8m by 12m.

34.
$$(x+2)^2 - 5x = 60$$
$$x^2 + 4x + 4 - 5x = 60$$
$$x^2 - x - 56 = 0$$
$$(x-8)(x+7) = 0$$
$$x - 8 = 0 \quad \text{or} \quad x + 7 = 0$$
$$x = 8 \quad \text{or} \quad x = -7$$
Discard -7 since it is not even, so integers are 8 and 10.

35.
$$28 = (2w-1)w = 2w^2 - w$$
$$0 = 2w^2 - w - 28$$
$$0 = (2w+7)(w-4)$$
$$0 = 2w + 7 \quad \text{or} \quad 0 = w - 4$$
$$-\frac{7}{2} = w \quad \text{or} \quad 4 = w$$
Width cannot be negative, so the dimensions are 4 by 7.

THINK CRITICALLY

36.
$$x = -2 \quad \text{or} \quad x = 8$$
$$x + 2 = 0 \quad \text{or} \quad x - 8 = 0$$
$$(x+2)(x-8) \to x^2 - 6x - 16 = 0$$

37.
$$x = 0 \quad \text{or} \quad x = -12$$
$$x = 0 \quad \text{or} \quad x + 12 = 0$$
$$(x+12) \to x^2 + 12x = 0$$

38.
$$x = \frac{1}{2} \quad \text{or} \quad x = -3$$
$$2x - 1 = 0 \quad \text{or} \quad x + 3 = 0$$
$$(2x-1)(x+3) \to 2x^2 + 5x - 3 = 0$$

39.
$$x = \frac{2}{3} \quad \text{or} \quad x = -1$$
$$3x - 2 = 0 \quad \text{or} \quad x + 1 = 0$$
$$(3x-2)(x+1) \to 3x^2 + x - 2 = 0$$

40. Factors of -10 are $1, -10$ and $-1, 10$ and $2, -5$ and $-2, 5$

Trial factors	Product
$(2x+1)(x-10)$	$2x^2 - 19x - 10$
$(2x-1)(x+10)$	$2x^2 + 19x - 10$
$(2x+10)(x-1)$	$2x^2 + 8x - 10$
$(2x-10)(x+1)$	$2x^2 - 8x - 10$
$(2x+2)(x-5)$	$2x^2 - 8x - 10$
$(2x-2)(x+5)$	$2x^2 + 8x - 10$
$(2x+5)(x-2)$	$2x^2 - x - 10$
$(2x-5)(x+2)$	$2x^2 + x - 10$

Possible values of n are $1, -1, 8, -8, 19, -19$; or n can be 0 which will factor as $2(x^2 - 5)$.

41. $b^2 - 4ac = (2)^2 - 4(n)(-4) = 4 + 16n$ must be a perfect square
So $n = 0, 2, 6, 12, 20, 30, 42, \ldots, k^2 - k$

MIXED REVIEW

42. $7x + 8 = 3x - 20$
$\quad\quad 4x = -28$
$\quad\quad x = -7$

43. $4(x-3) = 3x + 11$
$\quad\quad 4x - 12 = 3x + 11$
$\quad\quad x = 23$

44. $5x + 7 + 2x = 11 + 4x$
$\quad\quad 3x = 4$
$\quad\quad x = \dfrac{4}{3}$

45. Substitute members of $\{-2, 0, 3\}$ in $y = -3x$ to get $\{6, 0, -9\}$ respectively

46. Substitute members of $\{-2, 0, 3\}$ in $y = x + 5$ to get $\{3, 5, 8\}$ respectively

47. Substitute members of $\{-2, 0, 3\}$ in $y = x^2$ to get $\{4, 0, 9\}$ respectively

48. D; $2 - x > 7$
$\quad\quad -x > 5$
$\quad\quad x < -5$ Multiply by -1 and reverse the inequality symbol.

49. $(3x^2 + 4x - 2) + (4x^2 - 3) = (3x^2 + 4x^2) + 4x + (-2 - 3) = 7x^2 + 4x - 5$

50. $(4ab - 3a + b) - (2ab + a - 6b) = (4ab - 2ab) - (3a + a) + (b + 6b) = 4ab - 3a + b - 2ab - a + 6b = 2ab - 4a + 7b$

ALGEBRA WORKS

1. $lwx + lxd + lx^2$
$lx(w + d + x)$ Factor out the GCF of lx.

2. Since depth is only thing changing, change in volume is lwx from table:
$$lwx = 5280 \cdot 2640 \cdot \frac{1}{2} = 6{,}969{,}600 \text{ ft}^3$$

3. $6{,}969{,}600 \text{ ft}^3 \cdot \dfrac{1 \text{ gal}}{0.13 \text{ ft}^3} \approx 53{,}612{,}308 \text{ gal}$

4. $lwx = 5280 \cdot 2640 \cdot 1\frac{1}{2} \cdot \dfrac{1}{0.13} \approx 160{,}836{,}923 \text{ gal}$

5. $160{,}836{,}923 - 53{,}612{,}308 = 107{,}224{,}615 \text{ gal}$

6. Since a small rise in river depth can produce a great increase in volume of water, water resources technicians need to keep careful watch over changes in the water depth.

Lesson 12.7, page 602

EXPLORE THE PROBLEM

1. the width of bronze strip

2. length $= 80 - 2x$, width $= 60 - 2x$

3. $A = (80 - 2x)(60 - 2x) = 4800 - 280x + 4x^2$

4. $4800 - 280x + 4x^2 = \frac{1}{2}(60)(80)$ or $4x^2 - 280x + 2400 = 0$

5. $4(x^2 - 70x + 600) = 0$

$\qquad 4(x - 60)(x - 10) = 0$

$\qquad\qquad x - 60 = 0 \qquad$ or $\qquad x - 10 = 0$

$\qquad\qquad\qquad x = 60 \qquad$ or $\qquad\qquad x = 10$

6. $x = 60$ doesn't make sense because the strip would be larger than the actual buckle; the width is 10 mm.

7. Substitute 10 for x in the dimension of the silver rectangle; check that the area equals half the area of the buckle or 2400 in^2; $(80 - 20)(60 - 20) = (60)(40) = 2400$, which checks.

INVESTIGATE FURTHER

8. Three more than twice as long is $2y + 3$.

9. Sum of squares is 194, so

$\qquad y^2 + (2y + 3)^2 = 194$

$\qquad y^2 + 4y^2 + 12y + 9 = 194$

$\qquad 5y^2 + 12y - 185 = 0$

10. Yes; $(12)^2 - 4(5)(-185) = 3844$ which is 62^2, a perfect square

11. The factored form is $(5y + 37)(y - 5) = 0$

so $y = \dfrac{-37}{5}$ and $y = 5$

Since the number of years is not negative, only $y = 5$ makes sense in this problem.

12. Gloria = 5 years, Gustavo = 2(5) + 3 or 13 years; $5^2 + 13^2 = 194$

13. The answer would include the numbers 5 and 13 and the numbers $-\dfrac{37}{5}$ and $-\dfrac{59}{5}$, since negative numbers also satisfy the conditions of the problem.

14. Answers will vary. The solutions are 8 and −9, so only the positicve solution would make sense in an area problem; but both solutions would be acceptable for a problem that asked about consecutive integers.

APPLY THE STRATEGY

15a. $s^2 + 4s = 60$ or $s^2 + 4s - 60 = 0$
$\qquad$ (s^2 = Area, $4s$ = Perimeter of square)

b. $(s + 10)(s - 6) = 0$

$\qquad$ so $s + 10 = 0 \qquad$ or $\qquad s - 6 = 0$

$\qquad\qquad s = -10 \qquad$ or $\qquad\qquad s = 6$ ft

Discard -10 since length is not negative.

16. $1705 = (w + 15)w = w^2 + 15w$

$\qquad 0 = w^2 + 15w - 1750 = (w + 50)(w - 35)$

so $w + 50 = 0 \qquad$ or $\qquad w - 35 = 0$

$\qquad w = -50 \qquad$ or $\qquad\qquad w = 35$

Discard -50 since width is not negative.
Dimensions are 35m by 50m.

17a. $2l + 2w = 32 \;\rightarrow\; l = 16 - w$

$\qquad l \cdot w = (16 - w)w = 16w - w^2 = 63$

$\qquad 0 = w^2 - 16w + 63 = (w - 7)(w - 9)$

Either $w - 7 = 0$ or $w - 9 = 0$
Dimensions are 7 by 9.

b. The magazine page is $7 + 2\left(1\dfrac{1}{2}\right)$ by $9 + 2\left(1\dfrac{1}{2}\right)$ or 10 in. by 12 in.

18. $\qquad\qquad k(k - 1) = 756$

$\qquad\quad k^2 - k - 756 = 0$

$\qquad (k - 28)(k + 27) = 0$

$\qquad\qquad\qquad k = 28 \qquad$ or $\qquad k = -27$

So, Keiko is 28 and Yoshi is 27.

19. $\qquad\qquad n(n + 1) = 1560$

$\qquad\quad n^2 + n - 1560 = 0$

$\qquad (n + 40)(n - 39) = 0$

$\qquad\qquad\qquad n = -40 \qquad$ or $\qquad n = 39$

So, the integers are $-40, -39$ and $39, 40$.

20. $\qquad\qquad (n + 10)^2 = 9n^2$

$\qquad n^2 + 20n + 100 = 9n^2$

$\qquad 8n^2 - 20n - 100 = 0$

$\qquad 4(2n^2 - 5n - 25) = 0$

$\qquad 4(2n + 5)(n - 5) = 0$

so $2n + 5 = 0 \qquad$ or $\qquad n - 5 = 0$

$\qquad\qquad n = -\dfrac{5}{2} \qquad$ or $\qquad\qquad n = 5$

5 is the smaller integer.

21a. $(24 - 2w)(31 - 2w) = 450$

$\qquad 744 - 110w + 4w^2 = 450$

$\qquad 4w^2 - 110w + 294 = 0$

$\qquad 2(2w^2 - 55w + 147) = 0$

$\qquad 2(2w - 49)(w - 3) = 0$

so $2w - 49 = 0 \qquad$ or $\qquad w - 3 = 0$

$\qquad\qquad w = 24\dfrac{1}{2} \qquad$ or $\qquad\qquad w = 3$

So, the frame is 3 in. wide.

b. $24 - 2(3)$ by $31 - 2(3)$ or 18 in. by 25 in.

22. $4x$ = length of one piece,
then $56 - 4x$ = length of other piece

Sum of areas is $106 = \left(\dfrac{4x}{4}\right)^2 + \left(\dfrac{56 - 4x}{4}\right)^2$

$\qquad\qquad 106 = x^2 + (14 - x)^2$

(Solution continues on next page.)

$$x^2 + 196 - 28x + x^2 = 106$$

$$2x^2 - 28x + 90 = 0$$

$$2(x^2 - 14x + 45) = 0$$

$$2(x - 9)(x - 5) = 0$$

so $x - 9 = 0$ or $x - 5 = 0$

     $x = 9$ or $x = 5$

Wire is cut into $4(9) = 36$ in. and $4(5) = 20$ in.

23. $400\pi = \pi r^2 \rightarrow 400 = r^2$

                $\rightarrow 20 = r =$ radius of garden

Garden and path is

$$676\pi = \pi R^2$$

$$676 = R^2 \rightarrow R = 26$$

$$R - r = 26 - 20 = 6$$

The path is 6 ft wide.

24. $(10 - 2x)^2 = 4(x)(10 - 2x)$

$$100 - 40x + 4x^2 = 40x - 8x^2$$

$$12x^2 - 80x + 100 = 0$$

$$4(3x^2 - 20x + 25) = 0$$

$$4(3x - 5)(x - 5) = 0$$

$3x - 5 = 0$ or $x - 5 = 0$

   $x = \dfrac{5}{3}$ or $x = 5$

So, the squares that were cut out have length $1\dfrac{2}{3}$ in.

REVIEW PROBLEM SOLVING STRATEGIES

1a. There are six pairs; (G1, G2)(G2, G1)(Y1, Y2)
(Y2, Y1)(G1, Y2)(Y1, G2)

b. (G1, G2)(G2, G1)(G3, Y3)

c. $\dfrac{2}{3}$

d. $\dfrac{1}{3}$

2a. Cut the 19-in. cord into two pieces of 11 in. and 8 in.
Then the 11-in. piece is the average of the 8-in. and
the 14-in. pieces. The 11-in. piece is one-third the
sum of the original lengths $(19 + 14 = 33)$

b. If the two cords are unequal in length, cut from the
longer piece. If they are equal in length, cut from
either cord. Cut a piece that is one-third the sum of
the lengths of the original pieces. Let the lengths be x
and y $(y > x)$ and cut a length of b from y so you now
have lengths x, $y - b$, and b.

You want $\dfrac{x + (y - b)}{2} = b$, so $x + y - b = 2b$, $x + y =$

$3b$, and $b = \dfrac{x + y}{3}$.

3. 16 km; Use $d = rt$. Find the length of time the bird
flies. The bird's time equals the time it takes for the
two trains to cover 150 km.
So, $33t + 42t = 150$, $75t = 150$, $t = 2$. The bird's rate is
8 km/h, and $d = 8 \cdot 2 = 16$ km.

Chapter Review, page 606

1. c

2. a

3. d

4. b

5.

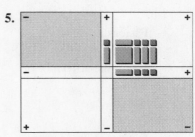

$$x^2 + 4x + 3 = (x + 3)(x + 1)$$

6.

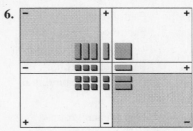

$$x^2 - 5x + 6 = (x - 2)(x - 3)$$

7.

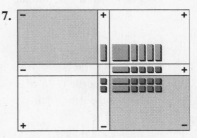

$$x^2 + 2x - 8 = (x + 4)(x - 2)$$

8.

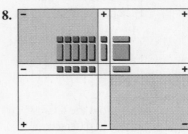

$$x^2 - 4x - 5 = (x - 5)(x + 1)$$

9. $15d + 25d^2$
$5d(3 + 5d)$ Factor out the GCF of $5d$.

10. $9xyz^2 - 3y^2z + 6x^2yz$
$3yz(3xz - y + 2x^2)$ Factor out the GCF of $3yz$.

11. $z^2 + 11z + 24$
$(z + 3)(z + 8)$ Factor the trinomial.

12. $y^2 + 8y - 9$
$(y + 9)(y - 1)$ Factor the trinomial.

13. $2x^2 + x - 15$
$(2x - 5)(x + 3)$ Factor the trinomial.

14. $c^2 + 8c + 16$
$(c + 4)^2$ Factor the trinomial.

15. $b^2 - 4ac = (-1)^2 - 4(1)(1) = -3$ not factorable

16. $r^2 - 6r + 9$
$(r - 3)^2$ Factor the trinomial.

17. $w^2 - 100$
$(w + 10)(w - 10)$ Factor the difference of squares.

18. $3(3) - 4x(x) = 9 - 4x^2 = (3 + 2x)(3 - 2x)$

19. $4(x - 3) + y(x - 3)$
$(4 + y)(x - 3)$ Factor out the binomial.

20. $(ab + 2a) + (4b + 8)$
$a(b + 2) + 4(b + 2)$ Factor each group.
$(a + 4)(b + 2)$ Factor out the binomial.

21. $(pr - ps) - (qr - qs)$
$p(r - s) - q(r - s)$ Factor each group.
$(p - q)(r - s)$ Factor out the binomial.

22. $(cd + 5c) - (5 + d)$
$c(d + 5) - 1(5 + d)$ Factor each group.
$c(d + 5) - 1(d + 5)$
$(c - 1)(d + 5)$ Factor out the binomial.

23. $(ab - 5a) + (3b - 15)$
$a(b - 5) + 3(b - 5)$ Factor each group.
$(a + 3)(b - 5)$ Factor out the binomial.

24. $3x^2 + 27x + 42$
$3(x^2 + 9x + 14)$ Factor out the GCF of 3.
$3(x + 7)(x + 2)$ Factor the trinomial.

25. $2ab^4 - 8ab^2 - 90a$
$2a(b^4 - 4b^2 - 45)$ Factor out the GCF of $2a$.
$2a(b^2 - 9)(b^2 + 5)$ Factor the trinomial.
$2a(b + 3)(b - 3)(b^2 + 5)$ Factor the difference of squares.

26. $6y^2 - 36y + 54$
$6(y^2 - 6y + 9)$ Factor out the GCF of 6.
$6(y - 3)^2$ Factor the trinomial.

27. $3xy^2 - 12x$
$3x(y^2 - 4)$ Factor out the GCF of $3x$.
$3x(y + 2)(y - 2)$ Factor the difference of squares.

28. $4x^2 - 144$
$4(x^2 - 36)$ Factor out the GCF of 4.
$4(x + 6)(x - 6)$ Factor the difference of squares.

29. $k(k + 5) = 0$
$k = 0$ or $k + 5 = 0$
             $k = -5$

30. $n(n - 9) = 0$
$n = 0$ or $n - 9 = 0$
             $n = 9$

31. $(y + 7)(y - 10) = 0$
$y + 7 = 0$ or $y - 10 = 0$
   $y = -7$ $y = 10$

32. $x^2 + 9x + 18 = 0$
$(x + 6)(x + 3) = 0$
$x + 6 = 0$ or $x + 3 = 0$
   $x = -6$ $x = -3$

33. $x(x + 2) = 224$
$x^2 + 2x - 224 = 0$
$(x + 16)(x - 14) = 0$
$x + 16 = 0$ or $x - 14 = 0$
   $x = -16$ $x = 14$
The integers are $-16, -14$ and $14, 16$.

34. $30(20) - (30 - x)(20 - x) = 264$
$600 - 600 + 50x - x^2 = 264$
$x^2 - 50x + 264 = 0$
$(x - 6)(x - 44) = 0$
$x - 6 = 0$ or $x - 44 = 0$
   $x = 6$ $x = 44$
44 is too big for the problem.

35. $x^2 + (x - 4)^2 = 170$
$x^2 + x^2 - 8x + 16 = 170$
$2x^2 - 8x - 154 = 0$
$2(x^2 - 4x - 77) = 0$
$2(x - 11)(x + 7) = 0$
$x - 11 = 0$ or $x + 7 = 0$
   $x = 11$ $x = -7$
Numbers are positive so -7 is inappropriate.
The numbers are 11 and 7.

Chapter Assessment, page 608

1. $2g^3h^2 - 8gh$
$2gh(g^2h - 4)$ Factor out the GCF.

2. $6a^3b^2 - 3a^2b + 18ab^2$
$3ab(2a^2b - a + 6b)$ Factor out the GCF.

3. $4ab - 12b + 4a$
$4(ab - 3b + a)$ Factor out the GCF.

4. $3x^3y - 12xy^3 + 6x^2y^2$
$3xy(x^2 - 4y^2 + 2xy)$ Factor out the GCF.

5. Factor out the common monomial $3x$: $3x(x^2 + 3x + 2)$. Then factor the trinomial into two binomial factors such that the product of the first two binomial terms equals the first term of the trinomial, the product of the last two binomial terms equals the last term of the trinomial, and the sum of the products of the inner and outer terms equals the middle term: $3x(x + 2)(x + 1)$

6.
$$x(x+2) = 195$$
$$x^2 + 2x = 195$$
$$x^2 + 2x - 195 = 0$$
$$(x-13)(x+15) = 0$$
So $x = 13$ and $x + 2 = 15$

7. $4r^2 - 16s^2 = 4(r^2 - 4s^2) = 4(r + 2s)(r - 2s)$

8. $x^2 - 81$
$(x + 9)(x - 9)$ Factor the difference of squares.

9. $x^3 - 16x$
$x(x^2 - 16)$ Factor out the GCF.
$x(x + 4)(x - 4)$ Factor the difference of squares.

10. $50x^2 - 32$
$2(25x^2 - 16)$ Factor out the GCF.
$2(5x + 4)(5x - 4)$ Factor the difference of squares.

11. $x^2 - 16x + 64$
$(x - 8)^2$ Factor the trinomial.

12. $3z^2 - 2z - 1$
$(z - 1)(3z + 1)$ Factor the trinomial.

13. $2x^2 - 28x + 98$
$2(x^2 - 14x + 49)$ Factor out the GCF.
$2(x - 7)^2$ Factor the trinomial.

14. $a^4 - 6a^2 + 5$
$(a^2 - 5)(a^2 - 1)$ Factor the trinomial.
$(a^2 - 5)(a + 1)(a - 1)$ Factor the difference of squares.

15. $2a^6a^6 - 7a^3b^3 - 4$
$(2a^3b^3 + 1)(a^3b^3 - 4)$ Factor the trinomial.

16. C; factors into $2(x^2 - 2)$ which is not a difference of squares

17. C;   **I.** $b^2 - 4ac = 13^2 - 4(3)(-10) = 289$, a perfect square so factorable
     II. $b^2 - 4ac = 11^2 - 4(2)(14) = 9$, a perfect square so factorable
     III. $b^2 - 4ac = (-4)^2 - 4(4)(-15) = 256$, a perfect square so factorable
     IV. $b^2 - 4ac = 19^2 - 4(2)(24) = 169$, a perfect square so factorable

18. $r(r + 6) = 0$
$r = 0$ or $r + 6 = 0$
                $r = -6$

19.
$$16x^2 - 9 = 0$$
$$(4x + 3)(4x - 3) = 0$$
$4x + 3 = 0$ or $4x - 3 = 0$
$x = -\dfrac{3}{4}$ or $x = \dfrac{3}{4}$

20. $y^2 - 10y - 56 = 0$
$(y + 4)(y - 14) = 0$
$y + 4 = 0$ or $y - 14 = 0$
   $y = -4$ or $y = 14$

21. $x^2 + x - 0.75 = 0$
$(x + 1.5)(x - 0.5) = 0$
$x + 1.5 = 0$ or $x - 0.5 = 0$
   $x = -1.5$ or $x = 0.5$

22. $k(k + 4) - 3(x + 4) = 0$
     $(k - 3)(k + 4) = 0$
$k - 3 = 0$ or $k + 4 = 0$
   $k = 3$ or $k = -4$

23. $n^2 - 7n + 10 = 0$
$(n - 5)(n - 2) = 0$
$n - 5 = 0$ or $n - 2 = 0$
   $n = 5$ or $n = 2$

24. $y^2 - 6y + 4 = 20$
$y^2 - 6y - 16 = 0$
$(y - 8)(y + 2) = 0$
$y - 8 = 0$ or $y + 2 = 0$
   $y = 8$ or $y = -2$

25. $4r(2r) - \pi r^2 = 8r^2 - \pi r^2 = r^2(8 - \pi)$

26. $(x^3 - 36x) + (3x^2 - 108)$
$x(x^2 - 36) + 3(x^2 - 36)$
$(x + 3)(x^2 - 36)$
$(x + 3)(x + 6)(x - 6)$

27. $(w + 7)w = 120$
$$w^2 + 7w = 120$$
$$w^2 + 7w - 120 = 0$$
$(w + 15)(w - 8) = 0$
$w + 15 = 0$ or $w - 8 = 0$
   $w = -15$ or $w = 8$
$w = 8$ cm and $l = w + 7 = 8 + 7 = 15$ cm

28. $2l + 2w = 38$ $\rightarrow l = 19 - w$
$(19 - w)w = 84$
$19w - w^2 = 84$
$w^2 - 19w + 84 = 0$
$(w - 12)(w - 7) = 0$
$w - 12 = 0$ or $w - 7 = 0$
   $w = 12$ or $w = 7$
So, dimensions are 7 m by 12 m.

29. $8(10) - 32 = (8 - 2x)(10 - 2x)$ where x = width of border

$$48 = 80 - 36x + 4x^2$$
$$0 = 4x^2 - 36x + 32$$
$$0 = 4(x^2 - 9x + 8)$$
$$0 = 4(x - 8)(x - 1)$$

$x - 8 = 0$ or $x - 1 = 0$

$x = 8$ too big   or $x = 1$ so border is 1 in.

30. $5(5) - 2(x)(2x) = 25 - 4x^2 = (5 + 2x)(5 - 2x)$

Cumulative Review, page 610

1. $12x^2y + 8xy^3$
$4xy(3x + 2y^2)$ Factor out the GCF.

2. $h^2 + 13h + 30$
$(h + 10)(h + 3)$ Factor the trinomial.

3. $n^2 - 9n + 20$
$(n - 5)(n - 4)$ Factor the trinomial.

4. $g^2 + 10g - 56$
$(g - 4)(g + 14)$ Factor the trinomial.

5. $\dfrac{x^8 \cdot x^7}{x^5} = x^{8 + 7 - 5} = x^{10}$

6. $(x^3y)^3(xy^4)^2 = x^{3 \cdot 3}y^3 \cdot x^2y^{4 \cdot 2} = x^{9 + 2}y^{3 + 8} = x^{11}y^{11}$

7. $0.24(3) + 1.00(x) = 0.28(3 + x)$
$$24(3) + 100(x) = 28(3 + x)$$
$$72 + 100x = 84 + 28x$$
$$72x = 12$$
$$x = \frac{12}{72} = \frac{1}{6} \text{ liter}$$

8. $f(-2) + g(-2)$
$(5(-2) - 2) + (2(-2)^2 + (-2) - 1)$
$-10 - 2 + 8 - 2 - 1$
-7

9. $\dfrac{f(4)}{g(2)} = \dfrac{5(4) - 2}{2(2)^2 + 2 - 1} = \dfrac{18}{9} = 2$

10. $f(x) + g(x)$
$(5x - 2) + (2x^2 + x - 1)$
$2x^2 + 6x - 3$

11. $g(x) - f(x)$
$(2x^2 + x - 1) - (5x - 2)$
$2x^2 + x - 1 - 5x + 2$
$2x^2 - 4x + 1$

12. $2x^2 + 7x + 3$
$(2x + 1)(x + 3)$ Factor the trinomial.

13. $r^2 - 14r + 49$
$(r - 7)^2$ Factor the trinomial.

14. $9t^2 - 4$
$(3t + 2)(3t - 2)$ Factor the difference of squares.

15. $16z^2 + 72z + 81$
$(4z + 9)^2$ Factor the trinomial.

16. $\begin{bmatrix} 4 + (-5) & -2 + 4 \\ -3 + 3 & 7 + (-9) \end{bmatrix} = \begin{bmatrix} -1 & 2 \\ 0 & -2 \end{bmatrix}$

17. $\begin{bmatrix} -5 - 4 & 4 - (-2) \\ 3 - (-3) & -9 - 7 \end{bmatrix} = \begin{bmatrix} -9 & 6 \\ 6 & -16 \end{bmatrix}$

18. $[4(7) - (-2)(-3)] - [(-5)(-9) - (4)(3)]$
$[28 - 6] - [45 - 12]$
$(22) - (33)$
-11

19. $27 + 0.16x < 24.50 + 0.18x$
$$2.50 < 0.02x$$
$$125 < x$$

20. $2x < -6$ or $x - 5 \geq -2$
$x < -3$ or $x \geq 3$

21. $-7 < 2x + 1 < 9$
$-8 < \quad 2x \quad < 8$
$-4 < \quad x \quad < 4$

22. $(2x^2 + 6xy) + (3x + 9y)$
$2x(x + 3y) + 3(x + 3y)$ Factor each group.
$(2x + 3)(x + 3y)$ Factor out the binomial.

23. $(x^3 - x^2) - (25x + 25)$
$x^2(x - 1) - 25(x - 1)$ Factor each group.
$(x^2 - 25)(x - 1)$ Factor out the binomial.
$(x + 5)(x - 5)(x - 1)$ Factor the difference of squares.

24. $2p^4 - 8p^3 - 24p^2$
$2p^2(p^2 - 4p - 12)$ Factor out the GCF.
$2p^2(p - 6)(p + 2)$ Factor the trinomial.

25. $2z^4 - 32$
$2(z^4 - 16)$ Factor out the GCF.
$2(z^2 + 4)(z^2 - 4)$.   Factor the difference of squares.
$2(z^2 + 4)(z + 2)(z - 2)$ Factor the difference of squares.

26. $0.15(280) = 42$

27. $0.12(x) = 45$
$$x = \frac{45}{0.12} = 375$$

28. Not necessarily; How many students did you ask? Was this a random sample, or possibly a cluster or convenience sample?

29.
$$x^2 + 8x = 20$$
$$x^2 + 8x - 20 = 0$$
$$(x + 10)(x - 2) = 0$$
$$x + 10 = 0 \quad \text{or} \quad x - 2 = 0$$
$$x = -10 \quad \text{or} \quad x = 2$$

30.
$$3x^2 + 2 = 7x$$
$$3x^2 - 7x + 2 = 0$$
$$(3x - 1)(x - 2) = 0$$
$$3x - 1 = 0 \quad \text{or} \quad x - 2 = 0$$
$$x = \frac{1}{3} \quad \text{or} \quad x = 2$$

31.
$$y^3 - y^2 - 2y = 0$$
$$y(y^2 - y - 2) = 0$$
$$y(y - 2)(y + 1) = 0$$
$$y = 0 \quad \text{or} \quad y - 2 = 0 \quad \text{or} \quad y + 1 = 0$$
$$y = 0 \quad \text{or} \quad y = 2 \quad \text{or} \quad y = -1$$

32.
$$2n^2 + 12n = -18$$
$$2n^2 + 12n + 18 = 0$$
$$2(n^2 + 6n + 9) = 0$$
$$2(n + 3)^2 = 0$$
So $\quad n + 3 = 0$
$$n = -3$$

33. Fred knew that the product of two factors would be zero only if one of the factors was zero. Where the graph crosses the x-axis (where $y = 0$) are the values that would make the product zero. Since the graph crosses the x-axis at -5 and 9, the factors must be $(x + 5)$ and $(x - 9)$.

Standardized Test, page 611

1. D
$$A = 2x^2 + 7x + 6$$
$$A = (2x + 1)(x + 3)$$

Side lengths cannot be compared without more information.

2. B
Col. 1: $\quad m < 0$
Col. 2: $-\dfrac{1}{m} > 0$
Col. 1 < Col. 2

3. C
Col. 1: $(2, -2); \; 2(-2) = -4$
Col. 2: $(-2, 2); \; -2(2) = -4$
Col. 1 = Col. 2

4. A
Col. 1: $a^5 \cdot a^3 = a^{5+3} = a^8 > 0$
Col. 2: $\left(a^5\right)^3 = a^{5 \cdot 3} = a^{15} < 0$
Col. 1 > Col. 2

5. C
Col. 1: $\dfrac{3}{6} = \dfrac{1}{2}$
Col. 2: $\dfrac{1}{2}$
Col. 1 = Col. 2

6. B
Col. 1: $(8 \div 4) \div 2 = 2 \div 2 = 1$
Col. 2: $8 \div (4 \div 2) = 8 \div 2 = 4$
Col. 1 < Col. 2

7. D
Col. 1: $-3x < 12 \rightarrow x > -4$
Col. 2: $-3(4) < 12 \rightarrow -12 < 12$

Col. 2 is a true statement and not a value that can be compared to Col. 1.

8. A
Col. 1: $\quad n + d = 20 \rightarrow d = 20 - n$
$$0.05n + 0.10d = 1.45$$
$$5n + 10d = 145$$
$$5n + 10(20 - n) = 145 \rightarrow n = 11$$
Col. 2: $\quad d = 20 - n = 20 - 11 = 9$
Col. 1 > Col. 2

9. B
Col. 1: parabola opens down, so $a < 0$
Col. 2: $c = 4$
Col. 1 < Col. 2

10. B
Col. 1: $83\% \rightarrow 0.83$
Col. 2: $\dfrac{5}{6} = 0.8\overline{3}$
Col. 1 < Col. 2

11. D
Col. 1: $-7 < x < 7$
Col. 2: $y < -7$ or $y > 7$

Values for x are between the two sets of values for y, so x and y cannot be compared.

12. C
Col. 1: a^{m-n}
Col. 2: $\dfrac{1}{a^{n-m}} = a^{-(n-m)} = a^{m-n}$
Col. 1 = Col. 2

13. B
Col. 1: $(x + y)(x - y) = x^2 - y^2 \rightarrow 2$ terms
Col. 2: $(x + y)(x + y) = x^2 + 2xy + y^2 \rightarrow 3$ terms
Col. 1 < Col. 2

14. B
Col. 1: $4(x + 5) < 12 \rightarrow x < -2$
Col. 2: $3(2 - n) < 12 \rightarrow n > -2$
Col. 1 < Col. 2

Data Activity, page 612

1. about 22 weeks

2. the difference between 23 weeks in 1991 and 19 weeks in 1971; about 4 more weeks

3. According to the table, the new car cost in week's earnings in 1994 was about 24 weeks out of a 52 week year. Therefore, the equation and solution to determine the figure that was used as the median family income for that year is

$$\frac{24}{52}x = 21{,}800$$
$$x \approx 47{,}233$$

4. If the income is below median, it will take more weeks of earnings to purchase a new car.

5. Answers will vary. One explanation is that the average price of a new car dropped due to an increase in the number of low-priced imports during this period. Import priceds have since risen.

Lesson 13.1, pages 615–617

THINK BACK/WORKING TOGETHER

1. The large square has sides of length 3 units and is made up of 9 smaller unit squares. Area = length of side squared

2. A square made up of 4 unit squares has a side of length 2 units. Square root of area = length of a side.

3. 1 square unit   4. 2 square units   5. 8 square units

EXPLORE

6. Activity, so no solution

7. $3^2 = 9$, $4^2 = 16$, $5^2 = 25$

8. $5^2 = 25$, $12^2 = 144$, $13^2 = 169$

9. $6^2 = 36$, $8^2 = 64$, $10^2 = 100$

10. $8^2 = 64$, $15^2 = 225$, $17^2 = 289$

11. Answers will vary.

MAKE CONNECTIONS

12. Area of large square = sum of areas of the other 2 squares.

13. $7^2 = 49$, $24^2 = 576$, $25^2 = 625$; $49 + 576 = 625$

14. the sum of

	Leg 1	Leg 2	Hypotenuse
15.	3	4	$\sqrt{3^2 + 4^2} = \sqrt{9 + 16} = \sqrt{25} = 5$
16.	5	12	$\sqrt{5^2 + 12^2} = \sqrt{25 + 144} = \sqrt{169} = 13$
17.	$\sqrt{15^2 - 12^2} = \sqrt{225 - 144} = \sqrt{81} = 9$	12	15
18.	8	$\sqrt{17^2 - 8^2} = \sqrt{289 - 64} = \sqrt{225} = 15$	17
19.	$\sqrt{41^2 - 40^2} = \sqrt{1681 - 1600} = \sqrt{81} = 9$	40	41
20.	m	n	$\sqrt{m^2 + n^2}$

21. $6^2 + 8^2 = 36 + 64 = 100 = 10^2$ Yes, right $\triangle$

22. $4^2 + 6^2 = 16 + 36 = 52 \neq 7^2$ No, not right $\triangle$

23. $2.5^2 + 6^2 = 6.25 + 36 = 42.25 = 6.5^2$ Yes, right $\triangle$

24. $12^2 + 35^2 = 144 + 1225 = 1369 = 37^2$ Yes, right $\triangle$

25. $14^2 + 22^2 = 196 + 484 = 680 \neq 26^2$ No, not right $\triangle$

SUMMARIZE

26. The square built on each leg of the shaded triangle consists of two triangles the same size and shape as the shaded triangle. The "sum" of these two squares consists of four triangles the same size and shape as the shaded triangle. The sum is equal to the square built on the hypotenuse, which also consists of four triangles the same size and shape as the shaded triangle.

27. Answers will vary.

28. The area of the large square is $(a + b)^2 = a^2 + 2ab + b^2$. The area of the four triangles is $\frac{1}{2}ab$. The small square in the middle with side c has an area of c^2. The combined area of the five polygons is

$$A = c^2 + 4\left(\frac{ab}{2}\right)$$
$$(a + b)^2 = c^2 + 2ab$$
$$a^2 + b^2 = c^2$$

29. Methods will vary, but multiples of pythagorean triples are also pythagorean triples.

EXPLORE

	Square of a Short Side (a^2)	Square of a Short Side (b^2)	Square of Longest Side (c^2)
1.	16	36	49
2.	16	25	64
3.	36	64	100
4.	25	64	81
5.	25	144	169
6.	9	25	49

7. right angle: 3 and 5; obtuse angle: 2 and 6; neither right nor obtuse: 1 and 4

8. $a^2 + b^2 = c^2$ for triangles 3 and 5, the ones with right angles

9. $a^2 + b^2 < c^2$ for triangles 2 and 6, those with obtuse angles

10. $a^2 + b^2 > c^2$ for triangles 1 and 4, those with neither right or obtuse angles

TRY THESE

1. No sides equal, 100.4° angle; so obtuse and scalene

2. Two sides equal, 90° angle; so right and isosceles

3. No sides equal, all angles < 90°, so acute and scalene

4. $180° - (42° + 81°) = 57°$ 5. $180° - (79° + 34°) = 67°$

6. $180° - (108° + 29°) = 43°$ 7. $180° - (44° + 46°) = 90°$

8. $\dfrac{180° - 34°}{2} = \dfrac{146°}{2} = 73°$

9. $x + (3x + 6°) = 90°$
$$4x = 84°$$
$$x = 21°$$
$3(21°) + 6° = 69°$, and right angle $= 90°$

10. $45^2 + 108^2 = 2025 + 11664 = 13689 = 117^2$ so, right triangle

11. $39^2 + 80^2 = 1521 + 6400 = 7921 < 90^2$ so, obtuse triangle

12. $15^2 + 23^2 = 225 + 529 = 754 > 24^2$ so, acute triangle

13. $80^2 + 192^2 = 6400 + 36864 = 43264 = 208^2$ so, right triangle

14. $200^2 + 75^2 = c^2 \rightarrow c^2 = 40{,}000 + 5{,}625 = 45{,}625$
$c = \pm\sqrt{45{,}625} \rightarrow c \approx 213.6$ ft

15. $4^2 + b^2 = 12^2 \rightarrow b^2 = 12^2 - 4^2 = 144 - 16 = 128$
$b = \pm\sqrt{128} \rightarrow b \approx 11.3$ ft

16. Possible answers: (1) Measure the angles at opposite corners. They should be right angles. (2) Measure the lengths of the sides of one of the triangular regions

formed by the frame and brace. If the triangle is a right triangle, the Pythagorean theorem should apply to those measures.

PRACTICE

1. Two sides equal, all angles < 90°, so acute and isosceles

2. No sides equal, angle > 90°, so obtuse and scalene

3. No sides equal, 90° angle, so right and scalene

4. $\dfrac{180° - 92°}{2} = \dfrac{88°}{2} = 44°$

5. Another must be 35° if each is a whole number and third is $180° - (35° + 35°) = 110°$

6. $x + x + 2x = 180°$ so $x = 45°$ and $2x = 90°$, it is an isosceles right triangle

7. They cannot because the sum of three odd numbers is odd and 180° is even.

EXTEND

8. $\overline{AB}$ is shortest, then $\overline{BC}$, and longest is $\overline{AC}$.

9. $\dfrac{1}{\sqrt{3}} = \dfrac{t}{50.2}$
$$t = \dfrac{50.2}{\sqrt{3}} \approx 29 \text{ ft}$$

10. Let $x = m\angle R$; $\dfrac{2}{3}x - 10 = m\angle P$; $2x + 3 = m\angle S$
$$x + \dfrac{2}{3}x - 10 + 2x + 3 = 180$$
$$3\dfrac{2}{3}x = 187$$
$x = 51°$; $\dfrac{2}{3}(51°) - 10° = 24°$; $2(51°) + 3° = 105°$

11. Yes; the longest diagonal is 66.52 in.; find this length by squaring 65, adding 200 (the square of the diagonal of the square face) and finding the square root of the sum.

THINK CRITICALLY

12. $x^2 + \left(\dfrac{x^2 - 1}{2}\right)^2 = \left(\dfrac{x^2 + 1}{2}\right)^2$
$$x^2 + \dfrac{(x^4 - 2x^2 + 1)}{4} = \dfrac{(x^4 + 2x^2 + 1)}{4}$$
$$4x^2 + x^4 - 2x^2 + 1 = x^4 + 2x^2 + 1$$
$$x^4 + 2x^2 + 1 = x^4 + 2x^2 + 1$$

13. Answers will vary, but there should be three different lengths and, if c is the longest side, $a^2 + b^2$ should be less than c^2.

14. Three different lengths, with $a^2 + b^2 > c^2$

15. Two lengths equal, with $a^2 + b^2 > c^2$; c can be either the longest side or one of two equal longer sides.

16. $|12| = 12$ **17.** $-|17| = -17$ **18.** $|-41| = 41$

19. $A:\ 0.25(150) = 37.5$

    $B:\ \mathbf{0.20(150) = 30}$

    $C:\ \mathbf{1.50(20) = 30}$

    $D:\ 1.50(25) = 37.5$

20. $\begin{array}{l} 3x - y = 2 \quad \rightarrow \quad 9x - 3y = 6 \\ 2x + 3y = 16 \quad \rightarrow \quad \underline{2x + 3y = 16} \\ \qquad\qquad\qquad\qquad\quad 11x \quad\quad = 22 \\ \qquad\qquad\qquad\qquad\qquad\quad x = 2 \end{array}$

$3(2) - y = 2 \quad \rightarrow \quad y = 4$

So, $(2, 4)$ solves the system.

21. $\begin{array}{ll} x - y = 1 & \text{Subtract Equation 1} \\ \underline{2x - y = 8} & \text{from Equation 2.} \\ x \quad\ = 7 & \end{array}$

    $7 - y = 1$

        $y = 6$

So, $(7, 6)$ solves the system.

22. $\begin{array}{l} 4x + 3y = 3 \quad \rightarrow \quad 8x + 6y = 6 \\ x - y = \dfrac{1}{6} \quad \rightarrow \quad \underline{6x - 6y = 1} \\ \qquad\qquad\quad \rightarrow \quad 14x \qquad\ = 7 \\ \qquad\qquad\quad \rightarrow \qquad\quad x = \dfrac{7}{14} = \dfrac{1}{2} \end{array}$

$\dfrac{1}{2} - y = \dfrac{1}{6}$

$\qquad y = \dfrac{1}{3}$

So, $\left(\dfrac{1}{2}, \dfrac{1}{3}\right)$ solves the system.

23. $3x^3 - 48x$

    $3x(x^2 - 16)$ Factor out GCF.

    $3x(x + 4)(x - 4)$ Factor difference of squares.

24. $4x^2 y - 4xy - 48y$

    $4y(x^2 - x - 12)$ Factor out GCF.

    $4y(x - 4)(x + 3)$ Factor trinomial.

25. $x^4 - 81$

    $(x^2 + 9)(x^2 - 9)$ Factor difference of squares.

    $(x^2 + 9)(x + 3)(x - 3)$ Factor difference of squares.

Lesson 13.3, pages 625–630

EXPLORE

1. $2^2 + 2^2 = l^2 \rightarrow l^2 = 8 \rightarrow l = \sqrt{8} = 2\sqrt{2}$

2. $1^2 + 1^2 = w^2 \rightarrow w^2 = 2 \rightarrow w = \sqrt{2}$

3. 4 unit squares

4. Area of $AFEB = l \bullet w = \sqrt{8} \bullet \sqrt{2} = 4 = \sqrt{16}$

TRY THESE

1. $\sqrt{50} = \sqrt{2 \bullet 5 \bullet 5} = \sqrt{5^2} \bullet \sqrt{2} = 5\sqrt{2}$

2. $\sqrt{27k^2} = \sqrt{3 \bullet 3 \bullet 3 \bullet k^2} = \sqrt{3^2} \bullet \sqrt{k^2} \bullet \sqrt{3} = 3|k|\sqrt{3}$

3. $2\sqrt{6} \bullet \sqrt{8} = 2\sqrt{6 \bullet 8} = 2\sqrt{3 \bullet 2 \bullet 2 \bullet 2 \bullet 2} = 2\sqrt{2^2} \bullet \sqrt{2^2} \bullet \sqrt{3} = 8\sqrt{3}$

4. $\sqrt{2}\left(\sqrt{3} - \sqrt{8}\right) = \sqrt{2 \bullet 3} - \sqrt{2 \bullet 2 \bullet 2 \bullet 2} = \sqrt{6} - \sqrt{2^2} \bullet \sqrt{2^2} = \sqrt{6} - 4$

5. $\dfrac{\sqrt{60}}{\sqrt{5}} = \sqrt{\dfrac{60}{5}} = \sqrt{12} = \sqrt{2 \bullet 2 \bullet 3} = \sqrt{2^2} \bullet \sqrt{3} = 2\sqrt{3}$

6. $\dfrac{3}{\sqrt{3}} \bullet \dfrac{\sqrt{3}}{\sqrt{3}} = \dfrac{3\sqrt{3}}{\sqrt{3^2}} = \dfrac{3\sqrt{3}}{3} = \sqrt{3}$

7. $\sqrt{\dfrac{5}{6}} = \dfrac{\sqrt{5}}{\sqrt{6}} \bullet \dfrac{\sqrt{6}}{\sqrt{6}} = \dfrac{\sqrt{5 \bullet 6}}{\sqrt{6^2}} = \dfrac{\sqrt{30}}{6}$

8. $\dfrac{4\sqrt{3}}{\sqrt{8}} \bullet \dfrac{\sqrt{2}}{\sqrt{2}} = \dfrac{4\sqrt{2 \bullet 3}}{\sqrt{2 \bullet 2 \bullet 2 \bullet 2}} = \dfrac{4\sqrt{6}}{\sqrt{2^2} \bullet \sqrt{2^2}} = \dfrac{4\sqrt{6}}{4} = \sqrt{6}$

9. $d = \sqrt{12 \bullet 375} = \sqrt{2 \bullet 2 \bullet 3 \bullet 3 \bullet 5 \bullet 5 \bullet 5} = \sqrt{2^2} \bullet \sqrt{3^2} \bullet \sqrt{5^2} \bullet \sqrt{5} = 2 \bullet 3 \bullet 5 \bullet \sqrt{5} = 30\sqrt{5}$ km

10. $d^2 = 90^2 + 90^2 = 90^2 \bullet 2 \rightarrow d = 90\sqrt{2}$

11. $30 = 2 \bullet 3 \bullet 5$, none of which is a perfect square so $\sqrt{30}$ is in simplest form

PRACTICE

1. $\sqrt{18} = \sqrt{2 \bullet 3 \bullet 3} = \sqrt{3^2} \bullet \sqrt{2} = 3\sqrt{2}$

2. $\sqrt{40} = \sqrt{2 \bullet 2 \bullet 2 \bullet 5} = \sqrt{2^2} \bullet \sqrt{2 \bullet 5} = 2\sqrt{10}$

3. $\sqrt{300} = \sqrt{2 \bullet 2 \bullet 3 \bullet 5 \bullet 5} = \sqrt{2^2} \bullet \sqrt{5^2} \bullet \sqrt{3} = 2 \bullet 5 \bullet \sqrt{3} = 10\sqrt{3}$

4. $\sqrt{147} = \sqrt{3 \bullet 7 \bullet 7} = \sqrt{7^2} \bullet \sqrt{3} = 7\sqrt{3}$

5. $\sqrt{9n^2} = \sqrt{n^2 \bullet 3^2} = 3|n|$

6. $\sqrt{50p^2} = \sqrt{2 \bullet 5 \bullet 5 \bullet p^2} = \sqrt{5^2} \bullet \sqrt{p^2} \bullet \sqrt{2} = 5|p|\sqrt{2}$

7. $3\sqrt{32c^3} = 3\sqrt{2 \bullet 2 \bullet 2 \bullet 2 \bullet 2 \bullet c^2 \bullet c} = 3 \bullet \sqrt{2^2} \bullet \sqrt{2^2} \bullet \sqrt{c^2} \bullet \sqrt{2c} = 3 \bullet 2 \bullet 2 \bullet c \bullet \sqrt{2c} = 12c\sqrt{2c}$

8. $-5\sqrt{24x^3y} = -5\sqrt{2 \cdot 2 \cdot 2 \cdot 3 \cdot x^2 \cdot x \cdot y} =$
$-5 \cdot \sqrt{2^2} \cdot \sqrt{x^2} \cdot \sqrt{2 \cdot 3 \cdot x \cdot y} =$
$-5 \cdot 2 \cdot x \cdot \sqrt{6xy} = -10x\sqrt{6xy}$

9. $\sqrt{3} \cdot \sqrt{3} = \sqrt{3^2} = 3$ **10.** $\sqrt{7} \cdot \sqrt{7} = \sqrt{7^2} = 7$

11. $5\sqrt{20} \cdot \sqrt{5} = 5\sqrt{2 \cdot 2 \cdot 5 \cdot 5} = 5\sqrt{2^2} \cdot \sqrt{5^2} =$
$5 \cdot 2 \cdot 5 = 50$

12. $-3\sqrt{18} \cdot \sqrt{2} = -3\sqrt{2 \cdot 3 \cdot 3 \cdot 2} = -3\sqrt{2^2} \cdot \sqrt{3^2} =$
$-3 \cdot 2 \cdot 3 = -18$

13. $6(2 - \sqrt{3}) = 6 \cdot 2 - 6 \cdot \sqrt{3} = 12 - 6\sqrt{3}$

14. $\sqrt{5}(5 + \sqrt{3}) = \sqrt{5} \cdot 5 + \sqrt{5} \cdot \sqrt{3} = 5\sqrt{5} + \sqrt{15}$

15. $\sqrt{12}(2\sqrt{3} - \sqrt{5}) = \sqrt{12} \cdot 2\sqrt{3} - \sqrt{12} \cdot \sqrt{5} =$
$2\sqrt{2 \cdot 2 \cdot 3 \cdot 3} - \sqrt{2 \cdot 2 \cdot 3 \cdot 5} =$
$2\sqrt{2^2} \cdot \sqrt{3^2} - \sqrt{2^2}\sqrt{3 \cdot 5} = 2 \cdot 2 \cdot 3 - 2\sqrt{15} =$
$12 - 2\sqrt{15}$

16. $\dfrac{\sqrt{30}}{\sqrt{6}} = \sqrt{\dfrac{30}{6}} = \sqrt{5}$

17. $\dfrac{\sqrt{72}}{\sqrt{8}} = \sqrt{\dfrac{72}{8}} = \sqrt{9} = \sqrt{3^2} = 3$

18. $\dfrac{\sqrt{28a^2}}{\sqrt{8}} \cdot \dfrac{\sqrt{2}}{\sqrt{2}} = \dfrac{\sqrt{2 \cdot 2 \cdot 7 \cdot a^2}}{\sqrt{2 \cdot 2 \cdot 2}} =$

$\dfrac{\sqrt{2^2} \cdot \sqrt{a^2}\sqrt{14}}{\sqrt{2^2} \cdot \sqrt{2^2}} = \dfrac{|a|\sqrt{14}}{2}$

19. $\dfrac{2}{\sqrt{5}} \cdot \dfrac{\sqrt{5}}{\sqrt{5}} = \dfrac{2\sqrt{5}}{\sqrt{5^2}} = \dfrac{2\sqrt{5}}{5}$

20. $\dfrac{6}{\sqrt{2}} \cdot \dfrac{\sqrt{2}}{\sqrt{2}} = \dfrac{6\sqrt{2}}{\sqrt{2^2}} = \dfrac{6\sqrt{2}}{2} = 3\sqrt{2}$

21. Area $= 2\sqrt{13} \cdot 2\sqrt{13} = 4\sqrt{13^2} = 4 \cdot 13 = 52$

22. Area $= 2\sqrt{8} \cdot 3\sqrt{6} = 6\sqrt{2 \cdot 2 \cdot 2 \cdot 2 \cdot 3} =$
$6\sqrt{2^2} \cdot \sqrt{2^2} \cdot \sqrt{3} = 6 \cdot 2 \cdot 2 \cdot \sqrt{3} = 24\sqrt{3}$

23. Area $= \dfrac{1}{2}bh = \dfrac{1}{2} \cdot 3\sqrt{18} \cdot 4\sqrt{2} =$
$\dfrac{1}{2} \cdot 3 \cdot 4\sqrt{2 \cdot 3 \cdot 3 \cdot 2} = 6\sqrt{2^2} \cdot \sqrt{3^2} =$
$6 \cdot 2 \cdot 3 = 36$

24. $\dfrac{\sqrt{3}}{\sqrt{2}} \cdot \dfrac{\sqrt{2}}{\sqrt{2}} = \dfrac{\sqrt{3 \cdot 2}}{\sqrt{2^2}} = \dfrac{\sqrt{6}}{2}$

25. Answers will vary. Samples:

   1. the radicand contains no perfect-square factors other than 1: $\sqrt{75} = 5\sqrt{3}$

   2. the radicand contains no fractions: $\sqrt{\dfrac{1}{3}} = \dfrac{\sqrt{3}}{3}$

   3. no denominator contains a radical: $\dfrac{3}{\sqrt{2}} = \dfrac{3\sqrt{2}}{2}$

EXTEND

26.

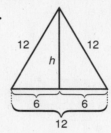

$h^2 + 6^2 = 12^2$
$h^2 + 36 = 144$
$h^2 = 108$
$h = \sqrt{108} = \sqrt{2 \cdot 2 \cdot 3 \cdot 3 \cdot 3}$
$ = \sqrt{2^2} \cdot \sqrt{3^2} \cdot \sqrt{3} = 2 \cdot 3 \cdot \sqrt{3} = 6\sqrt{3}$

27.

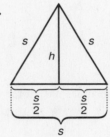

$h^2 + \left(\dfrac{s}{2}\right)^2 = s^2$

$h^2 + \dfrac{s^2}{4} = s^2$

$h^2 = s^2 - \dfrac{s^2}{4} = \dfrac{3s^2}{4}$

$h = \sqrt{\dfrac{3s^2}{4}} = \dfrac{\sqrt{3}\sqrt{s^2}}{\sqrt{2^2}} = \dfrac{s\sqrt{3}}{2}$

28. Conjugate of $4 + \sqrt{2}$ is $4 - \sqrt{2}$
Product is $(4 + \sqrt{2})(4 - \sqrt{2}) = 4^2 - (\sqrt{2})^2 = 16 - 2 = 14$

29. Conjugate of $\sqrt{5} - \sqrt{3}$ is $\sqrt{5} + \sqrt{3}$
Product is $(\sqrt{5} - \sqrt{3})(\sqrt{5} + \sqrt{3}) = (\sqrt{5})^2 - (\sqrt{3})^2 =$
$5 - 3 = 2$

30. Conjugate of $2\sqrt{6} - 3\sqrt{2}$ is $2\sqrt{6} + 3\sqrt{2}$

Product is $(2\sqrt{6} - 3\sqrt{2})(2\sqrt{6} + 3\sqrt{2}) =$
$$(2\sqrt{6})^2 - (3\sqrt{2})^2 = 4 \cdot 6 - 9 \cdot 2 =$$
$$24 - 18 = 6$$

31. Conjugate of $\sqrt{10} + 4$ is $\sqrt{10} - 4$

Product is $(\sqrt{10} + 4)(\sqrt{10} - 4) = (\sqrt{10})^2 - (4)^2 =$
$$10 - 16 = -6$$

32. $\dfrac{\sqrt{3} + \sqrt{2}}{\sqrt{3} - \sqrt{2}} \cdot \dfrac{\sqrt{3} + \sqrt{2}}{\sqrt{3} + \sqrt{2}} = \dfrac{(\sqrt{3})^2 + 2\sqrt{3} \cdot \sqrt{2} + (\sqrt{2})^2}{(\sqrt{3})^2 - (\sqrt{2})^2} =$

$\dfrac{3 + 2\sqrt{6} + 2}{3 - 2} = 5 + 2\sqrt{6}$

33. $\dfrac{1 + \sqrt{2}}{3 - \sqrt{3}} \cdot \dfrac{3 + \sqrt{3}}{3 + \sqrt{3}} = \dfrac{3 + \sqrt{3} + 3\sqrt{2} + \sqrt{2 \cdot 3}}{3^2 - (\sqrt{3})^2} =$

$\dfrac{3 + \sqrt{3} + 3\sqrt{2} + \sqrt{6}}{9 - 3} = \dfrac{3 + \sqrt{3} + 3\sqrt{2} + \sqrt{6}}{6}$

34. $\dfrac{1 + \sqrt{3}}{2 + \sqrt{5}} \cdot \dfrac{2 - \sqrt{5}}{2 - \sqrt{5}} = \dfrac{2 - 1\sqrt{5} + 2\sqrt{3} - \sqrt{3 \cdot 5}}{2^2 - (\sqrt{5})^2} =$

$\dfrac{2 - \sqrt{5} + 2\sqrt{3} - \sqrt{15}}{4 - 5} = -2 + \sqrt{5} - 2\sqrt{3} + \sqrt{15}$

THINK CRITICALLY

35.
$$\sqrt{x + y} = 10$$
$$x + y = 100$$
$$y = 100 - x$$
$$\sqrt{x \cdot y} = 48$$
$$xy = 48^2 = 2304$$
$$x(100 - x) = 2304 \qquad \text{Substitute } 100 - x \text{ for } y.$$
$$100x - x^2 = 2304$$
$$0 = x^2 - 100x + 2304$$
$$0 = (x - 36)(x - 64)$$

So $x = 36$ or $x = 64$; numbers are 36 and 64.

36.
$$\frac{\sqrt{7x}}{\sqrt{28x^2}} = \frac{1}{2}$$
$$\frac{7x}{28x^2} = \frac{1}{4} \qquad \text{Square both sides.}$$
$$7x = 7x^2 \qquad \text{Multiply by } 28x^2.$$
$$0 = 7x^2 - 7x \qquad \text{Subtract } 7x.$$
$$0 = 7x(x - 1) \qquad \text{Factor.}$$

So, $x = 0$ or 1, but since 0 makes $\dfrac{\sqrt{7x}}{28x^2}$ undefined,

x can only be 1.

37. $(m + \sqrt{n})(m - \sqrt{n}) = m^2 - n$

Since both m^2 and n are rational numbers, their difference is a rational number. Therefore, the product is a rational number.

MIXED REVIEW

38. $3x - 2 = 2x + 7$

$\quad x = 9 \qquad$ Subtract $2x$ and add 2.

39. $6 - 3x = -5x + 2$

$\quad 2x = -4 \qquad$ Add $5x$ and subtract 6.

$\quad x = -2 \qquad$ Divide by 2.

40. $5(2x - 5) = -(x - 8)$

$\quad 10x - 25 = -x + 8 \qquad$ Distribute.

$\quad 11x = 33 \qquad$ Add x and 25.

$\quad x = 3 \qquad$ Divide by 11.

41. $y = -4x + 3$; slope is -4; y-intercept is 3

42. $6x + 2y = 11$

$\quad 2y = -6x + 11$

$\quad y = -3x + \dfrac{11}{2}$

slope is -3

y-intercept is $\dfrac{11}{2}$

43. $3(y - 4) = 3x$

$\quad y - 4 = x$

$\quad y = x + 4$

slope is 1

y-intercept is 4

44. $\sqrt{20n^2} = \sqrt{2 \cdot 2 \cdot 5 \cdot n^2} = \sqrt{2^2} \cdot \sqrt{n^2}\sqrt{5} = 2|n|\sqrt{5}$; D

ALGEBRAWORKS

1. $\dfrac{306,000}{3,900,000} \approx 0.078$ or 7.8%

2. $\dfrac{3,900,000}{24,900} \approx 156.63$ or about 157 times

3. As r increases in $v = \sqrt{2.5r}$, v also increases

4.

r	10	20	30	40	50
$\sqrt{\ }$	5	$5\sqrt{2}$	$5\sqrt{3}$	10	$5\sqrt{5}$
r	60	70	80	90	100
$\sqrt{\ }$	$5\sqrt{6}$	$5\sqrt{7}$	$10\sqrt{2}$	15	$5\sqrt{10}$

5.

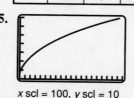

x scl = 100, y scl = 10

6. $v = \sqrt{2.5(4.1 \cdot 5280)} = \sqrt{54120} \approx 232.6$ mph

Lesson 13.4, pages 631–635

EXPLORE

1. $2^2 + 2^2 = AB^2 \rightarrow AB = 2\sqrt{2}$

2. $3^2 + 3^2 = BC^2 \rightarrow BC = 3\sqrt{2}$

3. $5^2 + 5^2 = AC^2 \rightarrow AC = 5\sqrt{2}$

4. $AB + BC = AC$; by substitution $2\sqrt{2} + 3\sqrt{2} = 5\sqrt{2}$

1. $(5+2)\sqrt{7} = 7\sqrt{7}$ 2. $(12-1)\sqrt{5} = 11\sqrt{5}$

3. $(10+9)\sqrt{2} = 19\sqrt{2}$

4. $\sqrt{4 \cdot 6} - \sqrt{6} = 2\sqrt{6} - \sqrt{6} = (2-1)\sqrt{6} = 1\sqrt{6} = \sqrt{6}$

5. $2\sqrt{10} - 3\sqrt{4 \cdot 10} + 4\sqrt{5} = 2\sqrt{10} - 6\sqrt{10} + 4\sqrt{5} =$
   $(2-6)\sqrt{10} + 4\sqrt{5} = -4\sqrt{10} + 4\sqrt{5}$

6. $(3-5)\sqrt{b} = -2\sqrt{b}$

7. $7\sqrt{9 \cdot 5} + 3\sqrt{4 \cdot 5} = 21\sqrt{5} + 6\sqrt{5} = (21+6)\sqrt{5} = 27\sqrt{5}$

8. $\sqrt{4 \cdot 2 \cdot x^2} + \sqrt{2 \cdot x^2} = 2|x|\sqrt{2} + |x|\sqrt{2} =$
   $(2|x| + |x|)\sqrt{2} = 3|x|\sqrt{2}$

9. $\sqrt{\dfrac{1}{2}} \cdot \sqrt{\dfrac{2}{2}} - \sqrt{\dfrac{1}{8}} \cdot \sqrt{\dfrac{2}{2}} = \dfrac{\sqrt{2}}{2} - \dfrac{\sqrt{2}}{4} = \dfrac{2\sqrt{2}}{4} - \dfrac{\sqrt{2}}{4} =$
   $(2-1)\left(\dfrac{\sqrt{2}}{4}\right) = \dfrac{\sqrt{2}}{4}$

10. $P = 2 \cdot \sqrt{54} + 2\sqrt{24} = 2 \cdot \sqrt{9 \cdot 6} + 2\sqrt{4 \cdot 6} =$
    $6\sqrt{6} + 4\sqrt{6} = (6+4)\sqrt{6} = 10\sqrt{6}$

11. In $7x^2 - 6x + 4x^2$, only $7x^2$ and $4x^2$ are like because they have the same exponent and base. So they can be combined to $11x^2$. In $7\sqrt{6} - 6\sqrt{2} + 4\sqrt{6}$, $7\sqrt{6}$ and $4\sqrt{6}$ are like because the radicands are the same, so they can be combined to $11\sqrt{6}$.

1. $(4+5)\sqrt{3} = 9\sqrt{3}$ 2. $(8+7)\sqrt{7} = 15\sqrt{7}$

3. $(13-5)\sqrt{5} = 8\sqrt{5}$ 4. $4\sqrt{10} - \sqrt{5}$

5. $3\sqrt{3} + 6\sqrt{2}$

6. $\sqrt{20} + \sqrt{80} - \sqrt{45} = \sqrt{4 \cdot 5} + \sqrt{16 \cdot 5} - \sqrt{9 \cdot 5} =$
   $2\sqrt{5} + 4\sqrt{5} - 3\sqrt{5} = 3\sqrt{5}$

7. $16\sqrt{h}$

8. $\sqrt{128y} - \sqrt{2y} = \sqrt{64 \cdot 2y} - \sqrt{2y} =$
   $8\sqrt{2y} - \sqrt{2y} = 7\sqrt{2y}$

9. $5\sqrt{18x} + 2\sqrt{8x} = 5\sqrt{9 \cdot 2x} + 2\sqrt{4 \cdot 2x} =$
   $15\sqrt{2x} + 4\sqrt{2x} = 19\sqrt{2x}$

10. $2\sqrt{3} + 3\sqrt{12} = 2\sqrt{3} + 3\sqrt{4 \cdot 3} = 2\sqrt{3} + 6\sqrt{3} = 8\sqrt{3}$

11. $5\sqrt{50} - 4\sqrt{32} = 5\sqrt{25 \cdot 2} - 4\sqrt{16 \cdot 2} =$
    $25\sqrt{2} - 16\sqrt{2} = 9\sqrt{2}$

12. $\sqrt{200c^2} - \sqrt{98c^2} = \sqrt{100 \cdot c^2 \cdot 2} - \sqrt{49 \cdot c^2 \cdot 2} =$
    $10|c|\sqrt{2} - 7|c|\sqrt{2} = 3|c|\sqrt{2}$

13. $\sqrt{8n+8} + \sqrt{2n+2} = \sqrt{4(2n+2)} + \sqrt{2n+2} =$
    $2\sqrt{2n+2} + \sqrt{2n+2} = 3\sqrt{2n+2}$

14. $\sqrt{2} + \sqrt{\dfrac{1}{2}} = \sqrt{2} + \sqrt{\dfrac{1}{2}} \cdot \sqrt{\dfrac{2}{2}} = \sqrt{2} + \dfrac{1}{2}\sqrt{2} = \dfrac{3\sqrt{2}}{2}$

15. $\sqrt{\dfrac{2}{3}} - \sqrt{\dfrac{1}{6}} = \sqrt{\dfrac{2}{3}} \cdot \sqrt{\dfrac{3}{3}} - \sqrt{\dfrac{1}{6}} \cdot \sqrt{\dfrac{6}{6}} = \dfrac{\sqrt{6}}{3} - \dfrac{\sqrt{6}}{6} =$
    $\dfrac{2\sqrt{6}}{6} - \dfrac{\sqrt{6}}{6} = \dfrac{\sqrt{6}}{6}$

16. $\text{Area} = (7\sqrt{3})(\sqrt{75})$
    $= 7\sqrt{225}$
    $= 7 \cdot 15$
    $= 105$
    $\text{Perimeter} = 2 \cdot 7\sqrt{3} + 2 \cdot \sqrt{75}$
    $= 14\sqrt{3} + 10\sqrt{3}$
    $= 24\sqrt{3}$

17. $\text{Area} = \dfrac{1}{2}(3 + \sqrt{200})(\sqrt{63})$
    $= \left(\dfrac{3}{2} + \dfrac{\sqrt{100 \cdot 2}}{2}\right)(\sqrt{9 \cdot 7})$
    $= \left(\dfrac{3}{2} + 5\sqrt{2}\right)(3\sqrt{7})$
    $= \dfrac{9\sqrt{7}}{2} + 15\sqrt{14}$
    $\text{Perimeter} = (2 + \sqrt{72}) + (\sqrt{98}) + (3 + \sqrt{200})$
    $= 2 + \sqrt{36 \cdot 2} + \sqrt{49 \cdot 2} + 3 + \sqrt{100 \cdot 2}$
    $= 2 + 6\sqrt{2} + 7\sqrt{2} + 3 + 10\sqrt{2}$
    $= 5 + 23\sqrt{2}$

18. $\text{Area} = \dfrac{1}{2}(\sqrt{15} + \sqrt{60})(\sqrt{20})$
    $= \dfrac{1}{2}(\sqrt{15} + \sqrt{4 \cdot 15})(\sqrt{4 \cdot 5})$
    $= \dfrac{1}{2}(\sqrt{15} + 2\sqrt{15})(2\sqrt{5})$
    $= \dfrac{1}{2}(3\sqrt{15})(2\sqrt{5})$
    $= 3\sqrt{75} = 3\sqrt{25 \cdot 3}$
    $= 15\sqrt{3}$
    $\text{Perimeter} = \sqrt{18} + \sqrt{15} + \sqrt{32} + \sqrt{60}$
    $= \sqrt{9 \cdot 2} + \sqrt{15} + \sqrt{16 \cdot 2} + \sqrt{4 \cdot 15}$
    $= 3\sqrt{2} + \sqrt{15} + 4\sqrt{2} + 2\sqrt{15}$
    $= 7\sqrt{2} + 3\sqrt{15}$

19. $s_1 = 2\sqrt{5(230.4)}$ $s_2 = 2\sqrt{5(160)}$

$\quad = 2\sqrt{1152}$ $\quad = 2\sqrt{800}$

$\quad = 2\sqrt{576 \cdot 2}$ $\quad = 2\sqrt{400 \cdot 2}$

$\quad = 48\sqrt{2}$ $\quad = 40\sqrt{2}$

The difference is $48\sqrt{2} - 40\sqrt{2} = 8\sqrt{2}$.

20. $d_1 = \sqrt{12 \cdot 320}$ $d_2 = \sqrt{12 \cdot 245}$

$\quad = \sqrt{3840}$ $\quad = \sqrt{2940}$

$\quad = \sqrt{256 \cdot 15}$ $\quad = \sqrt{196 \cdot 15}$

$\quad = 16\sqrt{15}$ $\quad = 14\sqrt{15}$

The difference is $16\sqrt{15} - 14\sqrt{15} = 2\sqrt{15}$ km.

21a. The shorter pendulum is faster.

b. $t_1 = 2\pi\sqrt{\dfrac{1}{32}}$ $t_2 = 2\pi\sqrt{\dfrac{4}{32}}$

$\quad = 2\pi\sqrt{\dfrac{1}{32}} \cdot \sqrt{\dfrac{2}{2}}$ $\quad = 2\pi\sqrt{\dfrac{4}{32}} \cdot \sqrt{\dfrac{2}{2}}$

$\quad = \dfrac{\pi}{4}\sqrt{2}$ $\quad = \dfrac{\pi}{4}\sqrt{8} = \dfrac{2\pi}{4}\sqrt{2}$

Difference in time is $\dfrac{2\pi}{4}\sqrt{2} - \dfrac{\pi}{4}\sqrt{2} = \dfrac{\pi}{4}\sqrt{2}$ seconds

22. Explanations will vary. Students should state that they can use a calculator to evaluate each expression and then compare the results.

EXTEND

23a. $3\sqrt{2} + 4\sqrt{2} + 7\sqrt{2} = 14\sqrt{2}$ miles

23b. $(3 + 3 + 4 + 4 + 7 + 7) - 14\sqrt{2} = 28 - 14\sqrt{2}$ miles

24. $\left(-2 + \sqrt{2}\right)^2 + 4\left(-2 + \sqrt{2}\right) + 2 =$

$4 - 4\sqrt{2} + 2 - 8 + 4\sqrt{2} + 2 = 0$

So, $-2 + \sqrt{2}$ is a solution of $x^2 + 4x + 2 = 0$.

25. $\left(2 - \sqrt{11}\right)^2 - 4\left(2 - \sqrt{11}\right) - 7 =$

$4 - 4\sqrt{11} + 11 - 8 + 4\sqrt{11} - 7 = 0$

So, $2 - \sqrt{11}$ is a solution of $x^2 - 4x - 7 = 0$.

26. $\left(1 + \sqrt{6}\right)^2 - 3\left(1 + \sqrt{6}\right) - 5 =$

$1 + 2\sqrt{6} + 6 - 3 - 3\sqrt{6} - 5 =$

$-1 - \sqrt{6} \neq 0$

So, $1 + \sqrt{6}$ is not a solution of $x^2 - 3x - 5 = 0$.

27. $\left(5 + 2\sqrt{5}\right)^2 - 10\left(5 + 2\sqrt{5}\right) + 5 =$

$25 + 20\sqrt{5} + 20 - 50 - 20\sqrt{5} + 5 = 0$

So, $5 + 2\sqrt{5}$ is a solution of $x^2 - 10x + 5 = 0$.

28a. $\sqrt{x} + \sqrt{\dfrac{1}{x}} = \sqrt{x} + \sqrt{\dfrac{1}{x}} \cdot \sqrt{\dfrac{x}{x}} = \sqrt{x} + \dfrac{\sqrt{x}}{x} =$

$\dfrac{x\sqrt{x}}{x} + \dfrac{\sqrt{x}}{x} = (x + 1)\dfrac{\sqrt{x}}{x} = \dfrac{x+1}{x}\sqrt{x}$

b. $\sqrt{5} + \sqrt{\dfrac{1}{5}} = \dfrac{5+1}{5}\sqrt{5} = \dfrac{6}{5}\sqrt{5}$

THINK CRITICALLY

29a. The expressions are equal.

b. $\left(\sqrt{5} - 1\right)^2 = 5 - 2\sqrt{5} + 1 = 6 - 2\sqrt{5} = \left(\sqrt{6 - 2\sqrt{5}}\right)^2$

c. Let $10 + 4\sqrt{6} = \left(\sqrt{6} + x\right)^2 = 6 + 2x\sqrt{6} + x^2$

$10 = 6 + x^2$ and $4\sqrt{6} = 2x\sqrt{6}$

So $x = 2$ and $\sqrt{10 + 4\sqrt{6}} = \sqrt{6} + 2$

d. Let $7 + 4\sqrt{3} = \left(\sqrt{3} + x\right)^2 = 3 + 2x\sqrt{3} + x^2$

$7 = 3 + x^2$ and $4\sqrt{3} = 2x\sqrt{3}$

So, $x = 2$ and $\sqrt{7 + 4\sqrt{3}} = \sqrt{3} + 2$

MIXED REVIEW

30. $3(-2) - 4(5) = -6 - 20 = -26$

31. $5(-3)^2 + 7(-3) = 5(9) - 21 = 45 - 21 = 24$

32. $6(2 - 2(-2)) + 5(2 + (-2)) = 6(2 + 4) + 5(0) = 6(6) = 36$

33. $(-1)^3 - (-1)^2 + 5(-1) - 2 = -1 - 1 - 5 - 2 = -9$

34. $f(3) = 4(3) - 5 = 12 - 5 = 7$

35. $f(-5.5) = 4(-5.5) - 5 = -22 - 5 = -27$

36. $f(0.5) = 4(0.5) - 5 = 2 - 5 = -3$

37. $f(x + 1) = 4(x + 1) - 5 = 4x + 4 - 5 = 4x - 1$

38. $|x| < 3$

$\quad -3 < x < 3$ Write a conjunction.

39. $|x| \geq 6$

$\quad x \leq -6$ or $x \geq 6$ Write a disjunction.

40. C; $\left(3 - \sqrt{5}\right)\left(5 + 2\sqrt{5}\right)$

$\quad 15 + 6\sqrt{5} - 5\sqrt{5} - 10$

$\quad 5 + \sqrt{5}$

41. $6\sqrt{7}$

42. $\sqrt{18} - \sqrt{8} = \sqrt{9 \cdot 2} - \sqrt{4 \cdot 2} = 3\sqrt{2} - 2\sqrt{2} = \sqrt{2}$

43. $\sqrt{27} + 5\sqrt{3} = \sqrt{9 \cdot 3} + 5\sqrt{3} = 3\sqrt{3} + 5\sqrt{3} = 8\sqrt{3}$

44. $5\sqrt{x} + 2\sqrt{9x} = 5\sqrt{x} + 6\sqrt{x} = 11\sqrt{x}$

PROJECT CONNECTION

1. Activity, so no solution.

2. Use the Pythagorean theorem twice, first to find the length of the diagonal along a surface of the carton, and again to find the corner-to-corner diagonal.

7. I: $\sqrt{31} \approx 5.5678$

    II: 35% or 0.35

    III: 6.01001001...

    IV: $\dfrac{23}{4} = 5.75$

8. B; $x + (x+1) = 351$

$$x = 175$$
$$x + 1 = 176$$

Since there are 40 pages per chapter, then

$\dfrac{175}{40} = 4.375$ or in the fifth chapter.

9. $f(6) + g(-2) = \dfrac{2}{3}(6) - 5 + 2(-2)^2 + 3(-2) - 5 =$

$4 - 5 + 8 - 6 - 5 = -4;$ B

10. D; $\dfrac{x_1 + 4}{2} = 1$ and $\dfrac{y_1 + (-10)}{2} = -3$

$$x_1 + 4 = 2 \qquad\qquad y_1 - 10 = -6$$
$$x_1 = -2 \qquad\qquad y_1 = 4$$

11. B; $\left[24 - \dfrac{1}{3}(24)\right] - x\left[24 - \dfrac{1}{3}(24)\right] = 12$

$$[24 - 8] - x[24 - 8] = 12$$
$$16 - 16x = 12$$
$$-16x = -4$$
$$x = \dfrac{-4}{-16} = \dfrac{1}{4}$$
$$x = 0.25 \text{ or } 25\%$$

Data Activity, page 673

1. $5.60\frac{3}{4} - 5.85\frac{1}{2} = 5.6075 - 5.855 = -0.24\frac{3}{4}$

2. $2.98(10,000) = \$29,800;$

   $2.92\frac{1}{2}(10,000) = 2.925(10,000) = \$29,250;$

   $29,800 - 29,250 = \$550$

3. $\dfrac{4.54 - 4.73}{4.73} \approx -0.04$ or 4% decrease

4. $\dfrac{2}{7}$; Only soybeans and oats had the best price on Wednesday.

5. Answers will vary.

GETTING STARTED

1. because its area is (8.5)(11) or 93.5 square units

2. The one with the $8\frac{1}{2}$ in. height has the greater volume.

3. One doubles while the other is halved.

4. Answers will vary.

Lesson 14.1, pages 675-678

THINK BACK/WORKING TOGETHER

1.
Number of hours	1	2	3	4	5	x
Miles traveled	50	100	150	200	250	50x

2. $y = 50x$; Yes this is a direct variation.

3.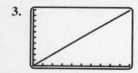

   Answers will vary.

   x min: 0 y min: 0

   x max: 10 y max: 500

   x scl: 1 y scl: 50

EXPLORE/WORKING TOGETHER

4.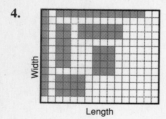

5.
Length	1	2	3	4	6	12
Width	12	6	4	3	2	1

6. $y = \dfrac{12}{x}$; As x increases, y decreases; As x decreases, y increases; No, it is an inverse variation.

7.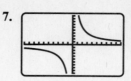

8. Yes, they make sense as possible dimensions of rectangles having an area 12 square units.

9. No; they make sense as real number factors of 12 but not as rectangle dimensions.

10. No; division by zero is undefined; no; a rectangle cannot have a dimension that is equal to zero.

MAKE CONNECTIONS

11.
Length, x	0.01	0.02	0.05	0.08	0.1	0.2	0.5	0.8	1
Width, y	1200	600	240	150	120	60	24	15	12

Length, x	12	24	36	48	60	72	84	96	100	120
Width, y	1	0.5	0.3	0.25	0.2	0.16	0.143	0.125	0.12	0.1

12. Small changes in x produce large changes in y.

13. Large changes in x produce small changes in y.

14. The rate of change in y is constant in this inverse variation, but is not a direct variation.

15. The value of y increases infinitely.

16. Values of y approach, but do not equal zero.

17. Direct variations pass through (0, 0) and are linear. Inverse variations are curved and do not include the point (0, 0).

SUMMARIZE

18. The equation is $y = \dfrac{k}{x}$. The graph shape is two curves in two quadrants (I and III or II and IV). y is undefined when x is zero. y becomes infinitely great as x becomes infinitely small. y becomes infinitely small as x becomes infinitely great. $k = xy$.

19. No. Setting $\dfrac{1}{x}$ equal to $\dfrac{10}{x}$ yields $x = 10x$. The only value that satisfies this equation is $x = 0$ and both equations are undefined for $x = 0$.

20. Branches are in Quadrants I and III when $k > 0$ and in Quadrants II and IV when $k < 0$.

21. For $k > 0$, the graph is in Quadrant I and Quadrant II; for $k < 0$, the graph is in Quadrant III and Quadrant IV.

ALGEBRAWORKS

1. Since the denominator increases, the value of t decreases.

2. $t = \dfrac{12,500}{175 - 167} = \dfrac{12,500}{8} = 1562.5$ tons

3. Profit = Revenue – Cost

$$= (2250 \cdot \$174) - (2250 \cdot \$167 + \$12,500)$$

$$= \$3250$$

Profit = Revenue – Cost, so

$$\$6500 = (\$174x) - (\$167x + 12,500)$$

$$19,000 = 7x$$

$$\frac{19,000}{7} = x$$

$$x \approx 2714.3 \text{ tons}$$

Lesson 14.2, pages 679–685

EXPLORE

1. $C = \pi d$; as d increases, C also increases so it is a direct variation.

2. Constant of variation is π; $\pi = \dfrac{C}{d}$

3. $d = 2r$; as r increases, d also increases so it is a direct variation.

4. Yes, since $C = \pi d$ and $d = 2r \to C = \pi \cdot 2r$ or $2\pi r$ which is also a direct variation with $k = 2\pi$.

5. Riding in a car is quicker. A car has a greater rate of speed. Speed varies inversely with time.

$4 = st$ or $\dfrac{4}{t} = s$ or $\dfrac{4}{s} = t$.

TRY THESE

1. s varies jointly as q and r; $k = 2$.

2. d varies directly as r; $k = 2$.

3. y varies inversely as x; $k = 10$.

4. As x increases, y decreases so y varies inversely as x.

Use substitution in $y = \dfrac{k}{x}$ and solve for k; $k = 8$

5. As x increases, y decreases so it is an inverse variation

$x = \dfrac{k}{y}$. Substitute values for x and y in $y = \dfrac{k}{x}$; $k = 60$

6. $w = \dfrac{k}{x}$ **7.** $c = \dfrac{k}{d^2}$ **8.** $f = kgh$.

9. $y = \dfrac{k}{x}$, $10 = \dfrac{k}{50}$

So, $k = 500$ and $y = \dfrac{500}{x}$;

When $x = 20$, then $y = \dfrac{500}{20} = 25$.

10. $y = \dfrac{k}{x}$, $8 = \dfrac{k}{12}$

So, $k = 96$ and $y = \dfrac{96}{x}$;

When $y = 6$, then $6 = \dfrac{96}{x} \to x = 16$

11. $y = kxz$, $30 = k \cdot 3 \cdot 5$

So, $k = 2$ and $y = 2xz$;

When $x = 12$ and $z = \dfrac{1}{2}$, then $y = 2 \cdot 12 \cdot \dfrac{1}{2} = 12$.

12. $y = kxz$, $75 = k \cdot 25 \cdot 6$

So, $k = \dfrac{1}{2}$ and $y = \dfrac{1}{2}xz$;

When $x = 30$ and $x = 4$, then $y = \dfrac{1}{2} \cdot 30 \cdot 4 = 60$.

13. $F = \dfrac{k}{l}$; $12 = \dfrac{k}{2}$

So, $k = 24$ and $F = \dfrac{24}{l}$;

When $l = 1.6$, then $F = \dfrac{2.4}{1.6} = 15 \text{ N}$.

14. $e = kt$; $2.3 = 2.1k$

So, $k \approx 0.1095$ and $e = 0.1095t$;

When $t = 80$, then $e = 0.1095 \cdot 80 = 8.76 \text{ kWh}$.

15. The formula for the area of a triangle is $A = 0.5bh$. When the area is 24 cm^2 then $0.5bh = 24$ and $bh = 48$. So, $k = 48$ and b varies inversely as h.

PRACTICE

1. a varies directly as b; $k = 9.5$

2. c varies jointly as d and e; $k = \dfrac{1}{2}$

3. l varies inversely as w; $k = 72$

4. g varies jointly as e and f; $k = 4$

5. u varies directly as v; $k = 0.14$

6. q varies inversely as r; $k = 16$

7. $y = \dfrac{k}{x}$, $12 = \dfrac{k}{60}$

So, $k = 720$ and $y = \dfrac{720}{x}$;

When $x = 15$, then $y = \dfrac{720}{15} = 48$.

8. $y = \dfrac{k}{x}$, $15 = \dfrac{k}{20}$

So, $k = 300$ and $y = \dfrac{300}{x}$;

When $x = 30$, then $y = \dfrac{300}{30} = 10$.

9. $y = \dfrac{k}{x}$, $5 = \dfrac{k}{14}$

So, $k = 70$ and $y = \dfrac{70}{x}$;

When $y = 10$, then $10 = \dfrac{70}{x} \to x = 7$.

10. $y = \dfrac{k}{x}$, $9 = \dfrac{k}{50}$

So, $k = 450$ and $y = \dfrac{450}{x}$;

When $y = 15$, then $15 = \dfrac{450}{x} \to x = 30$.

11. $y = kxz$, $45 = k \cdot 9 \cdot 6$

So $k = \dfrac{5}{6}$ and $y = \dfrac{5}{6}xz$;

When $x = 10$ and $z = 6$, then $y = \dfrac{5}{6} \cdot 10 \cdot 6 = 50$.

12. $y = kxz$, $75 = k \cdot 25 \cdot 6$

So, $k = \frac{1}{2}$ and $y = \frac{1}{2}xz$;

When $x = 30$ and $z = 4$, then $y = \frac{1}{2} \cdot 30 \cdot 4 = 60$.

13. $y = kxz$, $600 = k \cdot 50 \cdot 24$

So, $k = 24$ and $y = \frac{1}{2}xz$;

When $x = 50$ and $z = 84$, then $y = \frac{1}{2} \cdot 50 \cdot 84 = 2100$.

14. $y = kxz$, $100 = k \cdot 10 \cdot 40$

So, $k = \frac{1}{4}$ and $y = \frac{1}{4}xz$;

When $x = 28$ and $z = 14$, then $y = \frac{1}{4} \cdot 28 \cdot 14 = 98$.

15. $d = kr$ **16.** $A = kbh$ **17.** $w = \frac{k}{l}$

18. $y = \frac{k}{x}$; Possible example: the more weed killer used, the fewer weeds that will grow.

EXTEND

19. $y = kx^2 \rightarrow 245 = k \cdot 7^2 \rightarrow k = 5$;

$y = 5x^2 \rightarrow y = 5 \cdot 11^2 \rightarrow y = 605$

20. $y = \frac{k}{x^2} \rightarrow 16 = \frac{k}{9^2} \rightarrow k = 1296$;

$y = \frac{1296}{6^2} \rightarrow y = 36$

21. $n = kt^2 \rightarrow 5000 = k \cdot 25^2 \rightarrow k = 8$;

$n = 8t^2 \rightarrow 10{,}000 = 8t^2 \rightarrow t \approx 35.4$ minutes

THINK CRITICALLY

22. When $k > 0$, the graph is in Quadrant I and Quadrant III; if $k < 0$, the graph is in Quadrant II and Quadrant IV.

23. either $y = x$ or $y = -x$

24. The closest point lies on the line of symmetry; the intersection of $y = x$ and $xy = 1$ is $(1, 1)$.

25. The closest point lies on the line of symmetry; the intersection of $y = x$ and $xy = 4$ is $(2, 2)$.

26. The closest point lies on the line of symmetry; the intersection of $y = x$ and $xy = 10$ is $\left(\sqrt{10}, \sqrt{10}\right)$.

MIXED REVIEW

27. B; $0.30(80) = 0.15(80 + x)$

$2400 = 1200 + 15x$

$80 = x$

28. $6x + 3y - 6 = 12$

$3y = -6x + 18$

$y = -2x + 6$

29. $2x + 15 = 3y + 9$

$2x + 6 = 3y$

$\frac{2}{3}x + 2 = y$

30. $2x - 6 < -12$ or $2x - 6 > 12$ Write a disjunction.

$2x < -6$ or $2x > 18$

$x < -3$ or $x > 9$

31. $-16 < 4x + 8 < 16$

$-24 < 4x < 8$

$-6 < x < 2$

32. $y = \frac{k}{x}$, $15 = \frac{k}{5}$

So, $k = 75$ and $y = \frac{75}{x}$;

When $x = 25$, then $y = \frac{75}{25} = 3$.

33. $y = \frac{k}{x}$, $45 = \frac{k}{5}$

So, $k = 225$ and $y = \frac{225}{x}$;

When $y = 9$, then $9 = \frac{225}{x}$ so $x = 25$.

PROJECT CONNECTION

1. product always equals $8.5 \cdot 11 = 93.5$ in.^{2}

2. Area is constant. Changing the dimensions does not change the area.

3. $h = \frac{93.5}{C}$ for $C \neq 0$.

4. It represents an inverse variation.

ALGEBRA WORKS

1. A wider tube causes the pressure to increase at a slower rate. Since P is smaller, V must be greater to maintain a particular temperature. Thus, a longer tube must be used to hold a greater volume.

2. $C = \pi d$ so the inside circumference is $0.049\pi \approx 0.154$ in.; the outside circumference is $0.114\pi \approx 0.358$ in.

3. $T = kPV$ or $\frac{T}{k} = PV$ and $\frac{T}{k} = 30 \cdot 120 = 3600$.

If $P = 20$, then $3600 = 20 \cdot V$ and $V = 180$ cubic inches.

4. $T = kPV$ and $405 = 10 \cdot kP$ so $kP = 40.5$. If $T = 345$, then $345 = 40.5 \cdot V$ and $V \approx 8.52$ cubic inches.

Lesson 14.3, pages 686–691

EXPLORE

1. four layers of 6 x-blocks

$SA = 20x + 48$, $V = 24x$; $\dfrac{20x + 48}{24x}$

2. a stack of 7 x^2-blocks

$SA = 2x^2 + 28x$; $V = 7x^2$; $\dfrac{2x^2 + 28x}{7x^2}$

3. a stack of 5 xy-blocks

$SA = 10x + 2xy + 10y$; $V = 5xy$;

$\dfrac{10x + 2xy + 10y}{5xy}$

4. 6 x^2-blocks in 2 stacks of 3 x^2-blocks

$SA = 4x^2 + 18x$; $V = 6x^2$; $\dfrac{4x^2 + 16x}{6x^2}$

TRY THESE

1. $2x = 0$

$x = 0$

2. $3x + 6 = 0$

$x = -2$

3. $(x + 3)(x - 7) = 0$

$x = -3$ or $x = 7$

4. $x^2 - 2x - 35 = 0$

$(x - 7)(x + 5) = 0$

$x = 7$ or $x = -5$

5. $\dfrac{6x}{3x^2} = \dfrac{2 \cdot \cancel{3} \cdot \cancel{x}}{\cancel{3} \cdot x \cdot \cancel{x}} = \dfrac{2}{x}$; $3x^2 \neq 0$ so $x \neq 0$

6. $\dfrac{5x^2}{25} = \dfrac{\cancel{5} \cdot x^2}{\cancel{5} \cdot 5} = \dfrac{x^2}{5}$; no restrictions on x

7. $\dfrac{4x - 12}{24} = \dfrac{\cancel{4}(x - 3)}{\cancel{4} \cdot 6} = \dfrac{(x - 3)}{6}$; no restrictions on x

8. $\dfrac{6a - 12}{a - 2} = \dfrac{6\cancel{(a - 2)}}{\cancel{a - 2}} = 6$; $a - 2 \neq 0$ so $a \neq 0$

9. $\dfrac{a - 7}{7a - 49} = \dfrac{\cancel{a - 7}}{7\cancel{(a - 7)}} = \dfrac{1}{7}$; $7a - 49 \neq 0$ so $a \neq 7$

10. $\dfrac{3 - 2b}{6b - 9} = \dfrac{\cancel{3 - 2b}}{-3\cancel{(-2b + 3)}} = \dfrac{1}{-3}$; $6b - 9 \neq 0$ so $b \neq \dfrac{3}{2}$

11. $\dfrac{5 + 2b}{-8b - 20} = \dfrac{\cancel{5 + 2b}}{-4\ \cancel{(2b + 5)}} = \dfrac{1}{-4}$; $-8b - 20 \neq 0$ so $b \neq -\dfrac{5}{2}$

12. $\dfrac{3 + x}{x^2 - 2x - 15} = \dfrac{\cancel{3 + x}}{\cancel{(x + 3)}(x - 5)} = \dfrac{1}{x - 5}$;

$(x + 3)(x - 5) \neq 0$ so $x \neq -3$ or 5

13. $\dfrac{4 + x}{x^2 + 10x + 24} = \dfrac{\cancel{4 + x}}{\cancel{(x + 4)}(x + 6)} = \dfrac{1}{x + 6}$;

$(x + 4)(x + 6) \neq 0$ so $x \neq -4$ or -6

14. $\dfrac{x + x}{y + z} = \dfrac{2x}{y + z}$

15. $\dfrac{y^2 - 25}{y - 5}$ has the restriction that $y \neq 5$; $y + 5$ has no restrictions. The two functions are the same at every point except $y = 5$.

PRACTICE

1. $9mn = 0$

$m = 0$ or $n = 0$

2. $pqr = 0$

$p = 0$, $q = 0$, or $r = 0$

3. $5y + 15 = 0$

$y = -3$

4. $18x - 12 = 0$

$x = \dfrac{2}{3}$

5. $(z - 1)(z + 8) = 0$

$z = 1$ or $z = -8$

6. $(q + 6)(q - 4) = 0$

$q = -6$ or $q = 4$

7. $x^2 + 2x - 48 = 0$

$(x + 8)(x - 6) = 0$

$x = -8$ or $x = 6$

8. $y^2 - 10y - 56 = 0$

$(y - 14)(y + 4) = 0$

$y = 14$ or $y = -4$

9. $\dfrac{8x}{4x^2} = \dfrac{2 \cdot \cancel{4} \cdot \cancel{x}}{\cancel{4} \cdot x \cdot \cancel{x}} = \dfrac{2}{x}$; $4x^2 \neq 0$ so $x \neq 0$

10. $\dfrac{10x}{5x^2} = \dfrac{2 \cdot \cancel{5} \cdot \cancel{x}}{\cancel{5} \cdot x \cdot \cancel{x}} = \dfrac{2}{x}$; $5x^2 \neq 0$ so $x \neq 0$

11. $\dfrac{3y - 12}{15} = \dfrac{\cancel{3}(y - 4)}{\cancel{3} \cdot 5} = \dfrac{y - 4}{5}$; no restrictions on y

12. $\dfrac{6y + 15}{12} = \dfrac{\cancel{3}(2y + 5)}{\cancel{3} \cdot 4} = \dfrac{2y + 5}{4}$; no restrictions on y

13. $\dfrac{4a - 16}{a - 4} = \dfrac{4(a - 4)}{\cancel{a - 4}} = 4$; $a - 4 \neq 0$ so $a \neq 4$

14. $\dfrac{a - 9}{9a - 81} = \dfrac{\cancel{a - 9}}{9\cancel{(a - 9)}} = \dfrac{1}{9}$; $9a - 81 \neq 0$ so $a \neq 9$

15. $\dfrac{4 - 8b}{40b - 20} = \dfrac{\cancel{4}(1 - 2b)}{\cancel{4} \cdot (-5)(-2b + 1)} =$

$\dfrac{1}{-5}$; $40b - 20 \neq 0$ so $b \neq \dfrac{1}{2}$

16. $\dfrac{4 - 3b}{9b - 12} = \dfrac{\cancel{4 - 3b}}{-3\cancel{(-3b + 4)}} = \dfrac{1}{-3}$; $9b - 12 \neq 0$ so $b \neq \dfrac{4}{3}$

17. $\dfrac{2 + x}{x^2 - 5x - 14} = \dfrac{\cancel{2 + x}}{\cancel{(x + 2)}(x - 7)} = \dfrac{1}{x - 7}$; $(x + 2)(x - 7) \neq 0$

so, $x \neq -2$ or 7

18. $\dfrac{SA}{V} = \dfrac{2 \cdot 4x \cdot x + 2 \cdot x \cdot x + 2 \cdot x \cdot 4x}{x \cdot x \cdot 4x} = \dfrac{18x^2}{4x^3} =$

$\dfrac{\cancel{2} \cdot 9 \cdot \cancel{x} \cdot \cancel{x}}{\cancel{2} \cdot 2 \cdot \cancel{x} \cdot \cancel{x} \cdot x} = \dfrac{9}{2x}$

19. $\dfrac{4x^2 + 26x + 12}{2x + 1} = \dfrac{2(2x + 1)(x + 6)}{2x + 1} = 2(x + 6) = 2x + 12$

20. Both have restrictions on the variable. They are different in that some factors of $\dfrac{6x^2 - 24x - 30}{4x - 20}$ are binomials but the factors of $\dfrac{30x^3}{2x}$ are monomials.

EXTEND

21. $\dfrac{x^4 - 1}{x^2 - 2x + 1} = \dfrac{(x^2 + 1)(x + 1)\cancel{(x - 1)}}{(x - 1)\cancel{(x - 1)}} = \dfrac{(x^2 + 1)(x + 1)}{x - 1}$;

$(x - 1)^2 \neq 0$ so $x \neq 1$

22. $\dfrac{x^3 - 6x^2 + 9x}{x^3 - 9x} = \dfrac{\cancel{x}(x - 3)\cancel{(x - 3)}}{\cancel{x}(x + 3)\cancel{(x - 3)}} = \dfrac{x - 3}{x + 3}$;

$x(x - 3)(x + 3) \neq 0$ so $x \neq 0$, 3, or -3

23. $\dfrac{x^5 + 10x^4 + 25x^3}{x^3 + 4x^2 - 5x} = \dfrac{x \cdot x \cdot x(x+5)(x+5)}{x(x+5)(x-1)} = \dfrac{x^2(x+5)}{x-1}$;

$x(x+5)(x-1) \neq 0$ so $x \neq 0, -5,$ or 1

THINK CRITICALLY

24. Answers will vary.

Since $x \neq 0, x \neq \dfrac{1}{2}, x \neq -\dfrac{1}{4}$, then $x, (2x-1), (4x+1)$ are factors of the denominator; possible answer is $\dfrac{1}{x(2x-1)(4x+1)}$.

25. Answers will vary, possible answer is $\dfrac{x^3}{x^4}$; $x \neq 0$.

26. $\dfrac{x^{2n} + 7x^n + 6}{x^{2n} + 9x^n + 8} = \dfrac{(x^n+6)(x^n+1)}{(x^n+8)(x^n+1)} = \dfrac{x^n+6}{x^n+8}$

27. $\dfrac{(x+y)^q}{(x+y)^p} = (x+y)^{q-p}$ by the laws of exponents.

$x^2 + 2xy + y^2 = (x+y)^2$. Since $(x+y)^{q-p} = (x+y)^2$, then $q - p = 2$ or $q = p + 2$.

MIXED REVIEW

28. C; $(4 \times 10^5)(1.5 \times 10^4) = (4)(1.5)(10^{5+4}) = 6 \times 10^9$

29. $m = \dfrac{7 - (-3)}{4 - (-2)} = \dfrac{10}{6} = \dfrac{5}{3}$;

$y - 7 = \dfrac{5}{3}(x - 4)$ or $y = \dfrac{5}{3}x + \dfrac{1}{3}$

30. $m = \dfrac{-8 - 6}{4 - 3} = -14$;

$y - 6 = -14(x - 3)$ or $y = -14x + 48$

31. Since the coefficient of x^2 is negative, the parabola opens downward.

32. Since the coefficient of x^2 is positive, the parabola opens upward.

PROJECT CONNECTION

1. $C = 2\pi r$, so $\dfrac{C}{2\pi} = r$

$V = \pi r^2 h$, so $V = \pi\left(\dfrac{C}{2\pi}\right)^2 h = \dfrac{C^2 h}{4\pi}$

2. $V = \dfrac{93.5C}{4\pi}$

3. $C \neq 0$; the graph is a line with a hole at 0; the graph implies that the volume continues to increase as the cylinder gets wider and, therefore, shorter; C can't be negative.

4. Answers will vary. Although the volume may be greater, an extremely wide silo may not be useful, because it would occupy a large piece of land and would require a lot of material for a cover.

ALGEBRA WORKS

1. $1259 - 1208 = 51$
$1208 - 1162 = 46$
$1162 - 1120 = 42$
$1120 - 1081 = 39$
$1081 - 1046 = 35$
$1046 - 1014 = 32$
$C(N-1) - C(N) = D(N)$ where $N =$ row number

2. $0.51, 0.46, 0.42, 0.39, 0.35, 0.32$; $E(N) = \dfrac{D(N)}{B(N)}$.

3. 0.42 occurs when 1400 lbs of hay and 1120 lbs of grain are used.

4. $\dfrac{0.0725}{0.15} = 0.48\overline{3}$, combination occurs when 1200 lbs of hay and 1208 lbs of grain are used.

5. $\dfrac{0.06}{0.165} = 0.\overline{36}$ combination occurs when 1500 lbs of hay and 1081 lbs of grain are used.

6. If 1800 lbs of hay and 984 lbs of grain are used, then the change in the grain used is $1014 - 984 = 30$ lbs and the ratio is 0.30. So the price ratio is $0.30 = \dfrac{x}{0.175}$ where x is the price of hay and $x = 0.30(0.175) = 0.0525$ per lb.

Lesson 14.4, pages 692–696

EXPLORE

1. $\dfrac{5}{6} \cdot \dfrac{24}{5} = \dfrac{120}{30} = 4$

2. $\dfrac{3(5)^2}{4} \cdot \dfrac{8}{12(5)} = \dfrac{600}{240} = \dfrac{5}{2}$

3. $\dfrac{5+6}{5} \cdot \dfrac{2}{5+6} = \dfrac{11}{5} \cdot \dfrac{2}{11} = \dfrac{22}{55} = \dfrac{2}{5}$

4. $\dfrac{x}{6} \cdot \dfrac{24}{x} = \dfrac{24x}{6x} = 4$; same result as when $x = 5$ was used

5. $\dfrac{3(x)^2}{4} \cdot \dfrac{8}{12(x)} = \dfrac{24x^2}{6x} = \dfrac{x}{2}$; same result as when $x = 5$ was used

6. The multiplied expressions should result in $\dfrac{2}{x}$.

TRY THESE

1. $\dfrac{3x}{5} \cdot \dfrac{10}{12x} = \dfrac{3 \cdot x}{5} \cdot \dfrac{2 \cdot 5}{2 \cdot 2 \cdot 3 \cdot x} = \dfrac{1}{2}$

2. $\dfrac{4x^2}{7} \cdot \dfrac{14}{5x} = \dfrac{4 \cdot x \cdot x}{7} \cdot \dfrac{2 \cdot 7}{5 \cdot x} = \dfrac{8x}{5}$

3. $\dfrac{1}{4n} \div \dfrac{6n}{15} = \dfrac{1}{2 \cdot 2 \cdot n} \cdot \dfrac{3 \cdot 5}{2 \cdot 3 \cdot n} = \dfrac{5}{8n^2}$

4. $\dfrac{7r^2}{5} \div \dfrac{3r}{21} = \dfrac{7 \cdot r \cdot r}{5} \cdot \dfrac{\cancel{3} \cdot 7}{\cancel{3} \cdot \cancel{r}} = \dfrac{49r}{5}$

5. $\dfrac{ab^2}{c} \cdot \dfrac{3c^2}{b} = \dfrac{a \cdot \cancel{b} \cdot b}{\cancel{c}} \cdot \dfrac{3 \cdot \cancel{c} \cdot c}{\cancel{b}} = 3abc$

6. $\dfrac{2u^2}{v^2 w} \cdot \dfrac{vw^2}{5} = \dfrac{2 \cdot u \cdot u}{v \cdot \cancel{v} \cdot \cancel{w}} \cdot \dfrac{\cancel{v} \cdot \cancel{w} \cdot w}{5} = \dfrac{2u^2 w}{5v}$

7. $\dfrac{4d^2}{7e} \div \dfrac{8d}{e^2} = \dfrac{\cancel{2} \cdot \cancel{2} \cdot \cancel{d} \cdot d}{7 \cdot \cancel{e}} \cdot \dfrac{\cancel{e} \cdot e}{\cancel{2} \cdot \cancel{2} \cdot 2 \cdot \cancel{d}} = \dfrac{de}{14}$

8. $\dfrac{3q^2 r^2}{4} \div \dfrac{9qr}{3} = \dfrac{\cancel{3} \cdot q \cdot q \cdot \cancel{r} \cdot r}{4} \cdot \dfrac{\cancel{3}}{\cancel{3} \cdot 3 \cdot \cancel{q} \cdot \cancel{r}} = \dfrac{qr}{4}$

9. $\dfrac{m-3}{6(m+4)} \cdot \dfrac{3(m+4)}{m-3} = \dfrac{\cancel{m-3}}{2 \cdot \cancel{3}(m+4)} \cdot \dfrac{\cancel{3}(m+4)}{\cancel{m-3}} = \dfrac{1}{2}$

10. $\dfrac{x-6}{8x+12} \cdot \dfrac{10x+15}{3x-18} = \dfrac{\cancel{x-6}}{2 \cdot 2(2x+3)} \cdot \dfrac{5(2x+3)}{3(\cancel{x-6})} = \dfrac{5}{12}$

11. $\dfrac{(3x+6)(x-1)}{12x} \div \dfrac{x+2}{8} = \dfrac{3(x+2)(x-1)}{\cancel{2} \cdot \cancel{2} \cdot 3 \cdot x} \cdot \dfrac{\cancel{2} \cdot \cancel{2} \cdot 2}{x+2} =$

$\dfrac{2(x-1)}{x} = \dfrac{2x-2}{x}$

12. $\dfrac{3x^2 - 10x - 8}{6x} \div \dfrac{2x^2 - 32}{-5x - 20} =$

$\dfrac{(3x+2)(\cancel{x-4})}{2 \cdot 3 \cdot x} \cdot \dfrac{-5(x+4)}{2(\cancel{x+4})(\cancel{x-4})} = \dfrac{-15x-10}{12x}$

13. Answers will vary. In both products you can divide out common factors; no denominator can be zero. The product of rational expressions can be a rational number or rational expression; the product of rational numbers is always a rational number.

14. $W = F \cdot d = \dfrac{x+4}{x^2 - 16} \cdot \dfrac{x^2 + 13x + 12}{x+6} =$

$\dfrac{\cancel{x+4}}{(\cancel{x+4})(x-4)} \cdot \dfrac{(\cancel{x+6})(x+7)}{\cancel{x+6}} = \dfrac{x+7}{x-4}$ joules

PRACTICE

1. $\dfrac{5x}{7} \cdot \dfrac{14}{6x} = \dfrac{5 \cdot \cancel{x}}{\cancel{7}} \cdot \dfrac{\cancel{2} \cdot \cancel{7}}{\cancel{2} \cdot 3 \cdot \cancel{x}} = \dfrac{5}{3}$

2. $\dfrac{1}{6x} \cdot \dfrac{18x^2}{11} = \dfrac{1}{\cancel{2} \cdot \cancel{3} \cdot \cancel{x}} \cdot \dfrac{\cancel{2} \cdot \cancel{3} \cdot 3 \cdot \cancel{x} \cdot x}{11} = \dfrac{3x}{11}$

3. $\dfrac{1}{5k} \div \dfrac{3}{20k^2} = \dfrac{1}{\cancel{5} \cdot \cancel{k}} \cdot \dfrac{4 \cdot \cancel{5} \cdot \cancel{k} \cdot k}{3} = \dfrac{4k}{3}$

4. $\dfrac{8n}{5} \div \dfrac{5}{24n^2} = \dfrac{8n}{5} \cdot \dfrac{24n^2}{5} = \dfrac{192n^3}{25}$

5. $\dfrac{a^2 b}{c} \cdot \dfrac{2ac^2}{a} = \dfrac{a \cdot a \cdot b}{\cancel{c}} \cdot \dfrac{2 \cdot a \cdot \cancel{c} \cdot c}{\cancel{a}} = 2a^2 bc$

6. $\dfrac{wu^2}{v^2 w} \cdot \dfrac{v^2 w}{3u} = \dfrac{w \cdot \cancel{u} \cdot u}{\cancel{v} \cdot \cancel{v} \cdot \cancel{w}} \cdot \dfrac{\cancel{v} \cdot \cancel{v} \cdot \cancel{w}}{3 \cdot \cancel{u}} = \dfrac{u \cdot w}{3}$

7. $\dfrac{5c^2 d}{9ce^2} \div \dfrac{15c^2 d^2}{18e} = \dfrac{\cancel{5} \cdot \cancel{c} \cdot c \cdot \cancel{d}}{\cancel{3} \cdot 3 \cdot \cancel{c} \cdot e \cdot \cancel{e}} \cdot \dfrac{2 \cdot \cancel{3} \cdot \cancel{3} \cdot \cancel{e}}{3 \cdot \cancel{5} \cdot c \cdot \cancel{c} \cdot \cancel{d} \cdot d} =$

$\dfrac{2}{3cde}$

8. $\dfrac{8qr^2}{12qr} \div \dfrac{9qrs}{6s} = \dfrac{\cancel{2} \cdot \cancel{2} \cdot 2 \cdot \cancel{q} \cdot r \cdot \cancel{r}}{\cancel{2} \cdot \cancel{2} \cdot 3 \cdot q \cdot \cancel{r}} \cdot \dfrac{2 \cdot \cancel{3} \cdot \cancel{s}}{3 \cdot 3 \cdot q \cdot \cancel{r} \cdot \cancel{s}} = \dfrac{4}{9q}$

9. $\dfrac{m-5}{2(m+6)} \cdot \dfrac{4(m+6)}{8(m-5)} = \dfrac{4}{16} = \dfrac{1}{4}$

10. $\dfrac{x-3}{6(2x+1)} \cdot \dfrac{3(2x+1)}{4(x-3)} = \dfrac{3}{24} = \dfrac{1}{8}$

11. $\dfrac{4(x+3)(x-2)}{6x} \div \dfrac{(x+3)(x-3)}{3(x-3)} =$

$\dfrac{\cancel{2} \cdot 2(x+3)(x-2)}{\cancel{2} \cdot \cancel{3} \cdot x} \cdot \dfrac{\cancel{3}(x-3)}{(x+3)(x-3)} = \dfrac{2(x-2)}{x} = \dfrac{2x-4}{x}$

12. $\dfrac{(2x+3)(x-5)}{4x} \div \dfrac{3(x+5)(x-5)}{6(x+5)} =$

$\dfrac{(2x+3)(\cancel{x-5})}{\cancel{2} \cdot 2 \cdot x} \cdot \dfrac{\cancel{2} \cdot 3(x+5)}{3(x+5)(\cancel{x-5})} = \dfrac{2x+3}{2x}$

13. $\dfrac{7x+35}{3x^2 - 108} \cdot (6x+36) = \dfrac{7(x+5)}{\cancel{3}(x+6)(x-6)} \cdot \dfrac{[\cancel{2} \cdot \cancel{3}(x+6)]}{1} =$

$\dfrac{14(x+5)}{x-6} = \dfrac{14x+70}{x-6}$

14. $\dfrac{8x+32}{4x^2 - 100} \cdot (3x-15) = \dfrac{\cancel{2} \cdot \cancel{2} \cdot 2(x+4)}{\cancel{2} \cdot \cancel{2}(x+5)(\cancel{x-5})} \cdot \dfrac{[3(\cancel{x-5})]}{1} =$

$\dfrac{6(x+4)}{x+5} = \dfrac{6x+24}{x+5}$

15. $\dfrac{x^2 - 49}{x^2 - x - 42} \div (3x+21) = \dfrac{(x+7)(\cancel{x-7})}{(\cancel{x-7})(x+6)} \cdot \dfrac{1}{3(\cancel{x+7})} =$

$\dfrac{1}{3(x+6)} = \dfrac{1}{3x+18}$

16. $\dfrac{x^2 - 16}{x^2 - 7x + 12} \div (5x+20) = \dfrac{(\cancel{x+4})(\cancel{x-4})}{(x-3)(\cancel{x-4})} \cdot \dfrac{1}{5(\cancel{x+4})} =$

$\dfrac{1}{5(x-3)} = \dfrac{1}{5x-15}$

17. $\dfrac{3x^2 - 17x + 6}{3x^2 - 108} \cdot \dfrac{5x}{-21x - 7} =$

$\dfrac{(\cancel{3x+1})(\cancel{x-6})}{3(x+6)(\cancel{x-6})} \cdot \dfrac{5x}{-7(\cancel{3x+1})} = -\dfrac{5x}{21(x+6)} = -\dfrac{5x}{21x+126}$

18. $\dfrac{2x^2 + x - 10}{5x^2 - 20} \cdot \dfrac{5x}{-6x - 15} = \dfrac{(\cancel{2x+5})(\cancel{x-2})}{\cancel{5}(x+2)(\cancel{x-2})} \cdot \dfrac{\cancel{5}x}{-3(\cancel{2x+5})} =$

$-\dfrac{x}{3(x+2)} = -\dfrac{x}{3x+6}$

19. $\dfrac{x^2+x-6}{x^2-9} \div \dfrac{x^2-4}{7x-21} = \dfrac{(x+3)(x-2)}{(x+3)(x-3)} \cdot \dfrac{7(x-3)}{(x+2)(x-2)} =$

$\dfrac{7}{x+2}$

20. $\dfrac{x^2+5x-36}{x^2-81} \div \dfrac{x^2-16}{6x-54} = \dfrac{(x+9)(x-4)}{(x+9)(x-9)} \cdot \dfrac{6(x-9)}{(x+4)(x-4)} =$

$\dfrac{6}{x+4}$

21. $A = lw = \dfrac{6x-x^2}{2x+4} \cdot \dfrac{x^2-4}{36-x^2} = \dfrac{x(6-x)}{2(x+2)} \cdot \dfrac{(x+2)(x-2)}{(6-x)(6+x)} =$

$\dfrac{x(x-2)}{2(6+x)} = \dfrac{x^2-2x}{12+2x}$

22. $A = lw = \dfrac{a^2-1}{ab^2-b} \cdot \dfrac{b}{3-3a} =$

$\dfrac{(a+1)(a-1)}{b(ab-1)} \cdot \dfrac{b}{-3(-1+a)} = -\dfrac{a+1}{3(ab-1)} = -\dfrac{a+1}{3ab-3}$

23. Answers will vary, but should include changing the problem by multiplying by the reciprocal and canceling common factors.

EXTEND

24. $\dfrac{18-4x}{3x+2} \div \dfrac{6x-18}{-(6x+4)} \cdot \dfrac{3x-9}{81-4x^2} =$

$\dfrac{2(9-2x)}{3x+2} \cdot \dfrac{-2(3x+2)}{2 \cdot 3(x-3)} \cdot \dfrac{3(x-3)}{(9-2x)(9+2x)} = -\dfrac{2}{9+2x}$

25. $\dfrac{t^2-t}{t^2-2t-3} \cdot \dfrac{t^2+2t+1}{t^2+4t} \div \dfrac{t^2-3t-4}{2t^2-32} =$

$\dfrac{t(t-1)}{(t-3)(t+1)} \cdot \dfrac{(t+1)(t+1)}{t(t+4)} \cdot \dfrac{2(t-4)(t+4)}{(t-4)(t+1)} =$

$\dfrac{2(t-1)}{t-3} = \dfrac{2t-2}{t-3}$

26. $\dfrac{2x^2+4x-30}{2x^2-18} \div \dfrac{3x+15}{4x+12} =$

$\dfrac{2(x+5)(x-3)}{2(x+3)(x-3)} \cdot \dfrac{4(x+3)}{3(x+5)} = \dfrac{4}{3}$ or $4:3$

27. $\dfrac{2x^2-2x-4}{4x-8} \div \dfrac{x^2-1}{4x-4} =$

$\dfrac{2(x-2)(x+1)}{4(x-2)} \cdot \dfrac{4(x-1)}{(x-1)(x+1)} = \dfrac{2}{1} = 2$ or $2:1$

28. $\dfrac{x^3+3x^2+2x}{x} \div (10x^2+30x+20) =$

$\dfrac{x(x^2+3x+2)}{x} \cdot \dfrac{1}{10(x^2+3x+2)} = \dfrac{1}{10}$ or $1:10$

THINK CRITICALLY

For 29-32, examples will vary.

29. $\dfrac{x+2}{3x+3} \cdot \dfrac{x+5}{x-1}$ or $\dfrac{x+5}{3x+3} \cdot \dfrac{2x+4}{2x-2}$

30. $\dfrac{m-3}{2m-1} \cdot \dfrac{m+2}{m+3}$ or $\dfrac{m-3}{2m+6} \cdot \dfrac{2m+4}{2m-1}$

31. $\dfrac{3x^2+18x}{x+5} \div \dfrac{x^2+x-30}{x+5}$ or $\dfrac{6x^2+42x}{2x+16} \div \dfrac{x^2+2x-35}{x+8}$

32. $\dfrac{x^2+5x+4}{x^2-2x-3} \div \dfrac{x+6}{x-3}$ or $\dfrac{x^2+6x+8}{x^2+3x-18} \div \dfrac{x+2}{x-3}$

MIXED REVIEW

33. C; $3-4x = 12$ or $3-4x = -12$

34. $y = kx$; $8.1 = 9k$ so $0.9 = k$.

35. $y = kx$; $9.9 = 3.3k$ so $3 = k$.

36. $-3x > x - 16$

$\quad -4x > -16$

$\quad\quad x < 4$ Divide both sides by -4 and reverse the inequality symbol.

37. $6x + 12 < 3x + 6$

$\quad\quad 3x < -6$

$\quad\quad\; x < -2$

38. $\dfrac{x^2+2x-15}{25-x^2} \cdot \dfrac{x^2-4x-5}{x^2-2x-3} =$

$\dfrac{(x+5)(x-3)}{-(x+5)(x-5)} \cdot \dfrac{(x-5)(x+1)}{(x-3)(x+1)} = \dfrac{1}{-1} = -1$

39. $\dfrac{x-3}{x-6} \div \dfrac{x^2-9}{x^2-2x-24} = \dfrac{x-3}{x-6} \cdot \dfrac{(x-6)(x+4)}{(x-3)(x+3)} = \dfrac{x+4}{x+3}$

Lesson 14.5, pages 697–701

EXPLORE

1. 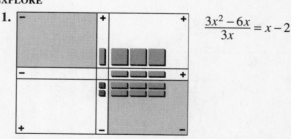 $\dfrac{3x^2-6x}{3x} = x - 2$

2. 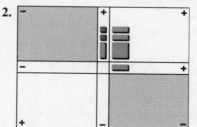 $\dfrac{x^2+2x}{x} = x + 2$

3.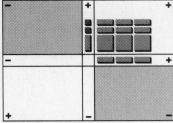

$$\frac{3x^2+6x}{x+2}=3x$$

4.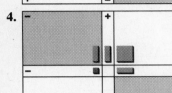

$$\frac{x^2-x}{x-1}=x$$

TRY THESE

1. Dividend is x^2+8, Divisor is $x-2$, Quotient is $x+2$, Remainder is 12.

2. Dividend is x^2+16, Divisor is $x+4$, Quotient is $x-4$, Remainder is 32.

3. $\left(6x^3+2x^2+x\right) \div 2x^2 = \dfrac{6x^3}{2x^2}+\dfrac{2x^2}{2x^2}+\dfrac{x}{2x^2} = 3x+1+\dfrac{1}{2x}$

4. $\left(8x^3+4x^2+2x\right) \div 4x^2 = \dfrac{8x^3}{4x^2}+\dfrac{4x^2}{4x^2}+\dfrac{2x}{4x^2} = 2x+1+\dfrac{1}{2x}$

5.
$$\begin{array}{r} a+\;5 \\ a-2\overline{)\,a^2+3a-10} \\ \underline{a^2-2a} \\ 5a-10 \\ \underline{5a-10} \\ 0 \end{array}$$

So $(a^2+3a-10) \div (a-2) = a+5$

6.
$$\begin{array}{r} 4m+10 \\ m-4\overline{)\,4m^2-\;6m-\;5} \\ \underline{4m^2-16m} \\ 10m-\;5 \\ \underline{10m-40} \\ 35 \end{array}$$

So $4m^2-6m-6) \div (m-4) = 4m+10+\dfrac{35}{m-4}$

7.
$$\begin{array}{r} p^2+\;p-3 \\ 2p+1\overline{)\,2p^3+3p^2-5p+1} \\ \underline{2p^3+\;p^2} \\ 2p^2-5p \\ \underline{2p^2+\;p} \\ -6p+\;1 \\ \underline{-6p-\;3} \\ 4 \end{array}$$

So $(2p^3+3p^2-5p+1) \div (2p+1) = p^2+p-3+\dfrac{4}{2p+1}$

8.
$$\begin{array}{r} d^2+\;4d+16 \\ d-4\overline{)\,d^3+0d^2+\;0d+12} \\ \underline{d^3-4d^2} \\ 4d^2+\;0d \\ \underline{4d^2-16d} \\ 16d+12 \\ \underline{16d-64} \\ 76 \end{array}$$

So $(d^3+12) \div (d-4) = d^2+4d+16+\dfrac{76}{d-4}$

9.
$$\begin{array}{r} 4t+\;3 \\ 2t+5\overline{)\,8t^2+26t+15} \\ \underline{8t^2+20t} \\ 6t+15 \\ \underline{6t+15} \\ 0 \end{array}$$

So, the width is $\dfrac{\text{Area}}{\text{Length}}=4t+3$

10. Answers will vary, but should include that the divisor is a factor of the dividend.

PRACTICE

1. Dividend is x^2+10, Divisor is $x-4$, Quotient is $x+4$, Remainder is 26.

2. Dividend is x^2+12, Divisor is $x-3$, Quotient is $x+3$, Remainder is 21.

3. $\left(12x^3+6x^2+x\right) \div 3x^2 = \dfrac{12x^3}{3x^2}+\dfrac{6x^2}{3x^2}+\dfrac{x}{3x^2} = 4x+2+\dfrac{1}{3x}$

4. $\left(15x^3+10x^2+x\right) \div 5x^2 = \dfrac{15x^3}{5x^2}+\dfrac{10x^2}{5x^2}+\dfrac{x}{5x^2} = 3x+2+\dfrac{1}{5x}$

5.
$$\begin{array}{r} a+\;7 \\ a-3\overline{)\,a^2+4a-21} \\ \underline{a^2-3a} \\ 7a-21 \\ \underline{7a-21} \\ 0 \end{array}$$

So $(a^2+4a-21) \div (a-3) = a+7$

6.
$$\begin{array}{r} b-\;2 \\ b+8\overline{)\,b^2+6b-16} \\ \underline{b^2+8b} \\ -2b-16 \\ \underline{-2b-16} \\ 0 \end{array}$$

So $(b^2+6b-16) \div (b+8) = b-2$

7.

$$\begin{array}{r} 4m-2 \\ 3m+1\overline{\smash{)}12m^2-2m-2} \\ \underline{12m^2+4m} \\ -6m-2 \\ \underline{-6m-2} \\ 0 \end{array}$$

So $(12m^2-2m-2) \div (3m+1) = 4m-2$

8.

$$\begin{array}{r} 9y^2+6y+1 \\ 3y+1\overline{\smash{)}27y^3+27y^2+9y+1} \\ \underline{27y^3+9y^2} \\ 18y^2+9y \\ \underline{18y^2+6y} \\ 3y+1 \\ \underline{3y+1} \\ 0 \end{array}$$

So $(27y^3+27y^2+9y+1) \div (3y+1) = 9y^2+6y+1$

9. Answers will vary, but should include that both are done by long division.

10.

$$\begin{array}{r} p^2+4p-5 \\ 2p+1\overline{\smash{)}2p^3+9p^2-6p+2} \\ \underline{2p^3+p^2} \\ 8p^2-6p \\ \underline{8p^2+4p} \\ -10p+2 \\ \underline{-10p-5} \\ 7 \end{array}$$

So $(2p^3+9p^2-6p+2) \div (2p+1) = p^2+4p-5+\dfrac{7}{2p+1}$

11.

$$\begin{array}{r} q^2+2q+2 \\ 3q-1\overline{\smash{)}3q^3+5q^2+4q-3} \\ \underline{3q^3-q^2} \\ 6q^2+4q \\ \underline{6q^2-2q} \\ 6q-3 \\ \underline{6q-2} \\ -1 \end{array}$$

So $(3q^3+5q^2+4q-3) \div (3q-1) = q^2+2q+2-\dfrac{1}{3q-1}$

12.

$$\begin{array}{r} 4x^3+16x^2+72x+288 \\ x-4\overline{\smash{)}4x^4+0x^3+8x^2+0x+16} \\ \underline{4x^4-16x^3} \\ 16x^3+8x^2 \\ \underline{16x^3-64x^2} \\ 72x^2+0x \\ \underline{72x^2-288x} \\ 288x+16 \\ \underline{288x-1152} \\ 1168 \end{array}$$

(Solution continues)

So $(4x^4+8x^2+16) \div (x-4) =$
$$4x^3+16x^2+72x+288+\dfrac{1168}{x-4}$$

13.

$$\begin{array}{r} 5y^3+25y^2+135y+675 \\ y-5\overline{\smash{)}5y^4+0y^3+10y^2+0y+5} \\ \underline{5y^4-25y^3} \\ 25y^3+10y^2 \\ \underline{25y^3-125y^2} \\ 135y^2+0y \\ \underline{135y^2-675y} \\ 675y+5 \\ \underline{675y-3375} \\ 3380 \end{array}$$

So $(5y^4+10y^2+5) \div (y-5) =$
$$5y^3+25y^2+135y+675+\dfrac{3380}{y-5}$$

14. $A = \dfrac{10,000}{r+1}(r^3+3r^2+3r+1)$

Divide (r^3+3r^2+3r+1) by $(r+1)$
and multiply by 10,000.

$$\begin{array}{r} r^2+2r+1 \\ r+1\overline{\smash{)}r^3+3r^2+3r+1} \\ \underline{r^3+r^2} \\ 2r^2+3r \\ \underline{2r^2+2r} \\ r+1 \\ \underline{r+1} \\ 0 \end{array}$$

So, the value is
$(r^2+2r+1) \cdot (10,000) = 10,000r^2+20,000r+10,000$

15. $10,000(-0.20)^2+20,000(-0.20)+10,000 =$
$400-4,000+10,000 = \$6,400$

16. $A = \dfrac{18,000}{-0.15+1}\left[(-0.15)^3+3(-0.15)^2+3(-0.15)+1\right]$

$= \dfrac{18,000}{0.85}(-0.003375+0.0675-0.45+1) = \$13,005$

EXTEND

17.

$$\begin{array}{r} 8m^2+12mn+24n^2 \\ m-n\overline{\smash{)}8m^3+4m^2n+12mn^2+16n^3} \\ \underline{8m^3-8m^2n} \\ 12m^2n+12mn^2 \\ \underline{12m^2n-12mn^2} \\ 24mn^2+16n^3 \\ \underline{24mn^2-24n^3} \\ 40n^3 \end{array}$$

So $(8m^3+4m^2n+12mn^2+16n^3) \div (m-n) =$
$$8m^2+12mn+24n^2+\dfrac{40n^3}{m-n}$$

Integrated Approach

18.

$$\begin{array}{r} 8n^2 + 10mn + 7m^2 \\ 2n-m{\overline{\smash{\big)}\,16n^3 + 12mn^2 + 4mn^2 + 8m^3}} \\ \underline{16n^3 - 8mn^2} \\ 20mn^2 + 4mn^2 \\ \underline{20mn^2 - 10mn^2} \\ 14mn^2 + 8m^3 \\ \underline{14mn^2 - 7m^3} \\ 15m^3 \end{array}$$

So $(16n^3 + 12mn^2 + 4m^2n + 8m^3) \div (2n-m) =$

$8n^2 + 10mn + 7m^2 + \dfrac{15m^3}{2n-m}$

19. $P = \left(\dfrac{336}{1+r}\right)\left(r^4 + 4r^3 + 6r^2 + 4r + 1\right)$

$= 336\left(\dfrac{r^4 + 4r^3 + 6r^2 + 4r + 1}{r+1}\right)$

$$\begin{array}{r} r^3 + 3r^2 + 3r + 1 \\ r+1{\overline{\smash{\big)}\,r^4 + 4r^3 + 6r^2 + 4r + 1}} \\ \underline{r^4 + r^3} \\ 3r^3 + 6r^2 \\ \underline{3r^3 + 3r^2} \\ 3r^2 + 4r \\ \underline{3r^2 + 3r} \\ r+1 \\ \underline{r+1} \\ 0 \end{array}$$

$P = 336(r^3 + 3r^2 + 3r + 1)$

$= 336r^3 + 1008r^2 + 1008r + 336$

20. $336(0.12)^3 + 1008(0.12)^2 + 1008(0.12) + 336 =$
$0.580608 + 14.5152 + 120.96 + 336 \approx 472$

21. $\left(\dfrac{460}{1+0.15}\right)\left[(0.15)^4 + 4(0.15)^3 + 6(0.15)^2 + 4(0.15) + 1\right] =$

$\left(\dfrac{460}{1.15}\right)(0.00050625 + 0.135 + 0.135 + 0.6 + 1) \approx 700$

THINK CRITICALLY

22.

$$\begin{array}{r} x^2 + 2x - 1 \\ x+3{\overline{\smash{\big)}\,x^3 + 5x^2 + 5x + k}} \\ \underline{x^3 + 3x^2} \\ 2x^2 + 5x \\ \underline{2x^2 + 6x} \\ -x+k \\ \underline{-x-3} \\ k+3 \end{array}$$

Remainder must be zero
to be a factor. So,

$3 + k = 0$

$k = -3$

23.

$$\begin{array}{r} x^2 + 7x + (k+35) \\ x-5{\overline{\smash{\big)}\,x^3 + 2x^2 + kx + 45}} \\ \underline{x^3 - 5x^2} \\ 7x^2 + kx \\ \underline{7x^2 - 35x} \\ (k+35)x + 45 \\ \underline{(k+35)x - 5(k+35)} \\ 5(k+35) + 45 \end{array}$$

To be a factor, the remainder must be 0.

So, $45 + 5(k+35) = 0 \rightarrow k = -44$

24.

$$\begin{array}{r} x^2 + \dfrac{k-3}{2}x - \dfrac{3k-9}{4} \\ 2x+3{\overline{\smash{\big)}\,2x^3 + kx^2 + \quad 0 \ x + \quad 18}} \\ \underline{2x^3 + 3x^2} \\ (k-3)x^2 + \quad 0 \ x \\ \underline{(k-3)x^2 + \dfrac{3k-9}{4}x} \\ -\dfrac{3k-9}{2}x + \quad 18 \\ \underline{-\dfrac{3k-9}{2}x - \dfrac{9k-27}{4}} \\ \dfrac{9k-27}{2} + 18 \end{array}$$

To be a factor, the remainder must be 0.

So, $18 + \dfrac{9k-27}{4} = 0 \rightarrow k = -5$

MIXED REVIEW

25. A; $3x - 5y + 6z = 54$

$2x - 5y + 8z = 61$

$2z = 10 \ \rightarrow z = 5$

Substitute 5 for z in Equations 1 and 2.

$3x - 5y = 24$

$\underline{2x - 5y = 21}$

$x \quad\quad = 3$ Subtract.

Substitute 3 for x and 5 for z in Equation 1.

$3(3) - 5y + 6(5) = 54 \rightarrow y = -3$

26. c; Coefficient of x^2 is positive so the parabola opens upward

27. b; Coefficient of x^2 is negative so the parabola opens downward, y-intercept is -2

28. a; Coefficient of x^2 is negative so the parabola opens downward, y-intercept is 2

EXPLORE

1. $\dfrac{10}{5}+\dfrac{4}{10}=\dfrac{10}{5}+\dfrac{2}{5}=\dfrac{12}{5}$

2. $\dfrac{3\cdot 10}{4}+\dfrac{1}{2\cdot 10}=\dfrac{30}{4}\cdot\dfrac{5}{5}+\dfrac{1}{20}=\dfrac{150}{20}+\dfrac{1}{20}=\dfrac{151}{20}$

3. $\dfrac{10-6}{10-1}+\dfrac{2}{10-4}=\dfrac{4}{9}+\dfrac{2}{6}=\dfrac{4}{9}+\dfrac{1}{3}\cdot\dfrac{3}{3}=\dfrac{4}{9}+\dfrac{3}{9}=\dfrac{7}{9}$

4. Use the product of the denominators or find the least common denominator (LCD).

5. LCD of 2 and 3 is 6. 6. LCD of 2 and 6 is 6.

7. LCD of x and x is x. 8. LCD of 4 and x is $4x$.

9. LCD of $3x$ and x is $3x$. 10. LCD of x^2 and $3x$ is $3x^2$.

11. LCD of $(x-5)$ and 8 is $8(x-5)$.

12. LCD of $2x$ and $2(x+1)$ is $2x(x+1)$.

13. LCD of 5 and x is $5x$
LCD of 4 and $2x$ is $4x$
LCD of $(x-1)$ and $(x-4)$ is $(x-1)(x-4)$.

TRY THESE

1. LCD of mn^2 and mn is mn^2.

2. LCD of a^2b and $2ac$ is $2a^2bc$.

3. LCD of $(x-4)$ and $(x-5)$ is $(x-4)(x-5)$.

4. $\dfrac{a}{11}+\dfrac{7a}{11}=\dfrac{8a}{11}$ 5. $\dfrac{5}{7y}-\dfrac{9}{7y}=\dfrac{-4}{7y}$

6. $\dfrac{3p}{p-4}-\dfrac{12}{p-4}=\dfrac{3p-12}{p-4}=\dfrac{3(p-4)}{p-4}=3$

7. $\dfrac{6}{5x}+\dfrac{-2}{15x}$ The LCD of $5x$ and $15x$ is $15x$.

$\dfrac{6}{5x}\cdot\dfrac{3}{3}+\dfrac{-2}{15x}=\dfrac{18}{15x}+\dfrac{-2}{15x}=\dfrac{16}{15x}$

8. $\dfrac{6}{5a^2}+\dfrac{2}{15a}-\dfrac{5}{a}$ LCD of $5a^2$, $15a$, and a is $15a^2$.

$\dfrac{6}{5a^2}\cdot\dfrac{3}{3}+\dfrac{2}{15a}\cdot\dfrac{a}{a}-\dfrac{5}{a}\cdot\dfrac{15a}{15a}=\dfrac{18}{15a^2}+\dfrac{2a}{15a^2}-\dfrac{75a}{15a^2}=$

$\dfrac{18-73a}{15a^2}$

9. $\dfrac{4}{a+4}+\dfrac{5}{a-9}$

The LCD of $(a+4)(a-9)$ is $(a+4)(a-9)$.

$\dfrac{4}{a+4}\cdot\dfrac{a-9}{a-9}+\dfrac{5}{a-9}\cdot\dfrac{a+4}{a+4}=$

$\dfrac{4a-36}{(a+4)(a-9)}+\dfrac{5a+20}{(a+4)(a-9)}=\dfrac{9a-16}{(a+4)(a-9)}$

10. $\dfrac{2x+5}{(x+2)(x-2)}+\dfrac{-2}{(x+2)}$

The LCD of $(x+2)(x-2)$ and $(x+2)$ is $(x+2)(x-2)$.

$\dfrac{2x+5}{(x+2)(x-2)}+\dfrac{-2}{(x+2)}\cdot\dfrac{(x-2)}{(x-2)}=$

$\dfrac{2x+5}{(x+2)(x-2)}+\dfrac{-2x+4}{(x+2)(x-2)}=\dfrac{9}{(x+2)(x-2)}$

11. $\dfrac{4z+2}{16-z^2}+\dfrac{4}{z-4}=\dfrac{4z+2}{16-z^2}+\dfrac{-4}{4-z}$

The LCD of $(16-z^2)$ and $(4-z)$ is $(16-z^2)$.

$\dfrac{4z+2}{16-z^2}+\dfrac{-4}{4-z}\cdot\dfrac{4+z}{4+z}=\dfrac{4z+2}{16-z^2}+\dfrac{-16-4z}{16-z^2}=\dfrac{-14}{16-z^2}$

12. $\dfrac{1}{x}+\dfrac{2}{x+2}$ The LCD of x and $(x+2)$ is $x(x+2)$.

$\dfrac{1}{x}\cdot\dfrac{x+2}{x+2}+\dfrac{2}{x+2}\cdot\dfrac{x}{x}=\dfrac{x+2}{x(x+2)}+\dfrac{2x}{x(x+2)}=\dfrac{3x+2}{x(x+2)}$

13. Find a common denominator; write equivalent expressions with the same denominator; add the numerators; combine like terms.

PRACTICE

1. LCD of ab^2 and b^3 is ab^3.

2. LCD of $(x-2)$ and $(x-3)$ is $(x-2)(x-3)$.

3. LCD of $3x$ and $x(x+1)$ is $3x(x+1)$.

4. $\dfrac{12}{5y}-\dfrac{7}{5y}=\dfrac{5}{5y}=\dfrac{1}{y}$ 5. $\dfrac{22}{7z}+\dfrac{6}{7z}=\dfrac{28}{7z}=\dfrac{4}{z}$

6. $\dfrac{3}{p+2}+\dfrac{p-3}{p+2}=\dfrac{p}{p+2}$

7. $\dfrac{8d}{d-6}-\dfrac{d+42}{d-6}=\dfrac{8d-d-42}{d-6}=\dfrac{7d-42}{d-6}=\dfrac{7(d-6)}{d-6}=7$

8. $\dfrac{13e}{e-12}-\dfrac{e+144}{e-12}=\dfrac{13e-e-144}{e-12}=\dfrac{12e-144}{e-12}=$

$\dfrac{12(e-12)}{e-12}=12$

9. $\dfrac{4}{b-4}-\dfrac{b}{b-4}=\dfrac{4-b}{b-4}=-1$

10. $\dfrac{9}{2x}+\dfrac{-5}{8x}$ The LCD of $2x$ and $8x$ is $8x$.

$\dfrac{9}{2x}\cdot\dfrac{4}{4}+\dfrac{-5}{8x}=\dfrac{36}{8x}+\dfrac{-5}{8x}=\dfrac{31}{8x}$

11. $\dfrac{8}{3x}+\dfrac{-1}{9x}$ The LCD of $3x$ and $9x$ is $9x$.

$\dfrac{8}{3x}\cdot\dfrac{3}{3}+\dfrac{-1}{9x}=\dfrac{24}{9x}+\dfrac{-1}{9x}=\dfrac{23}{9x}$

12. $\dfrac{5t}{4t^2}-\dfrac{3}{8t}-\dfrac{2}{t}$ The LCD of $4t^2$, $8t$, and t is $8t^2$.

$\dfrac{5}{4t^2}\cdot\dfrac{2}{2}-\dfrac{3}{8t}\cdot\dfrac{t}{t}-\dfrac{2}{t}\cdot\dfrac{8t}{8t}=\dfrac{10}{8t^2}-\dfrac{3t}{8t^2}-\dfrac{16t}{8t^2}=\dfrac{10-19t}{8t^2}$

13a. $\dfrac{1}{d_o}+\dfrac{1}{d_i}=\dfrac{1}{f}$ The LCD of d_o, d_i, and f is $d_o d_i f$.

$$d_o d_i f\left(\dfrac{1}{d_o}+\dfrac{1}{d_i}\right)=d_o d_i f\cdot\dfrac{1}{f}\qquad\text{Multiply by }d_o d_i f.$$

$$d_i f+d_o f=d_o d_i\qquad\text{Distribute.}$$

$$d_o f=d_o d_i-d_i f\qquad\text{Subtract }d_i f.$$

$$d_o f=d_i(d_o-f)\qquad\text{Factor.}$$

$$\dfrac{d_o f}{d_o-f}=d_i\qquad\text{Divide.}$$

b. $d_i=\dfrac{(1.65)(5000)}{5000-1.65}=\dfrac{8250}{4998.35}\approx1.65\text{ cm}$

c. $d_i=\dfrac{(1.65)(30)}{30-1.65}=\dfrac{49.5}{28.35}\approx1.75\text{ cm}$

14. $\dfrac{3}{a+6}+\dfrac{2}{a-5}$

The LCD of $(a+6)$ and $(a-5)$ is $(a+6)(a-5)$.

$$\dfrac{3}{a+6}\cdot\dfrac{a-5}{a-5}+\dfrac{2}{a-5}\cdot\dfrac{a+6}{a+6}=$$

$$\dfrac{3a-15}{(a+6)(a-5)}+\dfrac{2a+12}{(a+6)(a-5)}=\dfrac{5a-3}{(a+6)(a-5)}$$

15. $\dfrac{5}{a+7}+\dfrac{4}{a-4}$

The LCD of $(a+7)$ and $(a-4)$ is $(a+7)(a-4)$.

$$\dfrac{5}{a+7}\cdot\dfrac{a-4}{a-4}+\dfrac{4}{a-4}\cdot\dfrac{a+7}{a+7}=$$

$$\dfrac{5a-20}{(a+7)(a-4)}+\dfrac{4a+28}{(a+7)(a-4)}=\dfrac{9a+8}{(a+7)(a-4)}$$

16. $\dfrac{t}{2(t+3)}-\dfrac{2}{3(t+3)}$

LCD of $2(t+3)$ and $3(t+3)$ is $6(t+3)$.

$$\dfrac{t}{2(t+3)}\cdot\dfrac{3}{3}-\dfrac{2}{3(t+3)}\cdot\dfrac{2}{2}=\dfrac{3t-4}{6(t+3)}$$

17. $\dfrac{r}{3(r+3)}-\dfrac{3}{4(r+3)}$

The LCD of $3(r+3)$ and $4(r+3)$ is $12(r+3)$.

$$\dfrac{r}{3(r+3)}\cdot\dfrac{4}{4}-\dfrac{3}{4(r+3)}\cdot\dfrac{3}{3}=\dfrac{4r-9}{12(r+3)}$$

18. $\dfrac{4y+16}{(y+4)(y-4)}+\dfrac{-4}{y+4}$

The LCD of $(y+4)(y-4)$ and $y+4$ is $(y+4)(y-4)$.

$$\dfrac{4y+16}{(y+4)(y-4)}+\dfrac{-4}{y+4}\cdot\dfrac{y-4}{y-4}=$$

$$\dfrac{4y+16}{(y+4)(y-4)}+\dfrac{-4y+16}{(y+4)(y-4)}=\dfrac{32}{(y+4)(y-4)}$$

19. $\dfrac{5x+3}{(5-x)(5+x)}+\dfrac{5}{x-5}=\dfrac{5x+3}{(5-x)(5+x)}+\dfrac{-5}{5-x}$

The LCD of $(5-x)(5+x)$ and $5-x$ is $(5-x)(5+x)$.

$$\dfrac{5x+3}{(5-x)(5+x)}+\dfrac{-5}{5-x}\cdot\dfrac{5+x}{5+x}=$$

$$\dfrac{5x+3}{(5-x)(5+x)}+\dfrac{-25-5x}{(5-x)(5+x)}=\dfrac{-22}{(5-x)(5+x)}=\dfrac{-22}{25-x^2}$$

20. $\dfrac{6q-1}{81-q^2}-\dfrac{2q}{q+9}=\dfrac{6q-1}{(9-q)(9+q)}-\dfrac{2q}{9+q}$

The LCD of $(9-q)(9+q)$ and $(9+q)$ is $(9-q)(9+q)$.

$$\dfrac{6q-1}{(9-q)(9+q)}-\dfrac{2q}{9+q}\cdot\dfrac{9-q}{9-q}=$$

$$\dfrac{6q-1}{(9-q)(9+q)}-\dfrac{18q-2q^2}{(9-q)(9+q)}=\dfrac{2q^2-12q-1}{81-q^2}$$

21. $\dfrac{10v-1}{100-v^2}-\dfrac{5v}{v+10}=\dfrac{10v-1}{(10-v)(10+v)}-\dfrac{5v}{10+v}$

LCD of $(10-v)(10+v)$ and $(10+v)$ is $(10-v)(10+v)$.

$$\dfrac{10v-1}{(10-v)(10+v)}-\dfrac{5v}{10+v}\cdot\dfrac{10-v}{10-v}=$$

$$\dfrac{10v-1}{(10-v)(10+v)}-\dfrac{50v-5v^2}{(10-v)(10+v)}=$$

$$\dfrac{10v-1-50v+5v^2}{(10-v)(10+v)}=\dfrac{5v^2-40v-1}{100-v^2}$$

22. $\dfrac{11}{121-u^2}-\dfrac{u^2}{u+11}=\dfrac{11}{(11-u)(11+u)}-\dfrac{u^2}{11+u}$

LCD of $(11-u)(11+u)$ and $(11+u)$ is $(11-u)(11+u)$.

$$\dfrac{11}{(11-u)(11+u)}-\dfrac{u^2}{11+u}\cdot\dfrac{11-u}{11-u}=$$

$$\dfrac{11}{(11-u)(11+u)}-\dfrac{11u^2-u^3}{(11-u)(11+u)}=$$

$$\dfrac{11-11u^2+u^3}{(11-u)(11+u)}=\dfrac{u^3-11u^2+11}{121-u^2}$$

23. $\dfrac{3a}{6a^2+13a+2}+\dfrac{a+1}{a^2+5a+6}=$

$$\dfrac{3a}{(6a+1)(a+2)}+\dfrac{a+1}{(a+2)(a+3)}$$

The LCD is $(6a+1)(a+2)(a+3)$.

$$\dfrac{3a}{(6a+1)(a+2)}\cdot\dfrac{a+3}{a+3}+\dfrac{a+1}{(a+2)(a+3)}\cdot\dfrac{6a+1}{6a+1}=$$

$$\dfrac{3a^2+9a}{(6a+1)(a+2)(a+3)}+\dfrac{6a^2+7a+1}{(6a+1)(a+2)(a+3)}=$$

$$\dfrac{9a^2+16a+1}{(6a+1)(a+2)(a+3)}$$

24. $\dfrac{5b}{3b^2-b-2}+\dfrac{b+2}{b^2+4b-5}=$

$\dfrac{5b}{(3b+2)(b-1)}+\dfrac{b+2}{(b+5)(b-1)}$

The LCD of is $(3b+2)(b-1)(b+5)$.

$\dfrac{5b}{(3b+2)(b-1)}\cdot\dfrac{b+5}{b+5}+\dfrac{b+2}{(b+5)(b-1)}\cdot\dfrac{3b+2}{3b+2}=$

$\dfrac{5b^2+25b}{(3b+2)(b-1)(b+5)}+\dfrac{3b^2+8b+4}{(3b+2)(b-1)(b+5)}=$

$\dfrac{8b^2+33b+4}{(3b+2)(b-1)(b+5)}$

25. Multiplying polynomials can be more complicated, leaving room for errors in the computations.

EXTEND

26. $\left(\dfrac{2}{x+2}-\dfrac{1}{x+5}\right)\left(\dfrac{2x+10}{x^2+8x}\right)=$

$\left(\dfrac{2}{x+2}\cdot\dfrac{x+5}{x+5}-\dfrac{1}{x+5}\cdot\dfrac{x+2}{x+2}\right)\left(\dfrac{2x+10}{x^2+8x}\right)=$

$\left(\dfrac{2x+10}{(x+2)(x+5)}-\dfrac{x+2}{(x+2)(x+5)}\right)\left(\dfrac{2(x+5)}{x(x+8)}\right)=$

$\left(\dfrac{\cancel{x+8}}{(x+2)(\cancel{x+5})}\right)\left(\dfrac{2(\cancel{x+5})}{x(\cancel{x+8})}\right)=\dfrac{2}{x(x+2)}$

27. $\left(\dfrac{3}{x-5}-\dfrac{1}{x^2-25}\right)\div\left(\dfrac{3x+14}{x-8}\right)=$

$\left(\dfrac{3}{x-5}\cdot\dfrac{x+5}{x+5}-\dfrac{1}{(x+5)(x-5)}\right)\div\left(\dfrac{3x+14}{x-8}\right)=$

$\left(\dfrac{\cancel{3x+14}}{(x+5)(x-5)}\right)\cdot\left(\dfrac{x-8}{\cancel{3x+14}}\right)=\dfrac{x-8}{x^2-25}$

28. $x+\dfrac{3}{x}+\dfrac{4x}{1-x}$ The LCD is $x(1-x)$.

$\dfrac{x}{1}\cdot\dfrac{x(1-x)}{x(1-x)}+\dfrac{3}{x}\cdot\dfrac{1-x}{1-x}+\dfrac{4x}{1-x}\cdot\dfrac{x}{x}=$

$\dfrac{x^2-x^3}{x(1-x)}+\dfrac{3-3x}{x(1-x)}+\dfrac{4x^2}{x(1-x)}=\dfrac{-x^3+5x^2-3x+3}{x(1-x)}$

29. $\dfrac{1}{x}+\dfrac{2}{x+1}+\dfrac{x}{6x-1}$ The LCD is $x(x+1)(6x-1)$.

$\dfrac{1}{x}\cdot\dfrac{(x+1)(6x-1)}{(x+1)(6x-1)}+\dfrac{2}{x+1}\cdot\dfrac{x(6x-1)}{x(6x-1)}+\dfrac{x}{6x-1}\cdot\dfrac{x(x+1)}{x(x+1)}=$

$\dfrac{6x^2+5x-1}{x(x+1)(6x-1)}+\dfrac{12x^2-2x}{x(x+1)(6x-1)}+\dfrac{x^3+x^2}{x(x+1)(6x-1)}=$

$\dfrac{x^3+19x^2+3x-1}{x(x+1)(6x-1)}$

30. $\dfrac{2}{x}+\dfrac{3}{x+2}+\dfrac{x}{3x-1}$ The LCD is $x(x+2)(3x-1)$.

$\dfrac{2}{x}\cdot\dfrac{(x+2)(3x-1)}{(x+2)(3x-1)}+\dfrac{3}{x+2}\cdot\dfrac{x(3x-1)}{x(3x-1)}+\dfrac{x}{3x-1}\cdot\dfrac{x(x+2)}{x(x+2)}=$

$\dfrac{6x^2+10x-4}{x(x+2)(3x-1)}+\dfrac{9x^2-3x}{x(x+2)(3x-1)}+\dfrac{x^3+2x^2}{x(x+2)(3x-1)}=$

$\dfrac{x^3+17x^2+7x-4}{x(x+2)(3x-1)}$

31. $x+\dfrac{4}{x}+\dfrac{10x}{1-2x}$ The LCD is $x(1-2x)$.

$\dfrac{x}{1}\cdot\dfrac{x(1-2x)}{x(1-2x)}+\dfrac{4}{x}\cdot\dfrac{1-2x}{1-2x}+\dfrac{10x}{1-2x}\cdot\dfrac{x}{x}=$

$\dfrac{x^2-2x^3}{x(1-2x)}+\dfrac{4-8x}{x(1-2x)}+\dfrac{10x^2}{x(1-2x)}=\dfrac{-2x^3+11x^2-8x+4}{x(1-2x)}$

THINK CRITICALLY

32. An example is $\dfrac{2}{x+1}+\dfrac{6}{2x+2}=\dfrac{10}{2x+2}$, which can be simplified to $\dfrac{5}{x+1}$.

33. An example is $\dfrac{2}{x+1}+\dfrac{3}{2x+2}=\dfrac{7}{2x+2}$.

34. $2x-6=1\bullet 2\bullet(x-3)$ so $\dfrac{k}{1},\ \dfrac{k}{2},\ \dfrac{k}{x-3},\ \dfrac{k}{2(x-3)}$

MIXED REVIEW

35. D; $d=\dfrac{1}{2}|a|t^2$

$72=\dfrac{1}{2}|-9.8|t^2$

$72=4.9t^2$

$\dfrac{72}{4.9}=t^2$

$\sqrt{\dfrac{72}{4.9}}=t\approx 3.83$ seconds

36. 25 is the midpoint, 4 is the distance from the midpoint, so $|x-25|\le 4$.

37. $3\left(12-3^2\right)^3-2(7-1)^2$

$3(12-9)^3-2(6)^2$

$3(3)^3-2(36)$

$3(27)-72$

$81-72$

9

38. $\left[1260 \div \left(4 - 2^8\right)\right]^2 - 6 + 5^3$

$\left[1260 \div (4 - 256)\right]^2 - 6 + 125$

$\left[1260 \div (-252)\right]^2 + 119$

$[-5]^2 + 119$

$25 \quad + 119$

144

39. c; The coefficient of x^2 is positive so the parabola opens upward; there is no translation.

40. a; There is a vertical translation of -2.

41. b; There is a horizontal translation of -2.

42. $\dfrac{7}{6t^2} + \dfrac{12}{12t} - \dfrac{3}{t} = \dfrac{7}{6t^2} + \dfrac{1}{t} - \dfrac{3}{t} = \dfrac{7}{6t^2} - \dfrac{2}{t} =$

$\dfrac{7}{6t^2} - \dfrac{2}{t} \cdot \dfrac{6t}{6t} = \dfrac{7 - 12t}{6t^2}$

43. $\dfrac{3}{8r^2} + \dfrac{8}{16r} - \dfrac{2}{r} = \dfrac{3}{8r^2} + \dfrac{1}{2r} - \dfrac{2}{r} =$

$\dfrac{3}{8r^2} + \dfrac{1}{2r} \cdot \dfrac{4r}{4r} - \dfrac{2}{r} \cdot \dfrac{8r}{8r} = \dfrac{3}{8r^2} + \dfrac{4r}{8r^2} - \dfrac{16r}{8r^2} = \dfrac{3 - 12r}{8r^2}$

Lesson 14.7, pages 707–711

EXPLORE

1. $3 - \dfrac{1}{4} = \dfrac{12}{4} - \dfrac{1}{4} = \dfrac{11}{4}$

2. $4 + 3 + \dfrac{6}{4+3} = 7 + \dfrac{6}{7} = 7\dfrac{6}{7} = \dfrac{55}{7}$

3. $2 + \dfrac{4^2 - 4}{4^2 - 9} = 2 + \dfrac{16 - 4}{16 - 9} = 2 + \dfrac{12}{7} = 2\dfrac{12}{7} = 3\dfrac{5}{7} = \dfrac{26}{7}$

4. Answers will vary, but could include multiplying the whole number part of the mixed number by the denominator and adding the result to the numerator.

Example: $5\dfrac{2}{7} = \dfrac{5 \cdot 7 + 2}{7} = \dfrac{37}{7}$.

5. $\dfrac{5}{3} \div \dfrac{15}{4} = \dfrac{5}{3} \cdot \dfrac{4}{15} = \dfrac{4}{9}$ **6.** $\dfrac{1}{2} \div \dfrac{2}{3} = \dfrac{1}{2} \cdot \dfrac{3}{2} = \dfrac{3}{4}$

7. $8 \div \dfrac{1}{4} = 8 \cdot \dfrac{4}{1} = 32$

TRY THESE

1. $5 + \dfrac{9}{t} = \dfrac{5}{1} \cdot \dfrac{t}{t} + \dfrac{9}{t} = \dfrac{5t + 9}{t}$

2. $2x - \dfrac{x+5}{x} = \dfrac{2x}{1} \cdot \dfrac{x}{x} - \dfrac{x+5}{x} = \dfrac{2x^2 - x - 5}{x}$

3. $3n + \dfrac{2n+3}{4n+5} = \dfrac{3n}{1} \cdot \dfrac{4n+5}{4n+5} + \dfrac{2n+3}{4n+5} =$

$\dfrac{12n^2 + 15n}{4n+5} + \dfrac{2n+3}{4n+5} = \dfrac{12n^2 + 17n + 3}{4n+5}$

4. Answers will vary.

5. $\dfrac{100}{4x+1} \div \dfrac{80}{3x+2} = \dfrac{\overset{5}{\cancel{100}}}{4x+1} \cdot \dfrac{3x+2}{\underset{4}{\cancel{80}}} = \dfrac{15x+10}{16x+4}$

6. $\dfrac{5}{x} \div \dfrac{7}{y} = \dfrac{5}{x} \cdot \dfrac{y}{7} = \dfrac{5y}{7x}$ **7.** $\dfrac{8}{c} \div \dfrac{3}{d} = \dfrac{8}{c} \cdot \dfrac{d}{3} = \dfrac{8d}{3c}$

8. $\dfrac{13}{2u} \div \dfrac{6}{5z} = \dfrac{13}{2u} \cdot \dfrac{5z}{6} = \dfrac{65z}{12u}$

9. $\dfrac{\frac{1}{p} - \frac{3}{q}}{\frac{11}{2q}} \cdot \dfrac{2pq}{2pq} = \dfrac{\frac{1}{p} \cdot \frac{2p\cancel{q}}{1} - \frac{3}{\cancel{q}} \cdot \frac{2p\cancel{q}}{1}}{\frac{11}{2\cancel{q}} \cdot \frac{2\cancel{q}p}{1}} = \dfrac{2q - 6p}{11p}$

10. $\dfrac{\frac{1}{r} - \frac{1}{t}}{\frac{7}{3t}} \cdot \dfrac{3rt}{3rt} = \dfrac{\frac{1}{\cancel{r}} \cdot \frac{3\cancel{r}t}{1} - \frac{1}{\cancel{t}} \cdot \frac{3r\cancel{t}}{1}}{\frac{7}{3\cancel{t}} \cdot \frac{3r\cancel{t}}{1}} = \dfrac{3t - 3r}{7r}$

11. $\dfrac{\frac{6}{a} - 5}{-2 - \frac{3}{a}} \cdot \dfrac{a}{a} = \dfrac{\frac{6}{\cancel{a}} \cdot \frac{\cancel{a}}{1} - 5 \cdot a}{-2 \cdot a - \frac{3}{\cancel{a}} \cdot \frac{\cancel{a}}{1}} = \dfrac{6 - 5a}{-2a - 3} = \dfrac{5a - 6}{2a + 3}$

12. $\dfrac{7 - \frac{5}{g+3}}{14 - \frac{10}{g+3}} \cdot \dfrac{g+3}{g+3} = \dfrac{7(g+3) - \frac{5}{g+3} \cdot \frac{g+3}{1}}{14(g+3) - \frac{10}{g+3} \cdot \frac{g+3}{1}} =$

$\dfrac{7g + 21 - 5}{14g + 42 - 10} = \dfrac{\cancel{7g + 16}}{2(\cancel{7g + 16})} = \dfrac{1}{2}$

13. $\dfrac{60x}{60 - x} = \dfrac{60(12)}{60 - 12} = \dfrac{720}{48} = 15$

PRACTICE

1. $11 - \dfrac{13}{b} = \dfrac{11}{1} \cdot \dfrac{b}{b} - \dfrac{13}{b} = \dfrac{11b - 13}{b}$

2. $3x - \dfrac{x+2}{x} = \dfrac{3x}{1} \cdot \dfrac{x}{x} - \dfrac{x+2}{x} = \dfrac{3x^2 - x - 2}{x}$

3. $\dfrac{2z-5}{3z} + 3z - 1 = \dfrac{2z-5}{3z} + \dfrac{3z}{1} \cdot \dfrac{3z}{3z} - \dfrac{1}{1} \cdot \dfrac{3z}{3z} =$

$\dfrac{2z - 5 + 9z^2 - 3z}{3z} = \dfrac{9z^2 - z - 5}{3z}$

4. $n + \dfrac{n-4}{3n+4} = \dfrac{n}{1} \cdot \dfrac{3n+4}{3n+4} + \dfrac{n-4}{3n+4} = \dfrac{3n^2 + 4n}{3n+4} + \dfrac{n-4}{3n+4} =$

$\dfrac{3n^2 + 5n - 4}{3n+4}$

5. $t - 3 - \dfrac{5}{t+1} = \dfrac{t}{1} \cdot \dfrac{t+1}{t+1} - \dfrac{3}{1} \cdot \dfrac{t+1}{t+1} - \dfrac{5}{t+1} =$

$\dfrac{t^2 + t}{t+1} - \dfrac{3t + 3}{t+1} - \dfrac{5}{t+1} = \dfrac{t^2 - 2t - 8}{t+1}$

6. $b + 5 + \dfrac{5}{b-5} = \dfrac{b}{1} \cdot \dfrac{b-5}{b-5} + \dfrac{5}{1} \cdot \dfrac{b-5}{b-5} + \dfrac{5}{b-5} =$

$\dfrac{b^2 - 5b}{b-5} + \dfrac{5b - 25}{b-5} + \dfrac{5}{b-5} = \dfrac{b^2 - 20}{b-5}$

7. $\dfrac{3}{x} \div \dfrac{4}{y} = \dfrac{3}{x} \cdot \dfrac{y}{4} = \dfrac{3y}{4x}$

8. $\dfrac{5}{m} \div \dfrac{6}{n} = \dfrac{5}{m} \cdot \dfrac{n}{6} = \dfrac{5n}{6m}$

9. $\dfrac{\dfrac{1}{p} - \dfrac{3}{q}}{\dfrac{4}{q}} \cdot \dfrac{pq}{pq} = \dfrac{\dfrac{1}{\cancel{p}} \cdot \dfrac{p\cancel{q}}{1} - \dfrac{3}{\cancel{q}} \cdot \dfrac{p\cancel{q}}{1}}{\dfrac{4}{\cancel{q}} \cdot \dfrac{p\cancel{q}}{1}} = \dfrac{q - 3p}{4p}$

10. $\dfrac{\dfrac{2}{r} + \dfrac{3}{t}}{-\dfrac{5}{r}} \cdot \dfrac{rt}{rt} = \dfrac{\dfrac{2}{\cancel{r}} \cdot \dfrac{\cancel{r}t}{1} + \dfrac{3}{\cancel{t}} \cdot \dfrac{r\cancel{t}}{1}}{-\dfrac{5}{\cancel{r}} \cdot \dfrac{\cancel{r}t}{1}} = \dfrac{2t + 3r}{-5t}$

11. $\dfrac{7}{2u} \div \dfrac{8}{3v} = \dfrac{7}{2u} \cdot \dfrac{3v}{8} = \dfrac{21v}{16u}$

12. $\dfrac{5}{uv^2} \div \dfrac{2}{u^2 v} = \dfrac{5}{\cancel{u} \cdot v \cdot \cancel{v}} \cdot \dfrac{u \cdot u \cdot \cancel{v}}{2} = \dfrac{5u}{2v}$

13. $\dfrac{\dfrac{1}{r} - \dfrac{1}{t}}{-\dfrac{6}{5t}} \cdot \dfrac{-5rt}{-5rt} = \dfrac{\dfrac{1}{\cancel{r}} \cdot \dfrac{(-5r t)}{1} - \dfrac{1}{\cancel{t}} \cdot \dfrac{(-5 r \cancel{t})}{1}}{-\dfrac{6}{5\cancel{t}} \cdot \dfrac{(-5r\cancel{t})}{1}} = \dfrac{-5t + 5r}{6r}$

14. $\dfrac{\dfrac{1}{m} - \dfrac{1}{n}}{\dfrac{3}{2m} - \dfrac{3}{2n}} \cdot \dfrac{2mn}{2mn} = \dfrac{\dfrac{1}{\cancel{m}} \cdot \dfrac{2\cancel{m}n}{1} - \dfrac{1}{\cancel{n}} \cdot \dfrac{2m\cancel{n}}{1}}{\dfrac{3}{\cancel{2m}} \cdot \dfrac{\cancel{2m}n}{1} - \dfrac{3}{\cancel{2n}} \cdot \dfrac{2m\cancel{n}}{1}} = \dfrac{2n - 2m}{3n - 3m} =$

$\dfrac{2(\cancel{n - m})}{3(\cancel{n - m})} = \dfrac{2}{3}$

15. $\dfrac{\dfrac{4}{p} + 5}{\dfrac{5}{q}} \cdot \dfrac{pq}{pq} = \dfrac{\dfrac{4}{\cancel{p}} \cdot \dfrac{p q}{1} + 5 \cdot pq}{\dfrac{5}{\cancel{q}} \cdot \dfrac{p\cancel{q}}{1}} = \dfrac{4q + 5pq}{5p}$

16. $\dfrac{\dfrac{8}{a} - 9}{\dfrac{6}{b}} \cdot \dfrac{ab}{ab} = \dfrac{\dfrac{8}{\cancel{a}} \cdot \dfrac{\cancel{a}b}{1} - 9 \cdot ab}{\dfrac{6}{\cancel{b}} \cdot \dfrac{a\cancel{b}}{1}} = \dfrac{8b - 9ab}{6a}$

17. $\dfrac{\dfrac{c}{d^2} + \dfrac{3}{d}}{\dfrac{c}{d} - 1} \cdot \dfrac{d^2}{d^2} = \dfrac{\dfrac{c}{\cancel{d^2}} \cdot \dfrac{\cancel{d^2}}{1} + \dfrac{3}{\cancel{d}} \cdot \dfrac{d^{\cancel{2}}}{1}}{\dfrac{c}{\cancel{d}} \cdot \dfrac{d^{\cancel{2}}}{1} - 1 \cdot d^2} = \dfrac{c + 3d}{cd - d^2}$

18. $\dfrac{\dfrac{4}{e^2} + \dfrac{f}{e}}{\dfrac{2}{e} - 12} \cdot \dfrac{e^2}{e^2} = \dfrac{\dfrac{4}{\cancel{e^2}} \cdot \dfrac{\cancel{e^2}}{1} + \dfrac{f}{\cancel{e}} \cdot \dfrac{e^{\cancel{2}}}{1}}{\dfrac{2}{\cancel{e}} \cdot \dfrac{e^{\cancel{2}}}{1} - 12 \cdot e^2} = \dfrac{4 + ef}{2e - 12e^2}$

19. $\dfrac{\dfrac{5}{q} \cdot \dfrac{10}{q^2}}{\dfrac{5}{q} + \dfrac{10}{q^2}} \cdot \dfrac{q^3}{q^3} = \dfrac{\dfrac{50}{\cancel{q^3}} \cdot \dfrac{\cancel{q^3}}{1}}{\dfrac{5}{\cancel{q}} \cdot \dfrac{q^{\cancel{3}2}}{1} \cdot q + \dfrac{10}{\cancel{q^2}} \cdot \dfrac{q^{\cancel{3}}}{1}} = \dfrac{50}{5q^2 + 10q} =$

$\dfrac{\cancel{5} \cdot 10}{\cancel{5}(q^2 + 2q)} = \dfrac{10}{q^2 + 2q}$

20. $\dfrac{\text{Total distance}}{\text{Total time}} = \dfrac{m + m}{\dfrac{m}{40} + \dfrac{m}{60}} \cdot \dfrac{120}{120} =$

$\dfrac{2m \cdot 120}{\dfrac{m}{40} \cdot \dfrac{120}{1} + \dfrac{m}{60} \cdot \dfrac{120}{1}} = \dfrac{240m}{5m} = 48 \dfrac{\text{km}}{\text{hr}}$

21. $\dfrac{250{,}000}{3x - 10} \div \dfrac{4{,}000{,}000}{x^2} = \dfrac{250{,}000}{3x - 10} \cdot \dfrac{x^2}{4{,}000{,}000} =$

$\dfrac{x^2}{48x - 160}$

22. Answers will vary.

EXTEND

23. $\dfrac{\dfrac{4x}{(x+5)(x-5)} + \dfrac{6}{x+5}}{\dfrac{3}{2(x+5)}} \cdot \dfrac{2(x+5)(x-5)}{2(x+5)(x-5)} =$

$\dfrac{4x \cdot 2 + 6 \cdot 2(x-5)}{3(x-5)} = \dfrac{20x - 60}{3x - 15}$

24. $\dfrac{\dfrac{3x}{(x+7)(x-7)} + \dfrac{7}{x+7}}{\dfrac{1}{x-7} - \dfrac{7}{2(x+7)}} \cdot \dfrac{2(x+7)(x-7)}{2(x+7)(x-7)} =$

$\dfrac{3x \cdot 2 + 7 \cdot 2(x-7)}{2(x+7) - 7(x-7)} = \dfrac{20x - 98}{-5x + 63}$

25. $\dfrac{\dfrac{2b}{b-4} - \dfrac{3}{b^2}}{\dfrac{5}{5b-20} + \dfrac{3}{4b^2-16b}} = \dfrac{\dfrac{2b}{b-4} - \dfrac{3}{b^2}}{\dfrac{5}{5(b-4)} + \dfrac{3}{4b(b-4)}} \cdot \dfrac{4b^2(b-4)}{4b^2(b-4)} =$

$\dfrac{2b \cdot 4b^2 - 3 \cdot 4(b-4)}{4b^2 + 3 \cdot b} = \dfrac{8b^3 - 12b + 48}{4b^2 + 3b}$

26. $\dfrac{\dfrac{6}{(n+5)(n-4)} + \dfrac{3}{n-4}}{\dfrac{2}{3(n-4)} + \dfrac{3}{n+5}} \cdot \dfrac{3(n+5)(n-4)}{3(n+5)(n-4)} =$

$\dfrac{6 \cdot 3 + 3 \cdot 3(n+5)}{2(n+5) + 3 \cdot 3(n-4)} = \dfrac{9n + 63}{11n - 26}$

27. $\dfrac{x^2 + 2x + 1}{4} \div \dfrac{x+1}{x-1} = \dfrac{(x+1)(x+1)}{4} \cdot \dfrac{x-1}{\cancel{x+1}} = \dfrac{x^2 - 1}{4}$ ft

28. $\dfrac{x^2 + 14x + 45}{x^2 + 6x} \div \dfrac{2x + 10}{3x + 18} = \dfrac{(x+5)(x+9)}{x(\cancel{x+6})} \cdot \dfrac{3(\cancel{x+6})}{2(\cancel{x+5})} =$

$\dfrac{3x + 27}{2x}$

THINK CRITICALLY

29. $\dfrac{5}{6}$ is a complex rational expression since 5 and 6 are each rational expressions.

30. Answers may vary. Example: $\dfrac{\frac{x+1}{x^2-3}}{\frac{x+1}{x^2-3}}$

31. $\dfrac{\frac{1}{a+1}}{a-\frac{1}{a+\frac{1}{a}}} = \dfrac{\frac{1}{a+1}}{a-\frac{1}{\frac{a^2+1}{a}}} = \dfrac{\frac{1}{a+1}}{a-\frac{a}{a^2+1}} = \dfrac{\frac{1}{a+1}}{\frac{a^3}{a^2+1}} = \dfrac{1}{a+1} \div \dfrac{a^3}{a^2+1} =$

$\dfrac{1}{a+1} \cdot \dfrac{a^2+1}{a^3} = \dfrac{a^2+1}{a^4+a^3}$

32. $\dfrac{2}{c+\frac{1}{1+\frac{c+1}{5-c}}} = \dfrac{2}{c+\frac{1}{\frac{6}{5-c}}} = \dfrac{2}{c+\frac{5-c}{6}} = \dfrac{2}{\frac{5c+5}{6}} = \dfrac{12}{5c+5}$

Lesson 14.8, pages 712–717

EXPLORE

1a. 36   **b.** Find the LCD of all the fractions.

c. $108+9x = 4x$

$x = -\dfrac{108}{5}$

2. $\dfrac{x}{2}+\dfrac{x}{5} = 50$

$5x+2x = 500$ Multiply by 10.

$x = \dfrac{500}{7}$

3. $\dfrac{2x}{7} = \dfrac{3x}{8}-\dfrac{5}{4}$

$16x = 21x-70$ Multiply by 56.

$14 = x$

4. $\dfrac{3x}{10} = \dfrac{2x}{15}+\dfrac{3}{2}$

$9x = 4x+45$ Multiply by 30.

$x = 9$

5. Multiply both sides by the LCD of x and solve the resulting equation.

6. Multiply both sides of the equation by the LCD of all the denominators and solve the resulting equation.

TRY THESE

1. When you multiply both sides of an equation by a variable, it is possible you introduced extraneous solutions to the original equation.

2. $2x+\dfrac{4}{5} = \dfrac{17}{20}$

$40x+16 = 17$ Multiply both sides by 20.

$x = \dfrac{1}{40}$

3. $\dfrac{3t}{4}-\dfrac{5t}{8} = 8$

$6t-5t = 64$ Multiply both sides by 8.

$t = 64$

4. $\dfrac{7t}{3}-\dfrac{8t}{9} = 13$

$21t-8t = 117$ Multiply both sides by 9.

$t = 9$

5. $\dfrac{b}{b-2}-2 = \dfrac{2}{b-2}$

$b-2(b-2) = 2$ Multiply both sides by $(b-2)$.

$b = 2$,   but checking 2 in the original equation yields zero in the denominator, so there is no solution.

6. $4+\dfrac{5}{c} = -11$

$4c+5 = -11c$ Multiply both sides by c.

$c = -\dfrac{1}{3}$ Check $-\dfrac{1}{3}$ in the original equation and it works.

7. $\dfrac{3}{4e}-\dfrac{5}{2e} = \dfrac{7}{2}$

$3-10 = 14e$ Multiply both sides by $4e$.

$-\dfrac{1}{2} = e$ Check $-\dfrac{1}{2}$ in the original equation and it works.

8. $\dfrac{h+8}{h}+\dfrac{3}{4h} = 11$

$4(h+8)+3 = 4h(11)$ Multiply both sides by $4h$.

$\dfrac{7}{8} = h$ Check $\dfrac{7}{8}$ in the original equation and it works.

9. $\dfrac{2c+5}{c+10} = \dfrac{c-9}{c+10}+1$

$2c+5 = c-9+c+10$ Multiply by $(c+10)$.

$2c+5 = 2c+1$

$5 = 1$ Not true, so there is no solution.

10. $\dfrac{4d-6}{d-11} = \dfrac{d+8}{d-11}+2$

$4d-6 = d+8+2(d-11)$ Multiply both sides by $(d-11)$.

$d = -8$ Check -8 in the original equation and it works.

11. c = rate of the current; time is equal for each trip;

$$t = \frac{d}{r}$$

$$\frac{24}{16+c} = \frac{12}{16-c}$$

$24(16-c) = 12(16+c)$ Multiply both sides by $(16+c)(16-c)$.

$$5\frac{1}{3} = c$$

PRACTICE

1. $2m + \frac{2}{3} = \frac{23}{10}$

$60m + 20 = 69$ Multiply both sides by 30.

$$m = \frac{49}{60}$$

2. $6a - \frac{7}{8} = -\frac{12}{5}$

$240a - 35 = -96$ Multiply both sides by 40.

$$a = -\frac{61}{240}$$

3. $\frac{3u}{4} + \frac{4u}{6} = 17$

$9u + 8u = 204$ Multiply both sides by 12.

$u = 12$

4. $\frac{4v}{15} + \frac{v}{5} = 14$

$4v + 3v = 210$ Multiply both sides by 15.

$v = 30$

5. $\frac{j}{j+6} = j - \frac{6}{j+6}$

$j = 3(j+6) - 6$ Multiply both sides by $(j+6)$.

$-6 = j$ But checking -6 in the original equation yields zero in the denominator, so there is no solution.

6. $\frac{k}{k+2} - \frac{5}{k+2} = 8$

$k - 5 = 8(k+2)$ Multiply both sides by $(k+2)$.

$-3 = k$ Check -3 in the original equation and it works.

7. $\frac{m}{m-5} - \frac{13}{m-5} = 7$

$m - 13 = 7(m-5)$ Multiply both sides by $(m-5)$.

$\frac{11}{3} = m$ Check $\frac{11}{3}$ in the original equation and it works.

8. $4 - \frac{6}{n} = 5$

$4n - 6 = 5n$ Multiply both sides by n.

$-6 = n$ Check -6 in the original equation and it works.

9. $\frac{4}{5q} - \frac{2}{7q} = \frac{1}{35}$

$28 - 10 = q$ Multiply both sides by $35q$.

$18 = q$ Check 18 in the original equation and it works.

10. $\frac{w+3}{w} + \frac{2}{3w} = 12$

$3w + 9 + 2 = 36w$ Multiply both sides by $3w$.

$\frac{1}{3} = w$ Check $\frac{1}{3}$ in the original equation and it works.

11. $\frac{x-3}{x} + \frac{3}{4x} = 16$

$4x - 12 + 3 = 64x$ Multiply both sides by $4x$.

$-\frac{3}{20} = x$ Check $-\frac{3}{20}$ in the original equation and it works.

12. $\frac{3d-8}{d-5} = \frac{d+6}{d-5} + 2$

$3d - 8 = d + 6 + 2(d-5)$ Multiply both sides by $(d-5)$.

$3d - 8 = d + 6 + 2d - 10$

$3d - 8 = 3d - 4$

$-8 = -4$ false equation, so no solution

13. Determine the LCD and multiply both sides of the equation by the LCD.

14. Let x represent the tailwind rate; time for each trip is equal; $t = \frac{d}{r}$

$$\frac{2400}{550+t} = \frac{2000}{550-t}$$

$2400(550-t) = 2000(550+t)$ Multiply by $(550+t)(550-t)$.

$$50 = t$$

So, time $= \frac{2400}{550+50} = \frac{2400}{600} = 4$ hr

15. x = the amount in Keeno's account

$$\frac{46}{25} = \frac{x+1500}{x+450}$$

$46(x+450) = 25(x+1500)$ Multiply by $25(x+450)$.

$$x = 800$$

So, Keemo has $800 + 450 = \$1250$
and Rikuichi has $800 + 1500 = \$2300$.

16. $\dfrac{2t-5}{(t-2)(t+1)} = \dfrac{t-5}{(t-5)(t+1)}$

$2t-5 = t-2$ — Multiply both sides by $(t-2)(t+1)$.

$t = 3$ — Check 3 in the original equation and it works.

17. $\dfrac{x+5}{(x+5)(x-1)} = \dfrac{2(x+4)}{(x+4)(x-1)}$

$1 = 2$ — Multiply both sides by $(x-1)$.

false equation, so no solution

18. b = the boat's rate in still water; $t = \dfrac{d}{r}$

$$\dfrac{12}{b+4} = \dfrac{3}{4}\left(\dfrac{12}{b-4}\right)$$

$12 \cdot 4(b-4) = 3 \cdot 12(b+4)$ — Multiply both sides by $4(b+4)(b-4)$.

$b = 28$ — Check 28 in the original equation and it works.

19. Let x represent the time for the faster one to fill the tank, so it fills $\dfrac{1}{x}$ of the tank in one hour. The slower one fills $\dfrac{1}{4x}$ of the tank in one hour. Since it takes 12 hours to fill the tank, $\dfrac{1}{12}$ is done in one hour and:

$\dfrac{1}{12} = \dfrac{1}{x} + \dfrac{1}{4x}$

$x = 12 + 3$ — Multiply both sides by $12x$.

$x = 15$ hours

20. x = time for wallpaper hanger to do the job; $2x$ = time for the apprentice to do the job

$\dfrac{1}{x} + \dfrac{1}{2x} = \dfrac{1}{4}$

$4 + 2 = x$ — Multiply both sides by $4x$.

$6 = x$ and $2x = 12$

THINK CRITICALLY

For 21-23, examples will vary.

21. $\dfrac{x}{x+4} + \dfrac{8}{x-4} = \dfrac{x+60}{x^2-16}$ **22.** $\dfrac{2}{x+1} = \dfrac{-2x}{x+1} + 4$

23. $\dfrac{4}{x+1} + \dfrac{6}{x-2} = \dfrac{-2}{x^2-x-2}$

24. $\left(\dfrac{x-3}{x+6}\right)^2 \left(\dfrac{x+6}{2x-6}\right)^3 = 1$

$\dfrac{(x-3)(x-3)(x+6)(x+6)(x+6)}{(x+6)(x+6)(2)(x-3)(2)(x-3)(2)(x-3)} = 1$

$\dfrac{x+6}{8(x-3)} = 1$

$x+6 = 8x - 24$

$\dfrac{30}{7} = x$

25. $\left(\dfrac{x-2}{x+4}\right)^2 \div \left(\dfrac{3x-6}{x+4}\right)^3 = 1$

$\dfrac{(x-2)(x-2)}{(x+4)(x+4)} \cdot \dfrac{(x+4)(x+4)(x+4)}{(3)(x-2)(3)(x-2)(3)(x-2)} = 1$

$\dfrac{x+4}{27(x-2)} = 1$

$x+4 = 27x - 54$

$\dfrac{29}{13} = x$

PROJECT CONNECTION

1. $\dfrac{42}{1-0.50} = \dfrac{42}{0.50} = 84$ Tons

$\dfrac{42}{1-0.65} = \dfrac{42}{0.35} = 120$ Tons

2. $\dfrac{31}{1-x} = 77.5$

$31 = 77.5(1-x)$

$x = 0.6$ or 60%

3. 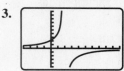 $0 \le x < 1$; Moisture cannot be greater than 1, or less than 0; $x \ne 1$ because the function is not defined at 1.

Lesson 14.9, pages 718-721

EXPLORE THE PROBLEM

1. Answer should be in grams of salt.

2. 28 students $\cdot \, 12.5 \dfrac{\text{g}}{\text{student}} = (28)(12.5)\text{g} = 350$ g

3. need to know the number of grams in a pound, a conversion factor

4. $1.5 \text{ lb} \cdot \dfrac{453.6 \text{ g}}{1 \text{ lb}} = (1.5)(453.6)\text{g} = 680.4$ g

5. $680.4 - 350 = 330.4$

INVESTIGATE FURTHER

6. $2.5 \text{ lb} \cdot \dfrac{453.6 \text{ g}}{\text{lb}} - \left(\dfrac{8.5 \text{ g}}{\text{student}} \cdot 24 \text{ students}\right) =$

$1134 \text{ g} - 204 \text{ g} = 930$ g

7. The value of any conversion factor is 1; multiplying by 1 does not change the value.

8. $\dfrac{1 \text{ km}}{1\times10^3 \text{ m}}$; $\dfrac{3\times10^8 \text{ m}}{1 \text{ s}} \cdot \dfrac{1 \text{ km}}{1\times10^3 \text{ m}} = 3\times10^5 \dfrac{\text{km}}{\text{s}}$

9. $\dfrac{60 \text{ s}}{1 \text{ min}}$; $\dfrac{60 \text{ min}}{1 \text{ h}}$; $\dfrac{24 \text{ h}}{1 \text{ day}}$

10. $\dfrac{3\times10^8 \text{ m}}{1 \text{ s}} \cdot \dfrac{1 \text{ km}}{1\times10^3 \text{ m}} \cdot \dfrac{60 \text{ s}}{1 \text{ min}} \cdot \dfrac{60 \text{ min}}{1 \text{ h}} \cdot \dfrac{24 \text{ h}}{1 \text{ da}} \cdot 365 \text{ da} =$

94608×10^{12} km

11. If units of measure or physical quantities are involved; many ratios and steps are involved; conversion of units is required.

APPLY THE STRATEGY

12a. yes   **b.** yes

  c. No, this is $\dfrac{\text{length}}{(\text{length})(\text{time})}$; it does not fit the model for acceleration.

  d. yes

13. need to know how many kilometers are in a mile; use 1 mile = 1.6 km

14. $46 \text{ cm} \cdot \dfrac{100 \text{ km}}{2.5 \text{ cm}} \cdot \dfrac{1 \text{ m}}{1.6 \text{ km}} \cdot \dfrac{1 \text{ gal}}{22 \text{ mi}} \cdot \dfrac{1.16 \text{ dollars}}{1 \text{ gal}} \approx$ 60.64 dollars or \$60.64.

15. $\dfrac{258{,}000 \text{ customers} \cdot \left(50 \dfrac{\text{calls}}{\text{customer}}\right) \cdot \left(0.8 \dfrac{\text{hours}}{\text{call}}\right)}{1500 \dfrac{\text{hours}}{\text{salesperson}}} =$

6880 hours $\div \dfrac{\text{hours}}{\text{salesperson}} = 6880$ salespersons

REVIEW PROBLEM SOLVING STRATEGIES

1a. $x^2 + y^2 = 9$ **b.** $(a-x)^2 + y^2 = 16$

  c. $(a-x)^2 + (b-y)^2 = 25$ **d.** $x^2 + (b-y)^2 = (PD)^2$

  e. Methods will vary. One method is to subtract Equation II from Equation III and add Equation I to the result.

  f. Since $x^2 + (b-y)^2 = (DP)^2$ and $x^2 + (b-y)^2 = 18$, it follows that $(DP)^2 = 18$. So, $DP = \sqrt{18}$

2. 1,312,432,004

3. Sonja can set both timers at the same time. As soon as the 5-min timer finishes, she turns it over. As soon as the 8-min timer finishes, she turns it over. As soon as the 5-min timer finishes the second time she immediately turns it over again. The amount of time from when the 5-min timer finishes for the third time to when the 8-min timer finishes for the second time is 1-min.

Chapter Review, pages 722-723

1. b   **2.** a   **3.** c

4. $y = \dfrac{k}{x}$; $16 = \dfrac{k}{20}$ so $k = 320$

$y = \dfrac{320}{x} = \dfrac{320}{64} = 5$

5. $y = kxz$; $60 = k \cdot 30 \cdot 4$ so $k = \dfrac{1}{2}$

$y = \dfrac{1}{2}xz = \dfrac{1}{2} \cdot 12 \cdot 3 = 18$

6. $\dfrac{12x}{4x^2} = \dfrac{\cancel{2} \cdot \cancel{2} \cdot 3 \cdot \cancel{x}}{\cancel{2} \cdot \cancel{2} \cdot x \cdot \cancel{x}} = \dfrac{3}{x}$ and $x \neq 0$

7. $\dfrac{6b+18}{b+3} = \dfrac{6(b+3)}{\cancel{b+3}} = 6$ and $b \neq -3$

8. $\dfrac{2y-1}{4-8y} = \dfrac{\cancel{2y-1}}{-4(\cancel{-1+2y})} = \dfrac{1}{-4}$ and $y \neq \dfrac{1}{2}$

9. $\dfrac{\cancel{p}q^2}{\cancel{r}} \cdot \dfrac{3p^2\cancel{r}}{\cancel{p}} = 3p^2q^2$

10. $\dfrac{4}{3y} \div \dfrac{2}{9y^2} = \dfrac{\cancel{2} \cdot 2}{\cancel{3} \cdot \cancel{y}} \cdot \dfrac{\cancel{3} \cdot 3 \cdot \cancel{y} \cdot y}{\cancel{2}} = 6y$

11. $\dfrac{n+7}{2n-6} \div \dfrac{10(n+7)}{5(n-3)} = \dfrac{\cancel{n+7}}{2(\cancel{n-3})} \cdot \dfrac{\cancel{5}(\cancel{n-3})}{2 \cdot \cancel{5}(\cancel{n+7})} = \dfrac{1}{4}$

12. $\dfrac{6x+24}{2x^2-8} \cdot (3x+6) = \dfrac{\cancel{2} \cdot 3(x+4)}{\cancel{2}(x+2)(x-2)} \cdot \dfrac{3(x+2)}{1} =$

$\dfrac{9(x+4)}{x-2} = \dfrac{9x+36}{x-2}$

13. $\dfrac{x^2-36}{x^2+4x-12} \div (3x-18) = \dfrac{(x+6)(\cancel{x-6})}{(\cancel{x+6})(x-2)} \cdot \dfrac{1}{3(\cancel{x-6})} =$

$\dfrac{1}{3(x-2)} = \dfrac{1}{3x-6}$

14. $\dfrac{5x^2+5x}{x^2-3x-4} \cdot (3x-12) = \dfrac{5x(\cancel{x+1})}{(\cancel{x-4})(\cancel{x+1})} \cdot 3(\cancel{x-4}) = 15x$

15. $\dfrac{6x^3}{2x^2} - \dfrac{8x^2}{2x^2} + \dfrac{x}{2x^2} = 3x - 4 + \dfrac{1}{2x}$

16. $(y^2+y-56) \div (y-7) = \dfrac{(y+8)(\cancel{y-7})}{1} \cdot \dfrac{1}{\cancel{y-7}} = y+8$

17.

$$
\begin{array}{r}
3x^3 + 3x^2 - 4x - 4 \\
x-1 \overline{)\,3x^4 + 0x^3 - 7x^2 + 0x + 5} \\
\underline{3x^4 - 3x^3} \\
3x^3 - 7x^2 \\
\underline{3x^3 - 3x^2} \\
-4x^2 + 0x \\
\underline{-4x^2 + 4x} \\
-4x + 5 \\
\underline{-4x + 4} \\
1
\end{array}
$$

So, $(3x^4 - 7x^2 + 5) \div (x-1) =$

$3x^3 + 3x^2 - 4x - 4 + \dfrac{1}{x-1}$

18.

$$
\begin{array}{r}
2x^2 - x + 2 \\
x+3 \overline{)\,2x^3 + 5x^2 - x + 2} \\
\underline{2x^3 + 6x^2} \\
-x^2 - x \\
\underline{-x^2 - 3x} \\
2x + 2 \\
\underline{2x + 6} \\
-4
\end{array}
$$

So, $(2x^2 + 5x^2 - x + 2) \div (x+3) = 2x^2 - x + 2 - \dfrac{4}{x+3}$

19. $\dfrac{7}{g+2} - \dfrac{g+6}{g+2} = \dfrac{1-g}{g+2}$

20. $\dfrac{8}{3x} - \dfrac{2}{9x} = \dfrac{8}{3x} \cdot \dfrac{3}{3} - \dfrac{2}{9x} = \dfrac{24}{9x} - \dfrac{2}{9x} = \dfrac{22}{9x}$

21. $\dfrac{5}{3z^2} - \dfrac{2}{z} + \dfrac{1}{6z}$ The LCD is $6z^2$.

$\dfrac{5}{3z^2} \cdot \dfrac{2}{2} - \dfrac{2}{z} \cdot \dfrac{6z}{6z} + \dfrac{1}{6z} \cdot \dfrac{z}{z} = \dfrac{10}{6z^2} - \dfrac{12z}{6z^2} + \dfrac{z}{6z^2} = \dfrac{10-11z}{6z^2}$

22. $\dfrac{t}{2t+8} - \dfrac{5}{3t+12} = \dfrac{t}{2(t+4)} - \dfrac{5}{3(t+4)}$

The LCD is $2 \cdot 3(t+4)$.

$\dfrac{t}{2(t+4)} \cdot \dfrac{3}{3} - \dfrac{5}{3(t+4)} \cdot \dfrac{2}{2} =$

$\dfrac{3t}{6(t+4)} - \dfrac{10}{6(t+4)} = \dfrac{3t-10}{6t+24}$

23. $\dfrac{4}{a} \div \dfrac{7}{b} = \dfrac{4}{a} \cdot \dfrac{b}{7} = \dfrac{4b}{7a}$

24. $\dfrac{\frac{1}{m} - \frac{2}{n}}{\frac{4}{m}} \cdot \dfrac{mn}{mn} = \dfrac{\frac{1}{m} \cdot \frac{mn}{1} - \frac{2}{n} \cdot \frac{mn}{1}}{\frac{4}{m} \cdot \frac{mn}{1}} = \dfrac{n-2m}{4n}$

25. $\dfrac{\frac{2}{x} + \frac{3}{y}}{\frac{1}{2x} - \frac{1}{2y}} \cdot \dfrac{2xy}{2xy} = \dfrac{\frac{2}{x} \cdot \frac{2xy}{1} + \frac{3}{y} \cdot \frac{2xy}{1}}{\frac{1}{2x} \cdot \frac{2xy}{1} - \frac{1}{2y} \cdot \frac{2xy}{1}} = \dfrac{4y+6x}{y-x}$

26. $\dfrac{\frac{3}{r} + 2}{\frac{4}{s}} \cdot \dfrac{rs}{rs} = \dfrac{\frac{3}{r} \cdot \frac{rs}{1} + 2 \cdot rs}{\frac{4}{s} \cdot \frac{rs}{1}} = \dfrac{3s+2rs}{4r}$

27. $2b + \dfrac{4}{5} = \dfrac{2}{3}$ Multiply both sides by 15.

$30b + 12 = 10$

$b = -\dfrac{1}{15}$

28. $\dfrac{y}{y-2} - \dfrac{14}{y-2} = 5$

$y - 14 = 5(y-2)$ Multiply by $(y-2)$.

$-1 = y$ Check -1 in the original equation and it works.

29. $\dfrac{k+5}{k} - \dfrac{2}{5k} = 10$

$5k + 25 - 2 = 50k$ Multiply both sides by $5k$.

$\dfrac{23}{45} = k$ Check $\dfrac{23}{45}$ in the original equation and it works.

30. $\left(6\dfrac{3}{8}\text{ in.}\right) \cdot \left(\dfrac{20 \text{ miles}}{\frac{3}{4}\text{ in.}}\right) \cdot \left(\dfrac{1 \text{ gallon}}{24 \text{ miles}}\right) \cdot \left(\dfrac{\$1.29}{\text{gallon}}\right) \approx \$9.14$

Chapter Assessment, pages 724-725

1. Examples will vary. Students should state that as the x-values increase, the y-values decrease, and as the x-values decrease, the y-values increase.

2. B; $y = \dfrac{k}{x}$; $14 = \dfrac{k}{2}$ so $k = 28$

3. inverse; $k = 4$ **4.** direct; $k = 5$ **5.** joint; $k = 2\pi$

6. The LCD of $2x$ and $x(x+2)$ is $2x(x+2)$.

7. The LCD of $2^3 \cdot d \cdot e \cdot f^2, 2 \cdot 3 \cdot d^2 \cdot e$, and f^2 is $2^3 \cdot 3 \cdot d^2 \cdot e \cdot f^2 = 24d^2ef^2$.

8. $\dfrac{5x+10}{15} = \dfrac{5(x+2)}{3 \cdot 5} = \dfrac{x+2}{3}$

9. $\dfrac{x+3}{x^2-10x-39} = \dfrac{x+3}{(x+3)(x-13)} = \dfrac{1}{x-13}$

and $x \neq -3$ or 13

10. $\dfrac{x^2+10x+25}{x^2+11x+30} = \dfrac{(x+5)(x+5)}{(x+5)(x+6)} = \dfrac{x+5}{x+6}$

and $x \neq -5$ or -6

11. $\dfrac{c^2-c-56}{c^2-16c+64} = \dfrac{(c+7)(c-8)}{(c-8)(c-8)} = \dfrac{c+7}{c-8}$ and $c \neq 8$

12. $\dfrac{3x^2-15x}{4x+2} \cdot \dfrac{8x+4}{125-5x^2} = \dfrac{3x(x-5)}{2(2x+1)} \cdot \dfrac{2 \cdot 2(2x+1)}{-5(x-5)(x+5)} =$

$-\dfrac{6x}{5x+25}$

13. $\dfrac{4x^2-8x}{3x+1} \cdot \dfrac{9x+3}{16-4x^2} = \dfrac{4x(x-2)}{3x+1} \cdot \dfrac{3(3x+1)}{-4(x-2)(x+2)} =$

$-\dfrac{3x}{x+2}$

14. $\dfrac{x^2+x-30}{x^2-36} \div \dfrac{x^2-25}{3x-18} =$

$\dfrac{(x+6)(x-5)}{(x+6)(x-6)} \cdot \dfrac{3(x-6)}{(x+5)(x-5)} = \dfrac{3}{x+5}$

15. $\dfrac{7}{5r^2} + \dfrac{9}{10r} - \dfrac{5}{r}$ The LCD is $10r^2$.

$= \dfrac{7}{5r^2} \cdot \dfrac{2}{2} + \dfrac{9}{10r} \cdot \dfrac{r}{r} - \dfrac{5}{r} \cdot \dfrac{10r}{10r}$

$= \dfrac{14}{10r^2} + \dfrac{9r}{10r^2} - \dfrac{50r}{10r^2} = \dfrac{14-41r}{10r^2}$

16. $\dfrac{6}{a+3} + \dfrac{14}{a-10}$ The LCD is $(a+3)(a-10)$.

$= \dfrac{6}{a+3} \cdot \dfrac{a-10}{a-10} + \dfrac{14}{a-10} \cdot \dfrac{a+3}{a+3}$

$= \dfrac{6a-60}{(a+3)(a-10)} + \dfrac{14a+42}{(a+3)(a-10)} = \dfrac{20a-18}{(a+3)(a-10)}$

17. $\dfrac{21}{16a^2} - \dfrac{5}{24ab} - \dfrac{1}{6b^2}$ The LCD is $48a^2b^2$.

$= \dfrac{21}{16a^2} \cdot \dfrac{3b^2}{3b^2} - \dfrac{5}{24ab} \cdot \dfrac{2ab}{2ab} - \dfrac{1}{6b^2} \cdot \dfrac{8a^2}{8a^2}$

$= \dfrac{63b^2 - 10ab - 8a^2}{48a^2b^2}$

18. $\dfrac{7q-5}{49-q^2} - \dfrac{7}{7-q}$ LCD is $(7-q)(7+q)$ or $49-q^2$.

$= \dfrac{7q-5}{49-q^2} - \dfrac{7}{7-q} \cdot \dfrac{7+q}{7+q} = \dfrac{7q-5}{49-q^2} - \dfrac{49+7q}{49-q^2} = \dfrac{-54}{49-q^2}$

19. When one polynomial is divided by another and the two expressions have no common factors other than 1, you must use long division.

20. $2p - \dfrac{4p-7}{4p-6}$

$= \dfrac{2p}{1} \cdot \dfrac{4p-6}{4p-6} - \dfrac{4p-7}{4p-6}$

$= \dfrac{8p^2 - 12p}{4p-6} - \dfrac{4p-7}{4p-6}$

$= \dfrac{8p^2 - 16p + 7}{4p-6}$

21. $q - 5 - \dfrac{4}{q+3}$

$= \dfrac{(q-5)}{1}\left(\dfrac{q+3}{q+3}\right) - \dfrac{4}{q+3}$

$= \dfrac{q^2 - 2q - 15}{q+3} - \dfrac{4}{q+3}$

$= \dfrac{q^2 - 2q - 19}{q+3}$

22. $\dfrac{2k-1}{k+10} - k$

$= \dfrac{2k-1}{k+10} - \dfrac{k}{1} \cdot \dfrac{k+10}{k+10}$

$= \dfrac{2k-1}{k+10} - \dfrac{k^2 + 10k}{k+10}$

$= \dfrac{-k^2 - 8k - 1}{k+10}$

23. $a + 4 + \dfrac{4}{a-4}$

$= \dfrac{(a+4)}{1}\left(\dfrac{a-4}{a-4}\right) + \dfrac{4}{a-4}$

$= \dfrac{a^2 - 16}{a-4} + \dfrac{4}{a-4}$

$= \dfrac{a^2 - 12}{a-4}$

24. $\dfrac{\dfrac{1}{5c^2} - \dfrac{16}{5d^2}}{\dfrac{1}{c^2d} + \dfrac{4}{cd^2}} \cdot \dfrac{5c^2d^2}{5c^2d^2} = \dfrac{d^2 - 16c^2}{5d + 20c} =$

$\dfrac{(d-4c)(d+4c)}{5(d+4c)} = \dfrac{d-4c}{5}$

25. $\dfrac{\dfrac{1}{x} - \dfrac{1}{y}}{\dfrac{3}{2x} - \dfrac{3}{2y}} \cdot \dfrac{2xy}{2xy} = \dfrac{2y-2x}{3y-3x} = \dfrac{2(y-x)}{3(y-x)} = \dfrac{2}{3}$

26. $\dfrac{\dfrac{6}{a^2} + \dfrac{6}{a}}{\dfrac{3}{a} - 18} \cdot \dfrac{a^2}{a^2} = \dfrac{6+6a}{3a-18a^2} = \dfrac{3(2+2a)}{3(a-6a^2)} = \dfrac{2+2a}{a-6a^2}$

27. $\dfrac{5 - \dfrac{6}{x+9}}{10 - \dfrac{12}{x+9}} \cdot \dfrac{x+9}{x+9} = \dfrac{5x+45-6}{10x+90-12} = \dfrac{5x+39}{10x+78} =$

$\dfrac{5x+39}{2(5x+39)} = \dfrac{1}{2}$

28. $2x^2 + 4x = 3x + 15$

$2x^2 + x - 15 = 0$

$(2x-5)(x+3) = 0$

$x = \dfrac{5}{2}$ or $x = -3$

29. $\dfrac{p^2-4}{p^2q - 2p} \cdot \dfrac{p^2 - p}{p^2 - 3p + 2} =$

$\dfrac{(p+2)(p-2)}{p(pq-2)} \cdot \dfrac{p(p-1)}{(p-2)(p-1)} = \dfrac{p+2}{pq-2}$

30. w = windspeed; t = time; $t = \dfrac{d}{r}$

$\dfrac{3000}{960+w} = \dfrac{2760}{960-w}$

$3000(960-w) = 2760(960+w) \rightarrow 40 = w$

31. $(9.3 \times 10^7 \text{ miles})\left(\dfrac{1 \text{ year}}{5.88 \times 10^{12} \text{ miles}}\right)\left(\dfrac{3.65 \times 10^2 \text{ days}}{1 \text{ year}}\right)$

$\left(\dfrac{2.4 \times 10^1 \text{ hours}}{1 \text{ day}}\right)\left(\dfrac{6.0 \times 10^1 \text{ minutes}}{1 \text{ hour}}\right) \approx 8.3 \text{ minutes}$

Cumulative Review, page 726

1. $y = kxz$; $36 = k \cdot 2 \cdot 9$, so $k = 2$
$y = 2xz \rightarrow 40 = 2x \cdot 4$, so $x = 5$

2. $\dfrac{2x^3 - 18x}{4x - 12} = \dfrac{2x(x+3)(x-3)}{2 \cdot 2(x-3)} = \dfrac{x^2 + 3x}{2}$ and $x \neq 3$

3. $\dfrac{x^2 + x - 20}{x^2 + 3x - 10} = \dfrac{(x+5)(x-4)}{(x+5)(x-2)} = \dfrac{x-4}{x-2}$
and $x \neq -5$ or 2

4. $12n^3 - 23n - 10$

5. $8y^3 + 36y^2 - 8y - 14y^2 - 63y + 14 =$
$8y^3 + 22y^2 - 71y + 14$

6. $4t^3 + 5t^2 - 6t + 12$

7.

$$x + 3 \overline{)\, x^4 + 5x^3 + 0x^2 - 7x - 2}$$

$$\begin{array}{r} x^3 + 2x^2 - 6x + 11 \\ \underline{x^4 + 3x^3} \\ 2x^3 + 0x^2 \\ \underline{2x^3 + 6x^2} \\ -6x^2 - 7x \\ \underline{-6x^2 - 18x} \\ 11x - 2 \\ \underline{11x + 33} \\ -35 \end{array}$$

So, $(x^4 + 5x^3 + 7x - 2) \div (x + 3) =$

$$x^3 + 2x^2 - 6x + 11 - \frac{35}{x+3}$$

8. In the first example, 4 is a factor of both the numerator and the denominator. In the second example, 4 is a term but is not a factor.

9. $4x + 15 = 2x - 9\left(\frac{1}{3}x - 2\right)$

$4x + 15 = 2x - 3x + 18$

$5x = 3$

$x = \frac{3}{5}$

10. $\frac{2}{3}x + 5 > 8 - \frac{5}{6}x$

$4x + 30 > 48 - 5x$ Multiply both sides by 6.

$9x > 18$ Add $5x$; subtract 30.

$x > 2$ Divide by 9.

11. $-3m \le 12$

$m \ge -4$ Divide by -3 and reverse inequality symbol.

12. $5n^2 + 27n - 18 = 0$

$(5n - 3)(n + 6) = 0$

$n = \frac{3}{5}$ or $n = -6$

13. $2y^2 + 5y - 4 = 0$

$$y = \frac{-5 \pm \sqrt{5^2 - 4(2)(-4)}}{2(2)} = \frac{-5 \pm \sqrt{57}}{4}$$

14. $\sqrt{x + 7} + 9 = x + 4$

$\sqrt{x + 7} = x - 5$ Isolate the radical.

$x + 7 = x^2 - 10x + 25$ Square both sides.

$0 = x^2 - 11x + 18$

$0 = (x - 9)(x - 2)$

$x = 9$ or $x = 2$

Only 9 checks.

15. $\dfrac{x^3 - 16x}{x^2 + 3x - 10} \div \dfrac{x^2 + 2x - 8}{x^2 - 4x + 4} =$

$$\frac{x(x-4)(x+4)}{(x+5)(x-2)} \bullet \frac{(x-2)(x-2)}{(x+4)(x-2)} = \frac{x^2 - 4x}{x + 5}$$

16. $\dfrac{y+4}{3y^2 - 27} + \dfrac{6}{y-3} - \dfrac{4}{3y+9}$

$$= \frac{y+4}{3(y+3)(y-3)} + \frac{6}{y-3} - \frac{4}{3(y+3)} \quad \begin{array}{l}\text{The LCD is} \\ 3(y-3)(y+3).\end{array}$$

$$= \frac{y+4}{3(y+3)(y-3)} + \frac{6}{y-3} \bullet \frac{3(y+3)}{3(y+3)} - \frac{4}{3(y+3)} \bullet \frac{y-3}{y-3}$$

$$= \frac{y+4}{3(y+3)(y-3)} + \frac{18y+54}{3(y+3)(y-3)} - \frac{4y-12}{3(y+3)(y-3)}$$

$$= \frac{15y+70}{3(y+3)(y-3)}$$

17. $4\sqrt{18} + 2\sqrt{72} - 4\sqrt{\dfrac{1}{2}}$

$$= 4\sqrt{9 \bullet 2} + 2\sqrt{36 \bullet 2} - 4\sqrt{\frac{1}{2} \bullet \frac{2}{2}}$$

$$= 12\sqrt{2} + 12\sqrt{2} - 2\sqrt{2} = 22\sqrt{2}$$

18. $\dfrac{\left(x^2 y^{-4}\right)^6}{\left(x^{-4} y^8\right)^{-3}} = \dfrac{x^{12} y^{-24}}{x^{12} y^{-24}} = x^0 y^0 = 1 \bullet 1 = 1$

19. $\dfrac{4ab^5 c^{-2}}{12a^3 b^{-4} c^{-3}} = \dfrac{1}{3} a^{1-3} b^{5-(-4)} c^{-2-(-3)} = \dfrac{1}{3} a^{-2} b^9 c = \dfrac{b^9 c}{3a^2}$

20. $x + 2 = 2$ or $x + 2 = -2$

     $x = 0$ or $x = -4$

21. $|2x + 3| + 11 = 12$

     $|2x + 3| = 1$

     $2x + 3 = 1$ or $2x + 3 = -1$

          $x = -1$ or $x = -2$

22. $2|x - 5| - 6 = 0$

     $2|x - 5| = 6$

     $|x - 5| = 3$

$x - 5 = 3$ or $x - 5 = -3$

$x = 8$ or $x = 2$

23. $-3|x + 4| = -12$

     $|x + 4| = 4$

$x + 4 = 4$ or $x + 4 = -4$

   $x = 0$ or $x = -8$

24. $\dfrac{112 + 75 + 98 + 140 + 52 + 101 + 110 + 94 + 82}{9} = \dfrac{864}{9} = 96$

25. Sorted data: 52, 75, 82, 94, **98**, 101, 110, 112, 140

Median $= Q_2 = 98$

$Q_1 = \dfrac{75 + 82}{2} = 78.5$ $Q_3 = \dfrac{110 + 112}{2} = 111$

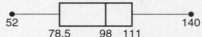

26. A line graph would best show the highs and lows and the difference from one game to the next.

27.
$$xy - 5 = x + 4$$
$$xy - x = 9$$
$$x(y - 1) = 9$$
$$x = \frac{9}{y - 1}$$

28.
$$450(0.25) = 0.2(450 + x)$$
$$x = 112.5$$

29. $x =$ the time it takes Donald
$x - 3 =$ the time it takes Mickey
$$\frac{1}{x} + \frac{1}{x - 3} = \frac{1}{2}$$
$2(x - 3) + 2x = x(x - 3)$ Multiply by $2x(x - 3)$.
$$2x - 6 + 2x = x^2 - 3x$$
$$0 = x^2 - 7x + 6$$
$$0 = (x - 6)(x - 1)$$
$$x = 6 \text{ or } x = 1$$
$x = 1$ does not make sense since Mickey's time would then be -2 hours. So, it takes Donald 6 hours alone.

30. Domain: all real numbers; Range: all real numbers

31. Domain: all real numbers; Range: $y \geq -2$

32. Domain: all real numbers; Range: $y \leq 1$

33. Domain: $x \geq -2$; Range: $y \geq 1$

34. A; $\begin{vmatrix} 5 & n \\ 2 & 4 \end{vmatrix} = 5(4) - 2(n) = 20 - 2n$

35. $x =$ pompoms ordered; $y =$ T-shirts ordered
Maximize: $P = 3x + 5y$
Subject to: $x + y = 1500$, $x \leq 800$, $y \leq 1000$

Vertex	$P = 3x + 5y$	Profit
$(0, 1000)$	$3(0) + 5(1000)$	5000
$(500, 1000)$	$3(500) + 5(1000)$	6500
$(800, 700)$	$3(800) + 5(700)$	5900
$(800, 0)$	$3(800) + 5(0)$	2400

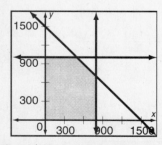

For a profit of $6500, order 500 pompoms, and 1000 T-shirts.

1. $\sin A = \dfrac{3}{5} = \dfrac{\text{opp}}{\text{hyp}}$

$\text{adj} = \sqrt{\text{hyp}^2 - \text{opp}^2} = \sqrt{5^2 - 3^2} = \sqrt{16} = 4$

$\tan A = \dfrac{\text{opp}}{\text{adj}} = \dfrac{3}{4}$

2. $x^2 - 4x - 2 = 0$
$$x = \frac{-(-4) \pm \sqrt{(-4)^2 - 4(1)(-2)}}{2(1)}$$
$$= \frac{4 \pm \sqrt{24}}{2} = \frac{4 \pm 2\sqrt{6}}{2} = 2 \pm \sqrt{6}$$
Product of the solutions is:
$$(2 + \sqrt{6})(2 - \sqrt{6}) = 2^2 - (\sqrt{6})^2 = 4 - 6 = -2$$

3. $x =$ number of nickels; $12 + x =$ number of dimes.
$$0.05x + 0.10(12 + x) = 2.70$$
$$x = 10, \; 12 + x = 22$$
There are 32 coins.

4. $m = \dfrac{4 - 1}{-3 - 5} = \dfrac{3}{-8}$, so the perpendicular slope is $\dfrac{8}{3}$.

5. $d_1 = \sqrt{(-8 - 4)^2 + (6 - 1)^2} = \sqrt{144 + 25} = \sqrt{169} = 13$

$d_2 = \sqrt{(4 - 1)^2 + [1 - (-3)]^2} = \sqrt{9 + 16} = \sqrt{25} = 5$

$d_3 = \sqrt{(-11 - 1)^2 + [2 - (-3)]^2} = \sqrt{144 + 25} = \sqrt{169} = 13$

$d_4 = \sqrt{[-8 - (-11)]^2 + (6 - 2)^2} = \sqrt{9 + 16} = \sqrt{25} = 5$

Perimeter $= d_1 + d_2 + d_3 + d_4 = 13 + 5 + 13 + 5 = 36$

6. $2 \cdot 6 \cdot 6 = 72$

7. $p = kmn$; $12 = k \cdot 8 \cdot 6$ so $k = \dfrac{1}{4}$

$p = \dfrac{1}{4}mn \to 20 = \dfrac{1}{4} \cdot m \cdot 5$ so $m = 16$

8.
$$
\require{enclose}
\begin{array}{r}
2x^2 + 6x + 18 \\
x - 3 \enclose{longdiv}{2x^3 + 0x^2 + 0x - 54} \\
\underline{2x^3 - 6x^2} \\
6x^2 + 0x \\
\underline{6x^2 - 18x} \\
18x - 54 \\
\underline{18x - 54} \\
0
\end{array}
$$

9. $\begin{cases} x + 2y = 10 \\ 2x + y = 5 \end{cases} \rightarrow x = 10 - 2y$

$2(10 - 2y) + y = 5$ Substitute $10 - 2y$ for x.

$y = 5$ and $x = 0$

So, the sum is 5.

10. The matrix has 4 rows and 2 columns.

11. $\dfrac{\frac{2}{3} - \frac{3}{5}}{\frac{5}{6} + \frac{1}{4}} \cdot \dfrac{60}{60} = \dfrac{40 - 36}{50 + 15} = \dfrac{4}{65}$

12. $(2x^3 y^5)^4 = 16x^{3 \cdot 4} y^{5 \cdot 4} = 16x^{12} y^{20}$

So, the degree is $12 + 20 = 32$

13. r = running rate, then $r - 6$ = walking rate

$$1 = \frac{8}{r} + \frac{2}{r - 6}$$

$r(r - 6) = 8(r - 6) + 2r$ Multiply by the LCD of $r(r - 6)$.

$r^2 - 16r + 48 = 0$

$(r - 4)(r - 12) = 0$

$r = 4$ or $r = 12$

4 would give a negative walking rate, so Sebastian's running rate is 12 mi/hr.

14. $\dfrac{18}{45 + 38 + 25 + 18 + 24} = \dfrac{18}{150} = 0.12$

15. $\dfrac{(4.2 \times 10^7)(2 \times 10^{-9})}{2.1 \times 10^{-4}} =$

$(4.2 \cdot 2 \div 2.1) \times 10^{\,7 + (-9) - (-4)} = 4 \times 10^2 = 400$

16. $\begin{vmatrix} x & 4 \\ -3 & 5 \end{vmatrix} + \begin{vmatrix} -4 & 2 \\ x & 1 \end{vmatrix} = 11$

$5x - (-12) + (-4) - 2x = 11$

$x = 1$

17. $y = \dfrac{k}{x}$; $5 = \dfrac{k}{12}$ so $k = 60$

$y = \dfrac{60}{x} \rightarrow 30 = \dfrac{60}{x}$ so $x = 2$

18. The mode is 22, since the value 22 occurs 5 times which is more frequent than any other value.

19. $\dfrac{76}{16} = \mathbf{4.75}$

$\sqrt{10} \approx 3.16$

$\dfrac{159}{50} = 3.18$

$3.\overline{15} = 3.151515\cdots$